ALGEBRAIC METHODS in the GLOBAL THEORY of COMPLEX SPACES

CONSTANTIN BĂNICĂ
and
OCTAVIAN STĂNĂȘILĂ

ALGEBRAIC METHODS in the GLOBAL THEORY of COMPLEX SPACES

EA

EDITURA ACADEMIEI
București

JOHN WILEY & SONS
London · New York · Sydney · Toronto

1976

Library of Congress Cataloging in Publication Data

Bănică, Constantin.
Algebraic methods in the global theory of complex spaces

Rev. English version of Metode algebrice în teoria globală a spațiilor complexe, published in 1974.

Bibliography: p.
1. Analytic spaces. 2. Homology theory. 3. Duality theory (Mathematics) I. Stănăşilă, Octavian, joint author. II. Title.

QA331.B25413 515′.7 76-5823

ISBN 0 471 01809 0

This edition is the revised English version of the Romanian book
"METODE ALGEBRICE ÎN TEORIA GLOBALĂ A SPAȚIILOR COMPLEXE"
published in 1974
by EDITURA ACADEMIEI, Calea Victoriei 125, BUCHAREST.

PRINTED IN ROMANIA

Preface

The theory of functions of several complex variables has been significantly developed in the last decades. The study of complex spaces, i.e. the geometric models associated to this theory, involves methods of analysis, algebra, differential equations, geometry and algebraic topology. The author intends to present some global problems of complex spaces, where the emphasis falls on algebraic methods.

Each chapter is preceded by a paragraph on preliminaries and by an introduction which summarises the main results of the text. These introductory passages will give the reader a rough picture of the book.

The proofs, except some of those in chapter seven, are complete apart from the assumption of basic results on complex spaces, algebra, sheaf theory, topological vector spaces, etc.

The book is intended mainly for experts in complex spaces who are not fully acquainted with the algebraic aspect, and for experts in algebraic geometry who wish to be introduced to the theory of complex spaces.

The authors would wish to express their profound gratitude to the Institute of Mathematics of the Romanian Academy.

Bucharest, October 1973.

C. BĂNICĂ
O. STĂNĂŞILĂ

We are under obligation to our colleagues V. Brînzănescu and Manuela Stoia for their valuable assistance in the preparation of the English edition.

To the Diary of Theodor Pallady

Contents

Chapter I

Cohomology with compact supports on Stein spaces

Introduction

We recall the following two theorems from the function theory of several complex variables:

"If D is an open subset of $\mathbb{C}^n$, $n \geqslant 2$ and $K \subset D$ is a compact such that $D \setminus K$ is connected, then any holomorphic function on $D \setminus K$ can be extended to a holomorphic function on D" (Hartogs' theorem, [41], Ch. II, § 3).

"If X is a Stein manifold of dimension $\geqslant 3$, and $U \subset X$ is a relatively compact Stein open subset, then the additive Cousin problem has a solution on $X \setminus U$; if in addition $H^2(X \setminus U; \mathbb{Z}) = 0$, then the multiplicative Cousin problem has also a solution on $X \setminus U$" (Serre's theorem, [74]).

The generalizations suggested by the first theorem are of the following nature:
– the substitution of D for a complex manifold or, more generally, for a complex space;
– the consideration of some entities more general than functions (sections in a sheaf, cohomology classes, divisors, meromorphic sections, subspaces, coherent sheaves).

All of these are subsumed to the problem of the extension of analytic entities defined out of a compact.

The second theorem belongs also to a general problem in connection with the previous one, namely, the study of the properties of the complementary of a compact (or a relatively compact open subset) in a complex space.

Let X be a topological space, $K \subset X$ a compact and $\mathcal{F}$ a sheaf of abelian groups on X. The exact cohomology sequence

$$\ldots \to H^q(X, \mathcal{F}) \to H^q(X \setminus K, \mathcal{F}) \to H_K^{q+1}(X, \mathcal{F}) \to \ldots$$

shows that the elements of the cohomology group with supports in K, $H_K^{q+1}(X, \mathcal{F})$ are just the obstructions for the extension to whole X of the cohomology q-classes on $X \setminus K$ with coefficients in $\mathcal{F}$.

If $U \subset X$ is a relatively compact open subset, then we have the exact sequence

$$\ldots \to H^q(X, \mathcal{F}) \to H^q(X \setminus U, \mathcal{F}) \to H_c^{q+1}(U, \mathcal{F}) \to \ldots .$$

Therefore, it results that the elements of the cohomology group with compact supports $H_c^{q+1}(U, \mathcal{F})$ represent the obstructions for the extension to whole X of the cohomology q-classes on $X \setminus U$ with coefficients in $\mathcal{F}$.

Thus one can notice the prominent part of the invariants $H_c^{\bullet}$, $H_K^{\bullet}$ in the investigation of the above stated problems. Our aim in this chapter is to study these invariants on Stein spaces and, in particular, to find vanishing theorems for them.

The first paragraph contains preliminaries of sheaf theory, topological vector spaces and local algebra. In the second one we prove two duality theorems (2.1 and 2.9) on Stein manifolds, which allow us to express the invariants $H_c^{\bullet}$, $H_K^{\bullet}$ in terms of global invariants $\mathrm{Ext}^{\bullet}(X; \ldots)$, $\mathrm{Ext}^{\bullet}(K; \ldots)$. The third paragraph deals with the algebraic notions of depth and dimension for a coherent analytic sheaf. Under the assumption that X is a Stein space and K is a holomorphically convex compact, the global Ext's determine completely the local *Ext*'s; this fact will allow us, *via* the duality proved in § 2, to reduce the invariants $H_c^{\bullet}$, $H_K^{\bullet}$ to invariants $\mathrm{Ext}^{\bullet}_{\mathcal{O}_x} \ldots$ ($\mathcal{O}_x$ is the local ring at $x \in X$). So we will make the connection with the depth and the dimension and will obtain the cohomological characterizations 3.6 and 3.7 for them: the cohomology groups with compact supports vanish out of the interval [prof, dim] and are $\neq 0$ at the ends of this interval. The fourth paragraph contains applications (in the frame of the above considered problems): results of type Hartogs or Cousin, properties of the boundary of a Stein space, applications regarding the compact analytic spaces. For instance, when X is a Stein space and $K \subset X$ is a holomorphically convex compact, the following assertions are proved:

— The restriction map $\Gamma(X, \mathcal{O}) \to \Gamma(X \setminus K, \mathcal{O})$ is bijective if and only if $\mathrm{prof}\, \mathcal{O}_x \geqslant 2$ for all $x \in K$ (corollary 4.4).

— If $\mathrm{prof}\, \mathcal{O}_x \geqslant 3$ for any $x \in K$, then the additive Cousin problem has solution on $X \setminus K$; if in addition $H^2(X \setminus K, \mathbb{Z}) = 0$, then the multiplicative Cousin problem has also solution on $X \setminus K$ (corollary 4.5).

— If $\mathrm{prof}\, \mathcal{O}_x \geqslant 2$ for all $x \in K$, then X is connected if and only if $X \setminus K$ is (corollary 4.8).

§ 1. Preliminaries

(a) We recall some facts in sheaf theory [26], [35].

Let X be a topological space. A family Φ of closed subsets of X is called a *family of supports* if any closed subset of an element of Φ belongs to Φ, and if any union of two elements of Φ is still an element of Φ. Let $\mathcal{F}$ be a sheaf of abelian groups on X (we will briefly write $\mathcal{F} \in \mathbf{Ab}(X)$). Denote by $\Gamma_\Phi(X, \mathcal{F})$ the subgroup of $\Gamma(X, \mathcal{F})$ of the sections whose supports belong to Φ. Thus, one obtains a left exact additive functor $\mathcal{F} \mapsto \Gamma_\Phi(X, \mathcal{F})$ from the category $\mathbf{Ab}(X)$ to the category $\mathbf{Ab}$ of abelian groups. Its right derived functors are denoted by $H_\Phi^{\bullet}(X, *)$ and are called *the cohomology groups with supports in* Φ. These invariants can be calculated by means of resolutions with flabby sheaves ([26], Ch. II, § 4; [35], Ch. III, § 3). We have $H_\Phi^0(X, \mathcal{F}) \simeq \Gamma_\Phi(X, \mathcal{F})$. In particular, if Φ is the family of all closed sets of X, we obtain the usual cohomology groups $H^{\bullet}(X, *)$.

Next suppose X paracompact. If Φ is the family of compact subsets, one obtains the invariants $H_c^{\bullet}(X, *)$, *the cohomology groups with compact supports*. For their calculation (and for the invariants $H^{\bullet}(X, *)$), one could make also use of resolutions with soft sheaves ([26], Ch. II, § 4). We have natural isomorphisms

$$\varinjlim H_c^{\cdot}(U, \mathfrak{F}) \xrightarrow{\sim} H_c^{\cdot}(X, \mathfrak{F}),$$

where the inductive limit is taken with respect to all open subsets U of X (or over a cofinal part of them).

If Φ is the family of closed subsets of a compact $K \subset X$, then the obtained invariants are denoted by $H_K^{\cdot}(X, *)$ and are called the *cohomology groups with supports in K*. For any open neighbourhood U of K there are canonical isomorphisms

$$H_K^{\cdot}(U, \mathfrak{F}) \xrightarrow{\sim} H_K^{\cdot}(X, \mathfrak{F}) \text{ (the excision property!).}$$

We also have canonical isomorphisms

$$\varinjlim_{K} H_K^{\cdot}(X, \mathfrak{F}) \xrightarrow{\sim} H_c^{\cdot}(X, \mathfrak{F}),$$

where the inductive limit is taken on all the compact subsets K of X (or over a cofinal part of them).

Let $0 \to \mathfrak{F} \to \mathfrak{J}^{\cdot}$ be an injective resolution of $\mathfrak{F}$. One thus obtains an exact sequence

$$0 \to \Gamma_K(X, \mathfrak{J}^{\cdot}) \to \Gamma(X, \mathfrak{J}^{\cdot}) \to \Gamma(X \setminus K, \mathfrak{J}^{\cdot}) \to 0$$

and passing to cohomology, an exact sequence

$$\ldots \to H^q(X, \mathfrak{F}) \to H^q(X \setminus K, \mathfrak{F}) \to H_K^{q+1}(X, \mathfrak{F}) \to \ldots .$$

This sequence shows that the invariants $H_K^{\cdot}(X, \mathfrak{F})$ represent the obstructions for the extension to whole X of the cohomology classes from $X \setminus K$. In the same manner one obtains the exact sequence

$$\ldots \to H_c^q(X \setminus K, \mathfrak{F}) \to H_c^q(X, \mathfrak{F}) \to H^q(K, \mathfrak{F}) \to H_c^{q+1}(X \setminus K, \mathfrak{F}) \to \ldots .$$

In particular, if X is compact and K a closed part of it, it will result the exact sequence

$$\ldots \to H_c^q(X \setminus K, \mathfrak{F}) \to H^q(X, \mathfrak{F}) \to H^q(K, \mathfrak{F}) \to H_c^{q+1}(X \setminus K, \mathfrak{F}) \to \ldots .$$

Let now U be a relatively compact open subset of X. From the exact sequence

$$0 \to \Gamma_c(U, \mathfrak{J}^{\cdot}) \to \Gamma(X, \mathfrak{J}^{\cdot}) \to \Gamma(X \setminus U, \mathfrak{J}^{\cdot}) \to 0$$

we deduce the exact cohomological sequence

$$\ldots \to H_c^q(U, \mathfrak{F}) \to H^q(X, \mathfrak{F}) \to H^q(X \setminus U, \mathfrak{F}) \to H_c^{q+1}(U, \mathfrak{F}) \to \ldots,$$

hence the invariants $H_c^{\cdot}(U, \mathcal{F})$ are the very obstructions for the extension of the cohomology classes on $X \setminus U$ to X.

Consider now $(X, \mathcal{O})$ a ringed space. For any two $\mathcal{O}$-modules $\mathcal{F}$, $\mathcal{G}$ denote $\mathrm{Hom}_{\Phi,\mathcal{O}}(\mathcal{F}, \mathcal{G}) = \Gamma_\Phi(X, \mathit{Hom}_{\mathcal{O}}(\mathcal{F}, \mathcal{G}))$ and this correspondence defines an exact left additive functor whose right derived functors are denoted by $\mathrm{Ext}^{\cdot}_{\Phi,\mathcal{O}}(*, *)$, and are called *the* Ext'*s with supports in* Φ. In particular, $\mathrm{Ext}^0_{\Phi,\mathcal{O}}(\mathcal{F}, \mathcal{G}) \simeq \simeq \mathrm{Hom}_{\Phi,\mathcal{O}}(\mathcal{F}, \mathcal{G})$. If $\mathcal{F}$ is a free $\mathcal{O}$-module of finite rank, then from the natural isomorphisms $\check{\mathcal{F}} \otimes_{\mathcal{O}} \mathcal{G} \rightrightarrows \mathit{Hom}_{\mathcal{O}}(\mathcal{F}, \mathcal{G})$ we get isomorphisms

$$\mathrm{Ext}^{\cdot}_{\Phi,\mathcal{O}}(\mathcal{F}, \mathcal{G}) \simeq H^{\cdot}_\Phi(X, \check{\mathcal{F}} \otimes_{\mathcal{O}} \mathcal{G}),$$

where $\check{\mathcal{F}} = \mathit{Hom}_{\mathcal{O}}(\mathcal{F}, \mathcal{O})$ is the dual of $\mathcal{F}$ (the proof follows easily since for any injective $\mathcal{O}$-module $\mathcal{J}$ and any $\mathcal{O}$-module $\mathcal{H}$, $\mathit{Hom}_{\mathcal{O}}(\mathcal{H}, \mathcal{J})$ is a flabby sheaf).

Recall now the definition of the functors *Ext*. Denote by $\mathit{Ext}^{\cdot}_{\mathcal{O}}(*, *)$ the derived functors of the functor $(*, *) \mapsto \mathit{Hom}_{\mathcal{O}}(*, *)$. One can easily check that for any two $\mathcal{O}$-modules $\mathcal{F}$ and $\mathcal{G}$, $\mathit{Ext}^{\cdot}_{\mathcal{O}}(\mathcal{F}, \mathcal{G})$ is the sheaf associated with the presheaf $U \mapsto \mathrm{Ext}^{\cdot}_{\mathcal{O}|U}(U; \mathcal{F}|U, \mathcal{G}|U)$ (for convenience, we sometimes write $\mathrm{Ext}^{\cdot}_{\mathcal{O}}(U; \mathcal{F}, \mathcal{G})$ instead of $\mathrm{Ext}^{\cdot}_{\mathcal{O}|U}(U; \mathcal{F}|U, \mathcal{G}|U)$). We have $\mathit{Ext}^0_{\mathcal{O}}(\mathcal{F}, \mathcal{G}) \simeq \mathit{Hom}_{\mathcal{O}}(\mathcal{F}, \mathcal{G})$. One can easily deduce that $\mathit{Ext}^q_{\mathcal{O}}(\mathcal{F}, \mathcal{G}) = 0$ for $q \geqslant 1$ and $\mathcal{F}$ a locally free $\mathcal{O}$-module of finite rank: accordingly, the *Ext*'s can be calculated by means of resolutions with free sheaves of finite rank in the first argument (in case that such resolutions do exist!). If $\mathcal{F}$ is locally an $\mathcal{O}$-module of finite presentation (locally there is an exact sequence of the form $\mathcal{O}^p \to \mathcal{O}^q \to \mathcal{F} \to 0$), then there are natural isomorphisms

$$\mathit{Hom}_{\mathcal{O}}(\mathcal{F}, \mathcal{G})_x \rightrightarrows \mathrm{Hom}_{\mathcal{O}_x}(\mathcal{F}_x, \mathcal{G}_x)$$

for any $\mathcal{O}$-module $\mathcal{G}$ and all $x \in X$. Moreover, if the stalks of the structural sheaf are noetherian rings, then the stalks of any injective $\mathcal{O}$-module are injective modules over these rings. Under these supplementary conditions on $\mathcal{O}$ and $\mathcal{F}$, we obtain natural isomorphisms

$$\mathit{Ext}^{\cdot}_{\mathcal{O}}(\mathcal{F}, \mathcal{G})_x \rightrightarrows \mathrm{Ext}^{\cdot}_{\mathcal{O}_x}(\mathcal{F}_x, \mathcal{G}_x)$$

for any point $x \in X$ and any $\mathcal{O}$-module $\mathcal{G}$.

We also recall the following property: if $\mathcal{O}$ is a coherent sheaf of rings and $\mathcal{F}$, $\mathcal{G}$ are coherent $\mathcal{O}$-modules, then the $\mathcal{O}$-modules $\mathit{Ext}^q_{\mathcal{O}}(\mathcal{F}, \mathcal{G})$ are coherent too. This fact can be easily proved by induction on q: the case $q = 0$ represents a well-known property of coherent sheaves and, for the general induction step, one can use the local existence of some exact sequences of the form $0 \to \mathcal{F}' \to \mathcal{O}^p \to \mathcal{F} \to 0$.

For details with respect to the functors Ext, *Ext*, one can see ([26], Ch. II, § 7; [35], Ch. IV).

(b) Let $\mathscr{C}_1, \mathscr{C}_2, \mathscr{C}_3$ be abelian categories, $\mathscr{C}_1$ and $\mathscr{C}_2$ having enough injective objects (for any object M there is a monomorphism $M \to I$, I an injective object). Let $\mathscr{C}_1 \xrightarrow{S} \mathscr{C}_2, \mathscr{C}_2 \xrightarrow{T} \mathscr{C}_3$ be left exact additive functors. Suppose in addition that S transforms any injective object in a T-acyclic object (an object M of $\mathscr{C}_2$ is called *T-acyclic* if $R^qT(M) = 0$ for any $q \geqslant 1$). Then there is a spectral sequence of term $E_2^{p,q} = (R^pT)(R^qS)$ which converges to the derived functors of the composition $R^{p+q}(TS)$ ([35], 2.4.1). Recall the construction of lateral morphisms

$$R^n(TS) = E^n \to E_2^{0,n} = T(R^nS), \ (R^nT)S = E_2^{n,0} \to E^n = R^n(TS).$$

Let M be an object of $\mathscr{C}_1$ and $I^\bullet$ one of its Cartan-Eilenberg resolutions ([16], XVII, 1.2). By the left exactness of T we get isomorphisms $\operatorname{Ker}(TS(I^n) \to TS(I^{n+1})) \xrightarrow{\sim} T(\operatorname{Ker}(S(I^n) \to S(I^{n+1})))$. From the commutative diagram

$$\begin{array}{ccc} TS(I^{n-1}) & \to & TS(I^n) \\ & \searrow \quad \nearrow & \\ & T(B^n) & \end{array}$$

where $B^n = \operatorname{Im}(S(I^{n-1}) \to S(I^n))$, we derive a morphism (in fact a monomorphisme) $\operatorname{Im}(TS(I^{n-1}) \to TS(I^n)) \to T(\operatorname{Im}(S(I^{n-1}) \to S(I^n)))$. Then we get morphisms

$$R^n(TS)(M) = H^n(TS(I^\bullet)) \to T(H^n(S(I^\bullet))) = T(R^nS)(M).$$

Their independence on the choice of the resolution $I^\bullet$ and the functoriality in M can be proved. One proceeds similarly for the other lateral morphisms.

We will use the following property: if for an object M of $\mathscr{C}_1$, $R^pT(R^qS(M)) = 0$ for all q and any $p \geqslant 1$, then the natural morphisms $R^n(TS)(M) \to T(R^nS(M))$ are isomorphisms for any integer $n \geqslant 0$. This is an immediate consequence of the general properties of spectral sequences [16], [26], [35]; however, one can straightforwardly prove it as follows

LEMMA 1.1. *Suppose $\mathscr{C}_1 \xrightarrow{S} \mathscr{C}_2 \xrightarrow{T} \mathscr{C}_3$ as above, and let M be an object of $\mathscr{C}_1$. If $R^pT(R^qS(M)) = 0$ for $p \geqslant 1$ and $q \geqslant 0$, then the natural morphisms $R^n(TS)(M) \to T(R^nS(M))$ are isomorphisms for any integer $n \geqslant 0$.*

Proof. Let $0 \to M \to I^0 \to I^1 \to \ldots$ be an injective resolution for M. $R^\bullet S(M)$ are the cohomology objects of the complex $S(I^\bullet)$ and $R^\bullet(TS)(M)$ the ones of $TS(I^\bullet)$. Denote $Z^n = \operatorname{Ker}(S(I^n) \to S(I^{n+1}))$, $B^n = \operatorname{Im}(S(I^{n-1}) \to S(I^n))$, $'Z^n = \operatorname{Ker}(TS(I^n) \to TS(I^{n+1}))$ and $'B^n = \operatorname{Im}(TS(I^{n-1}) \to TS(I^n))$. Obviously, $Z^0 = S(M)$. By the left exactness of functor T we get isomorphisms $'Z^n \xrightarrow{\sim} T(Z^n)$. If the exact sequences

$$0 \to Z^n \to S(I^n) \to B^{n+1} \to 0 \quad (n \geqslant 0),$$

$$0 \to B^n \to Z^n \to R^nS(M) \to 0 \quad (n \geqslant 1),$$

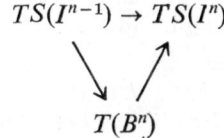

and the ascending induction on n are used one can verify that

$$R^pT(Z^n) = R^pT(B^{n+1}) = 0 (p \geqslant 1) \text{ and } 'B^{n+1} \xrightarrow{\sim} T(B^{n+1}).$$

The proof will be concluded by means of the exact commutative diagram:

$$\begin{array}{ccccccccc} 0 \to & 'B^n & \to & 'Z^n & \to & R^n(TS)(M) & \to 0 \\ & \downarrow\wr & & \downarrow\wr & & \downarrow & \\ 0 \to & T(B^n) & \to & T(Z^n) & \to & T(R^nS(M)) & \to 0. \end{array}$$

Remark. By the same argument one may establish the following result: if $R^pT(R^qS(M)) = 0$ for $p \geqslant 1$ and $0 \leqslant q \leqslant n_0$, then the natural morphisms $R^n(TS)(M) \to T(R^nS(M))$ are isomorphisms for each integer n, $0 \leqslant n \leqslant n_0 + 1$.

We will also use the following result, which is sometimes called "the de Rham abstract theorem".

LEMMA 1.2. *Let $\mathscr{C} \xrightarrow{T} \mathscr{C}'$ be a left exact additive functor between abelian categories, $\mathscr{C}$ having enough injective objects. Let M be an object of $\mathscr{C}$ and $I^\bullet$ a resolution of M with T-acyclic objects. Then there are natural isomorphisms $H^n(T(I^\bullet)) \to R^nT(M)$, $n \geqslant 0$.*

Proof. We proceed by induction on n. The case $n = 0$ will follow by the left exactness of T. Put $M' = \text{Ker}\,(I^1 \to I^2)$. From the exact sequence $0 \to M \to I^0 \to M' \to 0$ we get the exact sequence $0 \to T(M) \to T(I^0) \to T(M') \to R^1T(M) \to 0$ and isomorphisms $R^nT(M') \simeq R^{n+1}T(M)$, $n \geqslant 1$. Anyhow, by the exact sequence $0 \to T(M') \to T(I^1) \to T(I^2)$ the general induction step can be concluded.

We now consider two examples. Let $(X, \mathscr{O})$ be a ringed space and Φ a family of supports on X. Fix an $\mathscr{O}$-module and consider $\mathscr{C}_1 = \mathscr{C}_2 =$ the category of $\mathscr{O}$-modules, $\mathscr{C}_3 = \mathbf{Ab}$, $S = Hom_{\mathscr{O}}(\mathscr{F}, *)$ and $T = \Gamma_\Phi(X, *)$. Functor S carries injective objects in T-acyclic objects, since for any injective $\mathscr{O}$-module $\mathfrak{J}$, the sheaf $Hom_{\mathscr{O}}(\mathscr{F}, \mathfrak{J})$ is flabby. The following equality $TS = \text{Hom}_{\Phi,\mathscr{O}}(\mathscr{F}, *)$ holds. Therefore, we get a spectral sequence of term $E_2^{p,q} = H_\Phi^p(X, Ext_{\mathscr{O}}^q(\mathscr{F}, *))$ which converges to $\text{Ext}_{\Phi,\mathscr{O}}^{p+q}(X; \mathscr{F}, *)$. In particular, there is a spectral sequence of term $E_2^{p,q} = H^p(X, Ext_{\mathscr{O}}^q(\mathscr{F}, *))$ which converges to $\text{Ext}_{\mathscr{O}}^{p+q}(X; \mathscr{F}, *)$. In this case, the lateral morphisms $\text{Ext}_{\mathscr{O}}^n(X; \mathscr{F}, *) \to \Gamma(X, Ext_{\mathscr{O}}^n(\mathscr{F}, *))$ coincide with the morphisms given by passing from the presheaf to the associated sheaf.

Let now $(X, \mathscr{O}_X) \xrightarrow{f} (Y, \mathscr{O}_Y)$ be a morphism of ringed spaces. Consider $\mathscr{C}_1 =$ the category of $\mathscr{O}_X$-modules, $\mathscr{C}_2 =$ the category of $\mathscr{O}_Y$-modules, $\mathscr{C}_3 = \mathbf{Ab}$, $S = f_*$ (the direct image of sheaves) and $T = \Gamma(Y, *)$. Functor S carries injective objects to T-acyclic objects since any injective sheaf is flabby and f_* preserves the flabby sheaves. On the other hand, the equality $TS = \Gamma(X, *)$ holds. Thus, there is a spectral sequence of the term $E_2^{p,q} = H^p(Y, R^qf_*(*))$ which converges to $H^{p+q}(X, *)$. In this case one can show too that the lateral morphisms $H^n(X, *) \to \Gamma(Y, R^nf_*(*))$ enjoy remarkable interpretations. If in addition X and Y are paracompact and f is a proper map, then the equality of functors $\Gamma_c(Y, f_*(*)) = \Gamma_c(X, *)$ generates a spectral sequence of term $E_2^{p,q} = H_c^p(Y, R^q f_*(*))$ which converges to $H_c^{p+q}(X, *)$. The spectral sequences associated to the morphism f are called *the Leray's spectral sequences*.

In both previous examples, one can obtain remarkable consequences from lemma 1.1.

(c) In this section we recall some facts of topological vector spaces ([34], [68]), the basic field being the complex field $\mathbb{C}$.

A locally convex topological vector space, in addition metrisable and complete is called a *Fréchet space*. Such a space will be called of *type Schwartz* if for any neighbourhood of zero U there is a neighbourhood of zero V such that: for any $\varepsilon > 0$ there are $x_1, \ldots, x_n \in V$ with the property $V \subset \bigcup_{i=1}^{n} (x_i + \varepsilon U)$. Denote by FS (respectively DFS) the spaces Fréchet-Schwartz (respectively the strong duals of such spaces). These spaces are reflexive and have remarkable properties of stability: any closed subspace of an FS space is FS and the quotient of an FS space by a closed subspace is also FS and similar properties hold for DFS. We will often make use of the Banach theorem: a surjective continuous $\mathbb{C}$-linear map between two FS spaces (or DFS) is open.

Recall that a linear continuous map $u: E \to F$ between two topological vector spaces is called *strict* (or *topological homomorphism*) if the quotient topology on $u(E)$ coincides with that induced from F. The following result which is called "Serre's duality lemma" [75] will often be used in this chapter.

LEMMA 1.3. *Let* $E \xrightarrow{u} F \xrightarrow{v} G$ *be linear continuous maps between locally convex spaces, such that* v *is strict and* $vu = 0$. *Denote by* $E' \xleftarrow{u'} F' \xleftarrow{v'} G'$ *their transposition by topological duals. Then there is a natural algebraic isomorphism*

$$\operatorname{Ker} u'/\operatorname{Im} v' \xrightarrow{\sim} (\operatorname{Ker} v/\operatorname{Im} u)'.$$

Proof. Let $L \in \operatorname{Ker} u' \subset F'$. If $e \in E$, then $L(u(e)) = u'(L)(e) = 0$, hence $L|\operatorname{Im} u = 0$. Denote by $\dot{L}$ the functional determined by L on $\operatorname{Ker} v/\operatorname{Im} u$. If $L \in \operatorname{Im} v'$, then $L|\operatorname{Ker} v = 0$, hence $\dot{L} = 0$. In this way we get a $\mathbb{C}$-linear map $\operatorname{Ker} u'/\operatorname{Im} v' \to$ $\to (\operatorname{Ker} v/\operatorname{Im} u)'$ and we will prove its bijectivity. Let $L \in \operatorname{Ker} u'$ be such that $\dot{L} = 0$. Then L factorizes by a linear continuous functional $L': F/\operatorname{Ker} v \to \mathbb{C}$. Since v is strict, the topology of the quotient $F/\operatorname{Ker} v \xrightarrow{\sim} \operatorname{Im} v$ coincides with the topology induced from G. By Hahn-Banach theorem L' extends to a linear continuous functional $L'': G \to \mathbb{C}$. Since $v'(L'') = L$, then $L \in \operatorname{Im} v'$. It remains only to prove the surjectivity of the previous map. Suppose $T \in (\operatorname{Ker} v/\operatorname{Im} u)'$. By composition $\operatorname{Ker} v \to$ $\to \operatorname{Ker} v/\operatorname{Im} u \xrightarrow{T} \mathbb{C}$ we get a linear continuous functional on $\operatorname{Ker} v$ and let $L: F \to \mathbb{C}$ be an extension of it. We have $(u'(L))(e) = L(u(e)) = T(u(e) \bmod \operatorname{Im} u) = 0$, $e \in E$ an arbitrary element, hence $L \in \operatorname{Ker} u'$. It is easy to check that $\dot{L} = T$.

COROLLARY 1.4. *If* $\ldots \to E^{i-1} \to E^i \to E^{i+1} \to \ldots$ *is an exact sequence of FS spaces (or DFS) where the maps are linear and continuous, then the transposed sequence* $\ldots \leftarrow (E^{i-1})' \leftarrow (E^i)' \leftarrow (E^{i+1})' \leftarrow \ldots$ *is also exact.*

Proof. Since $\operatorname{Im}(E^{i-1} \to E^i) = \operatorname{Ker}(E^i \to E^{i+1})$ is a closed subspace of E^i, the differentials of the given complex are strict. The conclusion follows by lemma 1.3.

We will make also use in this chapter of the following facts:

— Any Banach space of at most countable dimension is finite-dimensional (consequence of Baire theorem).

— Any locally convex topological vector space, countable inductive limit of Fréchet spaces such that the maps of the inductive system are compact and injective, is separated, hence DFS taking into account the assertion which follows (recall that a linear continuous map $u: E \to F$ between locally convex topological vector spaces is called *compact* if there is a neighbourhood U of zero in E so that $u(U)$ is relatively compact in F).

— Any separated locally convex topological vector space, countable inductive limit of Fréchet spaces such that the maps of the inductive system are compact, is DFS.

— If $E = \varinjlim E_n$ is an inductive limit (in the category of locally convex topological vector spaces) of FS spaces such that the maps $E_n \to E_{n+1}$ are compact and in addition E is separated, then any bounded subset of E is the image of a bounded subset of some E_n (theorem of Raikov-Silva).

To conclude with, recall some things on complex manifolds. Let X be a complex manifold of dimension n. Denote by $\mathcal{E}$ the sheaf of germs of C^∞-differentiable functions on X. The space $\mathcal{E}(X) = \Gamma(X, \mathcal{E})$, endowed with the topology of the uniform convergence of functions and all their derivatives, is an FS space. Denote by $\mathcal{E}^{p,q}$(respectively $\mathcal{K}^{p,q}$) the sheaf of germs of differential forms of the type (p, q) with coefficients in $\mathcal{E}$ (distributions, respectively). The convergence of the coefficients of the forms defines FS topologies on the spaces $\mathcal{E}^{p,q}(X) = \Gamma(X, \mathcal{E}^{p,q})$. The space $\Gamma_c(X, \mathcal{K}^{p,q})$ coincides with the topological dual of $\mathcal{E}^{n-p,n-q}(X)$, hence, endowed with the strong dual topology, it becomes a DFS space. The differential $\mathcal{E}^{p,q}(X) \xrightarrow{d''} \mathcal{E}^{p,q+1}(X)$ is continuous with respect to these topologies and its transposition (*modulo* the sign) is the differential $\Gamma_c(X, \mathcal{K}^{n-p,n-q-1}) \xrightarrow{d''} \Gamma_c(X, \mathcal{K}^{n-p,n-q})$.

Denote by $\mathcal{O}$ the sheaf of germs of holomorphic functions on X. The space $\mathcal{O}(X) = \Gamma(X, \mathcal{O})$, endowed with the topology of the uniform convergence on compacts, is an FS space, more precisely a closed subspace of $\mathcal{E}(X)$. If U is a relatively compact open subset of X, then the restriction map $\mathcal{O}(X) \to \mathcal{O}(U)$ is compact. Let K be a compact of X and $(U_n)_{n \geq 0}$ a fundamental system of neighbourhoods of K such that $U_{n+1} \subset\subset U_n$ for any $n \geq 0$. Suppose in addition that any connected component of each U_n does intersect K. The restriction maps $\mathcal{O}(U_n) \to \mathcal{O}(U_{n+1})$ are compact and injective. It will result that the space $\mathcal{O}(K) = \varinjlim \mathcal{O}(U_n)$, endowed with the topology of inductive limit in the category of locally convex topological vector spaces, is DFS; its topology is independent on the above considered fundamental system. Moreover, it is a space of LF type (a separate locally convex topological vector space is called LF if it is a countable inductive limit of Fréchet spaces).

Any bounded subset of $\mathcal{O}(K)$ is the image of a bounded subset of some $\mathcal{O}(U_n)$. Similar considerations can be also made for the space $\Omega(K) = \Gamma(K, \Omega)$, where Ω

is the sheaf of germs of differential forms of the type $(n, 0)$ on X, with analytic coefficients.

(d) In this section we recall the most basic definitions and properties of the depth and of the dimension of a module ([10], [37] Ch. 0_{III}, [79]).

Let A be a ring (commutative and unitary). By Spec A one denotes the set of the prime ideals of A. If $\mathfrak{a}$ is an ideal of A, then put $V(\mathfrak{a}) = \{\mathfrak{p} | \mathfrak{p} \in \operatorname{Spec} A, \mathfrak{p} \supset \mathfrak{a}\}$. The sets by the form $V(\mathfrak{a})$ are the closed sets of a topology on Spec A, called the *Zariski topology*. A basis of open sets is given by the subsets $D(f) = \{\mathfrak{p} \in \operatorname{Spec} A, f \notin \mathfrak{p}\}$, for $f \in A$.

Let M be an A-module. The set $\operatorname{Supp} M = \{\mathfrak{p} | \mathfrak{p} \in \operatorname{Spec} A, M_\mathfrak{p} \neq 0\}$ is called the *support of M*.

Subsequently, we will assume that A is a noetherian ring and M an A-module of finite type. We have $\operatorname{Supp} M = V(\operatorname{Ann} M)$ where $\operatorname{Ann} M$ is the annihilator of M. An ideal $\mathfrak{p} \in \operatorname{Spec} A$ is said to be *associated to* M if there is a monomorphism of A-modules $A/\mathfrak{p} \to M$. The family of the prime ideals associated to A is finite and is denoted by Ass M. Obviously, $\operatorname{Ass} M \subset \operatorname{Supp} M$. The set of zero divisors for M is just $\bigcup_{\mathfrak{p} \in \operatorname{Ass} M} \mathfrak{p}$. The minimal elements of Ass M, Ass $(A/\operatorname{Ann} M)$ and Supp M are the same and one has

$$\dim M = \dim(A/\operatorname{Ann} M) = \sup_{\mathfrak{p} \in \operatorname{Ass} M} \dim(A/\mathfrak{p}).$$

We also would like to mention the following remarkable fact: there is a composition series $0 = M_0 \subset M_1 \subset \ldots \subset M_n = M$ such that each factor M_{i+1}/M_i is isomorphic to $A/\mathfrak{p}$ for some $\mathfrak{p} \in \operatorname{Supp} M$. For all the above considerations one could find details in ([79], Ch. III).

A sequence of elements $x_1, \ldots, x_r$ of A is called a *regular M-sequence* if any x_i is nonzerodivisor in $M \Big/ \sum_{j=1}^{i-1} x_j M$ (the hypothesis with respect to x_1 says that it is nonetheless a zerodivizor in M).

Suppose now A is a local ring and let $\mathfrak{m}$ be its maximal ideal and $k = A/\mathfrak{m}$ the residual field. Here is one further required result: "For any integer $r \geqslant 0$, $\operatorname{Ext}^i_A(k, M) = 0$ for $i < r$ if and only if there is a regular M-sequence formed by r elements of $\mathfrak{m}$". This can be seen in ([79], IV, prop. 6; [37], Ch. 0_{IV}, 16.4.4) or in the second chapter of this book, where a more general case will be considered. Denote

prof $M = \sup \{r |$ there is a regular M-sequence formed by r elements of $\mathfrak{m}\} = \inf \{i \mid \operatorname{Ext}^i_A(k, M) \neq 0\}$,

that is called *the depth* (or *profondeur*) *of M*. If $M = 0$, it results that prof $M = \infty$.

LEMMA 1.5. *If $x \in \mathfrak{m}$ is a M-regular element, then*

$$\dim M/xM = \dim M - 1.$$

Proof. The inequality $\dim M/xM \geqslant \dim M - 1$ is always true in virtue of the definition of the dimension via systems of parameters. If x is M-regular, then it is not contained in any ideal $\mathfrak{p} \in \operatorname{Ass} M$. The inequality $\dim M/xM \leqslant \dim M - 1$

easily results making use of the equality $\dim N = \sup_{\mathfrak{p} \in \mathrm{Ass} N} \dim A/\mathfrak{p}$, for $N = M$ and for $N = M/xM$.

PROPOSITION 1.6. *Let A be a noetherian local ring,* $\mathfrak{m}$ *its maximal ideal, and M an A-module of finite type. Then:*

(i) prof $M = 0$ *if and only if* $\mathfrak{m} \in \mathrm{Ass}\ M$.

(ii) *If* $x \in M$ *is M-regular, then* prof $M/xM = \text{prof } M - 1$.

(iii) *If* $M \neq 0$, *then* prof $M \leqslant \dim M$; *accordingly,* prof $M < \infty$ *for* $M \neq 0$.

(iv) *All maximal regular M-sequences of elements of* $\mathfrak{m}$ *have the same length, namely* prof M.

Proof. (i) prof $M = 0$ if and only if any element of $\mathfrak{m}$ is zerodivisor for M, hence if and only if $\mathfrak{m} \subset \bigcup_{\mathfrak{p} \in \mathrm{Ass} M} \mathfrak{p}$ and the conclusion results by use of ([10], Ch.IV, § 1, n$\overset{\circ}{=}$ 1, cor. 2, prop. 2).

(ii) If $(x_1, \ldots, x_r)$ is a regular M/xM-sequence formed by elements of $\mathfrak{m}$, then $(x, x_1, \ldots, x_r)$ is a regular M-sequence. Thus, prof $M/xM \leqslant \text{prof } M - 1$. Let i be an integer such that $\mathrm{Ext}^i_A(k, M/xM) \neq 0$. We will prove that $i + 1 \geqslant \text{prof } M$ and the proof is completed. Suppose the contrary, prof $M > i + 1$. Then $\mathrm{Ext}^i_A(k, M) = \mathrm{Ext}^{i+1}_A(k, M) = 0$. From the exact sequence

$$0 \to M \overset{x}{\to} M \to M/xM \to 0$$

we obtain the exact sequence

$$\ldots \to \mathrm{Ext}^i_A(k, M) \to \mathrm{Ext}^i_A(k, M/xM) \to \mathrm{Ext}^{i+1}_A(k, M) \to \ldots$$

and hence a contradiction.

(iii) If prof $M = 0$, then the inequality is clear. Assume prof $M > 0$ and let $x \in \mathfrak{m}$ be an M-regular element. By lemma 1.5, $\dim M/xM = \dim M - 1$ and the assertion will be concluded in accordance with (ii) by induction.

(iv) Again by (ii).

PROPOSITION 1.7. *Let* $A \overset{\rho}{\to} B$ *be a finite local morphism of noetherian local rings. If M is a B-module of finite type, then*

$$\mathrm{prof}_A M_{[\rho]} = \mathrm{prof}_B M.$$

($M_{[\rho]}$ is M, regarded as an A-module by means of ρ).

Proof. For $M = 0$ the conclusion is obvious, so we can suppose $M \neq 0$. Let $n = \mathrm{prof}_A M_{[\rho]}$ and $(x_i)_{1 \leqslant i \leqslant n}$ be a maximal regular $M_{[\rho]}$-sequence, all $x_i \in \mathfrak{m}_A$, $\mathfrak{m}_A$ being the maximal ideal of A. The elements $\rho(x_i)$ belong to the maximal ideal $\mathfrak{m}_B$ of B and constitute a regular M-sequence. Denote $N = M \Big/ \sum_{i=1}^{n} \rho(x_i) M$. Clearly,

$\mathrm{prof}_B N = \mathrm{prof}_B\, M - n,\ N_{[\rho]} = M_{]\rho]} \Big/ \sum_{i=1}^{n} x_i M_{[\rho]}$ and $\mathrm{prof}_A N_{[\rho]} = 0$. We thus reduce the problem to the case $n = 0$. Then $\mathfrak{m}_A \in \mathrm{Ass}_A\, M_{[\rho]}$, hence there is an injective morphism $A/\mathfrak{m}_A \to M_{[\rho]}$. Let M' be the submodule of M generated by the image of this morphism. Since the morphism ρ is finite, some power of the ideal $\mathfrak{m}_B$ is embedded in $\mathfrak{m}_A B$ and therefore annihilates M'. Then $\mathfrak{m}_B \in \mathrm{Ass}\ M' \subset \mathrm{Ass}\ M$, hence $\mathrm{prof}_B\, M = 0$.

PROPOSITION 1.8. *Let A be a noetherian local ring which is supposed to be normal of dimension $\geqslant 2$. Then prof $A \geqslant 2$.*

Proof. Let $\mathfrak{m}$ be the maximal ideal of A and x an element of $\mathfrak{m}$, $x \neq 0$. Since A is an integral domain, x is a nonzerodivizor. For any $\mathfrak{p} \in \mathrm{Ass}(A/xA)$, $\mathrm{ht}\mathfrak{p} = 1$ ([79], III, prop. 9). The hypothesis dim $A \geqslant 2$ implies that there is $y \in \mathfrak{m}$ which does not belong to any ideal of $\mathrm{Ass}(A/xA)$. Hence y is a nonzerodivisor in A/xA and the conclusion follows.

We recall that a module M (of finite type over a noetherian local ring) is called *a Cohen-Macauley module* if prof M = dim M. A noetherian local ring A is called *a Cohen-Macauley ring* if it is a Cohen-Macauley A-module, that is prof A = dim A.

COROLLARY 1.9. *Any normal noetherian local ring, of dimension 2, is a Cohen-Macauley ring.*

We now turn our attention to the characterization of the depth of the modules over regular rings. Recall some considerations of homological algebra ([16]; [79], IV C; [37], Ch. 0_{III}, 17.2).

Let A be a commutative unitary ring and M an A-module. The smallest n (integer or $+\infty$) such that there exists a left resolution of M which is projective and of length n, is called the *projective* (or *homological*) *dimension* of M and is denoted by dimproj M (or dh$M$). Equivalently, dh$M$ is the smallest n such that for any A-module N, $\mathrm{Ext}^i_A(M, N) = 0$ for $i > n$, or only for $i = n + 1$. The smallest n (integer or $+\infty$) such that dh$M \leqslant n$ for any A-module M, is called *the global cohomological dimension of the ring A* and is denoted by dimcoh A. This n results to be the smallest number such that for any two A-modules M, N, $\mathrm{Ext}^i_A(M, N) = 0$ for $i > n$ or $i = n + 1$ only.

Suppose now A a noetherian ring and M an A-module of finite type. Under these assumptions, in the definition of dh$M$ we can consider projective resolutions with modules of finite type only. Consequently, dh$M \leqslant n$ if and only if $\mathrm{Ext}^i_A(M, N) = 0$ for $i \geqslant n + 1$ (or $i = n + 1$ only), N being any A-module of finite type (by using induction on the number of generators one can consider here monogene modules N only). Since any such module N admits a composition series $0 = N_0 \subset \subset \ldots \subset N_k = N$ such that the succesive quotients N_i/N_{i-1} are isomorphic to modules of the form $A/\mathfrak{p}$, $\mathfrak{p} \in \mathrm{Spec}\ A$, we derive: dh$M \leqslant n$ if and only if $\mathrm{Ext}^i_A(M, A/\mathfrak{p}) = 0$ for any $i \geqslant n + 1$ (or $i = n + 1$ only) and all $\mathfrak{p} \in \mathrm{Spec}\ A$.

We now restrict ourselves to a more particular case, namely when A is a noetherian local ring of maximal ideal $\mathfrak{m}$. An A-module of finite type M is then free if and only if it is projective or equivalently, $\mathrm{Tor}^A_1(M, A/\mathfrak{m}) = 0$ ([10], Ch. II, § 3, n° 2, cor. 2, prop. 5; [79], IV, prop. 20). Therefore, one gets $\mathrm{dh}_A\ M \leqslant n$ if and only if $\mathrm{Tor}^A_{n+1}(M, A/\mathfrak{m}) = 0$ (M being an A-module of finite type) and dimcoh $A \leqslant n$

if and only if $\text{Tor}^A_{n+1}(A/\mathfrak{m}, A/\mathfrak{m}) = 0$ (for the proof we need only to interpret the definitions of dh and dimcoh by means of projective resolutions; for details, we send for instance to [79], Ch. IV).

Recall that a noetherian local ring A of maximal ideal $\mathfrak{m}$ and residual field $k = A/\mathfrak{m}$ is called *regular* if $\dim A = \dim_k \mathfrak{m}/\mathfrak{m}^2$ (the dimension theory gives rise to $\dim A \leqslant \dim_k \mathfrak{m}/\mathfrak{m}^2$). Henceforth we will make essential use of the following result ("Hilbert-Serre theorem" [79], IV, th. 9 or [37], Ch. 0_{III}, 17.3.1): "Let A be a noetherian local ring. A is regular if and only if its cohomological dimension is finite and in such case $\text{dimcoh } A = \dim A$".

We would point out some consequences of these facts concerning the notion of depth.

LEMMA 1.10. *Let A be a noetherian local ring, $\mathfrak{m}$ its maximal ideal, let M be an A-module of finite type, and x an element of $\mathfrak{m}$. If x is M-regular, then*

$$\text{dh } M/xM = \text{dh } M + 1.$$

Proof. From the exact sequence $0 \to M \to M \to M/xM \to 0$ induced by the homothety given by x, we get the exact sequence

$$\text{Tor}^A_i(M, A/\mathfrak{m}) \to \text{Tor}^A_i(M, A/\mathfrak{m}) \to \text{Tor}^A_i(M/xM, A/\mathfrak{m})$$
$$\to \text{Tor}^A_{i-1}(M, A/\mathfrak{m}) \to \text{Tor}^A_{i-1}(M, A/\mathfrak{m}),$$

where the first and the last arrow are both the homotheties given by x. Since $x \in \mathfrak{m}$, these two morphisms are null; hence we obtain the exact sequence

$$0 \to \text{Tor}^A_i(M, A/\mathfrak{m}) \to \text{Tor}^A_i(M/xM, A/\mathfrak{m}) \to \text{Tor}^A_{i-1}(M, A/\mathfrak{m}) \to 0$$

and the conclusion easily follows.

PROPOSITION 1.11. *Let A be a regular noetherian local ring of dimension n. Then for any A-module $M \neq 0$ of finite type,*

$$\text{prof } M + \text{dh} M = n.$$

Proof. We proceed by induction on $r = \text{prof } M$. If $r = 0$, then there exists an A-submodule N of M isomorphic to $A/\mathfrak{m}$ (1.6, i). The exact sequence $0 \to N \to M \to M/N \to 0$ yields the exact sequence

$$\text{Tor}^A_{n+1}(M/N, A/\mathfrak{m}) \to \text{Tor}^A_n(N, A/\mathfrak{m}) \to \text{Tor}^A_n(M, A/\mathfrak{m}).$$

The hypothesis on A shows that $\text{Tor}^A_{n+1}(M/N, A/\mathfrak{m}) = 0$ and $\text{Tor}^A_n(N, A/\mathfrak{m}) \simeq \simeq \text{Tor}^A_n(A/\mathfrak{m}, A/\mathfrak{m}) \neq 0$. Therefore, $\text{Tor}^A_n(M, A/\mathfrak{m}) \neq 0$, hence $\text{dh} M \geqslant n$; but on the other hand $\text{dh} M \leqslant \text{dimcoh } A = n$, hence $\text{dh} M = n$.

Assume now $r > 0$ and let x be an M-regular element, which belongs to $\mathfrak{m}$. As prof $M/xM = r - 1$ in accordance with (1.6, ii) and dh $M/xM =$ dh $M + 1$ (cf. 1.10), the general induction step is now clear.

Remark. For modules over regular rings, the depth is also called *cohomological codimension* and is denoted by codh. The above proposition asserts that dh + codh = = dim A and thus it justifies that terminology. If $B \overset{\rho}{\to} A$ is a surjective morphism of noetherian local rings, B regular of dimension n, then for any A-module M of finite type, $M \neq 0$, $\text{prof}_A M = n - \text{dh}_B M_{|\rho|}$. This fact will be often used.

COROLLARY 1.12. *Any regular noetherian local ring is a Cohen-Macauley ring.*

COROLLARY 1.13. *Suppose A is a regular noetherian local ring and $0 \to N \to$ $\to L \to M \to 0$ an exact sequence of A-modules of finite type where L is free. If M is not free, then*

$$\text{prof } M = \text{prof } N - 1.$$

Proof. We first remark that M, N are different from zero. The hypothesis on M then implies dh$M$ = dh$N$ + 1 and the proposition can be applied.

We must give, for future use, the characterization of the depth in terms of $\text{Ext}^{\bullet}_A(*, A)$.

LEMMA 1.14. *Let A be a regular noetherian local ring, M an A-module of finite type and $q \geqslant 0$ an integer. Then* dh$M \leqslant q$ *if and only if* $\text{Ext}^i_A(M, A) = 0$ *for any $i > q$.*

Proof. We proceed by descending induction on q. Since dh$M \leqslant$ dimcoh A = = dim A, the assertion of the lemma is obvious for $q >$ dim A. For the step $q + 1 \mapsto q$, assume $\text{Ext}^i_A(M, A) = 0$ when $i > q$ and prove that dh$M \leqslant q$ (the other implication is clear). It follows dh$M \leqslant q + 1$ If a module N would satisfy $\text{Ext}^{q+1}_A(M, N) \neq 0$, then take an exact sequence $0 \to P \to L \to N \to 0$ (L free) and got a contradiction.

COROLLARY 1.15. *Let A be a noetherian local ring, which is regular of dimension n. Then for any A-module M, $M \neq 0$ and for any integer q, we have* prof $M > q$ *if and only if* $\text{Ext}^i_A(M, A) = 0$ *for $i \geqslant n - q$.*

We are also going to give a characterization of the dimension in terms of $\text{Ext}^{\bullet}_A(*, A)$.

LEMMA 1.16. *Suppose A is a noetherian local ring, $\mathfrak{m}$ its maximal ideal, M and N are two A-modules of finite type such that the support of N contains only $\mathfrak{m}$, and q is an integer. Then* prof $M \geqslant q + 1$ *if and only if* $\text{Ext}^i_A(N, M) = 0$ *for any $i \leqslant q$.*

Proof. Choose a composition series $0 = N_0 \subset N_1 \subset \ldots \subset N_r = N$, with the factors of the form $A/\mathfrak{p}$, $\mathfrak{p} \in \text{Supp } N$. Since Supp $N = \{\mathfrak{m}\}$, all factors are in fact isomorphic to $A/\mathfrak{m}$. If prof $M \geqslant q + 1$, then $\text{Ext}^i_A(A/\mathfrak{m}, M) = 0$ for $i \leqslant q$ and the direct implication follows by applying the exact sequences of the Ext's associated to the exact sequences $0 \to N_j \to N_{j+1} \to N_{j+1}/N_j \to 0$ $(0 \leqslant j \leqslant r - 1)$.

Now we will prove the converse by induction. Look first at the case $q = 0$: if $\text{Hom}_A(N, M) = 0$ we derive prof $M \neq 0$; for otherwise there exists a monomorphism $0 \to A/\mathfrak{m} \to M$, hence a monomorphism $0 \to \text{Hom}_A(N, A/\mathfrak{m}) \to \text{Hom}_A(N, M)$, hence $\mathrm{Hom}_A(N, A/\mathfrak{m}) = 0$; this fact is a contradiction because there is a surjection $N = N_r \to N_r/N_{r+1} \xrightarrow{\sim} A/\mathfrak{m}$. We now assume $q \geqslant 1$ and the assertion proved for $q - 1$. Then prof $M \geqslant q \geqslant 1$ and fix $x \in \mathfrak{m}$ an M-regular element. From the exact sequence $0 \to M \xrightarrow{x} M \to M/xM \to 0$ we derive $\mathrm{Ext}^i_A(N, M/xM) = 0$ for $i \leqslant q - 1$, hence $\mathrm{prof}(M/xM) \geqslant q$. So, prof $M = \mathrm{prof}(M/xM) + 1 \geqslant q + 1$.

Let us fix a regular ring A and a prime ideal $\mathfrak{p}$ of A. By the Hilbert-Serre theorem it follows that the ring $A_\mathfrak{p}$ is regular too (as dimcoh $A_\mathfrak{p} \leqslant \dim A$!). In the proposition below we will use the relation

$$\dim A = \dim A_\mathfrak{p} + \dim A/\mathfrak{p}.$$

This formula is proved in ([79], IV, th. 6), in ([37], Ch. 0_{III}, 17.1.3 and 16.5.10) or in the case when A is a convergent power series ring (anyhow the only case when we apply 1.17), in Chapter II, cor. 1.28.

PROPOSITION 1.17. *Let A be a noetherian local ring, regular of dimension n. Then for any A-module of finite type M and for any integer q, $\dim M \leqslant q$ if and only if $\mathrm{Ext}^i_A(M, A) = 0$ for $i < n - q$.*

Proof. Suppose $\mathrm{Ext}^i_A(M, A) = 0$ for $i < n - q$. Let $\mathfrak{p}$ be an ideal, minimal within Supp M. We have

$$\mathrm{Ext}^{\cdot}_{A_\mathfrak{p}}(M_\mathfrak{p}, A_\mathfrak{p}) \simeq \mathrm{Ext}^{\cdot}_A(M, A)_\mathfrak{p}.$$

The ring $A_\mathfrak{p}$ is regular, hence a Cohen-Macauley ring and prof $A_\mathfrak{p} = \dim A_\mathfrak{p} = \dim A - \dim A/\mathfrak{p}$. One can see easily that Supp $M_\mathfrak{p}$ (in Spec $A_\mathfrak{p}$) contains only $\mathfrak{p}A_\mathfrak{p}$; by lemma 1.16 we conclude that prof $A_\mathfrak{p} \geqslant n - q$, hence $\dim A/\mathfrak{p} \leqslant q$. Accordingly, $\dim M \leqslant q$.

We are now going to prove the converse and proceed by induction on q. The case $q = 0$ follows by lemma 1.16, as Supp M contains at most the maximal ideal of A and prof $A = \dim A = n$. Let now $q \geqslant 1$ and the assertion is already verified for $q-1$. Consider a composition series $0 = M_0 \subset M_1 \subset \ldots \subset M_r = M$ such that the factors M_j/M_{j-1} are of the form $A/\mathfrak{p}$, $\mathfrak{p} \in \mathrm{Spec}\, A$. It is enough to prove that $\mathrm{Ext}^i_A(M_j/M_{j-1}, A) = 0$, for $i < n - q$ and $0 \leqslant j \leqslant r$; so we can assume M of the form $A/\mathfrak{p}$ (note that $\dim M_j/M_{j-1} \leqslant \dim M \leqslant q$). If $\mathfrak{p} = \mathfrak{m}$, then the conclusion occurs by the very definition of the depth. Otherwise, there exists an element x of $\mathfrak{m}$, which is M-regular. By induction hypothesis, the first morphism of the exact sequence $\mathrm{Ext}^i_A(M, A) \xrightarrow{x} \mathrm{Ext}^i_A(M, A) \to \mathrm{Ext}^{i+1}_A(M/xM, A)$ is surjective for $i < n - q$. The proof will be achieved by applying Nakayama lemma.

To conclude this section, we should mention a remarkable class of rings obtained from the regular rings. A ring is called *complete intersection* if it is isomorphic to a ring of the form $A \Big/ \sum_{i=1}^{r} x_i A$, where A is a regular noetherian local ring and $(x_1, \ldots, x_r)$ is an A-regular sequence.

PROPOSITION 1.18. *Any complete intersection ring is a Cohen-Macauley ring.*

Proof. Suppose $B = A \Big/ \sum_{i=1}^{r} x_i A$ as above. According to 1.5 and 1.6, $\dim B = \dim A - r$ and $\operatorname{prof} B = \operatorname{prof} A - r$. The conclusion follows from 1.12.

(e) Let X, Y be two separated topological spaces. We recall that a continuous map $f\colon X \to Y$ is said to be *finite* provided that it is a closed map (T closed in $X \Rightarrow f(T)$ closed in Y) and its fibers $f^{-1}(y)$, $y \in Y$, are finite sets.

LEMMA 1.19. *Let $f\colon X \to Y$ be a finite map and y a point of Y. If $\mathcal{V}$ is a fundamental system of neighbourhoods of y then $f^{-1}(\mathcal{V})$ is a fundamental system of neighbourhoods of the fiber $f^{-1}(y)$.*

The proof is a simple exercise of topology.

COROLLARY 1.20. *Let $f\colon X \to Y$ be a finite map between two separated topological spaces.*

(i) *If $\mathcal{F} \in \mathbf{Ab}(X)$, then the canonical morphism*

$$f_*(\mathcal{F})_y \to \prod_{f(x)=y} \mathcal{F}_x$$

is bijective for any $y \in Y$.

(ii) *The functor $f_*\colon \mathbf{Ab}(X) \to \mathbf{Ab}(Y)$ is exact.*

(iii) *If φ is a morphism in $\mathbf{Ab}(X)$ such that $f_*(\varphi)$ is an isomorphism then φ itself is an isomorphism.*

Proof. The assertion (i) immediately follows by the preceeding lemma and by the definition of f_*. The assertions (ii) and (iii) are obvious consequences of (i).

LEMMA 1.21. *Under the same assumptions as in 1.20, the canonical morphisms*

$$H^{\cdot}(Y, f_*(\mathcal{F})) \to H^{\cdot}(X, \mathcal{F}),\ H_c^{\cdot}(Y, f_*(\mathcal{F})) \to H_c^{\cdot}(X, \mathcal{F})$$

are bijective.

Proof. If we choose a flabby resolution $\mathcal{J}^{\cdot}$ for $\mathcal{F}$, then $f_*(\mathcal{J}^{\cdot})$ is a flabby resolution for $f_*(\mathcal{F})$. The conclusion follows by using the equalities:

$$\Gamma(Y, f_*(\mathcal{J}^{\cdot})) = \Gamma(X, \mathcal{J}^{\cdot}),\ \Gamma_c(Y, f_*(\mathcal{J}^{\cdot})) = \Gamma_c(X, \mathcal{J}^{\cdot}).$$

§ 2. Duality on Stein manifolds

The first result of the paragraph is the following:

THEOREM 2.1. *Let $(X, \mathcal{O})$ be a paracompact Stein manifold of dimension n and Ω the sheaf of germs of differential forms of the type $(n, 0)$ on X with analytic coefficients. Then, for any coherent analytic sheaf $\mathcal{F}$ on X and for any integer $q \geqslant 0$, $\operatorname{Ext}_{\mathcal{O}}^{n-q}(X; \mathcal{F}, \Omega)$ has a natural structure of Fréchet-Schwartz space with the topological dual algebraically isomorphic to $H_c^q(X, \mathcal{F})$.*

To prove it we need some preparations. Let $(X, \mathcal{O})$ be a complex space. We shall denote by $\mathbf{Coh}(X)$ the category of the coherent analytic sheaves on X. If $\mathcal{F} \in \mathbf{Coh}(X)$, then $\Gamma(X, \mathcal{F})$ has naturally a structure of topological vector space, which is even FS if X has a countable topology (a countable basis of open subsets); see [38], Ch. 8 and Chapter VII. Recall the definition of this structure. We confine ourselves to the case when X has a countable topology and consider a countable Stein open covering $\mathcal{U}$ of X, which is "sufficiently small". For any $U \in \mathcal{U}$, there is a closed immersion $U \xrightarrow{i} V$, V being a Stein open subset of some numerical space. We also assume the sheaf $i_*(\mathcal{F}|U)$ is the quotient of some sheaf $\mathcal{O}_V^p$ by a coherent subsheaf $\mathcal{G} \subset \mathcal{O}_V^p$. In virtue of theorem B, $\Gamma(U, \mathcal{F}) = \Gamma(V, i_*(\mathcal{F})) \simeq \Gamma(V, \mathcal{O}_V^p)/ /\Gamma(V, \mathcal{G})$. If we consider on $\Gamma(V, \mathcal{O}_V^p)$ the topology of uniform convergence on compact subsets, then one obtains an FS structure. By a theorem of Cartan, $\Gamma(V, \mathcal{G})$ is a closed subspace of $\Gamma(V, \mathcal{O}_V^p)$. In this way, we can endow $\Gamma(U, \mathcal{F})$ with an FS structure. $\Gamma(X, \mathcal{F})$ is a closed subspace of $\prod_{U \in \mathcal{U}} \Gamma(U, \mathcal{F})$ and endowed with the induced TVS structure, $\Gamma(X, \mathcal{F})$ has even an FS structure. This is independent on all choice. Thus $\mathcal{F}$ becomes a *Fréchet-Schwartz sheaf*, that is a sheaf such that for any open subset U, $\Gamma(U, \mathcal{F})$ has an FS structure and for any two open subsets $V \subset U$, the restriction maps $\Gamma(U, \mathcal{F}) \to \Gamma(V, \mathcal{F})$ are continuous.

If $\mathcal{F} \to \mathcal{G}$ is a morphism of coherent analytic sheaves on X, then the map $\Gamma(X, \mathcal{F}) \to \Gamma(X, \mathcal{G})$ is continuous in the above constructed topologies.

LEMMA 2.2. *Let X be a Stein space and $\mathcal{F} \to \mathcal{G}$ a morphism of coherent analytic sheaves on X. Then the map $\Gamma(X, \mathcal{F}) \to \Gamma(X, \mathcal{G})$ is strict.*

Proof. Denote $\mathcal{H} = \operatorname{Coker}(\mathcal{F} \to \mathcal{G})$. From the exact sequence $\mathcal{F} \to \mathcal{G} \to \mathcal{H}$, by applying theorem B, we get an exact sequence of linear continuous maps of FS spaces $\Gamma(X, \mathcal{F}) \to \Gamma(X, \mathcal{G}) \to \Gamma(X, \mathcal{H})$. Then the image of $\Gamma(X, \mathcal{F})$ in $\Gamma(X, \mathcal{H})$ is a closed subspace and in order to complete the proof we must make use of the Banach theorem.

Here is the convention we intend to use further. If X is a complex space, $K \subset X$ a compact, $\mathcal{F} \in \mathbf{Coh}(X)$, U a Stein neighbourhood of K, $U \xrightarrow{i} V$ a closed immersion, where V is a Stein open subset of a numerical space, L a compact in V such that $i^{-1}(L) = K$ and $\mathcal{O}_V^p \to i_*(\mathcal{F})$ is a sheaf epimorphism, then by taking the supremum on L, a seminorm on $\Gamma(V, \mathcal{O}_V^p)$ is so defined. By the surjection $\Gamma(V, \mathcal{O}_V^p) \to \to \Gamma(V, i_*(\mathcal{F})) = \Gamma(U, \mathcal{F})$ a seminorm on $\Gamma(U, \mathcal{F})$ is induced; the seminorm on $\Gamma(X, \mathcal{F})$ obtained by means of the restriction $\Gamma(X, \mathcal{F}) \to \Gamma(U, \mathcal{F})$ will be called "a seminorm of type '$\sup_K$' on $\Gamma(X, \mathcal{F})$". Such a seminorm has the following property: any Cauchy sequence with respect to "$\sup_K$" remains Cauchy by restriction to $\Gamma(D, \mathcal{F})$, whenever D is an open subset such that $D \subset K$ and $i(D) \subset \mathring{L}$.

The family "$\sup_K$", K compact in X, is a family of seminorms defining the topology of $\Gamma(X, \mathcal{F})$.

LEMMA 2.3. *Let X be a Stein space and $(U_r)_{r \geq 1}$ an exhaustion by relatively compact Stein open subsets for X. Suppose in addition that the maps* $\Gamma(X, \mathcal{O}) \to$

$\to \Gamma(U_r, \mathcal{O})$ *are dense for any* $r \geqslant 1$. *Then for any* $\mathcal{F} \in \mathbf{Coh}(X)$, *the canonical morphism*

$$\varinjlim_r (\Gamma(U_r, \mathcal{F}))' \to \Gamma(X, \mathcal{F})'$$

is bijective (the accent means as usual, the topological dual).

Proof. We first prove that the restriction maps $\Gamma(X, \mathcal{F}) \to \Gamma(U_r, \mathcal{F})$ are dense. Let $r \geqslant 1$ be an integer. By theorem *A* there is a sheaf morphism $\mathcal{O}_X^p \to \mathcal{F}$, which is surjective on U_r. The conclusion follows from the commutative diagram

$$\begin{array}{ccc} \Gamma(X, \mathcal{O}_X^p) & \to & \Gamma(X, \mathcal{F}) \\ \downarrow & & \downarrow \\ \Gamma(U_r, \mathcal{O}_X^p) & \to & \Gamma(U_r, \mathcal{F}), \end{array}$$

where the map $\Gamma(U_r, \mathcal{O}_X^p) \to \Gamma(U_r, \mathcal{F})$ is surjective by theorem *B*. Then it results that the maps $\Gamma(U_r, \mathcal{F})' \to \Gamma(X, \mathcal{F})'$ are injective, hence the asserted morphism is injective.

Next we will prove its surjectivity. Let $L\colon \Gamma(X, \mathcal{F}) \to \mathbb{C}$ be a linear continuous functional. There is a constant $\alpha > 0$, a compact $K \subset X$ and a seminorm p_K of the type "$\sup_K$" on $\Gamma(X, \mathcal{F})$, such that $|L(s)| \leqslant \alpha p_K(s)$ for any $s \in \Gamma(X, \mathcal{F})$. For r sufficiently large, one easily derives that L factorizes by any $\Gamma(U_r, \mathcal{F})$ (again according to the fact that the maps $\Gamma(X, \mathcal{F}) \to \Gamma(U_r, \mathcal{F})$ have dense images!).

LEMMA 2.4. *Let X be a Stein space and* $\mathcal{F}, \mathcal{G} \in \mathbf{Coh}$ (X). *Then the canonical morphism*

$$\mathrm{Ext}^q_{\mathcal{O}}(X; \mathcal{F}, \mathcal{G}) \to \Gamma(X, \mathcal{E}xt^q_{\mathcal{O}}(\mathcal{F}, \mathcal{G}))$$

is bijective, for any $q \geqslant 0$.

Proof. There exists a spectral sequence which converges to $\mathrm{Ext}^{\cdot}_{\mathcal{O}}(X; \mathcal{F}, \mathcal{G})$ for which $E_2^{p,q} = H^p(X, \mathcal{E}xt^q_{\mathcal{O}}(\mathcal{F}, \mathcal{G}))$, the morphisms from the statement being just the lateral morphisms $E^q \to E_2^{0,q}$. Since the $\mathcal{O}$-modules $\mathcal{E}xt^{\cdot}_{\mathcal{O}}(\mathcal{F}, \mathcal{G})$ are coherent, we obtain by theorem B, $E_2^{p,q} = 0$ for any $p \geqslant 1$. The conclusion follows from the properties of the spectral sequence.

We remark that one could avoid the considerations of spectral sequence by applying directly lemma 1.1 (according to section (**b**), § 1).

This lemma allows us to consider FS topologies on $\mathrm{Ext}^q_{\mathcal{O}}(X; \mathcal{F}, \mathcal{G})$, $q \geqslant 0$ arbitrary integer. In particular, one thus obtains the topologies required by the theorem.

LEMMA 2.5. *Let X be a Stein manifold of dimension n and* $\mathcal{F} \in \mathbf{Coh}(X)$. *Then* $H^q_c(X, \mathcal{F}) = 0$ *for* $q > n$.

Proof. Let $(U_r)_{r \geqslant 1}$ be an exhaustion of X by relatively compact Stein open subsets. We have $H^q_c(X, \mathcal{F}) \simeq \varinjlim_r H^q_c(U_r, \mathcal{F})$. It suffices to show that $H^q_c(U, \mathcal{F}) = 0$ for $q > n$ and $U \subset\subset X$ Stein open subset. The sheaf $\mathcal{F}$ admits on U a finite resolution by locally free sheaves of finite rank. By induction on the length of such a resolution, we are able to restrict the problem to the case $\mathcal{F}$ locally free.

Let

$$0 \to \mathcal{O} \to \mathcal{E} \xrightarrow{d''} \mathcal{E}^{0,1} \xrightarrow{d''} \ldots \to \mathcal{E}^{0,n} \to 0$$

be the Dolbeault resolution by forms C^∞. By tensoring $\otimes_{\mathcal{O}} \mathcal{F}$ we get a resolution of $\mathcal{F}$

$$0 \to \mathcal{F} \to \mathcal{F} \otimes_{\mathcal{O}} \mathcal{E} \to \mathcal{F} \otimes_{\mathcal{O}} \mathcal{E}^{0,1} \to \ldots \to \mathcal{F} \otimes_{\mathcal{O}} \mathcal{E}^{0,n} \to 0.$$

The sheaves $\mathcal{F} \otimes_{\mathcal{O}} \mathcal{E}^{0,*}$ are soft (as modules over the sheaf of germs of C^∞ functions on X). We thus conclude the lemma since $H_c^{\cdot}(X, \mathcal{F})$ are the cohomology groups of the complex $\Gamma_c(X, \mathcal{F} \otimes_{\mathcal{O}} \mathcal{E}^{0,*})$.

LEMMA 2.6. *Let $(X, \mathcal{O})$ be a Stein manifold of dimension n and let $\mathcal{L}$ be a free sheaf of finite rank. Then $H_c^q(X, \mathcal{L}) = 0$ for $q \neq n$ and $H_c^n(X, \mathcal{L})$ is isomorphic to the topological dual of* $\mathrm{Hom}_{\mathcal{O}}(\mathcal{L}, \Omega)$ *where* $\mathrm{Hom}_{\mathcal{O}}(\mathcal{L}, \Omega)$ *has the FS topology defined on the module of global sections of a coherent analytic sheaf,* $\mathrm{Hom}_{\mathcal{O}}(\mathcal{L}, \Omega) = \Gamma(X, Hom_{\mathcal{O}}(\mathcal{L}, \Omega))$. *Moreover, the duality isomorphisms agree with the $\mathcal{O}$-morphisms between free sheaves of finite rank.*

Proof. Let

$$0 \to \Omega \to \mathcal{E}^{n,0} \xrightarrow{d''} \mathcal{E}^{n,1} \xrightarrow{d''} \ldots \xrightarrow{d''} \mathcal{E}^{n,n} \to 0$$

and

$$0 \to \mathcal{O} \to \mathcal{K}^{0,0} \xrightarrow{d''} \mathcal{K}^{0,1} \xrightarrow{d''} \ldots \xrightarrow{d''} \mathcal{K}^{0,n} \to 0$$

the Dolbeault-Grothendieck resolutions of Ω ($\mathcal{O}$, respectively), by differential forms with coefficients C^∞ (distributions, respectively). Since $\mathcal{L}$ is of the form $\mathcal{O}^p$, we then obtain the resolutions

$$0 \to Hom_{\mathcal{O}}(\mathcal{L}, \Omega) \to$$

$$\to Hom_{\mathcal{O}}(\mathcal{L}, \mathcal{E}^{n,0}) \xrightarrow{d''} Hom_{\mathcal{O}}(\mathcal{L}, \mathcal{E}^{n,1}) \xrightarrow{d''} \ldots \xrightarrow{d''} Hom_{\mathcal{O}}(\mathcal{L}, \mathcal{E}^{n,n}) \to 0,$$

$$0 \to \mathcal{L} \to \mathcal{L} \otimes_{\mathcal{O}} \mathcal{K}^{0,0} \xrightarrow{d''} \mathcal{L} \otimes_{\mathcal{O}} \mathcal{K}^{0,1} \xrightarrow{d''} \ldots \xrightarrow{d''} \mathcal{L} \otimes_{\mathcal{O}} \mathcal{K}^{0,n} \to 0.$$

The sheaves $Hom_{\mathcal{O}}(\mathcal{L}, \mathcal{E}^{n,*})$ and $\mathcal{L} \otimes_{\mathcal{O}} \mathcal{K}^{0,*}$ are soft (as modules over the sheaf of germs of C^∞ functions on X). It will then result that the invariants $H^{\cdot}(X, Hom_{\mathcal{O}}(\mathcal{L}, \Omega))$ are the cohomology groups of the complex $\Gamma(X, Hom_{\mathcal{O}}(\mathcal{L}, \mathcal{E}^{n,*})) = \mathrm{Hom}_{\mathcal{O}}(\mathcal{L}, \mathcal{E}^{n,*})$ and the invariants $H_c^{\cdot}(X, \mathcal{L})$ are the cohomology groups of the complex $\Gamma_c(X, \mathcal{L} \otimes_{\mathcal{O}} \mathcal{K}^{0,*})$. The complex $\Gamma(X, \mathcal{E}^{n,*})$ is FS and its topological dual is isomorphic to $\Gamma_c(X, \mathcal{K}^{0,*})$ (the isomorphism changes the degree q in $n-q$). Since $\mathcal{L}$ is of the form $\mathcal{O}^p$, one can easily deduce that $\mathrm{Hom}_{\mathcal{O}}(\mathcal{L}, \mathcal{E}^{n,*})$ is an FS complex and its topological dual is isomorphic to $\Gamma_c(X, \mathcal{L} \otimes_{\mathcal{O}} \mathcal{K}^{0,*})$.

The FS topology obtained on $\mathrm{Hom}_{\mathcal{O}}(\mathcal{L}, \Omega)$, as the kernel of the map $\mathrm{Hom}_{\mathcal{O}}(\mathcal{L}, \mathcal{E}^{n,0}) \xrightarrow{d''} \mathrm{Hom}_{\mathcal{O}}(\mathcal{L}, \mathcal{E}^{n,1})$, coincides with that defined at the beginning of the paragraph (it is obviously finer and both of them are Fréchet topologies!). The conclusion of the lemma easily follows by theorem B and by 1.3.

Proof of the theorem 2.1. We consider on the invariants $\mathrm{Ext}^{\bullet}_{\mathcal{O}}(X; \mathcal{F}, \Omega)$ the FS topology given by lemma 2.4 and prove that it verifies the theorem. We first make the supplementary hypothesis that $\mathcal{F}$ admits a resolution (not necessarily finite!)

$$\ldots \to \mathcal{L}^2 \to \mathcal{L}^1 \to \mathcal{L}^0 \to \mathcal{F} \to 0$$

by free sheaves of finite rank. Consider the complex

$$(*) \qquad \ldots \to H^n_c(X, \mathcal{L}^2) \to H^n_c(X, \mathcal{L}^1) \to H^n_c(X, \mathcal{L}^0) \to 0.$$

Since $H^q_c(X, \mathcal{L}^i) = 0$ for any i and $q \neq n$ (by 2.6), $H^q_c(X, \mathcal{F})$ is just the homology group in dimension $n-q$ of this complex: one can easily verify this assertion by descending induction on q, by using the sheaves $\mathcal{F}^i = \mathrm{Coker}\,(\mathcal{L}^{i+1} \to \mathcal{L}^i)$, $i \geqslant 0$.

Consider now the complex

$$(**) \qquad \ldots \leftarrow \mathrm{Hom}_{\mathcal{O}}(\mathcal{L}^2, \Omega) \leftarrow \mathrm{Hom}_{\mathcal{O}}(\mathcal{L}^1, \Omega) \leftarrow \mathrm{Hom}_{\mathcal{O}}(\mathcal{L}^0, \Omega) \leftarrow 0.$$

Its cohomology groups are equal to $\mathrm{Ext}^{\bullet}_{\mathcal{O}}(X; \mathcal{F}, \Omega)$. By lemma 2.6, $(**)$ is an FS complex and its topological dual is algebraically isomorphic to $(*)$. By lemma 2.2 the maps of $(**)$ are strict. Then, by 1.3 we derive that $H^q_c(X, \mathcal{F})$ is algebraically isomorphic to the topological dual of $\mathrm{Ext}^{n-q}_{\mathcal{O}}(X; \mathcal{F}, \Omega)$ taking on the latter the FS topology deduced from $(**)$.

The homology sheaves of the complex

$$\ldots \leftarrow \mathcal{H}om_{\mathcal{O}}(\mathcal{L}^2, \Omega) \leftarrow \mathcal{H}om_{\mathcal{O}}(\mathcal{L}^1, \Omega) \leftarrow \mathcal{H}om_{\mathcal{O}}(\mathcal{L}^0, \Omega) \leftarrow 0$$

coincide with $\mathcal{E}xt^{\bullet}_{\mathcal{O}}(\mathcal{F}, \Omega)$. By applying lemma 2.2 it will easily result that the topology of $\mathrm{Ext}^{n-q}_{\mathcal{O}}(X; \mathcal{F}, \Omega)$ deduced from $(**)$ is identical with that deduced from $\mathcal{E}xt^{n-q}_{\mathcal{O}}(\mathcal{F}, \Omega)$ by means of lemma 2.4.

In this way, the theorem is proved in the supplementary hypothesis on $\mathcal{F}$.

We now pass to the general case. Let $(U_r)_{r\geqslant 1}$ be an exhaustion of X by relatively compact Stein open subsets such that the restriction maps $\Gamma(X, \mathcal{O}) \to \Gamma(U_r, \mathcal{O})$ are dense, $r \geqslant 1$. We have natural isomorphisms

$$H^q_c(X, \mathcal{F}) \simeq \varinjlim_r H^q_c(U_r, \mathcal{F}),$$

$q \geqslant 0$ being an arbitrary integer. According to lemmas 2.3, 2.4, we have also natural isomorphisms

$$(\mathrm{Ext}^{n-q}_{\mathcal{O}}(X; \mathcal{F}, \Omega))' \rightleftarrows \varinjlim_r (\mathrm{Ext}^{n-q}_{\mathcal{O}}(U_r; \mathcal{F}, \Omega))'.$$

Since every U_r is a relatively compact Stein open subset, $\mathcal{F}|U_r$ admits a resolution by free sheaves of finite rank.

To conclude with, it is enough to prove that the isomorphisms already built

$$H_c^q(U_r, \mathcal{F}) \simeq (\operatorname{Ext}_{\mathcal{O}}^{n-q}(U_r; \mathcal{F}, \Omega))'$$

agree with the morphisms given by the inclusions $U_r \subset U_{r+1}$. We must only show that the isomorphisms

$$H_c^q(X, \mathcal{F}) \simeq (\operatorname{Ext}_{\mathcal{O}}^{n-q}(X; \mathcal{F}, \Omega))',$$

constructed in the supplementary hypothesis on $\mathcal{F}$, do not depend on the resolution used. Let $\mathcal{L}^\bullet \to \mathcal{F}$ and $\mathcal{L}^{\bullet\bullet} \to \mathcal{F}$ be two such resolutions. Since X is Stein, one easily derives that $\mathcal{L}^\bullet$ and $\mathcal{L}^{\bullet\bullet}$ are homothopic (over $\mathcal{F}$). The conclusion then follows by lemmas 1.3, 2.6. The theorem is completely proved.

COROLLARY 2.7. *Let $(X, \mathcal{O})$ be a Stein manifold of dimension n, and $\mathcal{F}$ a locally free sheaf of finite rank on X. Then $H_c^q(X, \mathcal{F}) = 0$ for $q \neq n$ and $H_c^n(X, \mathcal{F})$ is algebraically isomorphic to the topological dual of* $\operatorname{Hom}_{\mathcal{O}}(\mathcal{F}, \Omega) \simeq \Gamma(X, \check{\mathcal{F}} \otimes_{\mathcal{O}} \Omega)$.

COROLLARY 2.8. *Let $(X, \mathcal{O})$ be a Stein space and $U \subset X$ a Stein open such that the restriction map $\Gamma(X, \mathcal{O}) \to \Gamma(U, \mathcal{O})$ is dense. Then, for any $\mathcal{F} \in \mathbf{Coh}(X)$ and for any integer $q \geqslant 0$, the natural maps*

$$H_c^q(U, \mathcal{F}) \to H_c^q(X, \mathcal{F})$$

are injective.

Proof. Let $(U_r)_{r\geqslant 1}$ be an exhaustion for U by relatively compact Stein open subsets, such that the maps $\Gamma(U, \mathcal{O}) \to \Gamma(U_r, \mathcal{O})$, $r \geqslant 1$, are dense. Since $H_c^q(U, \mathcal{F}) \simeq \simeq \varinjlim_r H_c^q(U_r, \mathcal{F})$, we have only to prove that every map $H_c^q(U_r, \mathcal{F}) \to H_c^q(X, \mathcal{F})$ is injective. In other words, we can assume from the very beginning that U is relatively compact.

Suppose now $(U_r)_{r\geqslant 1}$ is an exhaustion of X by relatively compact Stein open subsets such that the maps $\Gamma(X, \mathcal{O}) \to \Gamma(U_r, \mathcal{O})$, $r \geqslant 1$, are dense and $U_1 = U$. Suffice it to prove that every map $H_c^q(U, \mathcal{F}) \to H_c^q(U_r, \mathcal{F})$ is injective. In this way, we can assume in the statement of the corollary that there is a closed immersion $X \xrightarrow{i} Y$, Y being a Stein manifold. Again, using a suitable exhaustion of U, we can suppose in addition that there is a relatively compact Stein open V in Y such that $V \cap i(X) = i(U)$ and the map $\Gamma(Y, \mathcal{O}_Y) \to \Gamma(V, \mathcal{O}_Y)$ is dense.

So we have reduced the proof to the case $U \subset\subset X$ and X a Stein manifold. The conclusion follows by theorem 2.1, making use of the fact that the maps

$$\operatorname{Ext}^\bullet(X; \mathcal{F}, \Omega) = \Gamma(X, \mathit{Ext}^\bullet(\mathcal{F}, \Omega)) \to \operatorname{Ext}^\bullet(U; \mathcal{F}, \Omega) = \Gamma(U, \mathit{Ext}^\bullet(\mathcal{F}, \Omega))$$

are dense.

Remark. By their construction, the isomorphisms from the theorem do result functorial in $\mathcal{F}$, and are compatible with the short exact sequences and with the restrictions given by inclusions $U \subset X$, U Stein open. As a matter of fact, in

Chapter VII, which is dedicated to the analytic duality, we will show that the duality isomorphisms are naturally defined by means of the Yoneda map and of the "trace map". The statement 2.1 will be sufficient for the applications given in this chapter.

The following result refines 2.1.

THEOREM 2.9. *Let X be a complex manifold of dimension n, $K \subset X$ a Stein holomorphically convex compact, and Ω the sheaf of germs of differential forms of the type $(n, 0)$ with analytic coefficients. Then for any coherent analytic sheaf $\mathcal{F}$ and for any $q \geqslant 0$, $\operatorname{Ext}^{n-q}_{\mathcal{O}}(K; \mathcal{F}, \Omega)$ has a natural DFS structure and its topological dual is algebraically isomorphic to $H^q_K(X, \mathcal{F})$.*

$$(\text{By definition, } \operatorname{Ext}^{\cdot}_{\mathcal{O}}(K; \mathcal{F}, \Omega) = \operatorname{Ext}^{\cdot}_{\mathcal{O}K}(K; \mathcal{F}|K, \Omega|K)).$$

In order to prove this, we need some preparations. Recall that a compact subset of a complex space, admitting a fundamental system of Stein neighbourhoods is called a *Stein compact* (e.g. any holomorphically convex compact in a Stein space). Suppose X is a complex space, $K \subset X$ a Stein compact and $\mathcal{F} \in \mathbf{Coh}(X)$ and choose a countable fundamental system of Stein neighbourhoods U_r forK. Thus, $\Gamma(K, \mathcal{F}) = \varinjlim_r \Gamma(U_r, \mathcal{F})$ and consider on $\Gamma(K, \mathcal{F})$ the topolgy of inductive limit in the category of locally convex topological vector spaces. We will introduce another topology in the following way. We may clearly replace X by some neighbourhood of K such that there is a closed immersion $X \xrightarrow{i} Y$, Y being a Stein open subset of a numerical space such that there exists an epimorphism of $\mathcal{O}_Y$-modules $\mathcal{O}^p_Y \to i_*(\mathcal{F})$. By theorem B one finds a surjection $\Gamma(i(K), \mathcal{O}^p_Y) \to \Gamma(i(K), i_*(\mathcal{F})) = \Gamma(K, \mathcal{F})$. But $\Gamma(i(K), \mathcal{O}^p_Y)$ has a natural DFS structure (§ 1, (**c**)) and we can endow $\Gamma(K, \mathcal{F})$ with the quotient topology. It can be easily verified that this one coincides with the inductive limit topology defined above, in virtue of the definitions of these topologies and the fact that the topology on every $\Gamma(U_r, \mathcal{F})$ coincides with the quotient topology from $\Gamma(i(U_r), \mathcal{O}^p_Y)$.

PROPOSITION 2.10. *Let X be a complex space, $K \subset X$ a Stein compact and $\mathcal{F} \in$ **Coh** (X). The topology constructed above on $\Gamma(K, \mathcal{F})$ is LF and DFS. Moreover, it is independent on all choice.*

Proof. Fix $x \in X$ an arbitrary point. For any integer $n \geqslant 0$, $\mathcal{F}_x/\mathfrak{m}^n_x\mathcal{F}_x$ is an $\mathcal{O}_x/\mathfrak{m}^n_x$-module of finite type, hence it is of finite dimension over the complex field. Consider on it the Haussdorf topology of topological vector space. Let U be an open set containing the point x. We are going to prove that the composition

$$\Gamma(U, \mathcal{F}) \to \mathcal{F}_x \to \mathcal{F}_x/\mathfrak{m}^n_x\mathcal{F}_x$$

is continuous. We can assume that U is a Stein open set such that there is a closed immersion $U \xrightarrow{i} V$, V being a Stein open subset of a numerical space and, furthermore, $i_*(\mathcal{F}|U)$ is the quotient of some $\mathcal{O}^p_V$. Finally look at the commutative diagram

$$\begin{array}{ccc} \Gamma(V, \mathcal{O}^p_V) \to \Gamma(V, i_*(\mathcal{F}|U)) & = \Gamma(U, \mathcal{F}) \to 0 \\ \downarrow & \downarrow \\ \mathcal{O}^p_{V,i(x)}/\mathfrak{m}^n_{i(x)}\,\mathcal{O}^p_{V,i(x)} & \to \mathcal{F}_x/\mathfrak{m}^n_x\mathcal{F}_x \end{array}$$

and notice that $\Gamma(U, \mathcal{F})$ has the quotient-topology from $\Gamma(V, \mathcal{O}_V^p)$ and the first vertical arrow is obviously continuous.

In this way, we have proved that the maps $\Gamma(U, \mathcal{F}) \to \mathcal{F}_x/\mathfrak{m}_x^n\mathcal{F}_x$ are continuous, for any $x \in K$, $n \geqslant 1$. Accordingly, the maps $\Gamma(K, \mathcal{F}) \to \mathcal{F}_x/\mathfrak{m}_x^n\mathcal{F}_x$ are also continuous. By the Krull theorem, the continuous map

$$\Gamma(K, \mathcal{F}) \to \prod_{x\in K} (\prod_{n\geqslant 0} \mathcal{F}_x/\mathfrak{m}_x^n\mathcal{F}_x)$$

is injective, hence, $\Gamma(K, \mathcal{F})$ is separated. The topology on $\Gamma(K, \mathcal{F})$ turns out to be LF and DFS by the very way it has been defined. Its independence on the choice already made readily results by using the open map theorem for DFS spaces (or for LF spaces). The proposition is thus proved.

A remarkable case is provided by the reduction of K to one point. The topology thus obtained is called the canonical topology or the topology of the uniform convergence on neighbourhoods [33], [43].

LEMMA 2.11. *Let X be a complex space, $K \subset X$ a Stein compact, and $\mathcal{F} \to \mathcal{G}$ a morphism in* **Coh** *(X). Then the map $\Gamma(K, \mathcal{F}) \to \Gamma(K, \mathcal{G})$ is continuous and strict.*

Proof. The first assertion can easily derive from the continuity of $\Gamma(U, \mathcal{F}) \to$ $\to \Gamma(U, \mathcal{G})$, U being a Stein neighbourhood of K. In order to prove the second assertion one can proceed as in lemma 2.2.

LEMMA 2.12. *If $(X, \mathcal{O})$ is a complex space, $K \subset X$ a Stein compact and $\mathcal{F}, \mathcal{G} \in$* **Coh** *$(X)$, then for any $q \geqslant 0$ there exists a canonical isomorphism*

$$\mathrm{Ext}^q_{\mathcal{O}}(K; \mathcal{F}, \mathcal{G}) \xrightarrow{\sim} \Gamma(K, \mathcal{E}xt^q_{\mathcal{O}}(\mathcal{F}, \mathcal{G})).$$

Proof. We claim that the canonical morphism

$$\mathcal{E}xt^{\cdot}_{\mathcal{O}}(\mathcal{F}, \mathcal{G}) \mid K \to \mathcal{E}xt^{\cdot}_{\mathcal{O}|K}(\mathcal{F}|K, \mathcal{G}|K)$$

is an isomorphism: one can prove this by passing to stalks or computing the functors *Ext* by means of resolutions with free sheaves of finite rank for $\mathcal{F}$ around K, the existence of such resolutions being assured under the hypothesis. The morphism asserted in this lemma is given by this isomorphism and by the morphism

$$\mathrm{Ext}^q_{\mathcal{O}}(K; \mathcal{F}, \mathcal{G}) = \mathrm{Ext}^q_{\mathcal{O}|K}(K; \mathcal{F}|K, \mathcal{G}|K) \to \Gamma(K, \mathcal{E}xt^q_{\mathcal{O}|K}(\mathcal{F}|K, \mathcal{G}|K))$$

of passing from presheaves to associated sheaves. There is a spectral sequence by the term $E_2^{p,q} = H^p(K, \mathcal{E}xt^q_{\mathcal{O}|K}(\mathcal{F}|K, \mathcal{G}|K)) \simeq H^p(K, \mathcal{E}xt^q_{\mathcal{O}}(\mathcal{F}, \mathcal{G}))$, which converges to $\mathrm{Ext}^{\cdot}_{\mathcal{O}|K}(K; \mathcal{F}|K, \mathcal{G}|K)$. Since the sheaves $\mathcal{E}xt^{\cdot}_{\mathcal{O}}(\mathcal{F}, \mathcal{G})$ are coherent, theorem B shows that the sequence degenerates and the conclusion follows.

This lemma allows us to define a DFS topology on every $\mathrm{Ext}^q_{\mathcal{O}}(K; \mathcal{F}, \mathcal{G})$.

LEMMA 2.13. *Let X be a paracompact topological space with countable topology, $K \subset X$ a compact and $\mathcal{F}$ a sheaf of abelian groups on X. Then the canonical morphism*

$$H^q(X\setminus K, \mathcal{F}) \to \varprojlim_{U\supset K} H^q(X\setminus U, \mathcal{F})$$

is an epimorphism for any q. If, moreover, the restriction maps $H^{q-1}(X, \mathcal{F}) \to H^{q-1}(X\setminus U, \mathcal{F})$ are surjective for any open U which belongs to a fundamental system of neighbourhoods of K, then this morphism is an isomorphism.

Proof. Consider a resolution $\mathcal{N}^\bullet$ of $\mathcal{F}|X\setminus K$ by injective sheaves (or soft ones!). This allows us to calculate the invariants $H^\bullet(X\setminus K, \mathcal{F})$ and $H^\bullet(X\setminus U, \mathcal{F})$, U neighbourhood of K. Choose a countable fundamental system of neighbourhoods $(U_r)_{r\geqslant 1}$ for K (and for the second assertion, a system verifying the supplementary hypothesis!). The maps $\Gamma(X\setminus U_{r+1}, \mathcal{N}^\bullet) \to \Gamma(X\setminus U, \mathcal{N}^\bullet)$ are surjective, $r \geqslant 1$. The conclusion actually follows by an elementary reasoning of projective systems by using a suitable diagram.

The proof of the theorem 2.9. The problem being around K, we can assume X Stein. On the invariants $\mathrm{Ext}^{\bullet}_{\mathcal{O}}(K; \mathcal{F}, \Omega)$ we shall consider the DFS topology given by lemma 2.12 and proposition 2.10.

There is a canonical morphism (given by restriction)

$$(*) \qquad H^q_K(X, \mathcal{F}) \to \varprojlim_{U\supset K} H^q_c(U, \mathcal{F}),$$

$q \geqslant 0$ being an arbitrary integer. Since K is a Stein compact, the projective limit can be taken over a countable fundamental system of relatively compact Stein neighbourhoods U of K. According to theorem 2.1, every $\mathbb{C}$-linear space $\mathrm{Ext}^{n-q}_{\mathcal{O}}(U; \mathcal{F}, \Omega)$ has a natural FS structure, such that its topological dual is algebraically isomorphic to $H^q_c(U, \mathcal{F})$. Moreover, these isomorphisms agree with the maps induced by inclusions $U' \subset U$. We thus obtain a natural isomorphism

$$\varprojlim_{U\supset K} H^q_c(U, \mathcal{F}) \simeq \varprojlim_{U\supset K} (\mathrm{Ext}^{n-q}_{\mathcal{O}}(U; \mathcal{F}, \Omega))' \simeq (\mathrm{Ext}^{n-q}_{\mathcal{O}}(K; \mathcal{F}, \Omega))',$$

the last isomorphism being a consequence of lemmas 2.3, 2.4, 2.12 and of the manner of considering the topologies. By composition with $(*)$ one thus obtains the following morphism

$$(**) \qquad H^q_K(X, \mathcal{F}) \to (\mathrm{Ext}^{n-q}_{\mathcal{O}}(K; \mathcal{F}, \Omega))',$$

$q \geqslant 0$ being an arbitrary integer. These morphisms are functorial in $\mathcal{F}$ and compatible with short exact sequences. The maps $(**)$ are bijective if and only if the maps $(*)$ are. Thereby we can assume X a numerical space.

We first prove that $(**)$ is an isomorphism for $\mathcal{F}$ locally free. In this case it is enough to prove that $(*)$ is an isomorphism. The case $q = 0$ is obvious. For $q = 1$ we apply the exact sequences

$$0 \to \Gamma_K(X, \mathcal{F}) = 0 \to \Gamma(X, \mathcal{F}) \to \Gamma(X\setminus K, \mathcal{F}) \to H^1_K(X, \mathcal{F}) \to 0$$

$$0 \to \Gamma_c(U, \mathcal{F}) = 0 \to \Gamma(X, \mathcal{F}) \to \Gamma(X\setminus U, \mathcal{F}) \to H^1_c(U, \mathcal{F}) \to 0$$

($\Gamma_K(X, \mathcal{F})$ and $\Gamma_c(U, \mathcal{F})$ are subspaces in $\Gamma_c(X, \mathcal{F})$, which is null by corollary 2.7). Passing to projective limit (this can be considered countably indexed), one gets the exact sequence

$$0 \to \Gamma(X, \mathcal{F}) \to \varprojlim_{U \supset K} \Gamma(X \setminus U, \mathcal{F}) \to \varprojlim_{U \supset K} H_c^1(U, \mathcal{F}) \to 0.$$

Since the restriction map $\Gamma(X \setminus K, \mathcal{F}) \to \varprojlim_{U \supset K} \Gamma(X \setminus U, \mathcal{F})$ is bijective, the conclusion follows for $q = 1$.

Suppose now $q \geqslant 2$. We have the exact sequences

$$\ldots \to H^{q-1}(X, \mathcal{F}) \to H^{q-1}(X \setminus K, \mathcal{F}) \to H_K^q(X, \mathcal{F}) \to H^q(X, \mathcal{F}) \to \ldots,$$

$$\ldots \to H^{q-1}(X, \mathcal{F}) \to H^{q-1}(X \setminus U, \mathcal{F}) \to H_c^q(U, \mathcal{F}) \to H^q(X, \mathcal{F}) \to \ldots$$

For $q \neq n+1$, the conclusion follows by lemma 2.13, making use of corollary 2.7. It only remains to show that $H_K^{n+1}(X, \mathcal{F}) = 0$. To this aim it is sufficient to prove that the space $H_K^{n+1}(X, \mathcal{O}) \simeq H^n(X \setminus K, \mathcal{O})$ is null. At this point we use the properties of the Laplace operator (B. Malgrange, Ann. Inst. Fourier, 6, 1955): let $\omega = \alpha \, d\bar{z}_1 \wedge \ldots \wedge d\bar{z}_n$ be a form of type $(0, n)$ on $X \setminus K$ with C^∞ coefficients; there is a C^∞-function β on $X \setminus K$ such that $1/4 \, \Delta\beta = \sum_{i,j} \frac{\partial^2 \beta}{\partial z_i \partial \bar{z}_j} = \alpha$, hence the form $\bar{\omega} = \sum_i (-1)^{i+1} \frac{\partial \beta}{\partial z_i} d\bar{z}_1 \wedge \ldots \wedge \widehat{d\bar{z}_i} \ldots \wedge d\bar{z}_n$ verifies the relation $d''\bar{\omega} = \omega$.

Another argument for proving that $H^n(X \setminus K, \mathcal{O}) = 0$ can be given by using the theory of hyperfunctions (P. Schapira, Lecture notes, Vol. 126, 1970).

We now prove that (**) is an isomorphism in the general case $\mathcal{F} \in \mathbf{Coh}(X)$. We may replace X by a suitable neighbourhood of K (the invariants of interest do not change), thus we can assume that $\mathcal{F}$ admits a finite resolution by locally free sheaves of finite rank. By induction on the length of such a resolution, all that remains to end the proof is to show the following assertion: if $0 \to \mathcal{F} \to \mathcal{G} \to \mathcal{H} \to 0$ is an exact sequence in $\mathbf{Coh}(X)$ and the morphism (**) is an isomorphism for $\mathcal{F}$ and $\mathcal{G}$, then it is an isomorphism for $\mathcal{H}$ too. But we have the exact sequences

$$\ldots \to H_K^q(X, \mathcal{F}) \to H_K^q(X, \mathcal{G}) \to H_K^q(X, \mathcal{H}) \to H_K^{q+1}(X, \mathcal{F}) \to H_K^{q+1}(X, \mathcal{G}) \to \ldots,$$

$$\ldots \leftarrow \mathrm{Ext}_{\mathcal{O}}^{n-q}(K; \mathcal{F}, \Omega) \leftarrow \mathrm{Ext}_{\mathcal{O}}^{n-q}(K; \mathcal{G}, \Omega) \leftarrow \mathrm{Ext}_{\mathcal{O}}^{n-q}(K; \mathcal{H}, \Omega) \leftarrow$$

$$\leftarrow \mathrm{Ext}_{\mathcal{O}}^{n-q-1}(K; \mathcal{F}, \Omega) \leftarrow \mathrm{Ext}_{\mathcal{O}}^{n-q-1}(K; \mathcal{G}, \Omega) \leftarrow \ldots$$

The maps of the second sequence are strict by 2.11 and 2.12. In virtue of 1.4, one obtains a new exact sequence passing to topological duals. The proof ends from the lemma of the five homomorphisms.

By the above proof it results that the dualities of 2.1 and 2.9 agree with the natural maps

$$H^{\bullet}_K(X, \mathcal{F}) \to H^{\bullet}_c(X, \mathcal{F}), \ \mathrm{Ext}^{\bullet}_{\mathcal{O}}(K; \mathcal{F}, \Omega) \leftarrow \mathrm{Ext}^{\bullet}_{\mathcal{O}}(X; \mathcal{F}, \Omega).$$

Remark. If X is an analytic set in some $\mathbb{C}^n$, then one can prove that any Stein compact of X is a Stein compact in $\mathbb{C}^n$. By using this fact, in the statement of the theorem as well as in the following results (except 2.17) we can assume K to be Stein only.

In the meantime let us give some immediate consequences of theorems 2.1 and 2.9.

COROLLARY 2.14. *Let X be a complex manifold of dimension n, $K \subset X$ a holomorphically convex compact and $\mathcal{F}$ a locally free sheaf of finite rank on X. Then $H^q_K(X, \mathcal{F}) = 0$ for $q \neq n$ and $H^n_K(X, \mathcal{F})$ is algebraically isomorphic to the topological dual of the space $\Gamma(K, \check{\mathcal{F}} \otimes_{\mathcal{O}} \Omega)$.*

COROLLARY 2.15. *Suppose $(X, \mathcal{O})$ is a complex manifold of dimension n, $\mathcal{F} \in$ **Coh** X and $x \in X$. For any integer $q \geqslant 0$, the topological dual of the analytic module $\mathrm{Ext}^{n-q}_{\mathcal{O}_x}(\mathcal{F}_x, \mathcal{O}_x)$, which is endowed with the topology of the uniform convergence on neighbourhoods, is algebraically isomorphic to $H^q_x(X, \mathcal{F})$* ($H^{\bullet}_x(X, \mathcal{F}) = H^{\bullet}_{\{x\}}(X, \mathcal{F})$ are the cohomology groups of $\mathcal{F}$ with supports in $\{x\}$).

COROLLARY 2.16. *Let X be a Stein space, $K \subset X$ a holomorphically convex compact and $\mathcal{F} \in$ **Coh**(X). Then the canonical morphism*

$$H^q_K(X, \mathcal{F}) \to \varprojlim_{U \supset K} H^q_c(U, \mathcal{F})$$

is an isomorphism for any integer $q \geqslant 0$.

Proof. If X is a manifold, we proved this fact in the course of the proof of 2.9 (and it is easy to be obtained *a posteriori* from the dualities given by 2.1 and 2.9). The general case may be reduced to manifolds, putting instead of X a Stein neighbourhood of K and making use of a suitable immersion.

COROLLARY 2.17. *Let X be a Stein space, $K \subset X$ a holomorphically convex compact and $\mathcal{F} \in$ **Coh** (X). Then the canonical morphisms*

$$H^q_K(X, \mathcal{F}) \to H^q_c(X, \mathcal{F})$$

are injective, $q \geqslant 0$.

Proof. If X is a Stein manifold, the theorems 2.1 and 2.9 give the isomorphisms

$$H^q_c(X, \mathcal{F}) \simeq (\mathrm{Ext}^{n-q}_{\mathcal{O}}(X; \mathcal{F}, \Omega))', \ H^q_K(X, \mathcal{F}) \simeq (\mathrm{Ext}^{n-q}_{\mathcal{O}}(K; \mathcal{F}, \Omega))'.$$

Since $\mathrm{Ext}^{n-q}_{\mathcal{O}}(X; \mathcal{F}, \Omega) \simeq \Gamma(X, \mathcal{E}xt^{n-q}_{\mathcal{O}}(\mathcal{F}, \Omega))$ and $\mathrm{Ext}^{n-q}_{\mathcal{O}}(K; \mathcal{F}, \Omega) \simeq \Gamma(K, \mathcal{E}xt^{n-q}_{\mathcal{O}}(\mathcal{F}, \Omega))$, the corollary then follows by the fact that the map $\Gamma(X, \mathcal{H}) \to \Gamma(K, \mathcal{H})$ is dense for any $\mathcal{H} \in$ **Coh** (X).

For the general case, consider an exhaustion $(U_r)_{r\geqslant 1}$ of X by Stein relatively compact open sets containing K, each U_r being embedded in some numerical space. Obviously, $H^q_c(X, \mathcal{F}) \simeq \varinjlim_r H^q_c(U_r, \mathcal{F})$, hence it is enough to prove the corollary for U_r's; by immersions we reduce the proof to the above treated case of manifolds.

Let now X be a complex space with countable topology and $\mathcal{F} \in \mathbf{Coh}\,(X)$. Recall how the invariants $H^\bullet(X, \mathcal{F})$ can be topologized ([38] or Ch. VII). Let $\mathfrak{U}$ be a countable Stein open covering of X. We have a canonical isomorphism $H^\bullet(\mathfrak{U}, \mathcal{F}) \simeq H^\bullet(X, \mathcal{F})$ and if we consider on any $\Gamma(U, \mathcal{F})$, U an arbitrary open set in X, the usual FS topology, then $C^\bullet(\mathfrak{U}, \mathcal{F})$ becomes an FS complex and so, by passing to cohomology, one gets a topology on $H^\bullet(X, \mathcal{F})$, generally nonseparated. Its independence on the covering $\mathfrak{U}$ can be shown. If U is an open set in X, then it can be proved that the maps $H^\bullet(X, \mathcal{F}) \to H^\bullet(U, \mathcal{F})$, deduced by restriction, are continuous. If $\mathcal{F} \to \mathcal{G}$ is a morphism of coherent sheaves, then the maps $H^\bullet(X, \mathcal{F}) \to H^\bullet(X, \mathcal{G})$ are continuous too. Now, consider an exact sequence $0 \to \mathcal{F}' \to \mathcal{F} \to \mathcal{F}'' \to 0$ in $\mathbf{Coh}\,(X)$; then the "coboundary" maps $H^q(X, \mathcal{F}'') \to H^{q+1}(X, \mathcal{F}')$ are also continuous (the proof easily follows in virtue of the exact sequence of FS complexes $0 \to C^\bullet(\mathfrak{U}, \mathcal{F}') \to C^\bullet(\mathfrak{U}, \mathcal{F}) \to C^\bullet(\mathfrak{U}, \mathcal{F}'') \to 0$ and by the construction of the coboundary map, where $\mathfrak{U}$ is a countable Stein open covering of X).

LEMMA 2.18. *Let X be a complex space with countable topology, $\mathcal{F} \to \mathcal{G}$ a morphism in* $\mathbf{Coh}\,(X)$ *and $q \geqslant 0$ an integer. If the map $H^q(X, \mathcal{F}) \to H^q(X, \mathcal{G})$ is surjective, then it is strict.*

Proof. Take a countable covering $\mathfrak{U}$ of X, by Stein open subsets and consider the commutative diagram

$$\begin{array}{ccc} Z^q(\mathfrak{U}, \mathcal{F}) \oplus C^{q-1}(\mathfrak{U}, \mathcal{F}) & \to & Z^q(\mathfrak{U}, \mathcal{G}) \\ \downarrow & & \downarrow \\ H^q(X, \mathcal{F}) \simeq H^q(\mathfrak{U}, \mathcal{F}) & \to & H^q(\mathfrak{U}, \mathcal{G}) \simeq H^q(X, \mathcal{G}). \end{array}$$

The notations are the usual ones and the morphisms are obtained as follows: the first horizontal arrow is that given by the sum of the canonical maps $Z^q(\mathfrak{U}, \mathcal{F}) \to Z^q(\mathfrak{U}, \mathcal{G})$ and $C^{q-1}(\mathfrak{U}, \mathcal{G}) \to Z^q(\mathfrak{U}, \mathcal{G})$, the first vertical arrow is the composition of the projection on $Z^q(\mathfrak{U}, \mathcal{F})$ with the canonical surjection $Z^q(\mathfrak{U}, \mathcal{F}) \to H^q(\mathfrak{U}, \mathcal{F})$; finally, the maps $Z^q(\mathfrak{U}, \mathcal{G}) \to H^q(\mathfrak{U}, \mathcal{G})$ and $H^q(\mathfrak{U}, \mathcal{F}) \to H^q(\mathfrak{U}, \mathcal{G})$ are the canonical ones. By hypothesis the first horizontal arrow is surjective (and obviously continuous!), hence by Banach theorem it is strict. The conclusion easily follows.

THEOREM 2.19. *Suppose X is a Stein space, $K \subset X$ a holomorphically convex compact and $\mathcal{F}$ a coherent analytic sheaf on X. Then the topological vector spaces $H^\bullet(X \setminus K, \mathcal{F})$ are separated.*

Proof. Since $\Gamma(X \setminus K, \mathcal{F})$ is separated, it only remains to prove the separation of spaces $H^q(X \setminus K, \mathcal{F})$, $q \geqslant 1$. We have the exact sequence

$$\ldots \to H^q(X, \mathcal{F}) \to H^q(X \setminus K, \mathcal{F}) \to H^{q+1}_K(X, \mathcal{F}) \to H^{q+1}(X, \mathcal{F}) \to \ldots.$$

hence we get a canonical isomorphism $H^q(X\setminus K, \mathcal{F}) \xrightarrow{\sim} H_K^{q+1}(X, \mathcal{F})$ for any $q \geqslant 1$. If U is a relatively compact Stein neighbourhood of K, then we also derive a canonical isomorphism $H^q(U\setminus K, \mathcal{F}) \simeq H_K^{q+1}(U, \mathcal{F})$. Since $H_K^{q+1}(U, \mathcal{F}) \simeq H_K^{q+1}(X, \mathcal{F})$, we deduce the bijectivity of the natural map

$$H^q(X\setminus K, \mathcal{F}) \to H^q(U\setminus K, \mathcal{F}),$$

$q \geqslant 1$, which, moreover, is continuous. Therefore, if the spaces $H^q(U\setminus K, \mathcal{F})$ were separated, the same would be true for $H^q(X\setminus K, \mathcal{F})$.

So it suffices to prove the theorem for U and further, by a suitable immersion, we reduce the problem to the case of manifolds.

We can suppose henceforth X a Stein manifold. Denote $n = \dim X$. Then $\operatorname{Ext}_{\mathcal{O}}^{n-q-1}(K; \mathcal{F}, \Omega)$ has a DFS structure such that its topological dual is algebraically isomorphic to $H_K^{q+1}(X, \mathcal{F})$. We shall consider on $H_K^{q+1}(X, \mathcal{F})$ the strong topology on duals which is an FS topology (as FS spaces are reflexive). We are going to prove the continuity of the canonical map $H^q(X\setminus K, \mathcal{F}) \to H_K^{q+1}(X, \mathcal{F})$; since it is bijective for $q \geqslant 1$ and $H_K^{q+1}(X, \mathcal{F})$ is separated, the separation of $H^q(X\setminus K, \mathcal{F})$ follows as required.

We first consider $\mathcal{F}$ locally free. In a Stein neighbourhood U of K, consider an exact sequence of the form $0 \to \mathcal{G} \to \mathcal{O}^p \to \mathcal{F} \to 0$. The map $H^q(X\setminus K, \mathcal{F}) \to$ $\to H_K^{q+1}(X, \mathcal{F})$ is just the composition of maps $H^q(X\setminus K, \mathcal{F}) \to H^q(U\setminus K, \mathcal{F}) \to$ $\to H_K^{q+1}(U, \mathcal{F}) \simeq H_K^{q+1}(X, \mathcal{F})$ and thus we can replace X by U, that is we can suppose the above exact sequence defined on whole X. Since it splits in any point, the sequence $0 \to \mathit{Hom}(\mathcal{F}, \mathcal{G}) \to \mathit{Hom}(\mathcal{F}, \mathcal{O}^p) \to \mathit{Hom}(\mathcal{F}, \mathcal{F}) \to 0$ is exact. By theorem B, $H^1(X, \mathit{Hom}(\mathcal{F}, \mathcal{G})) = 0$, we have therefore an exact sequence

$$0 \to \operatorname{Hom}(\mathcal{F}, \mathcal{G}) \to \operatorname{Hom}(\mathcal{F}, \mathcal{O}^p) \to \operatorname{Hom}(\mathcal{F}, \mathcal{F}) \to 0$$

and deduce the existence of a morphism $\mathcal{F} \to \mathcal{O}^p$ such that its composition with the map $\mathcal{O}^p \to \mathcal{F}$ is the identity. Thus, in order to prove the continuity of the maps $H^q(X\setminus K, \mathcal{F}) \to H_K^{q+1}(X, \mathcal{F})$ we could suppose, without loss of generality, $\mathcal{F}$ a direct factor in some $\mathcal{O}^p$. Then $H_K^{q+1}(X, \mathcal{F})$ is a closed subspace of $H_K^{q+1}(X, \mathcal{O}^p)$. From the following commutative diagram given by the morphism $\mathcal{F} \to \mathcal{O}^p$

$$\begin{array}{ccc} H^q(X\setminus K, \mathcal{F}) & \to & H_K^{q+1}(X, \mathcal{F}) \\ \downarrow & & \downarrow \\ H^q(X\setminus K, \mathcal{O}^p) & \to & H_K^{q+1}(X, \mathcal{O}^p) \end{array}$$

it only remains to prove the continuity of the maps $H^q(X\setminus K, \mathcal{F}) \to H_K^{q+1}(X, \mathcal{F})$ for $\mathcal{F} = \mathcal{O}^p$ and by additivity for $\mathcal{F} = \mathcal{O}$. But $H_K^i(X, \mathcal{O}) = 0$ for $i \neq n$ and $H_K^n(X, \mathcal{O})$ is the topological dual of the space $\Omega(K)$. Hence what remains to demonstrate is the continuity of the map

$$H^{n-1}(X\setminus K, \mathcal{O}) \to H_K^n(X, \mathcal{O}).$$

If U is a Stein open in X, then $H^n_c(U, \mathcal{O})$ is algebraically isomorphic to the topological dual of the space $\Gamma(U, \Omega)$; endowed with the strong topology, $H^n_c(U, \mathcal{O})$ becomes an FS space. Let $(U_r)_{r\geqslant 1}$ be a fundamental system of Stein neighbourhoods of K such that $U_{r+1} \subset\subset U_r$, for any r. Then we have an isomorphism of locally convex topological vector spaces $\Gamma(K, \Omega) \leftarrow \varinjlim_r \Gamma(U_r, \Omega)$. Passing to the strong topological duals we get a topological isomorphism

$$H^{n-1}(X, \mathcal{O}) \rightarrow \varprojlim_r H^n_c(U_r, \mathcal{O})$$

since any bounded subset of $\Gamma(K, \Omega)$ is the image of a bounded subset of some $\Gamma(U_r, \Omega)$. Therefore, we must show that the canonical morphism

$$H^{n-1}(X \setminus K, \mathcal{O}) \xrightarrow{\delta} H^n_c(U, \mathcal{O}),$$

is continuous for a relatively compact Stein neighbourhood U of K. We shall explicit this morphism by means of Dolbeault resolution. Consider two open sets U', U'' such that $K \subset U' \subset\subset U'' \subset U$ and choose $\varphi \in C^\infty(X)$ such that $\varphi = 0$ on U' and $\varphi = 1$ on $X \setminus U''$. If ξ_n is a sequence of cycles from $\Gamma(X \setminus K, \mathcal{E}^{0,n-1})$ which tends to a cycle $\xi \in \Gamma(X \setminus K, \mathcal{E}^{0,n-1})$, we have to show that $\delta(\hat{\xi}_n) \rightarrow \delta(\hat{\xi})$ where "$\wedge$" denotes the associated cohomology class. The elements $d''(\varphi\xi_n)$, $d''(\varphi\xi)$ belong to $\Gamma(X, \mathcal{E}^{0,n})$ and their supports are contained in the closure of U''; hence they belong to $\Gamma_c(U, \mathcal{E}^{0,n})$. The morphism δ is the composition $H^{n-1}(X \setminus K, \mathcal{O}) \rightarrow$ $\rightarrow H^{n-1}(X \setminus U, \mathcal{O}) \rightarrow H^n_c(U, \mathcal{O})$, where the first map is given by restriction and the second by the coboundary map. The latter can be constructed by means of the exact sequnce of complexes

$$0 \rightarrow \Gamma_c(U, \mathcal{E}^{0,*}) \rightarrow \Gamma(X, \mathcal{E}^{0,*}) \rightarrow \Gamma(X \setminus U, \mathcal{E}^{0,*}) \rightarrow 0$$

which connects the invariants $H^\bullet_c(U, \mathcal{O})$, $H^\bullet(X, \mathcal{O})$, $H^\bullet(X \setminus U, \mathcal{O})$. It will then result that $\delta(\hat{\xi}_n)$, $\delta(\hat{\xi})$ are just the cohomology classes associated to the elements $d''(\varphi\xi_n)$, $d''(\varphi\xi)$ and the required result easily follows.

We analyse now the general case, namely we shall prove that for any sheaf $\mathcal{F} \in \mathbf{Coh}(X)$, the maps

$$H^q(X \setminus K, \mathcal{F}) \rightarrow H^{q+1}_K(X, \mathcal{F})$$

are continuous. That will be done by induction on the number prof $\mathcal{F} =$ $= \inf_{x \in X} (\text{prof}_{\mathcal{O}_x} \mathcal{F}_x)$. The case prof $= n$ is already treated. We can write an exact sequence of the form $0 \rightarrow \mathcal{G} \rightarrow \mathcal{O}^{n_0} \rightarrow \mathcal{F} \rightarrow 0$. By 1.13, prof $\mathcal{G} =$ prof $\mathcal{F} + 1$ (prof $\mathcal{F} < n$). The diagram

$$(*) \quad \begin{array}{ccccccccc} \ldots \rightarrow & H^q(X \setminus K, \mathcal{O}^{n_0}) & \rightarrow & H^q(X \setminus K, \mathcal{F}) & \rightarrow & H^{q+1}(X \setminus K, \mathcal{G}) & \rightarrow & H^{q+1}(X \setminus K, \mathcal{O}^{n_0}) & \rightarrow \ldots \\ & \downarrow & & \downarrow & & \downarrow & & \downarrow & \\ \ldots \rightarrow & H^{q+1}_K(X, \mathcal{O}^{n_0}) & \rightarrow & H^{q+1}_K(X, \mathcal{F}) & \rightarrow & H^{q+2}_K(X, \mathcal{G}) & \rightarrow & H^{q+2}_K(X, \mathcal{O}^{n_0}) & \rightarrow \ldots \end{array}$$

is commutative (modulo the signs) and the horizontal arrows are continuous, those of the line below being even strict (this line is an exact sequence of FS spaces). Let $q \leqslant n-2$, then by 2.14, $H_K^{q+1}(X, \mathcal{C}^{n_0}) = 0$ and the assertion follows by means of the diagram (*) by induction hypothesis. Finally, if $q \geqslant n-1$, then $H_K^{q+1}(X \setminus K, \mathcal{G}) \simeq H_K^{q+2}(X, \mathcal{G}) = 0$ and the assertion follows again from the diagram (*) by using 2.18. This completes the proof.

COROLLARY 2.20. *Let X be a Stein space, $K \subset X$ a holomorphically convex compact and $0 \to \mathcal{F} \to \mathcal{G} \to \mathcal{H} \to 0$ an exact sequence in* **Coh**(X). *Then the sequence*

$$\ldots \to H^q(X \setminus K, \mathcal{F}) \to H^q(X \setminus K, \mathcal{G}) \to H^q(X \setminus K, \mathcal{H}) \to H^{q+1}(X \setminus K, \mathcal{F}) \to \ldots$$

is exact in the category of all topological vector spaces.

Proof. The maps of the sequence are continuous; since the corresponding spaces are FS, these maps are strict.

COROLLARY 2.21. *Let X be a Stein space, x any point of X and $\mathcal{F} \in$* **Coh**(X). *Then the topological vector spaces $H^{\bullet}(X \setminus \{x\}, \mathcal{F})$ are separated (hence Fréchet-Schwartz).*

§ 3. Dimension and depth of a coherent analytic sheaf

Let $(X, \mathcal{O})$ be a complex space and $\mathcal{F} \in$ **Coh**(X). Following Andreotti and Grauert [2] we will denote

$$\operatorname{prof} \mathcal{F} = \inf_{x \in X} \operatorname{prof}_{\mathcal{O}_x} \mathcal{F}_x$$

(we will often write $\operatorname{prof} \mathcal{F}_x$ instead of $\operatorname{prof}_{\mathcal{O}_x} \mathcal{F}_x$). This number is called *the depth* (or *profondeur*) *of the sheaf* $\mathcal{F}$. Recall some of its simple properties:

– $\operatorname{prof} \mathcal{F} = \infty$ if and only if $\mathcal{F} = 0$.

– If $X \xrightarrow{i} Y$ is a closed immersion of complex spaces and if $\mathcal{F} \in$ **Coh**(X), then $\operatorname{prof} \mathcal{F} = \operatorname{prof} i_*(\mathcal{F})$ (cf. 1.7).

– If X is a manifold of dimension n and $\mathcal{F} \in$ **Coh**(X), then $\operatorname{prof} \mathcal{F} = n - \operatorname{dh}\mathcal{F}$, where $\operatorname{dh}\mathcal{F} = \sup_{x \in X} \operatorname{dh}_{\mathcal{O}_x}\mathcal{F}_x$(cf. 1.11).

Recall that a complex space is said to be *perfect space* if the stalks of the structural sheaf are Cohen-Macauley rings. Special cases of such spaces are the *locally complete intersections*, the stalks of the structural sheaf being then complete intersection rings. If X a perfect space and $\mathcal{F} \in$ **Coh**(X) is locally free of positive rank, then the formula $\operatorname{prof} \mathcal{F}_x = \operatorname{prof} \mathcal{O}_x = \dim \mathcal{O}_x$, x any point in X, allows us to compute $\operatorname{prof} \mathcal{F}$.

THEOREM 3.1. *Let X be a Stein space, $K \subset X$ a holomorphically convex compact, $\mathcal{F} \in$* **Coh**(X) *and $N \geqslant 0$ an integer. Then*

(*a*) $\operatorname{prof} \mathcal{F}_x \geqslant N+1$ *for any $x \in K$ if and only if $H_K^q(X, \mathcal{F}) = 0$ for $q \leqslant N$;*

(*b*) *there is a neighbourhood U of K such that* $\operatorname{prof}(\mathcal{F}|U \setminus K) \geqslant N+1$ *if and only if the vectorial $\mathbb{C}$-spaces $H_K^q(X, \mathcal{F})$ are finite-dimensional for any $q \leqslant N$.*

Proof. (*a*) We first suppose X a manifold of dimension n. According to theorem 2.9, lemma 2.12 and Cartan's theorem A, we have the equivalences $H^q_K(X, \mathcal{F}) = 0 \Leftrightarrow \mathrm{Ext}^{n-q}_{\mathcal{O}}(K; \mathcal{F}, \Omega) = 0 \Leftrightarrow \mathcal{E}xt^{n-q}_{\mathcal{O}}(\mathcal{F}, \Omega)|K = 0 \Leftrightarrow \mathrm{Ext}^{n-q}_{\mathcal{O}_x}(\mathcal{F}_x, \Omega_x) = 0$ for any $x \in K \Leftrightarrow \mathrm{Ext}^{n-q}_{\mathcal{O}_x}(\mathcal{F}_x, \mathcal{O}_x) = 0$ for any $x \in K$. Since the rings $\mathcal{O}_{X,x}$ are regular, the conclusion is clearly forthcoming by 1.15.

For the general case, eventually replacing X by a suitable Stein neighbourhood of K, we can set up a closed immersion $X \xrightarrow{i} \mathbb{C}^p$ for some p. The image $K' = i(K)$ is a holomorphically convex compact in $\mathbb{C}^p$. If we put $\mathcal{F}^* = i_*(\mathcal{F})$, then $H^\bullet_K(X, \mathcal{F}) \simeq H^\bullet_{K'}(\mathbb{C}^p, \mathcal{F}^*)$ and $\mathrm{prof}\, \mathcal{F}_x = \mathrm{prof}\, \mathcal{F}^*_{i(x)}$, for all $x \in X$. The question is reduced to the case of manifolds.

The assertion (*b*) can be similarly proved by applying to the sheaves *Ext* the following

LEMMA 3.2. *Let X be a Stein space, $K \subset X$ a compact and $\mathcal{H} \in$ **Coh**(X). Then $\dim_{\mathbb{C}} \Gamma(K, \mathcal{H}) < \infty$ if and only if there exists a neighbourhood U of K such that $\mathcal{H}|U \setminus K = 0$.*

Proof. Let U be a neighbourhood of K and $\mathcal{H}|U \setminus K = 0$. Then $\mathrm{Supp}(\mathcal{H}|U) \subset K$ is a finite set, be it $\{x_1, \ldots, x_p\}$. By Nullstellensatz, any stalk $\mathcal{H}_{x_i}$ is annihilated by a power of the maximal ideal $\mathfrak{m}_{x_i}$, hence $\dim \mathcal{H}_{x_i} < \infty$. Since $\Gamma(K, \mathcal{H}) = \prod_{k=1}^{p} \mathcal{H}_{x_i}$, then $\dim_{\mathbb{C}} \Gamma(K, \mathcal{H}) < \infty$.

We now prove the converse implication. For any point $x \in K$ denote by $\mathfrak{m}(x)$ the assigned maximal ideal of $\Gamma(X, \mathcal{O})$. The hypothesis $\dim_{\mathbb{C}} \Gamma(K, \mathcal{H}) < \infty$ implies the existence of an integer r such that $\mathfrak{m}(x)^r \Gamma(K, \mathcal{H}) = \mathfrak{m}(x)^{r+1} \Gamma(K, \mathcal{H}) = \ldots$ By theorem A we get $\mathfrak{m}^r_x \mathcal{H}_x = \mathfrak{m}^{r+1}_x \mathcal{H}_x = \ldots$, hence by Krull's theorem $\mathfrak{m}^r_x \mathcal{H}_x = 0$. Then the points of $K \cap \mathrm{Supp}\, \mathcal{H}$ are isolated. Hence one can find a neighbourhood U of K so that $\mathcal{H}|U \setminus K = 0$.

COROLLARY 3.3. *Let X be a complex space, $x \in X$, $\mathcal{F} \in$ **Coh**(X) and N an integer. Then*

(*a*) prof $\mathcal{F}_x \geqslant N + 1$ *if and only if* $H^q_x(X, \mathcal{F}) = 0$ *for* $q \leqslant N$;

(*b*) *there exists a neighbourhood U of x such that* $\mathrm{prof}\, (\mathcal{F}|U \setminus \{x\}) \geqslant N + 1$ *if and only if the spaces $H^q_x(X, \mathcal{F})$ are finite-dimensional for $q \leqslant N$.*

Remark. If U is neighbourhood of K, then $H^\bullet_K(U, \mathcal{F}) \simeq H^\bullet_K(X, \mathcal{F})$. From the exact sequence of cohomology

$$\ldots \to H^q_K(U, \mathcal{F}) \to H^q(U, \mathcal{F}) \to H^q(U \setminus K, \mathcal{F}) \to H^{q+1}_K(U, \mathcal{F}) \to \ldots$$

one obtains an interpretation of the assertion (*a*) of 3.1, in terms of extension of cohomology classes.

For the next theorem we need two lemmas.

LEMMA 3.4. *If X is a Stein space and $\mathcal{H} \in$* **Coh**(X), *then* $\dim_{\mathbb{C}} \Gamma(X, \mathcal{H}) < \infty$ *if and only if Supp $\mathcal{H}$ is a finite set.*

The proof is analogous to that for 3.2.

LEMMA 3.5. *If X is a Stein space and $\mathcal{H} \in$* **Coh**(X), *then the topological dual of the Fréchet space $\Gamma(X, \mathcal{H})$ has at most countable complex dimension if and only if* Supp $\mathcal{H}$ *is a discrete set.*

Proof. Let us fix $(U_r)_{r\geqslant 1}$ an exhaustion of X by relatively compact Stein open subsets such that the restriction maps $\Gamma(X, \mathcal{O}) \to \Gamma(U_r, \mathcal{O})$ are dense, $r \geqslant 1$. If Supp $\mathcal{H}$ is discrete, then the previous lemma shows that every $\Gamma(U_r, \mathcal{H})$ is finite-dimensional. Since $\Gamma(X, \mathcal{H})' \simeq \varinjlim_r \Gamma(U_r, \mathcal{H})'$, an implication is immediate.

Conversely, by lemma 3.4 we should prove only that $\dim_{\mathbf{C}}\Gamma(U_r, \mathcal{H}) < \infty$ for all r. For some fixed integer r, denote $U_r = U$ and let K be a holomorphically convex compact, $U \subset K$. Take p a seminorm on $\Gamma(X, \mathcal{H})$ of type "$\sup_K$", enjoying the property that any Cauchy sequence with respect to p remains Cauchy by restriction at $\Gamma(U, \mathcal{H})$. By $\Gamma(X, \mathcal{H})^\wedge$ we mean the completion of the space $\Gamma(X, \mathcal{H})/p^{-1}(0)$ in this seminorm. $\Gamma(X, \mathcal{H})^\wedge$ is a Banach space and since the map $\Gamma(X, \mathcal{H}) \to \Gamma(X, \mathcal{H})^\wedge$ is dense, the map $(\Gamma(X, \mathcal{H})^\wedge)' \to \Gamma(X, \mathcal{H})'$ is injective. But $(\Gamma(X, \mathcal{H})^\wedge)'$ has a structure of Banach space, and by hypothesis it has at most countable dimension. Accordingly, it is finite-dimensional and $\dim \Gamma(X, \mathcal{H})^\wedge < \infty$. On the other hand, by the choice of K and p, we easily deduce that the map $\Gamma(X, \mathcal{H}) \to \Gamma(U, \mathcal{H})$ factorizes by $\Gamma(X, \mathcal{H})^\wedge \to \Gamma(U, \mathcal{H})$ and that the latter has dense image. Therefore $\dim_{\mathbf{C}}\Gamma(U, \mathcal{H}) < \infty$ and the lemma is proved.

THEOREM 3.6. *Let X be a Stein space, $\mathcal{F}$ a coherent analytic sheaf on X and $N \geqslant 0$ an integer. Then*

(*a*) prof $\mathcal{F} \geqslant N + 1$ *if and only if* $H^q_c(X, \mathcal{H}) = 0$ *for* $q \leqslant N$;

(*b*) *there is a finite set* $A \subset X$ *such that* prof $(\mathcal{F}|X \setminus A) \geqslant N + 1$ *if and only if the spaces* $H^q_c(X, \mathcal{F})$ *are finite-dimensional for* $q \leqslant N$;

(*c*) *there is a discrete set* $A \subset X$ *such that* prof $(\mathcal{F}|X \setminus A) \geqslant N + 1$ *if and only if the spaces* $H^q_c(X, \mathcal{F})$ *have at most countable dimension for* $q \leqslant N$.

Proof. (*a*) The nonsingular case follows exactly as the assertion 3.1 (*a*), by using the duality theorem 2.1. Actually, the general case may be reduced to this case by a suitable exhaustion and by immersions in Stein manifolds.

(*b*) We first consider the particular case when X is a manifold and let $n = \dim X$. Let $A \subset X$ be a finite set such that prof $(\mathcal{F}|X \setminus A) \geqslant N + 1$. It follows $\mathrm{Ext}^{n-q}_{\mathcal{O}_x}(\mathcal{F}_x, \Omega_x) = 0$ for any $x \notin A$, $q \leqslant N$. Hence $\mathcal{E}xt^{n-q}_{\mathcal{O}}(\mathcal{F}, \Omega)|X \setminus A = 0$ $(q \leqslant N)$ and Supp $(\mathcal{E}xt^{n-q}_{\mathcal{O}}(\mathcal{F}, \Omega))$ will be a finite set. By 2.4, 3.4, $\mathrm{Ext}^{n-q}_{\mathcal{O}}(X; \mathcal{F}, \Omega)$ is a $\mathbf{C}$-linear space of finite dimension and the assertion follows from 2.1. The converse implication can be proved in the same way using again 3.4. Suppose now X is an embeddable Stein space; this means there is a closed immersion $X \xrightarrow{i} \mathbf{C}^p$ for a suitable p. If we put $\mathcal{F}^* = i_*(\mathcal{F})$, then $H^\bullet_c(X, \mathcal{F}) \simeq H^\bullet_c(\mathbf{C}^p, \mathcal{F}^*)$ and for all $x \in X$, prof $\mathcal{F}_x =$ prof $\mathcal{F}^*_{i(x)}$; we thus reduce the proof at the nonsingular case.

We finally consider the general case. We first prove the direct implication. Let $A \subset X$ be a finite set such that prof $(\mathcal{F}|X \setminus A) \geqslant N + 1$ and U a relatively compact Stein open set containing A, chosen such that the map $\Gamma(X, \mathcal{O}) \to \Gamma(U, \mathcal{O})$ has a dense image. We shall show that the map $H^q_c(U, \mathcal{F}) \to H^q_c(X, \mathcal{F})$ is bijective $(q \leqslant N)$ and the conclusion will follow from the above considered case of embeddable Stein spaces. We must only show the following: for any relatively compact Stein open subset $V \supset U$, the map $H^q_c(U, \mathcal{F}) \to H^q_c(V, \mathcal{F})$ is bijective. Let $V \xrightarrow{i} \mathbf{C}^n$ be a closed immersion. We have $H^q_c(V, \mathcal{F}) \simeq H^q_c(\mathbf{C}^p, i_*(\mathcal{F})) \simeq$

$\simeq (\mathrm{Ext}^{n-q}_{\mathcal{O}_{\mathbf{C}^n}}(\mathbf{C}^n; i_*(\mathcal{F}), \Omega_{\mathbf{C}^n})' \simeq \prod_{x \in A} (\mathrm{Ext}^{n-q}_{\mathcal{O}_{\mathbf{C}^n, i(x)}}(\mathcal{F}_x, \mathcal{O}_{\mathbf{C}^n, i(x)}))'$, $q \leqslant N$ that proves the assertion.

Next look at the converse implication. Denote $A = \{x \in X \mid \mathrm{prof}\, \mathcal{F}_x \leqslant N\}$. Choose a relatively compact Stein open set U in X such that the map $\Gamma(X, \mathcal{O}) \to \Gamma(U, \mathcal{O})$ has a dense image. The maps $H^{\bullet}_c(U, \mathcal{O}) \to H^{\bullet}_c(X, \mathcal{O})$ are injective. From the above considered case of Stein embeddable spaces we deduce that $A \cap U$ is a finite set. If $(U_r)_{r \geqslant 1}$ is a Runge exhaustion of X by relatively compact Stein open sets, then $H^q_c(X, \mathcal{F}) \simeq \varinjlim_r H^q_c(U_r, \mathcal{F})$ and exactly as above, $A \cap U_r = A \cap U_{r+1} = \ldots$, as soon as the maps $H^q_c(U_r, \mathcal{F}) \to H^q_c(U_{r+1}, \mathcal{F})$, $H^q_c(U_{r+1}, \mathcal{F}) \to H^q_c(U_{r+2}, \mathcal{F}), \ldots$ are bijective, $q \leqslant N$. So A is a finite set.

The assertion (*c*) can be proved as (*b*) by using lemma 3.5.

Remark. The theorem may be proved straightforwardly by means of 3.1, which is a consequence of the duality theorem 2.9. The above given proof shows that it is but a consequence of 2.1.

We shall now give some results in connexion with the dimension of sheaves. Let $(X, \mathcal{O})$ be a complex space and $\mathcal{F} \in \mathbf{Coh}(X)$. Denote

$$\dim \mathcal{F} = \sup_{x \in X} \dim_{\mathcal{O}_x} \mathcal{F}_x.$$

This number is to be said *the dimension of the sheaf* $\mathcal{F}$. If $\mathrm{Ann}\, \mathcal{F}$ is the ideal-sheaf annihilator of $\mathcal{F}$, then $(\mathrm{Ann}\, \mathcal{F})_x = \mathrm{Ann}\, \mathcal{F}_x$ where $\mathrm{Ann}\, \mathcal{F}_x$ is the annihilator of the $\mathcal{O}_x$-module $\mathcal{F}_x$. Accordingly, $\dim \mathcal{F} = \dim (\mathrm{Supp}\, \mathcal{F})$. If $\mathcal{F} \neq 0$, then $\mathrm{prof}\, \mathcal{F} \leqslant \dim \mathcal{F}$.

Remark. One can easily construct examples where the above inequality is strict. A less trivial example is the following [71].

Let $X \subset \mathbf{C}^4$ be the reduced complex space which corresponds to the analytic set

$$\{z_1 = z_2 = 0\} \cup \{z_3 = z_4 = 0\}.$$

Obviously, X is connected, of pure dimension 2 at any point and it has a unique singularity in origin. The local ring at origin is

$$A = \mathbf{C}\{z_1, z_2, z_3, z_4\}/(z_1, z_2) \cap (z_3, z_4).$$

The class of $z_1 - z_3$ in A is a nonzerodivisor, since $z_1 - z_3$ does not belong to any of the prime ideals (z_1, z_2), (z_3, z_4). Accordingly

$$\mathrm{prof}\, A = \mathrm{prof}\, A/(z_1 - z_3)A + 1.$$

The element z_1 has a nonnull image in $B = A/(z_1 - z_3)A$ and one easily verifies that $z_1 z_i \in ((z_1, z_2) \cap (z_3, z_4)) + (z_1 - z_3)$, $i = 1, 2, 3, 4$. Thus, all the elements of the maximal ideal of B are zerodivisors, hence $\mathrm{prof}\, B = 0$. So $\mathrm{prof}\, A = 1$ and therefore

$$\mathrm{prof}\, \mathcal{O}_X = 1 < 2 = \dim \mathcal{O}_X.$$

THEOREM 3.7. *Let X be a Stein space, $\mathcal{F} \in \mathbf{Coh}(X)$ and $N \geqslant 0$ an integer. Then the following assertions are equivalent*:

(i) $\dim \mathcal{F} \leqslant N$;

(ii) $H_c^q(X, \mathcal{F}) = 0$ *for* $q > N$;

(iii) *the complex vectorial spaces* $H_c^q(X, \mathcal{F})$ *are finite-dimensional for* $q > N$;

(iv) *the complex vectorial spaces* $H_c^q(X, \mathcal{F})$ *have at most countable dimension for* $q > N$.

Proof. We proceed exactly as for theorem 3.6. It is sufficient to prove the implications (i) $\Rightarrow$ (ii) and (iv) $\Rightarrow$ (i).

(i) $\Rightarrow$ (ii). Let $(U_r)_{r \geqslant 1}$ a Runge exhaustion of X by relatively compact Stein open sets. By 2.8 it follows that the maps $H_c^{\bullet}(U_r, \mathcal{F}) \to H_c^{\bullet}(X, \mathcal{F})$ are injective. We must only prove the implication for all U_r. By using immersions in numerical spaces we can furthermore suppose X a manifold. The conclusion then follows by duality theorem 2.1, lemma 2.4, theorem A and by the characterization of the dimension of modules of finite type over regular local noetherian rings, given in 1.17.

(iv) $\Rightarrow$ (i). As shown above we reduce the question to the case X a Stein manifold. Put $n = \dim X$. Then the topological duals of the spaces $\mathrm{Ext}_{\mathcal{O}}^{n-q}(X; \mathcal{F}, \Omega) = \Gamma(X, \mathcal{E}xt_{\mathcal{O}}^{n-q}(\mathcal{F}, \Omega))$ have at most countable complex dimension for $q > N$. By 3.5 there exists a discrete set $A \subset X$ so that $\mathcal{E}xt_{\mathcal{O}}^{n-q}(\mathcal{F}, \Omega)|X \setminus A = 0$ for $q > N$. For all $x \notin A$, $\mathrm{Ext}_{\mathcal{O}_x}^{n-q}(\mathcal{F}_x, \mathcal{O}_x) = 0$ for $q > N$, hence $\dim \mathcal{F}_x \leqslant N$. Therefore, $\dim(\mathcal{F}|X \setminus A) \leqslant N$ and, since A is discrete and $N \geqslant 0$, it follows $\dim \mathcal{F} \leqslant N$.

By virtue of 3.6 and 3.7, the cohomology groups with compact supports for coherent analytic sheaves on Stein spaces vanish out of the interval [prof, dim]. For the extremities of this interval the following two corollaries hold:

COROLLARY 3.8. *Let* X *be a Stein space and* $\mathcal{F} \in \mathbf{Coh}(X)$. *Then* $H_c^{\mathrm{prof}\mathcal{F}}(X, \mathcal{F})$ *has a finite (resp. at most countable) complex dimension if and only if the set* $\{x| \operatorname{prof} \mathcal{F}_x = \operatorname{prof} \mathcal{F}\}$ *is finite (resp. discrete).*

This derives from theorem 3.6.

COROLLARY 3.9. *Let* X *be a Stein space and* $\mathcal{F} \in \mathbf{Coh}(X)$. *Then* $H_c^{\dim\mathcal{F}}(X, \mathcal{F})$ *has at most countable complex dimension if and only if* $\dim \mathcal{F} = 0$.

Proof. If $\dim \mathcal{F} = 0$, then $\mathcal{F}$ has discrete support and $H_c^{\dim\mathcal{F}}(X, \mathcal{F}) = \Gamma_c(X, \mathcal{F})$ has at most countable dimension. If $\dim \mathcal{F} \geqslant 1$, then the conclusion follows from 3.7, applied for $N = \dim \mathcal{F} - 1$.

Recall that a coherent analytic sheaf $\mathcal{F} \neq 0$ is said to be *a Cohen-Macauley sheaf* if $\operatorname{prof} \mathcal{F} = \dim \mathcal{F}$.

From corollaries 3.8, 3.9 it results:

COROLLARY 3.10. *Let* X *be a Stein space and* $\mathcal{F} \neq 0$ *a coherent analytic sheaf on* X. *Then*

(*a*) $\mathcal{F}$ *is Cohen-Macauley if and only if there is an integer* $N \geqslant 0$ *such that* $H_c^q(X, \mathcal{F}) = 0$ *for any* $q \neq N$;

(*b*) *there exists a finite set* $A \subset X$ *so that* $\mathcal{F}|X \setminus A$ *is Cohen-Macauley if and only if there is an integer* $N \geqslant 0$ *such that the spaces* $H_c^q(X, \mathcal{F})$ *are finite-dimensional for* $q \neq N$;

(*c*) *the assertion (b) is still true if* A *need not be a finite but a discrete set.*

This corollary gives particularly a cohomological characterization of the perfect spaces.

§ 4. Applications

(a) *The interpretation of the depth in small dimensions: results of the Hartogs and Cousin type.*

The first two corollaries characterize the coherent analytic sheaves (on Stein spaces) of depth $\geqslant 1$ or, respectively, $\geqslant 2$.

COROLLARY 4.1. *Let X be a Stein space and* $\mathcal{F} \in \mathbf{Coh}(X)$. *Then* $\operatorname{prof} \mathcal{F} \geqslant 1$ *if and only if any section of* $\Gamma(X, \mathcal{F})$ *with compact support is null.*

The proof follows from 3.6. (a), for $N = 0$.

COROLLARY 4.2. *Let X be a Stein space and* $\mathcal{F} \in \mathbf{Coh}(X)$. *Then the following assertions are equivalent:*

(i) $\operatorname{prof} \mathcal{F} \geqslant 2$;

(ii) *for any relatively compact Stein open set* $U \subset X$, *the restriction map* $\Gamma(X, \mathcal{F}) \to \Gamma(X \setminus U, \mathcal{F})$ *is bijective;*

(iii) *for any holomorphically-convex compact* $K \subset X$, *the restriction map* $\Gamma(X, \mathcal{F}) \to \Gamma(X \setminus K, \mathcal{F})$ *is bijective.*

Proof. It is sufficient to apply 3.1 and 3.6 for $N = 1$ by using the exact sequences

$$0 \to \Gamma_c(U, \mathcal{F}) \to \Gamma(X, \mathcal{F}) \to \Gamma(X \setminus U, \mathcal{F}) \to H^1_c(U, \mathcal{F}) \to H^1(X, \mathcal{F})$$

$$0 \to \Gamma_K(X, \mathcal{F}) \to \Gamma(X, \mathcal{F}) \to \Gamma(X \setminus K, \mathcal{F}) \to H^1_K(X, \mathcal{F}) \to H^1(X, \mathcal{F}).$$

The implication (i) $\Rightarrow$ (ii) does not require the hypothesis X Stein.

The following two corollaries are results of Hartogs type.

COROLLARY 4.3. *Suppose X is a Stein space,* $K \subset X$ *a holomorphically convex compact and* $\mathcal{F} \in \mathbf{Coh}(X)$. *Then the restriction map* $\Gamma(X, \mathcal{F}) \to \Gamma(X \setminus K, \mathcal{F})$ *is bijective if and only if* $\operatorname{prof} \mathcal{F}_x \geqslant 2$ *for any* $x \in K$.

Proof. Apply 3.1 for $N = 1$.

COROLLARY 4.4. *Suppose* $(X, \mathcal{O})$ *is a reduced Stein space and K a compac such that* $X \setminus K$ *has no relatively compact irreducible components (branches) in X. If* $\operatorname{prof} \mathcal{O}_x \geqslant 2$ *for all* $x \in K$, *then the map* $\Gamma(X, \mathcal{O}) \to \Gamma(X \setminus K, \mathcal{O})$ *is bijective. Conversely, if X is Stein and* $K \subset X$ *a holomorphically convex compact such that the map* $\Gamma(X, \mathcal{O}) \to \Gamma(X \setminus K, \mathcal{O})$ *is bijective, then* $\operatorname{prof} \mathcal{O}_x \geqslant 2$ *for all* $x \in K$.

Proof. The second assertion follows by the previous corollary.

We are going to prove the first assertion. Let $\varphi \in \Gamma(X, \mathcal{O})$, null on $X \setminus K$. The set of zeroes of φ is a closed analytic subset of X and since it is closed in K, it will be finite. As X is reduced and of positive dimension, it follows $\varphi = 0$. We then consider a holomorphically convex compact K' containing K and $\varphi' \in \Gamma(X \setminus K, \mathcal{O})$. By the previous corollary there is $\varphi \in \Gamma(X, \mathcal{O})$ so that $\varphi | X \setminus K = \varphi' | X \setminus K'$. The function $\varphi - \varphi' \subset \Gamma(X \setminus K, \mathcal{O})$ vanishes on $X \setminus K'$ and therefore is by hypothesis null. This completes the proof.

Remarks. 1) Corollaries 4.2, 4.3, 4.4 (as well as those from the next section) may be applied to the structural sheaf of a normal space of dimension $\geqslant 2$

(respectively normal of dimension $\geqslant 2$ in the points of K), because in this case prof $\geqslant 2$ (by 1.8). We can also consider the case of locally free sheaves of finite rank over perfect spaces (for instance locally complete intersections) of dimension $\geqslant 2$.

2) Let X be a paracompact, noncompact topological space. An open subset Uf is called *neighbourhood of the boundary of* X if $X \setminus U$ is compact. If $\mathcal{F}$ is a sheain $\mathbf{Ab}(X)$ we denote $\Gamma(\partial X, \mathcal{F}) = \varinjlim_{U} \Gamma(U, \mathcal{F})$, where $\varinjlim$ is taken over the neighbourhoods U of the boundary ([38], Ch. VII.D).

Suppose now X is a Stein space, $\mathcal{F} \in \mathbf{Coh}(X)$ and prof $\mathcal{F} \geqslant 2$. For any compact K there is a compact K' (for instance $K' = \hat{K}$, the holomorphically convex envelope) such that any section $s \in \Gamma(X \setminus K, \mathcal{F})$ admits a unique extension $s' \in \Gamma(X, \mathcal{F})$ and $s' = s$ on $X \setminus K'$. This follows from 4.3. In this way we have proved the following assertion: if X is a Stein space, $\mathcal{F} \in \mathbf{Coh}(X)$ and prof $\mathcal{F} \geqslant 2$, then the canonical morphism $\Gamma(X, \mathcal{F}) \to \Gamma(\partial X, \mathcal{F})$ is bijective.

3) As we have seen in the extension of Hartogs' theorem to the case of complex spaces, the condition dim $\geqslant 2$ must be substituted by prof $\geqslant 2$. The following example due to Harvey [40] points out this phenomenon.

Let X be the image of the morphism $h: \mathbb{C}^2 \to \mathbb{C}^4$, $h(x, y) = (x^2, x^3, y, xy)$. X is a subspace of dimension 2 of $\mathbb{C}^4$ with a single singularity in the origin and, moreover, $X \setminus \{0\}$ is connected. The function $f(z)$, $z \in X \setminus \{0\}$, equal to z_2/z_1 if $z_1 \neq 0$ and to z_4/z_3 if $z_3 \neq 0$, is a holomorphic function on $X \setminus \{0\}$. It can not be extended to whole X. For, otherwise, the holomorphic map $X \to \mathbb{C}^2$, $z = (z_1, z_2, z_3, z_4) \mapsto (f(z), z_3)$, is an inverse of $h: \mathbb{C}^2 \to X$, hence a contradiction. In this example X is irreducible and

$$1 = \text{prof } \mathcal{O}_X < \dim \mathcal{O}_X = 2.$$

The next applications regard the Cousin problems. First of all, restate them on complex spaces (not necessarily reduced).

Let $(X, \mathcal{O})$ be a complex space. For any open set $U \subset X$ denote $S_U = \{f \in \Gamma(U, \mathcal{O}_X) | f_x$ is nonzerodivisor in $\mathcal{O}_x$, for all $x \in U\}$. S_U is a multiplicative system and the sheaf $\mathcal{M}$ defined by the presheaf which assigns to an open subset $U \subset X$ the ring $\Gamma(U, \mathcal{O}_X)_{S_U}$ (the ring of quotients with denominators in S_U) is called *the sheaf of germs of meromorphic sections on* X. There is obviously a canonical inclusion of sheaves $\mathcal{O} \subset \mathcal{M}$. Let $x \in X$. If $s_x \in \mathcal{O}_x$ is nonzerodivisor, then the multiplication by s_x defines an injective morphism $\mathcal{O}_x \to \mathcal{O}_x$. Pick a section $s \in \Gamma(U, \mathcal{O})$, U being a neighbourhood of x, such that its germ in x is s_x. The morphism $\mathcal{O}|U \xrightarrow{s} \mathcal{O}|U$ is injective in x and since $\mathcal{O}$ is coherent, there is a neighbourhood $V \subset U$ of x such that $\mathcal{O}|V \xrightarrow{s} \mathcal{O}|V$ is injective. Thus $s|V \in S_V$. Accordingly, one will be able to verify that $\mathcal{M}_x$ is canonically identified with the total ring of quotients of $\mathcal{O}_x$. If X is reduced, the above multiplicative systems can be thus characterized: $S_U = \{f \in \Gamma(U, \mathcal{O}) | f$ does not vanish on any nonempty open set in $U\}$.

The Cousin problems can be formulated exactly as for manifolds. The additive (respectively multiplicative) problem consists in giving conditions of surjectivity of the map $\Gamma(X, \mathcal{M}) \to \Gamma(X, \mathcal{M}/\mathcal{O})$ (respectively $\Gamma(X, \mathcal{M}^*) \to \Gamma(X, \mathcal{M}^*/\mathcal{O}^*)$). As usual, $\mathcal{O}^*$ (resp. $\mathcal{M}^*$) is the subsheaf of $\mathcal{O}$ (resp. $\mathcal{M}$) formed by the invertible sections.

COROLLARY 4.5. *Let X be a Stein space and $K \subset X$ a holomorphically convex compact. If* prof $\mathcal{O}_x \geqslant 3$ *for all $x \in K$, then the additive Cousin problem has solution on $X \setminus K$. If moreover $H^2(X \setminus K, \mathbb{Z}) = 0$, then the multiplicative Cousin problem has also solution on $X \setminus K$.*

Proof. We have an exact sequence

$$H^1(X, \mathcal{O}) \to H^1(X \setminus K, \mathcal{O}) \to H^2_K(X, \mathcal{O}).$$

By 3.1 we deduce $H^1(X \setminus K, \mathcal{O}) = 0$ and the first assertion follows from the exact sequence

$$\Gamma(X \setminus K, \mathcal{M}) \to \Gamma(X \setminus K, \mathcal{M}/\mathcal{O}) \to H^1(X \setminus K, \mathcal{O}).$$

For the second part of the corollary consider the exact sequence

$$H^1(X \setminus K, \mathcal{O}) \to H^1(X \setminus K, \mathcal{O}^*) \to H^2(X \setminus K, \mathbb{Z}),$$

associated to the exact sequence

$$0 \to \mathbb{Z} \to \mathcal{O} \to \mathcal{O}^* \to 0$$

(recall that $\mathbb{Z}$ is the simple sheaf of stalk the ring of integers and that the morphism $\mathcal{O} \to \mathcal{O}^*$ is given by $\varphi \mapsto e^{2\pi i \varphi}$). Then there follows $H^1(X \setminus K, \mathcal{O}^*) = 0$ and the exact sequence

$$\Gamma(X \setminus K, \mathcal{M}^*) \to \Gamma(X \setminus K, \mathcal{M}^*/\mathcal{O}^*) \to H^1(X \setminus K, \mathcal{O}^*)$$

completes the proof.

COROLLARY 4.6. *Suppose X is a Stein space and $U \subset X$ a relatively compact Stein open set. If* prof $(\mathcal{O} \mid U) \geqslant 3$, *then the additive Cousin problem has solution on $X \setminus U$. If in addition $H^2(X \setminus U, \mathbb{Z}) = 0$, then the multiplicative Cousin problem has also a solution on $X \setminus U$.*

The proof is similar to that of the previous corollary.

(b) *Some topological properties of the boundary of a Stein space.*

A complex space X (supposed paracompact but noncompact) is called *connected at boundary* if for any compact $K \subset X$ the complementary in $X \setminus K$ of the union of relatively compact connected components of $X \setminus K$ is connected ([38], Ch. VII. D).

Recall the notion of *germ of analytic set around the boundary:* it is a pair (U, S) where U is a neighbourhood of the boundary and S a closed analytic subset in U; we shall identify two such pairs (U_1, S_1), (U_2, S_2) whenever there is a neighbourhood of the boundary $U \subset U_1 \cap U_2$ for which $S_1 \cap U = S_2 \cap U$.

Denote by ∂X the germ defined by any neighbourhood of the boundary. We shall say that a complex space X (paracompact, noncompact) is *irreducible at boundary* if the germ ∂X can not be written as a union of two germs distinct of it (the unions, intersections, etc. of germs as above are naturally defined). Any such space is connected at boundary: indeed, for any compact K we have $X = K \cup (\bigcup_i V_i) \cup (\bigcup_\alpha U_\alpha)$, where V_i, U_α are the connected components of $X \setminus K$, V_i are those relatively compact, and U_α the others. For any compact $K' \supset K$ and any α, $U_\alpha \cap (X \setminus K') \neq \emptyset$; by irreducibility it results that the index-set $\{\alpha\}$ contains only one element.

COROLLARY 4.7. *If $(X, \mathcal{O})$ is a connected Stein space such that* prof $\mathcal{O} \geqslant 2$, *then X is connected at boundary.*

Proof. Let K be an arbitrary compact and $X = K \cup (\bigcup_i V_i) \cup (\bigcup_\alpha U_\alpha)$ as above. Since for any α, $\hat{K} \cap U_\alpha \neq \emptyset$, it is sufficient to show that $X \setminus \hat{K}$ is connected. If this happens, then the index-set $\{\alpha\}$ would be reduced to one element and therefore X is connected at boundary. So we need

COROLLARY 4.8. *If $(X, \mathcal{O})$ is a Stein space and $K \subset X$ a holomorphically convex compact such that* prof $\mathcal{O}_x \geqslant 2$ *for all $x \in K$, then X is connected if and only if $X \setminus K$ is connected.*

Proof. By 4.4 the map $\Gamma(X, \mathcal{O}) \to \Gamma(X \setminus K, \mathcal{O})$ is bijective. The corollary will be concluded by the following

LEMMA 4.9. *A ringed space in local rings $(X, \mathcal{O})$ is connected if and only if the ring $\Gamma(X, \mathcal{O})$ can not be written as product of two nonnull rings (commutative and unitary).*

Proof. If $X = U \cup V$, U and V disjoint open sets, then $\Gamma(X, \mathcal{O}) = \Gamma(U, \mathcal{O}) \times \Gamma(V, \mathcal{O})$. Conversely, suppose $\Gamma(X, \mathcal{O}) = A \times B$ (A, B commutative unitary rings). The unity of $\Gamma(X, \mathcal{O})$ can be written $1 = e_1 + e_2$, $e_1 \in A$, $e_2 \in B$. Let $U = \{x \in X | e_1(x) \neq 0\}$ and $V = \{x \in X | e_2(x) \neq 0\}$. As usual, for a section $f \in \Gamma(X, \mathcal{O})$, $f(x)$ designes its image by the composition $\Gamma(X, \mathcal{O}) \to \mathcal{O}_x \to \mathcal{O}_x/\mathfrak{m}_x$ ($\mathfrak{m}_x$ is the maximal ideal of $\mathcal{O}_x$). The condition $f(x) \neq 0$ means $f_x \notin \mathfrak{m}_x$, hence it is equivalent to the fact that f is invertible in x. It will result that U and V are open sets. On the other hand, it is not difficult to see that $X = U \cup V$ and $U \cap V = \emptyset$.

COROLLARY 4.10. *If X is an irreducible Stein space of dimension $\geqslant 2$, then X is irreducible at boundary and in particular, connected at boundary.*

Proof. We need to show that $X \setminus K$ is an irreducible space for any holomorphically convex compact $K \subset X$. Let X' be the normalization of X and K' the inverse image of K by the normalization morphism. The space X' verifies the conditions of the previous corollary and K' is holomorphically convex. Hence $X' \setminus K'$ is connected. The conclusion follows since $X' \setminus K'$ is the normalization of $X \setminus K$.

Let $(X, \mathcal{O}_X)$ be a complex space, not necessarily reduced. The algebra of the global sections $\Gamma(X, \mathcal{O}_X)$ has a natural structure of topological $\mathbb{C}$-algebra. If X has a countable topology, then $\Gamma(X, \mathcal{O}_X)$ is a Fréchet algebra. If $(X, \mathcal{O}_X) \to (Y, \mathcal{O}_Y)$ is a morphism of complex spaces, then the corresponding morphism $\Gamma(Y, \mathcal{O}_Y) \to \Gamma(X, \mathcal{O}_X)$ is a morphism of topological algebras. We shall use the following result of Forster [22], whose proof given below belongs to the authors:

THEOREM 4.11. *If Y is Stein space, then the canonical map*

$$\mathrm{Hom}\,(X, Y) \to \mathrm{Hom}_{\mathrm{top.}\,\mathbb{C}\text{-alg.}}\,(\Gamma(Y, \mathcal{O}_Y), \Gamma(X, \mathcal{O}_X))$$

is bijective for any complex space X.

Proof. Denote by θ the map we are concerned with; two cases may be distinguished.

(1) *Y embeddable*. Let $i: Y \to \mathbb{C}^n$ be a closed immersion and $\mathfrak{J} \subset \mathcal{O}_{\mathbb{C}^n}$ the coherent ideal-sheaf defining $i(Y)$. We first prove θ is injective. Let $f, g \in \mathrm{Hom}\,(X,Y)$ be such that $\theta(f) = \theta(g)$. To show that $f = g$ it is sufficient to show $i.f = i.g$, that is we can suppose $Y = \mathbb{C}^n$. Hereupon the maps f, g are completely determined by the systems of maps $(\mathrm{pr}_j.f)_{1 \leqslant j \leqslant n}$, $(\mathrm{pr}_j.g)_{1 \leqslant j \leqslant n}$, where pr_j: $\mathbb{C}^n \to \mathbb{C}$ are the canonical projections [15]. But the system $(\mathrm{pr}_j.f)$ is identified with the system of images in $\Gamma(X, \mathcal{O}_X)$ of the coordinates $z_1, \ldots, z_n \in \Gamma(\mathbb{C}^n, \mathcal{O}_{\mathbb{C}^n})$ through the map $\theta(f)$. A similar assertion holds for the system $(\mathrm{pr}_j.\, g)$. Therefore $f = g$.

We now check the surjectivity of θ. Let $\alpha: \Gamma(Y, \mathcal{O}_Y) \to \Gamma(X, \mathcal{O}_X)$ be a continuous morphism of topological $\mathbb{C}$-algebras. By theorem *B* the following sequence $0 \to \Gamma(\mathbb{C}^n, \mathfrak{J}) \to \Gamma(\mathbb{C}^n, \mathcal{O}_{\mathbb{C}^n}) \to \Gamma(Y, \mathcal{O}_Y) \to 0$ is exact. The composition β: $\Gamma(\mathbb{C}^n, \mathcal{O}_{\mathbb{C}^n}) \to \Gamma(Y, \mathcal{O}_Y) \to \Gamma(X, \mathcal{O}_X)$ gives rise to n elements $\beta(z_1), \ldots, \beta(z_n) \in \Gamma(X, \mathcal{O}_X)$, hence yields a morphism $f: X \to \mathbb{C}^n$. The morphism $\theta(f): \Gamma(\mathbb{C}^n, \mathcal{O}_{\mathbb{C}^n}) \to \Gamma(X, \mathcal{O}_X)$ which corresponds to f coincides with β, because both morphisms coincide on coordinate functions $z_1, \ldots, z_n$ and by continuity, on all entire functions. We are going to prove that f factorizes by Y and the surjectivity of θ will follow. In this respect we must verify the equality $f^*(\mathfrak{J})\mathcal{O}_X = 0$, that is $\mathfrak{J}_{f(x)}\mathcal{O}_{X,x} = 0$ for all $x \in X$. But by theorem A, $\mathfrak{J}_{f(x)}$ is generated as $\mathcal{O}_{\mathbb{C}^n, f(x)}$-module by $\Gamma(\mathbb{C}^n, \mathfrak{J})$ and from the definition of β and the commutative diagram

$$\begin{array}{ccc} \Gamma(\mathbb{C}^n, \mathcal{O}_{\mathbb{C}^n}) & \to & \Gamma(X, \mathcal{O}_X) \\ \downarrow & & \downarrow \\ \mathcal{O}_{\mathbb{C}^n, f(x)} & \to & \mathcal{O}_{X,x} \end{array}$$

it will follow that $\Gamma(\mathbb{C}^n, \mathfrak{J})$ has null image in $\mathcal{O}_{X,x}$ and hereof the conclusion.

(2) *The general case*. We check the injectivity of θ. Let $f, g \in \mathrm{Hom}\,(X, Y)$ be such that $\theta(f) = \theta(g)$. In order to prove that $f = g$, it is enough to find, for any $x \in X$, a neighbourhood of x on which the two morphisms coincide. Choose an embeddable Stein open subset V of Y containing $f(x)$, $g(x)$ and such that the map $\Gamma(Y, \mathcal{O}_Y) \to \Gamma(V, \mathcal{O}_Y)$ is dense. Then $U = f^{-1}(V) \cap g^{-1}(V)$ is a neighbourhood of x. Denote by $f', g' \in \mathrm{Hom}\,(U, V)$ the morphisms deduced from f, g, respectively. Since $\theta(f) = \theta(g)$, $\theta(f') = \theta(g')$ hence $f' = g'$ by the particular case (1).

We now prove the surjectivity of θ.

Let $\alpha \in \mathrm{Hom}_{\mathrm{top.}\mathbb{C}\text{-alg}}\,(\Gamma(Y, \mathcal{O}_Y), \Gamma(X, \mathcal{O}_X))$. Since θ is injective, it suffices to find for any $x \in X$ a neighbourhood U and a morphism $U \to X$ such that the corresponding morphism $\Gamma(Y, \mathcal{O}_Y) \to \Gamma(U, \mathcal{O}_X)$ coincides with the composition $\Gamma(Y, \mathcal{O}_Y) \to \Gamma(X, \mathcal{O}_X) \xrightarrow{\text{restriction}} \Gamma(U, \mathcal{O}_X)$. Fix $x \in X$ and K a holomorphically convex compact in a neighbourhood of x such that $x \in \overset{\circ}{K}$. Denote by $\| \ \|_K$ a suitable seminorm of type "sup" on $\Gamma(X, \mathcal{O}_X)$. By the very definition of the topology on $\Gamma(Y, \mathcal{O}_Y)$, the seminorms $\| \ \|_L$ of type "sup", L holomorphically convex compact in Y, constitute a cofinal family. By the continuity of α it then follows there is such a seminorm $\| \ \|_L$ and a constant $C > 0$ such that $\|\alpha(s)\|_K \leqslant C \cdot \|s\|_L$ for all $s \in \Gamma(Y, \mathcal{O}_Y)$. Denote $U = \overset{\circ}{K}$. Let V be an embeddable Stein neighbourhood of L such that the map $\Gamma(Y, \mathcal{O}_Y) \to \Gamma(V, \mathcal{O}_Y)$ is dense. If $(s_n)_n$ is a sequence of elements of $\Gamma(Y, \mathcal{O}_Y)$ which converges in $\Gamma(V, \mathcal{O}_Y)$, then the sequence $(\alpha(s_n))_n$ of elements of $\Gamma(X, \mathcal{O}_X)$ is Cauchy with respect to the seminorm $\| \ \|_K$, hence it converges in $\Gamma(U, \mathcal{O}_X)$. Thus we get a continuous map $\Gamma(V, \mathcal{O}_Y) \to \Gamma(U, \mathcal{O}_X)$ which agrees with α. The corresponding morphism $U \to V$ deduced by the particular case (1) and composed with the inclusion $V \to Y$, verifies the assertion and the theorem is completely proved.

COROLLARY 4.12. *Let X, Y be Stein spaces. If* prof $\mathcal{O}_X \geqslant 1$ *(respectively $\geqslant 2$), then the natural map*

$$\operatorname{Hom}(X, Y) \to \operatorname{Hom}(\partial X, Y)$$

is injective (bijective, respectively) where $\operatorname{Hom}(\partial X, Y) = \varinjlim_U \operatorname{Hom}(U, Y)$ *with respect to the neighbourhoods U of the boundary of X.*

Proof. Let K be a holomorphically convex compact in X and $U = X \setminus K$. Consider the commutative diagram

$$\begin{array}{ccc} \operatorname{Hom}(X, Y) & \longrightarrow & \operatorname{Hom}(U, Y) \\ \downarrow & & \downarrow \\ \operatorname{Hom}(\Gamma(Y, \mathcal{O}_Y), \Gamma(X, \mathcal{O}_X)) & \to & \operatorname{Hom}(\Gamma(Y, \mathcal{O}_Y), \Gamma(U, \mathcal{O}_X)), \end{array}$$

the vertical arrows being bijective by the theorem. If prof $\mathcal{O}_X \geqslant 1$ (resp. $\geqslant 2$), then by 3.1 the undermost horizontal arrow is injective (resp. bijective). The conclusion follows because in any Stein space the complementaries of the holomorphically-convex compacts give a cofinal system of neighbourhoods of the boundary.

(c) *Applications to compact complex spaces.*

THEOREM 4.13. *Let X be a compact complex space, $Y \subset X$ a closed subset such that $X \setminus Y$ is a Stein open subset and $\mathcal{F} \in$* **Coh** *(X). Then* prof $(\mathcal{F}|X \setminus Y) \geqslant N + 1$ *if and only if the canonical map* $H^q(X, \mathcal{F}) \to H^q(Y, \mathcal{F})$ *is bijective for $q \leqslant N - 1$ and injective for $q = N$.*

Proof. The assertion follows from theorem 3.6 by means of the exact sequence

$$\ldots \to H^q_c(X \setminus Y, \mathcal{F}) \to H^q(X, \mathcal{F}) \to H^q(Y, \mathcal{F}) \to H^{q+1}_c(X \setminus Y, \mathcal{F}) \to \ldots$$

COROLLARY 4.14. *Let X be a compact complex space and Y a closed subset of X such that $X \setminus Y$ is Stein. If* prof $\mathcal{O}_x \geqslant 2$ *for all* $x \in X \setminus Y$, *then X is connected if and only if Y is connected.*

The proof is immediate by theorem and lemma 4.9.

Remark. These facts can be applied to the case when X is a projective complex manifold and Y a hypersurface of it.

Similar results can be obtained if Y is substituted for a holomorphically convex compact in a neighbourhood of it; for instance, $X = \mathbb{P}^r$ and Y a holomorphically convex compact in an affine cart.

PROPOSITION 4.15. *Let X be a complex space such that it has a covering by $d + 1$ Stein open subsets. If $\mathcal{F} \in$* **Coh** (X), *$N \geqslant 0$ is an integer and* prof $\mathcal{F} \geqslant N + 1$, *then $H_c^q(X, \mathcal{F}) = 0$ for any $q \leqslant N - d$.*

For the proof we proceed by induction on d when using 3.6 and the following result of Mayer-Vietoris type:

LEMMA 4.16. *Suppose X is a paracompact space and $\mathcal{F}$ a sheaf of abelian groups on it. If X is a union of two open subsets U, V then there exists the following exact cohomology-sequence*

$$\ldots \to H_c^q(U \cap V, \mathcal{F}) \to H_c^q(U, \mathcal{F}) \oplus H_c^q(V, \mathcal{F}) \to H_c^q(X, \mathcal{F}) \to$$
$$\to H_c^{q+1}(U \cap V, \mathcal{F}) \to \ldots$$

Proof. For any sheaf of abelian groups $\mathfrak{I}$ on X we have an exact sequence

$$0 \to \Gamma_c(U \cap V, \mathfrak{I}) \to \Gamma_c(U, \mathfrak{I}) \oplus \Gamma_c(V, \mathfrak{I}) \xrightarrow{\theta} \Gamma_c(X, \mathfrak{I})$$

where the second arrow is induced by the diagonal map and the third is defined by the formula $\theta(s, t) = s - t$, $s \in \Gamma_c(U, \mathfrak{I})$, $t \in \Gamma_c(V, \mathfrak{I})$ (we use in both cases the trivial extension of sections).

We now suppose $\mathfrak{I}$ flabby. Let $v \in \Gamma_c(X, \mathfrak{I})$ and $K =$ Supp v. Consider the compact sets $K_1 = K \cap (X \setminus U)$, $K_2 = K \cap (X \setminus V)$. They are obviously disjoint and one can choose U_1, U_2 as disjoint neighbourhood of theirs. We put

$$s = \begin{cases} 0 \text{ on } (X \setminus K) \cup U_1 \\ v \text{ on } U_2 \end{cases}$$

and thus obtain a section of $\mathfrak{I}$ over the open $(X \setminus K) \cup U_1 \cup U_2$. If $s' \in \Gamma(X, \mathfrak{I})$ is an extension of its then $s'|U \in \Gamma_c(U, \mathfrak{I})$ and if we put $s_1 = s'|U$, $s_2 = (s' - v)|V$, it is easy to check that $s_2 \in \Gamma_c(V, \mathfrak{I})$ and $s_1 - s_2 = v$, hence $\theta(s_1, s_2) = v$. Therefore the map θ is surjective.

Let now $\mathfrak{I}^\bullet$ be a flabby resolution for $\mathcal{F}$. We then get an exact sequence of complexes

$$0 \to \Gamma_c(U \cap V, \mathfrak{I}^\bullet) \to \Gamma_c(U, \mathfrak{I}^\bullet) \oplus \Gamma_c(V, \mathfrak{I}^\bullet) \to \Gamma_c(X, \mathfrak{I}^\bullet) \to 0.$$

The lemma follows by passing to cohomology.

For a compact complex space X, in conformity with Andreotti-Vesentini [3], we shall denote by $d(X)$ the minimal number of Stein open subsets which can cover together the space X.

COROLLARY 4.17. *Let $(X, \mathcal{O})$ be a compact complex space. Then*

$$d(X) \geqslant \operatorname{prof} \mathcal{O} + 1.$$

In particular, X can not be covered by less than prof $\mathcal{O} + 1$ *Stein open subsets.*

Proof. Otherwise, $\Gamma(X, \mathcal{O}) = \Gamma_c(X, \mathcal{O}) = 0$, hence a contradiction.

COROLLARY 4.18. *A normal compact complex space of dimension $\geqslant 2$ can not be covered by two Stein open subsets.*

The assertion follows since the depth of the structural sheaf is $\geqslant 2$ (by 1.8).

(d) *The invariance of the depth with respect to finite morphisms.*

Recall that a morphism of complex spaces $X \xrightarrow{f} Y$ is said to be *finite* if the corresponding topological map is finite.

PROPOSITION 4.19. *Let $X \xrightarrow{f} Y$ be a finite morphism of complex spaces and $\mathcal{F} \in$ **Coh** (X). Then*

$$\operatorname{prof} \mathcal{F} = \operatorname{prof} (f_*(\mathcal{F})).$$

Proof. The problem being local on Y, we can suppose Y Stein. By lemma 1.21 and theorem B we get $H^q(X, \mathcal{G}) \simeq H^q(Y, f_*(\mathcal{G})) = 0$ for any $q \geqslant 1$ and $\mathcal{G} \in$ **Coh** (X), $f_*(\mathcal{G})$ being also coherent [15], [56]. Accordingly, X is Stein. Then we shall use the characterization of the depth on Stein spaces in terms of cohomology with compact supports given in 3.6. Since the canonical morphisms $H_c^{\bullet}(Y, f_*(\mathcal{F})) \simeq H_c^{\bullet}(X, \mathcal{F})$ are isomorphisms, the conclusion follows.

The particular case when f is the normalization morphism was proved without cohomological methods in [94].

Bibliographical indications

The idea of using cohomology with compact supports in the problems recalled at the beginning of the introduction of this chapter, as well as the informations about this cohomology by means of duality, appear in Serre's papers [74], [75].

Theorem 2.1 belongs to the authors [8] and it is a particular case of the general Ramis-Ruget duality on complex spaces [61].

Theorem 2.9 for X domain of holomorphy, K polycompact (product of compacts of the complex plane) and $\mathcal{F} = \mathcal{O}$ was proved by Frenkel by means of Laurent expansions [24]. The general case (for locally free sheaves) was proved by Martineau [55], by using Malgrange's result on the triviality of the cohomology in the highest dimension on noncompact manifolds. The extension to coherent sheaves was done by Harvey [40] and by the authors (the proof given here is that from [8]); in [40] the result applies for a class of compacts larger than Stein compacts.

The separation theorem 2.19 was proved by Trautmann in [92] for the case when the compact is reduced to one point (2.21); the proof given here by means of duality is due to the first author [5].

Theorem 3.1 belongs to Harvey [40] and the authors [8]. Its particular case 3.3 was previously proved by Trautmann [90].

The inverse implication from theorem 3.6 (*a*) is a simple case of the Andreotti-Grauert finiteness theorems [2]. The implication (*i*) ⇒ (ii) of theorem 3.7 is a particular case of a result of Reiffen's [64]. The characterizations of the depth and dimension of coherent analytic sheaves on Stein spaces in terms of cohomology with compact supports (the direct implication of 3.6 (*a*) and the implication (ii) ⇒ (i) of 3.7) were obtained by Y. T. Siu in [81] by means of Thimm's gap sheaves; the characterization of the depth was also independently found by the authors [8]. The proofs given here, which make use of duality, are those from [8], [40]. The results 3.6 and 3.7 can be fairly exposed by means of the dualizing sheaves introduced by Andreotti and Kas ([4] and Ch. VII, § 5 of this book).

Paragraph 4 is conceived following papers [74], [8], [9] and [40], the results being straightforward consequences of the previous paragraph.

Chapter II

Analytic local cohomology

Introduction

An important problem of the theory of complex functions is the extension of the analytic entities defined beyond an analytic set. Recall some results in this context:

"Let Ω be an open set in $\mathbb{C}^n$ and $A \subset \Omega$ an analytic subset of codimension $\geqslant 2$. Then any holomorphic function on $\Omega \setminus A$ extends uniquely to Ω" (der 2. Riemannsche Hebbarkeitssatz [56]).

"The additive Cousin data in $\mathbb{C}^3 \setminus \{0\}$ extend uniquely to $\mathbb{C}^3$." This result of H. Cartan [13] was generalized for superior dimensions by Rothstein [67], namely:

"If G is an open in a numerical space and $A \subset G$ is an analytic subset of codimension $\geqslant 3$, then the additive Cousin data on $G \setminus A$ extend to whole G."

To these examples of extensions of functions and Cousin data one can add examples of extensions of cohomology classes, analytic subsets, sheaves,...

Let X be a topological space, $A \subset X$ a closed part and $\mathcal{F}$ a sheaf of abelian groups on X. The exact sequence

$$\ldots \to H^q(X, \mathcal{F}) \to H^q(X \setminus A, \mathcal{F}) \to H^{q+1}_A(X, \mathcal{F}) \to \ldots$$

shows that the cohomology group with supports in A, $H^{q+1}_A(X, \mathcal{F})$ stands for the obstruction to the extension to whole X of cohomology q-classes from $X \setminus A$. Beside the global invariants $H^\bullet_A(X, \mathcal{F})$ one can consider local invariants: the sheaves $\mathcal{H}^\bullet_A \mathcal{F}$ associated to the presheaves $U \mapsto H^\bullet_{A \cap U}(U, \mathcal{F})$. This chapter is dedicated to the study of the invariants $H^\bullet_A(X, \mathcal{F})$ and $\mathcal{H}^\bullet_A \mathcal{F}$ (X complex space, A analytic subset, $\mathcal{F} \in$ **Coh** (X)) and in this way, to the extension of cohomology classes.

The first paragraph contains some preliminaries of local cohomology and algebra (the depth of a module with respect to an ideal). In the second one, to any coherent analytic sheaf we associate a family of analytic subsets, which play a decisive part in the formulation and the proof of the results in this chapter. In the third paragraph we introduce the notion of depth of a coherent sheaf with respect to an analytic set and prove the vanishing theorem: $\mathcal{H}^i_A \mathcal{F} = 0$ for $i \leqslant q$ if and only if $\mathrm{prof}_A \mathcal{F} \geqslant q + 1$. Among the applications one can mention:

– If X is a perfect complex space and A an analytic subset of codimension $\geqslant 2$, then the restriction $\Gamma(X, \mathcal{O}) \to \Gamma(X \setminus A, \mathcal{O})$ is bijective.

— Let X be a perfect reduced complex space. Then X is normal if and only if the codimension of its singular locus is $\geqslant 2$.

— If X is a normal complex space and A an analytic subset of codimension $\geqslant 2$, then the restriction $\Gamma(X, \mathcal{O}) \to \Gamma(X \setminus A, \mathcal{O})$ is bijective.

In the fourth paragraph we give necessary and sufficient conditions regarding the coherence of the sheaves $\mathcal{H}^{\bullet}_A\mathcal{F}$. The next paragraph deals with the absolute local cohomological invariants: the sheaves $\mathcal{H}^i_d\,\mathcal{F}$ associated to the presheaves $U \mapsto \varinjlim \{H^i_A(U, \mathcal{F}) | \dim A \leqslant d\}$. A vanishing theorem and a finiteness theorem are proved for them. In the last paragraph some topologies on the spaces $H^{\bullet}_A(X, \mathcal{F})$ are defined and the following separation theorem is proved: if X is Stein and the sheaves $\mathcal{H}^i_A\,\mathcal{F}$ are coherent for $i \leqslant q$, then the spaces $H^i_A(X, \mathcal{F})$ are separated for $i \leqslant q + 1$.

§ 1. Preliminaries

(a) *Some elements of local cohomology* ([36], SGA 2). Let us consider a topological space X. A part A of X is said to be *locally closed* if there exists on open subset U of X such that A is a closed subset of U. Let $\mathcal{F} \in \mathbf{Ab}(X)$ be a sheaf of abelian groups on X. Denote by $\Gamma_A(X, \mathcal{F})$ the subgroup of the elements of $\Gamma(U, \mathcal{F})$ whose supports are contained in A. One easily verifies that $\Gamma_A(X, \mathcal{F})$ is independent on U. If $\mathcal{F} \to \mathcal{G}$ is a morphism of sheaves of abelian groups, then a natural map $\Gamma_A(X, \mathcal{F}) \to \Gamma_A(X, \mathcal{G})$ is induced. An additive functor

$$\Gamma_A(X, *) : \mathbf{Ab}(X) \to \mathbf{Ab}$$

is thus obtained and its right derived functors are denoted by $H^{\bullet}_A(X, *)$ and are called *the groups of local cohomology with supports in* A. Since the functor $\Gamma_A(X, *)$ is left exact, $H^0_A(X, *) \simeq \Gamma_A(X, *)$. If $\mathcal{F} \in \mathbf{Ab}\,(X)$ and

$$0 \to \mathcal{F} \to \mathcal{J}^0 \to \mathcal{J}^1 \to \mathcal{J}^2 \to \dots$$

is a resolution of $\mathcal{F}$ by injective sheaves in the category $\mathbf{Ab}\,(X)$, then $H^{\bullet}_A\,(X, \mathcal{F})$ are the cohomology groups of the complex of abelian groups

$$0 \to \Gamma_A(X, \mathcal{J}^0) \to \Gamma_A(X, \mathcal{J}^1) \to \Gamma_A(X, \mathcal{J}^2) \to \dots$$

LEMMA 1.1. (excision theorem). *If Y is an open subset of X and A is a closed subset of Y, then*

$$H^{\bullet}_A\,(X, \mathcal{F}) \simeq H^{\bullet}_A\,(Y, \mathcal{F} \mid Y), \quad \mathcal{F} \in \mathbf{Ab}\,(X).$$

The proof follows by definitions since the restriction of an injective sheaf to an open subset remains injective ([35], prop. 3.1.3.).

In particular, it follows by this lemma that $H_A^{\bullet}(X, \mathfrak{F}) \simeq H^{\bullet}(A, \mathfrak{F})$ for any open subset A of X.

LEMMA 1.2. *If* $0 \to \mathfrak{F}' \to \mathfrak{F} \to \mathfrak{F}'' \to 0$ *is an exact sequence in* $\mathbf{Ab}(X)$ *and* A *is locally closed in* X, *then we have the exact long cohomology sequence*

$$0 \to H_A^0(X, \mathfrak{F}') \to H_A^0(X, \mathfrak{F}) \to H_A^0(X, \mathfrak{F}'') \to H_A^1(X, \mathfrak{F}') \to H_A^1(X, \mathfrak{F}) \to \dots$$

The proof follows in virtue of general properties of derived functors.

Let A be a locally closed subset of X and A' a closed subset in A; $A'' = A \setminus A'$ is also locally closed in X.

PROPOSITION 1.3. *For any* $\mathfrak{F} \in \mathbf{Ab}\,(X)$ *the sequence*

$$0 \to H_{A'}^0(X, \mathfrak{F}) \to H_A^0(X, \mathfrak{F}) \to H_{A''}^0(X, \mathfrak{F}) \to H_{A'}^1(X, \mathfrak{F}) \to \dots$$

is exact, functorial in $\mathfrak{F}$ *and agrees with the short exact sequences in* $\mathbf{Ab}(X)$. *The maps* $H_{A'}^{\bullet}(X, \mathfrak{F}) \to H_A^{\bullet}(X, \mathfrak{F})$ *and* $H_A^{\bullet}(X, \mathfrak{F}) \to H_{A''}^{\bullet}(X, \mathfrak{F})$ *are the natural ones*; *the first is given by the inclusion* $A' \subset A$ *and the second* $H_A^{\bullet}(X, \mathfrak{F}) \to H_{A\setminus A'}^{\bullet}(X \setminus A', \mathfrak{F}) \simeq H_{A''}^{\bullet}(X, \mathfrak{F})$ *is obtained by restriction.*

Proof. If $\mathfrak{J}$ is a sheaf on X, then we have the following exact sequence given by inclusion $A' \subset A$ and by restriction

$$0 \to \Gamma_{A'}(X, \mathfrak{J}) \to \Gamma_A(X, \mathfrak{J}) \to \Gamma_{A\setminus A'}(X \setminus A', \mathfrak{J}) = \Gamma_{A''}(X, \mathfrak{J}).$$

If $\mathfrak{J}$ is injective (hence flabby!), then the map

$$\Gamma_A(X, \mathfrak{J}) = \Gamma_A(U, \mathfrak{J}) \to \Gamma_{A''}(X, \mathfrak{J}) = \Gamma_A(U \setminus A', \mathfrak{J})$$

is surjective (U is an open such that A is a closed subset of U). Let $\mathfrak{J}^{\bullet}$ be an injective resolution for $\mathfrak{F}$. We then derive the exact sequence

$$0 \to \Gamma_{A'}(X, \mathfrak{J}^{\bullet}) \to \Gamma_A(X, \mathfrak{J}^{\bullet}) \to \Gamma_{A''}(X, \mathfrak{J}^{\bullet}) \to 0$$

and the assertion follows.

COROLLARY 1.4. *If* A *is a closed subset of* X *and* $\mathfrak{F} \in \mathbf{Ab}\,(X)$, *then we get a natural exact sequence*

$$0 \to H_A^0(X, \mathfrak{F}) \to H^0(X, \mathfrak{F}) \to H^0(X \setminus A, \mathfrak{F}) \to H_A^1(X, \mathfrak{F}) \to H^1(X, \mathfrak{F}) \to \dots$$

From this sequence we can remark that the invariants $H_A^{i+1}(X, \mathfrak{F})$ are just the obstructions concerning the surjectivity of the restriction maps $H^i(X, \mathfrak{F}) \to H^i(X \setminus A, \mathfrak{F})$.

COROLLARY 1.5. *Let* A *be a locally closed subset of* X *and* $\mathfrak{F}$ *a flabby sheaf on* X. *Then* $H_A^i(X, \mathfrak{F}) = 0$ *for any* $i \geqslant 1$.

Proof. If U is an open such that A is a closed subset of U, then the sheaf $\mathfrak{F}|U$ is flabby on U. By substituting X for U we can assume A closed in X. The map $H^0(X, \mathcal{F}) \to H^0(X \setminus A, \mathcal{F})$ is surjective and $H^i(X, \mathcal{F}) = H^i(X \setminus A, \mathcal{F}) = 0$ for $i \geqslant 1$. The conclusion follows by the previous corollary.

COROLLARY 1.6. *The groups of local cohomology with supports in a locally closed subset can be calculated by means of resolutions by flabby sheaves.*

Apply 1.5 in accordance with a general property of derived functors ("de Rham's abstract theorem", I. 1.2).

Remark. In the investigation of the invariants $H_A^{\bullet}(X, *)$ as well as the other invariants of this chapter, we can avoid the injective sheaves. For a sheaf $\mathcal{F} \in \mathbf{Ab}(X)$ we can consider the flabby resolution of Godement ([26], Ch. II, 4.3)

$$0 \to \mathcal{F} \to \mathcal{C}^0 \to \mathcal{C}^1 \to \mathcal{C}^2 \to \dots$$

and define $H_A^{\bullet}(X, \mathcal{F})$ as the cohomology groups of the complex

$$0 \to \Gamma_A(X, \mathcal{C}^0) \to \Gamma_A(X, \mathcal{C}^1) \to \Gamma_A(X, \mathcal{C}^2) \to \dots$$

Starting from this definition we obtain again the preceeding properties of local cohomology. By corollary 1.5 there results that in order to calculate the invariants $H_A^{\bullet}(X, \mathcal{F})$ we can replace Godement's resolution by any flabby resolution.

To conclude with the global cohomological considerations, we introduce a new invariant. If $\Phi \subset \Psi$ are two families of supports in X (not necessarily paracompactifying), then we denote

$$\Gamma_{\Psi/\Phi}(X, \mathcal{F}) = \Gamma_{\Psi}(X, \mathcal{F})/\Gamma_{\Phi}(X, \mathcal{F})$$

for any sheaf $\mathcal{F} \in \mathbf{Ab}(X)$.

One thus obtains a functor and its right derived functors are denoted by $H_{\Psi/\Phi}^{\bullet}(X, *)$. Generally, $\Gamma_{\Psi/\Phi}(X, \mathcal{F})$ is not isomorphic to $H_{\Psi/\Phi}^0(X, \mathcal{F})$. From the properties of derived functors we get the following

LEMMA 1.7. *Under the above conditions, there is a naturally exact sequence*

$$0 \to H_{\Phi}^0(X, \mathcal{F}) \to H_{\Psi}^0(X, \mathcal{F}) \to H_{\Psi/\Phi}^0(X, \mathcal{F}) \to H_{\Phi}^1(X, \mathcal{F}) \to H_{\Psi}^1(X, \mathcal{F}) \to \dots$$

We are now going to study the local cohomological invariants, defined by Grothendieck. Preserve the above notations; hence X is a topological space, $A \subset X$ a locally closed subset of X and $\mathcal{F} \in \mathbf{Ab}(X)$.

For an integer i denote by $\mathcal{H}_A^i\mathcal{F}$ the sheaf associated to the presheaf

$$U \mapsto H_A^i(U, \mathcal{F}),\ U \subset V \mapsto \text{the natural map } H_A^i(U, \mathcal{F}) \leftarrow H_A^i(V, \mathcal{F})$$

(by $H_A^i(U, \mathcal{F})$ we mean $H_{A\cap U}^i(U, \mathcal{F}|U)\dots$). A family of functors $(\mathcal{H}_A^i(*))_{i\geqslant 0}$ is thus obtained and one can easily check that this is just the family of the derived functors of the functor $\mathcal{F} \mapsto \mathcal{H}_A^0\mathcal{F}$. The invariants $\mathcal{H}_A^{\bullet}\mathcal{F}$ are called the *sheaves of local cohomology with supports in A associated to* $\mathcal{F}$. If A is a closed subset of X, then

$\mathcal{H}_A^0\mathcal{F}$ is the subsheaf of $\mathcal{F}$, formed by the sections whose support is in A; to be more precise, for any open U of X,

$$\Gamma(U, \mathcal{H}_A^0\mathcal{F}) = \Gamma_A(U, \mathcal{F}) \subset \Gamma(U, \mathcal{F}).$$

It is easy to see that Supp $\mathcal{H}_A^i\mathcal{F} \subset A$ for all i.

LEMMA 1.8. *If* $0 \to \mathcal{F}' \to \mathcal{F} \to \mathcal{F}'' \to 0$ *is an exact sequence in* **Ab**(X), *then we have the long cohomology sequence*

$$0 \to \mathcal{H}_A^0\mathcal{F}' \to \mathcal{H}_A^0\mathcal{F} \to \mathcal{H}_A^0\mathcal{F}'' \to \mathcal{H}_A^1\mathcal{F}' \to \mathcal{H}_A^1\mathcal{F} \to \dots$$

The proof follows by lemma 1.2.

By using the proposition 1.3 we deduce

PROPOSITION 1.9. *If A is a locally closed subset of X, A' a closed subset in A, $A'' = A \setminus A'$* and *$\mathcal{F} \in$* **Ab**(X), *then we have an exact sequence*

$$0 \to \mathcal{H}_{A'}^0\mathcal{F} \to \mathcal{H}_A^0\mathcal{F} \to \mathcal{H}_{A''}^0\mathcal{F} \to \mathcal{H}_{A'}^1\mathcal{F} \to \mathcal{H}_A^1\mathcal{F} \to \dots$$

enjoying the same properties as the sequence from 1.3.

We shall now consider the case when A is a closed subset of X. Denote by $\mathcal{R}_A^i\mathcal{F}$ the sheaf of abelian groups defined by the presheaf

$$U \to H^i(U \setminus A, \mathcal{F}).$$

The sheaves $\mathcal{R}_A^\bullet\mathcal{F}$ are called the *gap-sheaves of $\mathcal{F}$ with respect to A*. By the above notations, $\mathcal{R}_A^\bullet\mathcal{F} = \mathcal{H}_{X\setminus A}^\bullet\mathcal{F}$. These sheaves coincide also with the direct image sheaves given by the inclusion $X \setminus A \subset X$.

Since the inductive limit is an exact functor and since for all $x \in X$, $\varinjlim_{U \ni x} H^i(U, \mathcal{F}) = 0$ $(i \geqslant 1)$, we get by 1.4 the following

COROLLARY 1.10. *Let $A \subset X$ be a closed subset and $\mathcal{F} \in$* **Ab**(X). *Then we have the exact sequence*

$$0 \to \mathcal{H}_A^0\mathcal{F} \to \mathcal{F} \to \mathcal{R}_A^0\mathcal{F} \to \mathcal{H}_A^1\mathcal{F} \to 0$$

and, for any $i \geqslant 1$, we have the isomorphisms

$$\mathcal{R}_A^i\mathcal{F} \simeq \mathcal{H}_A^{i+1}\mathcal{F}.$$

COROLLARY 1.11. *If $A \subset X$ is a locally closed subset of X and $\mathcal{F}$ a flabby sheaf, then $\mathcal{H}_A^i\mathcal{F} = 0$ for $i \geqslant 1$. If in addition A is closed, then $\mathcal{H}_A^0\mathcal{F}$ is flabby.*

Proof. Only the second assertion is not trivial. Let U be an arbitrary open subset of X and $s \in \Gamma(U, \mathcal{H}^0_A\mathcal{F}) = \Gamma_A(U, \mathcal{F}) \subset \Gamma(U, \mathcal{F})$. Consider

$$\tilde{s} = \begin{cases} 0 \text{ on } X \setminus A \\ s \text{ on } U. \end{cases}$$

We have $\tilde{s} \in \Gamma(U \cup (X \setminus A), \mathcal{F})$; since $\mathcal{F}$ is flabby, $\tilde{s}$ can be extended to a section $\hat{s} \in \Gamma(X, \mathcal{F})$. As $\hat{s}\,|X \setminus A = 0$ and $\hat{s}\,|U = s$, the conclusion follows.

COROLLARY 1.12. *The sheaves of local cohomology with supports in a locally closed subset can be calculated by means of flabby resolutions.*

COROLLARY 1.13. *Let* $X \xrightarrow{i} Y$ *be a closed immersion of topological spaces,* $A \subset X$ *a closed subset and* $\mathcal{F} \in \mathbf{Ab}(X)$. *Then*

$$i_*(\mathcal{H}^{\cdot}_A(\mathcal{F})) \simeq \mathcal{H}^{\cdot}_{i(A)}(i_*(\mathcal{F})).$$

Proof. Apply the previous corollary by virtue of the exactness of i_* and of the fact that $i_*(\mathcal{J})$ is flabby for any flabby sheaf $\mathcal{J}$ on X.

Before closing our considerations on local cohomology we shall establish the connexion between the local and global invariants.

Let A be a closed subset of X. We have the equality of functors $\Gamma_A(X, *) = \Gamma(X, \mathcal{H}^0_A(*))$. By corollary 1.11, the functor $\mathcal{H}^0_A$ transforms flabby sheaves (so much the more injective) into flabby sheaves, hence $\Gamma(X, *)$-acyclic. Applying ([26], Ch. II, 2.4) we get the following:

PROPOSITION 1.14. *There exists a spectral sequence of term* $E_2^{p,q} = H^p(X, \mathcal{H}^q_A\mathcal{F})$ *which converges to* $H^{p+q}_A(X, \mathcal{F})$, $\mathcal{F} \in \mathbf{Ab}(X)$.

Recall that this spectral sequence can be built in the following way: consider an injective (or flabby) resolution $0 \to \mathcal{F} \to \mathcal{J}^{\bullet}$. For the complex $0 \to \mathcal{H}^0_A(\mathcal{J}^0) \to \mathcal{H}^0_A(\mathcal{J}^1) \to \mathcal{H}^0_A(\mathcal{J}^2) \to \dots$, we consider again an injective (or flabby) resolution $\mathcal{J}^{\bullet\bullet}$. The spectral sequence we look for is one of the spectral sequences attached to the bicomplex $(\Gamma(X, \mathcal{J}^{p,q}))_{p,q}$. From the properties of spectral sequences there follows:

COROLLARY 1.15. *Let A be a closed subset in the topological space X and $\mathcal{F}$ a sheaf of abelian groups on X. Let $q \geqslant 0$ be an integer such that $H^i(X, \mathcal{H}^j_A\mathcal{F}) = 0$ for $j \leqslant q$ and $i \geqslant 1$. Then the natural morphism*

$$H^i_A(X, \mathcal{F}) \to \Gamma(X, \mathcal{H}^i_A\mathcal{F})$$

is an isomorphism for any $i \leqslant q + 1$.

Remark. The spectral sequence of the proposition is used only to prove this corollary. By applying I.1.1 to the composition $\Gamma_A(X, *) = \Gamma(X, \mathcal{H}^0_A(*))$, we obtain a straightforward proof of the corollary and the considerations of spectral sequences can thus be avoided (one can also see [84], lemma 0.6).

By the above corollary and by 1.4 we get:

COROLLARY 1.16. *Let A be a closed subset in the topological space X and $\mathcal{F} \in \mathbf{Ab}(X)$. For the integer $q \geqslant 0$ the following three conditions are equivalent:*

(*a*) $\mathcal{H}^i_A\mathcal{F} = 0$ *for* $i \leqslant q$.

(*b*) *For any open subset U of X,* $H^i_A(U, \mathcal{F}) = 0$ *for* $i \leqslant q$.

(*c*) *For any open subset U of X, the restrictions*

$$H^i(U, \mathcal{F}) \to H^i(U \setminus A, \mathcal{F})$$

are bijective for $i < q$ and injective for $i = q$.

We also remark that if $\mathcal{O}$ is a sheaf of rings on X and $\mathcal{F}$ is an $\mathcal{O}$-module, then the sheaves $\mathcal{H}^i_A\mathcal{F}$ and $\mathcal{R}^i_A\mathcal{F}$ have a natural structure of $\mathcal{O}$-modules.

(b) *The depth of a module with respect to an ideal* ([36], SGA 2).

THEOREM 1.17. *Let A be a noetherian ring, $\mathfrak{a}$ an ideal of A, M an A-module of finite type, and q an integer. Then the following three conditions are equivalent:*

(*a*) $\mathrm{Ext}^i_A(N, M) = 0$ *for any A-module of finite type N so that* $\mathrm{Supp}\, N \subset V(\mathfrak{a})$, *and for any $i < q$.*

(*b*) *There exists an A-module of finite type N with* $\mathrm{Supp}\, N = V(\mathfrak{a})$ *such that* $\mathrm{Ext}^i_A(N, M) = 0$ *for all $i < q$.*

(*c*) *There exists a regular M-sequence formed by q elements of the ideal $\mathfrak{a}$.*

In order to prove the theorem we need two lemmas.

LEMMA 1.18. *Let A be a noetherian ring and let M, N be two A-modules of finite type. Then* $\mathrm{Ass}\,(\mathrm{Hom}_A(N, M)) = \mathrm{Supp}\, N \cap \mathrm{Ass}\, M$.

Proof. Obviously, $\mathrm{Ass}\,(\mathrm{Hom}_A(N, M)) \subset \mathrm{Supp}\,(\mathrm{Hom}_A(N, M)) \subset \mathrm{Supp}\, N$. If we consider an epimorphism $A^r \to N \to 0$, we get thereby an exact sequence $0 \to \mathrm{Hom}_A(N, M) \to \mathrm{Hom}_A(A^r, M) = M^r$ and, therefore, $\mathrm{Ass}\,(\mathrm{Hom}_A(N, M)) \subset \mathrm{Ass}\, M^r = \mathrm{Ass}\, M$. We have thus proved the inclusion $\mathrm{Ass}\,(\mathrm{Hom}_A(N, M)) \subset \mathrm{Supp}\, N \cap \mathrm{Ass}\, M$.

The inverse inclusion is more difficult. Let $\mathfrak{p} \in \mathrm{Supp}\, N$. We first show that there exists a non-null morphism of A-modules $N \to A/\mathfrak{p}$. Indeed, $N_\mathfrak{p}/\mathfrak{p}N_\mathfrak{p}$ is a linear $A_\mathfrak{p}/\mathfrak{p}A_\mathfrak{p}$-space $\neq 0$ (by Nakayama lemma) and it is finite-dimensional. Let $v: N_\mathfrak{p}/\mathfrak{p}N_\mathfrak{p} \to A_\mathfrak{p}/\mathfrak{p}A_\mathfrak{p}$ be a non-null $A_\mathfrak{p}/\mathfrak{p}A_\mathfrak{p}$-morphism. Choose an element $a \in A/\mathfrak{p} \subset A_\mathfrak{p}/\mathfrak{p}A_\mathfrak{p}$ such that the morphism av satisfies $av(N_\mathfrak{p}/\mathfrak{p}N_\mathfrak{p}) \subset A/\mathfrak{p}$. The composition $N \to N_\mathfrak{p} \to N_\mathfrak{p}/\mathfrak{p}N_\mathfrak{p} \xrightarrow{av} A/\mathfrak{p}$ is a non-null A-morphism. Let now $\mathfrak{p} \in \mathrm{Supp}\, N \cap \mathrm{Ass}\, M$. Then there is an A-monomorphism $A/\mathfrak{p} \to M$ and by composition a morphism $u: N \to M$ is thus determined. One can easily check that $\mathfrak{p} = \mathrm{Ann}\, u$, hence $\mathfrak{p} \in \mathrm{Ass}\,(\mathrm{Hom}_A(N, M))$.

LEMMA 1.19. *Let A be a noetherian ring, $\mathfrak{a}$ an ideal of A, and M an A-module of finite type. The following assertions are equivalent:*

(*a*) $\mathrm{Hom}_A(N, M) = 0$ *for any A-module of finite type N such that* $\mathrm{Supp}\, N \subset V(\mathfrak{a})$.

(*b*) *There exists an A-module of finite type N such that* $\mathrm{Supp}\, N = V(\mathfrak{a})$ *and* $\mathrm{Hom}_A(N, M) = 0$.

(*c*) *The ideal $\mathfrak{a}$ is not contained in any ideal of* $\mathrm{Ass}\, M$.

(*d*) *The ideal $\mathfrak{a}$ contains a nonzerodivisor of M.*

Proof. The implication $(a) \Rightarrow (b)$ is utterly trivial. The implications $(b) \Rightarrow (c)$, $(c) \Rightarrow (a)$ easily follow from the above lemma (the set Ass associated to a module of finite type is empty if and only if the module is null!). The equivalence $(c) \Leftrightarrow (d)$ is clear since the set of zerodivisors of M is just the union of the ideals of Ass M and apply ([10], Ch. IV, § 1, n^{o} = 1, cor.2, prop. 2).

The proof of the theorem. The implication $(a) \Rightarrow (b)$ is patent.

$(b) \Rightarrow (c)$. The case $q \leqslant 0$ is clear; hence we can assume $q > 0$. As $\mathrm{Hom}(N, M) = 0$, there is, by 1.19, an element $f \in \mathfrak{a}$ which is M-regular. We then have the exact sequence $0 \to M \xrightarrow{f} M \to M/fM \to 0$. From the hypothesis and the exact sequence of Ext's it follows that $\mathrm{Ext}^i_A(N, M/fM) = 0$ for $i < q - 1$. One easily deduces the assertion by induction on q.

$(c) \Rightarrow (a)$. Again we proceed by induction on q. The case $q \leqslant 0$ is clear and so we can suppose $q > 0$ and prove the general induction step.

Let $f \in \mathfrak{a}$ be a M-regular element. From the exact sequence $0 \to M \xrightarrow{f} M \to \to M/fM \to 0$ and the induction hypothesis, we deduce $\mathrm{Ext}^i_A(N, M/fM) = 0$ for any A-module of finite type N such that $\mathrm{Supp}\, N \subset V(\mathfrak{a})$ and for all $i < q - 1$. Accordingly, the homotethy given by f, $\mathrm{Ext}^i_A(N, M) \xrightarrow{f} \mathrm{Ext}^i_A(N, M)$ is injective. Since $f \in \mathfrak{a}$, some power of it annihilates N, hence $\mathrm{Ext}^i_A(N, M)$ too. It will follow $\mathrm{Ext}^i_A(N, M) = 0$ for $i < q$; this completes the proof.

Let A, $\mathfrak{a}$, M be as in the statement of the theorem. The largest number q which satisfies the equivalent assertions of 1.17 is called *the depth of* M *with respect to* $\mathfrak{a}$ and it is denoted by $\mathrm{prof}\,_{\mathfrak{a}}M$. If A is a local ring with maximal ideal $\mathfrak{a}$, one finds again the usual notion of depth.

COROLLARY 1.20. *All maximal regular M-sequences formed by elements of* $\mathfrak{a}$ *have the same length, namely* $\mathrm{prof}\,_{\mathfrak{a}}M$.

COROLLARY 1.21. *If $f \in \mathfrak{a}$ is M-regular, then*

$$\mathrm{prof}_{\mathfrak{a}}\, M = \mathrm{prof}_{\mathfrak{a}}\, M/fM + 1.$$

The proof is in the same manner as I.1.6, (ii).

COROLLARY 1.22. $\mathrm{prof}\,_{\mathfrak{a}}M = \inf\limits_{\mathfrak{p} \in V(\mathfrak{a})} (\mathrm{prof}_{A_{\mathfrak{p}}} M_{\mathfrak{p}})$.

Proof. If the elements $f_1, \ldots, f_q$ of $\mathfrak{a}$ give a regular M-sequence, then for any $\mathfrak{p} \in V(\mathfrak{a})$ their images in $A_{\mathfrak{p}}$ are in $\mathfrak{p}A_{\mathfrak{p}}$ and form a regular $M_{\mathfrak{p}}$-sequence. Consequently, $\mathrm{prof}\,_{\mathfrak{a}}M \leqslant \inf\limits_{\mathfrak{p} \in V(\mathfrak{a})} (\mathrm{prof}\, M_{\mathfrak{p}})$.

In order to prove the inverse inequality we shall first consider the case when $\mathrm{prof}\,_{\mathfrak{a}}M = 0$. It will follow that $\mathfrak{a}$ is formed by zerodivisors for M. Then we have $\mathfrak{p} \in \mathrm{Ass}_A M$ such that $\mathfrak{a} \subset \mathfrak{p}$. Since $\mathfrak{p}A_{\mathfrak{p}}$ belongs to $\mathrm{Ass}_{A_{\mathfrak{p}}} M_{\mathfrak{p}}$, $\mathrm{prof}_{A_{\mathfrak{p}}} M_{\mathfrak{p}} = 0$ (I. 1.6) and the conclusion follows. Consider now the case when $\mathrm{prof}\,_{\mathfrak{a}}M > 0$ and let $f \in \mathfrak{a}$ be an M-regular element. By 1.21 and I. 1.6 we have $\mathrm{prof}\,_{\mathfrak{a}}(M/fM) = \mathrm{prof}\,_{\mathfrak{a}}M - 1$ and $\mathrm{prof}(M_{\mathfrak{p}}/fM_{\mathfrak{p}}) = \mathrm{prof}\, M_{\mathfrak{p}} - 1$, $\mathfrak{p} \in V(\mathfrak{a})$. To conclude the proof we apply induction.

COROLLARY 1.23. $\mathrm{prof}\,_{\mathfrak{a}}M < \infty$ *if and only if* $V(\mathfrak{a}) \cap \mathrm{Supp}\, M \neq \emptyset$.

Apply I. 1.6 (iii) and 1.22.

COROLLARY 1.24. *Let A be a normal ring and $\mathfrak{a} \subset A$ an ideal of height $\geqslant 2$. Then* prof $_{\mathfrak{a}}A \geqslant 2$.

Proof. Let $\mathfrak{p} \in V(\mathfrak{a})$. The ring $A_{\mathfrak{p}}$ is normal, of dimension $\geqslant 2$. By I. 1.8, prof $A_{\mathfrak{p}} \geqslant 2$ and to conclude we must apply 1.22.

The assertion may be proved straightforwardly as in I. 1.8.

We close this section by the following lemma.

LEMMA 1.25. *Suppose $A \xrightarrow{\varphi} B$ is a surjective morphism of noetherian rings, $\mathfrak{a} \subset A$ an ideal and M a B-module of finite type. Then*

$$\operatorname{prof}_{\varphi(\mathfrak{a})}(M) = \operatorname{prof}_{\mathfrak{a}} M_{[\varphi]}.$$

Proof. Let $f_1, \ldots, f_q$ be elements of $\mathfrak{a}$. They constitute a regular $M_{[\varphi]}$-sequence if and only if their images in B form a regular M-sequence.

(c) Further on we shall use the following

LEMMA 1.26. *Let A be a regular noetherian local ring, $n = \dim A$, and M an A-module of finite type. Then for any integer q,*

$$\dim(\operatorname{Ext}^q_A(M, A)) \leqslant n - q.$$

Proof. Let $\mathfrak{p} \in \operatorname{Supp}(\operatorname{Ext}^q_A(M, A))$. Suppose the contrary, $\dim A/\mathfrak{p} > n - q$. Since A is regular, we can write

$$\dim A = n = \dim A/\mathfrak{p} + \dim A_{\mathfrak{p}}.$$

The ring $A_{\mathfrak{p}}$ is regular and $\dim A_{\mathfrak{p}} < q$. By applying the Hilbert-Serre theorem ([79], IV, th. 9), we get $(\operatorname{Ext}^q_A(M, A))_{\mathfrak{p}} \simeq \operatorname{Ext}^q_{A_{\mathfrak{p}}}(M_{\mathfrak{p}}, A_{\mathfrak{p}}) = 0$, a contradiction.

Remark. We shall apply the lemma for a ring of convergent power series. The relation $\dim A = \dim A_{\mathfrak{p}} + \dim A/\mathfrak{p}$ will be proved for this case in the next section (1.28).

(d) *The lemma of normalization for analytic algebras*. Recall some well-known facts about analytic algebras [15], [33].

A $\mathbb{C}$-algebra isomorphic to a non-null factor-ring of a convergent power series ring is called an *analytic algebra*. The analytic algebras are noetherian local rings of residual field $\mathbb{C}$, the complex field. If A and B are two analytic algebras, then any morphism of $\mathbb{C}$-algebras $A \to B$ is local. We will use the following result: for any analytic algebra A and for any integer n, the map

$$\operatorname{Hom}_{\mathbb{C}\text{-alg.}}(\mathbb{C}\{X_1, \ldots, X_n\}, A) \to \overbrace{\mathfrak{m}_A \times \ldots \times \mathfrak{m}_A}^{n \text{ times}}, \varphi \mapsto (\varphi(X_i))_{1 \leqslant i \leqslant n}$$

is bijective ($\mathfrak{m}_A$ is the maximal ideal of A).

Consider a morphism of analytic $\mathbb{C}$-algebras $\varphi: A \to B$. The morphism φ is called *finite* if $B_{[\varphi]}$ is an A-module of finite type. The morphism φ is called *quasi-finite* if $B/\mathfrak{m}_A B$ is a finite-dimensional linear space over $A/\mathfrak{m}_A \simeq \mathbb{C}$; this means that $\mathfrak{m}_A B$ contains some power of the maximal ideal $\mathfrak{m}_B$ of B. We shall also use the following form of Weierstrass preparation theorem: a local morphism of analytic algebras is finite if and only if it is quasi-finite.

Let A be an analytic algebra and $x_1,\ldots, x_n$ a system of parameters for $\mathfrak{m}_A$. From the above, it follows that the correspondence $X_i \mapsto x_i$, $1 \leq i \leq n$ defines a finite morphism $\mathbb{C}\{X_1,\ldots, X_n\} \to A$. As $\dim A = n$, by the first Cohen-Seidenberg theorem ([79], III, prop. 3) this morphism is injective. In this way, we have shown how the lemma of normalization can be obtained: for any analytic algebra A there is a finite injective morphism $\mathbb{C}\{X_1,\ldots, X_n\} \to A$ ($n = \dim A$).

Next, by using this lemma of normalization we prove a more precise result.

THEOREM 1.27. *Let A be an analytic algebra and*

$$\mathfrak{a}_1 \subset \mathfrak{a}_2 \subset \ldots \subset \mathfrak{a}_p$$

a sequence of proper ideals of A. Then there exists a finite and injective morphism of $\mathbb{C}$*-algebras* $\mathbb{C}\{X_1,\ldots, X_n\} \to A$ *and, for any* i, $1 \leq i \leq p$, *a number* $h(i)$, $0 \leq h(i) \leq n$, *such that*

$$\mathfrak{a}_i \cap \mathbb{C}\{X_1,\ldots, X_n\} = (X_1,\ldots, X_{h(i)})\, \mathbb{C}\{X_1,\ldots, X_n\},$$

for any $i = 1,\ldots, p$.

Proof. We proceed exactly as in the case of polynomials ([79], III, th. 2). There exists a finite injective morphism $\mathbb{C}(z_1,\ldots, z_n\} \to A$ and thus it is sufficient to prove the theorem for $A = \mathbb{C}\,\{z_1,\ldots, z_n\}$.

We first consider the case $p = 1$ and $\mathfrak{a}_1$ a non-null principal ideal $\mathfrak{a}_1 = x_1A$. Since A is an integral domain, x_1 is nonzerodivisor and can be completed to a system of parameters x_1, $x_2,\ldots, x_n$ for A ([79], Ch. III). The morphism $\mathbb{C}\{\underline{X}\} \to A$, $X_i \mapsto x_i$, $\underline{X} = (X_1,\ldots, X_n)$, is finite and injective. We need the equality $\mathfrak{a}_1 \cap \mathbb{C}\{\underline{X}\} = X_1\mathbb{C}\{\underline{X}\}$. An inclusion is obvious. For the other, let $\alpha \in \mathfrak{a}_1 \cap \mathbb{C}\{\underline{X}\}$, $\alpha = x_1S$, where $S \in A$. Since the element S is integral over the ring $\mathbb{C}\{\underline{X}\}$ and lies in its quotient field, there results that $S \in \mathbb{C}\{\underline{X}\}$($\mathbb{C}\{\underline{X}\}$ is a normal ring!), therefore $\alpha \in X_1\mathbb{C}\{\underline{X}\}$. The theorem is thus proved in this particular case, if we put $h(1) = 1$.

Further, we will make induction on p. In the case $p = 1$ use induction on n. If $n = 1$, then the ideal $\mathfrak{a}_1$ is principal; hence we have the same previous situation. For the general induction step let us fix $x_1 \in \mathfrak{a}_1$, $x_1 \neq 0$. There exists a finite injective morphism $\mathbb{C}\{X_1, Y_2,\ldots, Y_n\} \to A$ such that $x_1A \cap \mathbb{C}\{X_1, Y_2,\ldots, Y_n\} = X_1\mathbb{C}\{X_1, Y_2,\ldots, Y_n\}$. The induction hypothesis applied to the ideal $\mathfrak{a}_1 \cap \mathbb{C}\{Y_2,\ldots$ $\ldots, Y_n\}$ gives rise to a finite injective morphism $\mathbb{C}\{X_2,\ldots, X_n\} \to \mathbb{C}\{Y_2,\ldots, Y_n\}$ and to a number h' such that $\mathfrak{a}_1 \cap \mathbb{C}\{X_2,\ldots, X_n\} = \{X_2,\ldots, X_{h'})\mathbb{C}\{X_2, \ldots, X_n\}$. It is easy to check that the morphism $\mathbb{C}\{\underline{X}\} \to \mathbb{C}\{X_1, Y_2,\ldots, Y_n\}$ is finite and injective; the composition $\mathbb{C}\{\underline{X}\} \to \mathbb{C}\{X_1, Y_2,\ldots. Y_n\} \to A$ has the same property. Moreover, $\mathfrak{a}_1 \cap \mathbb{C}\{\underline{X}\} = (X_1, \ldots, X_{h'})\; \mathbb{C}\{\underline{X}\}$ and the case $p = 1$ is over.

Suppose the theorem proved for $p - 1$ and prove it for p. There exists a finite injective morphism $\mathbb{C}\{Y_1,\ldots, Y_n\} \xrightarrow{u} A$ and some numbers $h(i)$, $1 \leq i \leq p - 1$ which satisfy the assertion of the theorem with respect to the sequence of ideals $\mathfrak{a}_1 \subset \ldots \subset \mathfrak{a}_{p-1}$. Denote $d = h(p - 1)$. Then we can find a finite injective morphism

$\mathbb{C}\{X_{d+1},\ldots, X_n\} \xrightarrow{v} \mathbb{C}\{Y_{d+1},\ldots, Y_n\}$ and a number $r \geqslant d+1$ such that $\mathfrak{a}_p \cap \cap \mathbb{C}\{X_{d+1},\ldots, X_n\} = (X_{d+1},\ldots, X_r)\, \mathbb{C}\{X_{d+1},\ldots, X_n\}$. The morphism $\mathbb{C}\{X_1,\ldots ,X_d, X_{d+1},\ldots, X_n\} \to \mathbb{C}\{Y_1,\ldots, Y_d, Y_{d+1},\ldots, Y_n\}$, $X_i \mapsto Y_i$ for $1 \leqslant i \leqslant d$, $X_i \mapsto v(X_i)$ for $d+1 \leqslant i \leqslant n$ is finite and by composition with u it determines a finite injective morphism $\mathbb{C}\{X_1,\ldots, X_n\} \to A$. It only remains to put $h(p) = r$. This completes the proof.

Remark. The hypothesis that the analytic algebras are considered over the complex field $\mathbb{C}$ is not esential; $\mathbb{C}$ can be replaced by an arbitrary complete valued field.

COROLLARY 1.28. *Let A be an integral analytic algebra and $\mathfrak{p}$ a prime ideal of A. Then*

$$\dim A = \dim A/\mathfrak{p} + \dim A_{\mathfrak{p}}.$$

Proof. There exists a finite injective morphism $\mathbb{C}\{\underline{X}\} = \mathbb{C}\{X_1,\ldots, X_n\} \to A$ and a number h, $0 \leqslant h \leqslant n$, such that $\mathfrak{p} \cap \mathbb{C}\{\underline{X}\} = (X_1,\ldots X_h)\mathbb{C}\{\underline{X}\}$. By applying the Cohen-Seidenberg theorems ([79], Ch. III), we get $\dim A = \dim \mathbb{C}\{\underline{X}\} = n$, $\dim A/\mathfrak{p} = \dim (\mathbb{C}\{\underline{X}\}/\mathfrak{p} \cap \mathbb{C}\{\underline{X}\}) = \dim \mathbb{C}\{X_{h+1},\ldots, X_n\} = n - h$ and $\dim A_{\mathfrak{p}} = = \operatorname{ht} \mathfrak{p} = \operatorname{ht}(\mathfrak{p} \cap \mathbb{C}\{\underline{X}\}) = h$.

COROLLARY 1.29. *Let A be an analytic algebra and $\mathfrak{p} \subset \mathfrak{q}$ prime ideals of A. Then all maximal chains of prime ideals which link $\mathfrak{p}$ with $\mathfrak{q}$ have the same length.*

This fact follows by the preceding corollary, applied to $A/\mathfrak{p}$ and $\mathfrak{q}/\mathfrak{p}$.

(e) We shall use the following result: "If $X \xrightarrow{f} Y$ is a finite morphism of complex spaces and if $\mathcal{F} \in \mathbf{Coh}\,(X)$, then $f_*(\mathcal{F}) \in \mathbf{Coh}\,(Y)$".

For a proof see [15] or [56], Ch. IV, th. 7, or the next chapter, where the more general case of proper morphisms is considered.

We shall also use the following result: "If X is a complex space and x is one of its points, then there is a neighbourhood U of x and a finite morphism $U \to U'$, U' being an open subset of a numerical space ([15], exp. 19)".

PROPOSITION 1.30. *Let X be a complex space and $\mathcal{F} \in \mathbf{Coh}(X)$. Then any ascending chain of coherent subsheaves of $\mathcal{F}$ is locally stationary.*

Proof. As the problem is of local nature, we may assume X finite-dimensional. We proceed by induction on $n = \dim X$. The case $n < 0$ is utterly trivial, since X is void. Let $n \geqslant 0$ and let $\mathcal{F}_1 \subset \mathcal{F}_2 \subset \ldots$ be a sequence as in the statement of the proposition. We must show that any point $x \in X$ admits a neighbourhood where the above sequence is stationary. Let $x \in X$. There exist a neighbourhood U of x, an open V in a numerical space and a finite morphism $\pi\colon U \to \to V$. The sheaves $\pi_*(\mathcal{F})$, $\pi_*(\mathcal{F}_i)$ are coherent. Moreover $\pi_*(\mathcal{F}_1) \subset \pi_*(\mathcal{F}_2) \subset \ldots \ldots \subset \pi_*(\mathcal{F})$ and any equality $\pi_*(\mathcal{F}_i) = \pi_*(\mathcal{F}_j)$ implies $\mathcal{F}_i = \mathcal{F}_j$ (by I. 1.20).

We have thus reduced the problem to the case when X is a manifold, which can also be supposed connected. As $\mathcal{F}$ is locally the quotient of a sheaf of the form $\mathcal{O}^r$, it is not difficult to reduce the proof to the case $\mathcal{F} = \mathcal{O}^r$, hence to the case $\mathcal{F} = \mathcal{O}$. Here, $\mathcal{F}_i$ are coherent ideal-sheaves. If $\mathcal{F}_i = \mathcal{O}$ for all i, we have nothing to prove. Otherwise, choose an index i such that $\mathcal{F}_i \neq \mathcal{O}$. The sheaf $\mathcal{F}_i$ defines a subspace X_i of X, which is distinct of X and, consequently, of dimension $\leqslant n - 1$.

The sheaves $\mathscr{F}_j, j \geqslant 1$ then correspond to ideal-sheaves on X_i. The conclusion follows by the induction hypothesis.

The above given proof is taken from [80]; another proof can be found in [29], th. 8.

§ 2. The singular sets of the coherent sheaves

Let X be a complex space and $\mathscr{O}$ its structural sheaf. For any sheaf $\mathscr{F} \in \mathbf{Coh}(X)$ we make the notations

$$\text{prof } \mathscr{F} = \inf_{x \in X} \text{prof } \mathscr{F}_x \text{ and } \dim \mathscr{F} = \sup_{x \in X} \dim \mathscr{F}_x (= \dim \text{Supp } \mathscr{F}).$$

For an integer m denote

$$S_m(\mathscr{F}) = \{x \in X | \text{prof } \mathscr{F}_x \leqslant m\}.$$

These sets are called the *singular sets of the sheaf* $\mathscr{F}$. The main result in this paragraph is the following theorem of Scheja [71]:

THEOREM 2.1. *Let X be a complex space and $\mathscr{F}$ an analytic coherent sheaf on X. The singular sets $S_m(\mathscr{F})$ are closed analytic subsets of X and* $\dim S_m(\mathscr{F}) \leqslant m$ *for any integer m.*

Proof. The problem is local in nature; thus, we may assume that there exists a closed immersion $X \xrightarrow{i} V$ where V is an open subset of some numerical space. By I. 1.7, $S_m(\mathscr{F}) = S_m(i_*(\mathscr{F}))$; thus, we should consider only the case X a manifold. Let $n = \dim X$. By I. 1.15,

$$S_m(\mathscr{F}) = \bigcup_{r \geqslant n-m} \text{Supp}\,(\mathcal{E}xt^r_{\mathscr{O}}(\mathscr{F}, \mathscr{O})).$$

The sheaves $\mathcal{E}xt^r$ are coherent and vanish for $r > n$, hence the first assertion is proved.

In order to prove the second assertion, we must only prove that $\dim \mathcal{E}xt^r_{\mathscr{O}}(\mathscr{F}, \mathscr{O}) \leqslant n - r$ for any integer r, that is $\dim (\mathcal{E}xt^r_{\mathscr{O}}(\mathscr{F}, \mathscr{O})_x) \leqslant n - r$ for all $x \in X$. But we have the isomorphism $\mathcal{E}xt^r_{\mathscr{O}}(\mathscr{F}, \mathscr{O})_x \simeq \text{Ext}^r_{\mathscr{O}_x}(\mathscr{F}_x, \mathscr{O}_x)$ and in virtue of 1.26, we can conclude.

If $A \subset X$ is a closed analytic subset, then $\text{codim } A = \inf_{x \in X} \text{codim}_x A$ where $\text{codim}_x A$ is the height of the ideal defined in $\mathscr{O}_x$ by A (if $x \notin A$ we put $\text{codim}_x A = \infty$). Generally, for $x \in A$, $\text{codim}_x A \neq \dim_x X - \dim_x A$, but the following remarkable fact is anyhow true: there is a neighbourhood U of x such that $\text{codim}_x A = \inf_{y \in U} (\dim_y X - \dim_y A)$ ([15]).

If $A \subset B$ are two closed analytic subsets, then $\text{codim}\,(A, B)$ is the codimension of A in the reduced analytic subspace of X given by B (the codimension of A

in X and in X_{red} is the same and this definition of the codimension agrees with the previous one).

For any sheaf $\mathcal{F} \in \mathbf{Coh}(X)$ and any point $x \in X$ we set

$$\operatorname{def}_x \mathcal{F} = \begin{cases} 0 & \text{if } \mathcal{F}_x = 0 \\ \dim \mathcal{F}_x - \operatorname{prof} \mathcal{F}_x & \text{if } \mathcal{F}_x \neq 0. \end{cases}$$

For any integer m denote $D_m(\mathcal{F}) = \{x \in X | \operatorname{def}_x(\mathcal{F}) \geqslant m\}$. These sets are called *the defect sets of* $\mathcal{F}$ [15], [71]. If $m \geqslant 1$, then $D_m(\mathcal{F}) \subset \operatorname{Supp} \mathcal{F}$ and $D_1(\mathcal{F}) = \{x \in X | \dim \mathcal{F}_x \neq \operatorname{prof} \mathcal{F}_x\}$ coincides with the set of the points of X where the stalk of $\mathcal{F}$ is not Cohen-Macauley.

COROLLARY 2.2. *The sets $D_m(\mathcal{F})$ are closed analytic subsets of X and, for any $m \geqslant 1$,* codim $(D_m(\mathcal{F}), \operatorname{Supp} \mathcal{F}) \geqslant m$.

Proof. For any $m \leqslant 0$, $D_m(\mathcal{F}) = X$. Replacing X by $(\operatorname{Supp} \mathcal{F}, \mathcal{O}/\operatorname{Ann} \mathcal{F} | \operatorname{Supp} \mathcal{F})$, we can suppose $\operatorname{Supp} \mathcal{F} = X$. The problem being local we may also assume $\dim X = d < \infty$. If $m > d$, then $D_m(\mathcal{F})$ is empty and the stated assertions are obvious. Suppose thereby $1 \leqslant m \leqslant d$. Then we can write the decomposition

$$D_m(\mathcal{F}) = \bigcup_{m \leqslant k \leqslant d} A_k \text{ where } A_k = \{x \in X | \dim_x X \geqslant k\} \cap S_{k-m}(\mathcal{F}).$$

Since $\{x | \dim_x X \geqslant k\}$ is the union of all irreducible components of X of dimension $\geqslant k$, by the theorem, $D_m(\mathcal{F})$ are analytic closed subsets. Moreover, $\dim_x A_k \leqslant \dim_x S_{k-m}(\mathcal{F}) \leqslant k - m \leqslant \dim_x X - m$ for any k and whatever $x \in A_k$. Then it will result that $\dim_x D_m(\mathcal{F}) \leqslant \dim_x X - m$, for all $x \in X$. Therefore codim $D_m(\mathcal{F}) \geqslant m$ and the corollary follows.

COROLLARY 2.3. *Let X be a complex space and $\mathcal{F} \in \mathbf{Coh}(X)$. The set of all points $x \in X$, such that $\mathcal{F}_x$ is not a Cohen-Macauley $\mathcal{O}_x$-module, is analytic and closed.*

COROLLARY 2.4. *For any complex space X, the set of all points of X where X is not perfect is a closed analytic subset of codimension* $\geqslant 1$.

§ 3. The vanishing theorem

Let X be a complex space, $\mathcal{O}$ the structural sheaf, $A \subset X$ a closed analytic subset, and $\mathcal{A} = \mathfrak{I}(A)$ the maximal ideal-sheaf defining A. For a sheaf $\mathcal{F} \in \mathbf{Coh}(X)$ we denote

$$\operatorname{prof}_{A,x} \mathcal{F} = \operatorname{prof}_{\mathcal{A}_x} \mathcal{F}_x (x \in A) \text{ and } \operatorname{prof}_A \mathcal{F} = \inf_{x \in A} \operatorname{prof}_{A,x} \mathcal{F}.$$

By 1.23, $\operatorname{prof}_{A,x} \mathcal{F} = \infty$ if and only if $\mathcal{F}_x = 0$. The number $\operatorname{prof}_A \mathcal{F}$ is called *the depth of $\mathcal{F}$ with respect to A.*

LEMMA 3.1. *The function $x \mapsto \operatorname{prof}_{A,x} \mathcal{F}$ is lower semi-continuous on A.*

Proof. The assertion is obvious in the points $x \in A$ where $\mathcal{F}_x = 0$. Assume $x \in A$ and $\operatorname{prof}_{A,x} \mathcal{F} = q < \infty$. Then there are elements $(f_1)_x, \ldots, (f_q)_x \in \mathcal{A}_x$ which give a regular $\mathcal{F}_x$-sequence. Let $f_1, \ldots, f_q \in \Gamma(U, \mathcal{A})$ be sections in some neighbourhood U of x, which represent these germs. The sheaf-morphisms on U

$$\mathcal{F} \Big/ \sum_{j=1}^{i-1} f_j \mathcal{F} \to \mathcal{F} \Big/ \sum_{j=1}^{i-1} f_j \mathcal{F}$$

defined by multiplication by $f_i (1 \leqslant i \leqslant q)$ are injective in x; hence they remain injective in a neighbourhood V of x. For any $y \in V \cap A$, $(f_1)_y, \ldots, (f_q)_y$ are elements of $\mathcal{A}_y$ and form a regular $\mathcal{F}_y$-sequence, therefore $\operatorname{prof}_{A, y} \mathcal{F} \geqslant q$.

PROPOSITION 3.2. *Let X be a complex space, A a closed analytic subset of X, and $\mathcal{F} \in$ **Coh**(X). Then the sheaf $\mathcal{H}^0_A \mathcal{F}$ is coherent and is equal to the subsheaf of all sections of $\mathcal{F}$ annihilated by suitable powers of the ideal $\mathcal{A} = \mathcal{J}(A)$.*

Proof. Let $\mathcal{F}_\infty$ denote the subsheaf of $\mathcal{F}$ formed by those sections which are locally annihilated by suitable powers of $\mathcal{A}$. If $\mathcal{F}_k$ is the subsheaf of $\mathcal{F}$ of all sections annihilated by $\mathcal{A}^k$, we get an ascending chain

$$\mathcal{F}_1 \subset \mathcal{F}_2 \ldots \subset \mathcal{F} \text{ and } \mathcal{F}_\infty = \bigcup_{k=1}^{\infty} \mathcal{F}_k.$$

Every $\mathcal{F}_k$ is coherent, as the kernel of the canonical map

$$\mathcal{F} \to Hom_{\mathcal{O}}(\mathcal{A}^k, \mathcal{F}), \text{ defined by } s \mapsto (f \mapsto fs).$$

By 1.30 the above chain is locally stationary; hence $\mathcal{F}_\infty$ is coherent. The proposition will follow if we can prove that $\mathcal{H}^0_A \mathcal{F} = \mathcal{F}_\infty$. Obviously $\mathcal{F}_\infty \subset \mathcal{H}^0_A \mathcal{F}$. Conversely, let $s \in \Gamma_A(U, \mathcal{F})$, U an open subset of X. Let $\mathcal{B}$ denote the kernel of the map $\mathcal{O}|U \to \mathcal{F}|U, f \mapsto fs$; $\mathcal{B}$ is the annihilator sheaf of s and Supp s is the associated analytic subset $V(\mathcal{B})$. Since $V(\mathcal{B}) \subset A \cap U$, the conclusion easily follows.

PROPOSITION 3.3. *Let A be a closed analytic subset of a complex space X and $\mathcal{F} \in$ **Coh**(X). For a point $x \in A$ the following two conditions are equivalent*:

(*a*) $\operatorname{prof}_{A, x} \mathcal{F} \geqslant 1$;

(*b*) $(\mathcal{H}^0_A \mathcal{F})_x = 0$.

Proof. (*a*) ⇒ (*b*). Let $f_x \in \mathcal{A}_x$ be a nonzerodivisor for $\mathcal{F}_x (\mathcal{A} = \mathcal{J}(A))$. Consider an arbitrary element $s_x \in (\mathcal{H}^0_A \mathcal{F})_x$. By the above proposition, $\mathcal{A}^k_x s_x = 0$ for some integer k. In particular, $f^k_x s_x = 0$, hence $s_x = 0$.

(*b*) ⇒ (*a*). On the contrary, if $\operatorname{prof}_{\mathcal{A}_x} \mathcal{F}_x = 0$, then all elements of $\mathcal{A}_x$ will be zerodivisors for $\mathcal{F}_x$, hence there exists an ideal $\mathfrak{p} \in \operatorname{Ass} \mathcal{F}_x$ such that $\mathcal{A}_x \subset \mathfrak{p}$. The ideal $\mathfrak{p}$ is the annihilator of some element $s_x \neq 0$ in $\mathcal{F}_x$. We have $\mathcal{A}_x s_x = 0$, hence $\mathcal{A}s = 0$ for a representing section s of s_x in some neighbourhood U of x. We then derive Supp $s \subset A \cap U$, therefore $s_x \in (\mathcal{H}^0_A \mathcal{F})_x$, contradiction.

We shall use the following particular case of the Frenkel's lemma [24].

LEMMA 3.4. *Let $\Delta \subset \mathbb{C}^n$ be an open polydisc centred in the origin, $d \geqslant 0$ an integer and $A = \Delta \cap \{z_{d+1} = \ldots = z_n = 0\}$. Then the canonical map $H^0(\Delta, \mathcal{O}) \to$*

$\to H^0(\Delta \setminus A, \mathcal{O})$ *is injective for* $d \leqslant n-1$, *bijective for* $d \leqslant n-2$ *and, in addition,* $H^i(\Delta \setminus A, \mathcal{O}) = 0$ *for* $1 \leqslant i < n-d-1$.

Proof [69]. Denote $U_l = \{z \in \Delta | \, z_l \neq 0\}$ for $l = d+1, \ldots, n$. Thus one obtains a Stein open covering $\mathfrak{U}$ of $\Delta - A$, therefore

$$H^\bullet(\Delta \setminus A, \mathcal{O}) \simeq H^\bullet(\mathfrak{U}, \mathcal{O}).$$

Let $U = U_{d+1} \cap \ldots \cap U_n$. Any element $f \in \mathcal{O}(U)$ can be expanded in Laurent series

$$f = \sum_\alpha f_\alpha(z_1, \ldots, z_d) z_{d+1}^{\alpha_{d+1}} \ldots z_n^{\alpha_n} \qquad \alpha = (\alpha_{d+1}, \ldots, \alpha_n),$$

f_α being holomorphic functions. For $S \subset \{d+1, \ldots, n\}$ denote by $\theta_S(f)$ the sum of terms whose negative exponents appear in the very variables that have indexes from S. In this way we get an endomorphism

$$\theta_S : \mathcal{O}(U) \to \mathcal{O}(U).$$

It can be easily checked that the sum of all of these is just the identity If $f \in \mathcal{O}(U_{l_0 \ldots l_i})$, then $\theta_S(f) \in \mathcal{O}(U_{l_0 \ldots l_i})$ and so any endomorphism θ_S can be extended to cochains. We also notice that whenever $l \notin S$ and $f \in \mathcal{O}(U_{ll_0 \ldots l_{i-1}})$, $\theta_S(f) \in \mathcal{O}(U_{l_0 \ldots l_{i-1}})$.

Let now $1 \leqslant i < n-d-1$; we are going to prove that any cocycle. $f \in Z^i(\mathfrak{U}, \mathcal{O})$ is a coboundary. It is sufficient to prove that any $\theta_S(f)$ is a coboundary. The condition $i < n-d-1$ shows that $\theta_S(f) = 0$ for $S = \{d+1, \ldots, n\}$. Thus we may assume that there is an index $l \in \{d+1, \ldots, n\}$ so that $l \notin S$. Define $g \in C^{i-1}(\mathfrak{U}, \mathcal{O})$ by the formula

$$g_{l_0 \ldots l_{i-1}} = \theta_S(f_{ll_0 \ldots l_{i-1}}).$$

We have

$$(\delta g)_{l_0 \ldots l_i} = \sum_{k=0}^{i} (-1)^k g_{l_0 \ldots \hat{l}_k \ldots l_i} =$$

$$= \theta_S \left(\sum_{k=0}^{i} (-1)^k f_{ll_0 \ldots \hat{l}_k \ldots l_i} \right) = \theta_S(f_{l_0 \ldots l_i}).$$

For the last equality we have applied the identity theorem for holomorphic functions. Therefore, $H^i(\Delta \setminus A, \mathcal{O}) = 0$ for $1 \leqslant i < n-d-1$.

The other assertions of the lemma can be easily proved.

Remark. The above lemma can be easily deduced by studying the cohomology of the complementary of a point (I. 3.3), by means of topological tensor products (a simple argument of Künneth type!).

PROPOSITION 3.5. *Let* $D \subset \mathbb{C}^n$ *be a domain,* $A \subset D$ *a closed analytic subset of dimension* d *and* $\mathcal{F} \in \mathbf{Coh}(D)$. *If* prof $\mathcal{F} \geqslant d+q$, *then* $\mathcal{H}_A^i\mathcal{F} = 0$ *for* $0 \leqslant i < q$.

Proof. We first prove that $\mathcal{H}^i_A\mathcal{O} = 0$ for $0 \leqslant i < n - d$. Consider in the beginning the case A nonsingular. Let x bc a fixed arbitrary point of A. By a biholomorphic transformation around x we can assume $x = 0$ and $A \cap U = \{z_{d+1} = \ldots = z_n = 0\}$, U being some neighbourhood of x. For any open polydisc Δ which is contained in U, and which is centred in x, by Frenkel's lemma and theorem B, we get $H^i_A(\Delta, \mathcal{O}) = 0$ for $i < n - d$. Whence $(\mathcal{H}^i_A\mathcal{O})_x = 0$ for $i < n - d$.

We now consider an arbitrary analytic subset A and proceed by induction on $d = \dim A$. The case $d = 0$ follows from above. For the general induction step, let A' be the singular locus of A and $A'' = A \setminus A'$. We can write the exact sequence, given by (1.9)

$$\ldots \to \mathcal{H}^i_{A'}\mathcal{O} \to \mathcal{H}^i_A\mathcal{O} \to \mathcal{H}^i_{A''}\mathcal{O} \to \ldots$$

Since $\dim A' \leqslant d - 1$, the induction hypothesis shows that $\mathcal{H}^i_{A'}\mathcal{O} = 0$ for $i < n - d + 1$. Let $D' = D \setminus A'$. By the first part of the proof, $\mathcal{H}^i_{A''}\mathcal{O} = 0$ on D' for $i < n - d$. Let $U \subset D$ be an arbitrary open set. By 1.1 and 1.16, $H^i_{A''}(U, \mathcal{O}) = H^i_{A''}(D' \cap U, \mathcal{O}) = 0$ for $i < n - d$. Hence $\mathcal{H}^i_{A''}\mathcal{O} = 0$ for $i < n - d$; it follows that $\mathcal{H}^i_A\mathcal{O} = 0$ for $i < n - d$.

We shall now prove the assertion of the statement and we shall proceed by descending induction on prof $\mathcal{F}$.

If prof $\mathcal{F} > n$, then $\mathcal{F} = 0$ and the assertion is clear. For the general induction step, let x be an arbitrary point of D. If $\mathcal{F}_x$ is free, then $\mathcal{F}$ is free in a neighbourhood of x; it will follow $q \leqslant n - d$, hence $(\mathcal{H}^i_A\mathcal{F})_x = 0$ for $0 \leqslant i < q$ by the above. Suppose now $\mathcal{F}_x$ is not a free $\mathcal{O}_x$-module. Then $n > d + q$. In some neighbourhood U of x we have an exact sequence of the form

$$0 \to \mathcal{G} \to \mathcal{O}^s \,|\, U \to \mathcal{F} \,|\, U \to 0.$$

We deduce the inequality prof $\mathcal{G} \geqslant$ prof $(\mathcal{F}|U) + 1 \geqslant$ prof $\mathcal{F} + 1$. By induction hypothesis, $\mathcal{H}^i_A\mathcal{G} = 0$ for $0 \leqslant i < q + 1$. Since $n \geqslant d + q + 1$, $\mathcal{H}^i_A\mathcal{O} = 0$ for $i < q + 1$. From the above exact sequence we derive that $\mathcal{H}^i_A\mathcal{F} = 0$ for $i < q$ and the proposition is proved.

Remark. The proposition remains true if we assume that A is an analytic subset which is not necessarily closed. Indeed, let $D' \subset D$ be an open subset where A is closed. We have $\mathcal{H}^i_A\mathcal{F} = 0$ on D' for $i < q$. Let U be an arbitrary open set in D. By applying 1.1 and 1.16 we obtain $H^i_A(U, \mathcal{F}) = H^i_A(D' \cap U, \mathcal{F}) = 0$, therefore $\mathcal{H}^i_A\mathcal{F} = 0$ for $i < q$.

The main result of the paragraph is the following *vanishing theorem for local cohomology* of Scheja and Trautmann [71], [90], [84].

THEOREM 3.6. *Let X be a complex space, $A \subset X$ a closed analytic subset and $\mathcal{F}$ a coherent analytic sheaf on X. Then for any integer $q \geqslant 0$ the following* *four conditions are equivalent*:

(*a*) $\mathrm{prof}_A \mathcal{F} \geqslant q + 1$.

(*b*) $\dim (A \cap S_{k+q+1}(\mathcal{F})) \leqslant k$ *for any* k.

(*c*) $\mathcal{H}^i_A \mathcal{F} = 0$ for $i \leqslant q$.

(*d*) *For any open subset* U *of* X, *the restriction maps*

$$H^i(U, \mathcal{F}) \to H^i(U \setminus A, \mathcal{F})$$

are bijective for $i < q$ *and injective for* $i = q$.

Proof. The equivalence (*c*) $\Leftrightarrow$ (*d*) follows from 1.16.

(*a*) $\Rightarrow$ (*b*). Let $\mathcal{a}$ be the ideal-sheaf given by A and x an arbitrary point of A. Since $\mathrm{prof}_{\mathcal{a}_x} \mathcal{F}_x \geqslant q + 1$, there is a neighbourhood U of x and sections $f_1, \ldots, f_{q+1}$ of $\Gamma(U, \mathcal{a})$ such that for any $y \in U \cap A$, the germs $(f_i)_y$, $1 \leqslant i \leqslant q + 1$, form a regular $\mathcal{F}_y$-sequence (by the proof of lemma 3.1). By applying 1.21 one obtains

$$A \cap U \cap S_{k+q+1}(\mathcal{F}) = A \cap U \cap S_k \left(\mathcal{F} \Big/ \sum_{i=1}^{q+1} f_i \mathcal{F} \right).$$

Then we derive by theorem 2.1, $\dim_x (A \cap S_{k+q+1}(\mathcal{F})) \leqslant k$.

(*b*) $\Rightarrow$ (*c*). We use induction on $\dim A$. Let $\dim A = 0$ and x be an arbitrary point of A. We have $\dim (A \cap S_q(\mathcal{F})) \leqslant -1$, hence the intersection $U \cap S_q(\mathcal{F})$ is empty for some neighbourhood U of x. Then $\mathrm{prof}\, (\mathcal{F}|U) \geqslant q + 1$. In virtue of 3.5 (and by the assertions of 1.13, 1.25!), we get $\mathcal{H}^i_A \mathcal{F}|U = 0$ for $i \leqslant q$.

We now prove the general induction step. Let $d = \dim A$ and suppose the implication proved for analytic subsets of dimension $< d$. Put $A' = A \cap S_{q+d}(\mathcal{F})$ and $A'' = A \setminus A'$. By 3.5, $\mathcal{H}^i_{A''} \mathcal{F} = 0$ for $i \leqslant q$. The set A' has dimension $\leqslant d - 1$ and satisfies the condition (*b*), hence by induction hypothesis, $\mathcal{H}^i_{A'} \mathcal{F} = 0$ for $i \leqslant q$. The conclusion follows from the exact sequence

$$\ldots \to \mathcal{H}^i_{A'} \mathcal{F} \to \mathcal{H}^i_A \mathcal{F} \to \mathcal{H}^i_{A''} \mathcal{F} \to \ldots$$

(*c*) $\Rightarrow$ (*a*). We will prove a more precise result: if $(\mathcal{H}^i_A \mathcal{F})_x = 0$ for a point $x \in A$ and for $i \leqslant q$, then $\mathrm{prof}_{A,x} \mathcal{F} \geqslant q + 1$. By 3.3 we can find a neighbourhood U of x and an element $f \in \Gamma(U, \mathcal{a})$ such that the sequnce

$$0 \to \mathcal{F}|U \to \mathcal{F}|U \to \mathcal{F}/f\mathcal{F}\,|U \to 0$$

is exact; the arrow $\mathcal{F}|U \to \mathcal{F}|U$ is the multiplication by f. We then obtain the exact sequence on U

$$\ldots \to \mathcal{H}^i_A \mathcal{F} \to \mathcal{H}^i_A(\mathcal{F}/f\mathcal{F}) \to \mathcal{H}^{i+1}_A \mathcal{F} \to \ldots$$

hence $\mathcal{H}^i_A(\mathcal{F}/f\mathcal{F})_x = 0$ for $i \leqslant q - 1$. To conclude the proof we make use of induction on q by applying corollary 1.21.

Remark. For $q = 0$ we obtain some necessary and sufficient conditions for the maps $H^0(U, \mathcal{F}) \to H^0(U \setminus A, \mathcal{F})$ to be injective. For $q = 1$ we obtain conditions equivalent to the fact that the maps $H^0(U, \mathcal{F}) \to H^0(U \setminus A, \mathcal{F})$ are bijective and $H^1(U, \mathcal{F}) \to H^1(U \setminus A, \mathcal{F})$ injective.

In the following we shall draw some consequences of the vanishing theorem.

COROLLARY 3.7. *Let x be a point of A such that $(\mathcal{H}^i_A\mathcal{F})_x = 0$ for $i \leqslant q$. Then there exists a neighbourhood U of x such that $\mathcal{H}^i_A\mathcal{F}|U = 0$ for $i \leqslant q$.*

Apply the proof of the implication $(c) \Rightarrow (a)$, lemma 3.1 and the very equivalence $(a) \Leftrightarrow (c)$.

COROLLARY 3.8. *Let $(X, \mathcal{O})$ be a complex space, $f \in \Gamma(X, \mathcal{O})$, $V(f) = \mathrm{Supp}\,(\mathcal{O}/f\mathcal{O})$ be the set of the zeros of f and $\mathcal{F} \in$ **Coh**(X). Then the following conditions are equivalent:*

(*a*) $\dim V(f) \cap S_{k+1}(\mathcal{F}) \leqslant k$ *for all* k;

(*b*) f_x *is nonzerodivisor for* $\mathcal{F}_x$, *whatever* $x \in X$.

Proof. $(a) \Rightarrow (b)$. Apply the theorem for $A = V(f)$ and $q = 0$. Then $\mathrm{prof}_{V(f)} \mathcal{F} \geqslant 1$. Let now x be a point in $V(f)$. There exists an element $g_x \in \mathcal{I}(V(f))_x$ which is nonzerodivisor for $\mathcal{F}_x$. One of its convenient powers is $f_x h_x$, $h_x \in \mathcal{O}_x$. It follows that f_x is nonzerodivisor for $\mathcal{F}_x$. If x does not lie in $V(f)$, then f_x is an inversible element and (*b*) holds too.

The inverse implication is an immediate consequence of the theorem.

We now reformulate 3.5 for complex spaces as a consequence of the theorem.

COROLLARY 3.9. *Let X be a complex space, $A \subset X$ a closed analytic subset, $\mathcal{F} \in$ **Coh**(X), and q an integer. If* $\mathrm{prof}\,\mathcal{F} \geqslant \dim A + q + 1$, *then* $\mathcal{H}^i_A\mathcal{F} = 0$ *for any $i \leqslant q$.*

Another consequence can be obtained if we consider that A is an analytic set of dimension zero in the theorem.

COROLLARY 3.10. *Let X be a complex space, x a point of X, $\mathcal{F} \in$ **Coh**(X), and q an integer. Then* $\mathrm{prof}\,\mathcal{F}_x \geqslant q + 1$ *if and only if for any neighbourhood U of x the canonical maps*

$$H^i(U, \mathcal{F}) \to H^i(U \setminus \{x\}, \mathcal{F})$$

are bijective for $i < q$ and injective for $i = q$.

COROLLARY 3.11. *Suppose $(X, \mathcal{O})$ is a complex space and A a closed analytic subset of X such that* $\mathrm{prof}_A \mathcal{O} \geqslant 2$. *Then X is connected if and only if $X \setminus A$ is.*

The proof follows from I. 4.9.

From the above we obtain remarkable results for perfect spaces and implicitly for manifolds, for locally complete intersection spaces, for the pure $(n-1)$-dimensional analytic subsets of n-dimensional manifolds.

In particular, we get the following result of Thimm and Scheja [88], [70].

COROLLARY 3.12. *Let $(X, \mathcal{O})$ be a perfect complex space and A a closed analytic subset of codimension $\geqslant 2$. Then the restriction map*

$$\Gamma(X, \mathcal{O}) \to \Gamma(X \setminus A, \mathcal{O})$$

is bijective.

For the next corollary suppose that X is a reduced complex space and let A be one of its closed analytic subsets. A holomorphic function f on $X \setminus A$ is called *bounded at the points of* A if for any point $x \in A$, there exists a neighbourhood U of x such that the function $f|U \setminus A$ is bounded.

COROLLARY 3.13. (Scheja [70]). *Suppose* X *is a perfect reduced analytic space. Then* X *is normal if and only if* codim $S(X) \geqslant 2$ ($S(X)$ *is the singular locus of* X).

Proof. The assertion X normal $\Rightarrow$ codim $S(X) \geqslant 2$ is a general property of complex spaces [1], [15], [56]. Suppose now conversely, codim $S(X) \geqslant 2$. Let $x \in X$ be an arbitrary point and f_x/g_x an element of the total ring of quotients of $\mathcal{O}_x$, integral over $\mathcal{O}_x$. Then there are a neighbourhood U of x and holomorphic functions f, g, $a_1, \ldots a_r$ on U such that the germs of f, g in x are just f_x, g_x and the following equality is true

$$(f/g)^r + a_1(f/g)^{r-1} + \ldots + a_r = 0.$$

f/g is a holomorphic function on $U \setminus A$ where A is the set of all zeros of g. By restricting eventually U, one derives that f/g is bounded at the points of A. From the first Riemann extension theorem [56], it will follow that there exists a holomorphic function h on $U \setminus S(X)$ such that $h = f/g$ on $(U \setminus S(X)) \setminus A$. By corollary 3.12, this function extends to a holomorphic function on U; hence $f_x/g_x \in \mathcal{O}_x$, and $\mathcal{O}_x$ is normal ring.

COROLLARY 3.14. (Scheja [70]). *Let* $(X, \mathcal{O})$ *be a normal space and* A *a closed analytic subset of codimension* $\geqslant 2$. *Under these assumptions, the restriction map*

$$\Gamma(X, \mathcal{O}) \to \Gamma(X \setminus A, \mathcal{O})$$

is bijective.

According to the theorem by means of 1.24.

Recall now that an $\mathcal{O}$-module $\mathcal{F}$ is said to be *reflexive* if the canonical morphism

$$\mathcal{F} \to (\check{\mathcal{F}})^{\vee} = \mathcal{H}om_{\mathcal{O}}(\mathcal{H}om_{\mathcal{O}}(\mathcal{F}, \mathcal{O}), \mathcal{O})$$

is an isomorphism.

COROLLARY 3.15 (Serre [80]). *Suppose* X *is a normal space,* $A \subset X$ *a closed analytic subset of codimension* $\geqslant 2$, *and* $\mathcal{F} \in \mathbf{Coh}(X)$ *a reflexive sheaf. Then* $\operatorname{prof}_A \mathcal{F} \geqslant 2$; *in particular, the canonical map*

$$\Gamma(X, \mathcal{F}) \to \Gamma(X \setminus A, \mathcal{F})$$

is bijective.

Proof. Denote by i the inclusion $X \setminus A \subset X$. Let $\mathcal{G} = \mathcal{H}om_{\mathcal{O}}(\mathcal{F}, \mathcal{O})$, hence $\mathcal{F} = \mathcal{H}om_{\mathcal{O}}(\mathcal{G}, \mathcal{O})$. We have $i^*\mathcal{F} = \mathcal{H}om(i^*\mathcal{G}, \mathcal{O}_{X \setminus A})$, hence $i_* i^* \mathcal{F} = i_*(\mathcal{H}om(i^*\mathcal{G}, \mathcal{O}_{X\setminus A})) = \mathcal{H}om(\mathcal{G}, i_*\mathcal{O}_{X \setminus A}) = \mathcal{H}om(\mathcal{G}, \mathcal{O}_X) = \mathcal{F}$ (the second equality follows by the adjoint functors i_* and i^* and the equality $i_*\mathcal{O}_{X\setminus A} = \mathcal{O}_X$ holds by the previous corollary). From the exact sequence

$$0 \to \mathcal{H}^0_A\mathcal{F} \to \mathcal{F} \to i_* i^* \mathcal{F} \to \mathcal{H}^1_A\mathcal{F} \to 0$$

we get $\mathcal{H}^0_A\mathcal{F} = \mathcal{H}^1_A\mathcal{F} = 0$, and the required conclusion follows now from the theorem.

§ 4. The finiteness theorem

We still use the notation from the preceding paragraphs. Denote by $\bar{S}_m(\mathcal{F}|X \setminus A)$ the topological closure in X of $S_m(\mathcal{F}|X \setminus A)$. It is easy to check that this coincides with the union of the irreducible components of $S_m(\mathcal{F})$which are not contained in A.

The main result of the paragraph is the following *finiteness theorem for local cohomology* of Siu and Trautmann [83], [92], [84], which is analogous to a theorem of Grothendieck in algebraic geometry.

THEOREM 4.1. *Let X be a complex space, $A \subset X$ a closed analytic subset and $\mathcal{F} \in$ **Coh**(X). For an integer $q \geqslant 0$, the following two conditions are equivalent*:

(*a*) $\dim A \cap \bar{S}_{k+q+1}(\mathcal{F}|X \setminus A) \leqslant k$ *for all* k;

(*b*) *the sheaves $\mathcal{H}^i_A\mathcal{F}$ are coherent for* $0 \leqslant i \leqslant q$.

In order to prove the implication (*a*) ⇒ (*b*) we need some preliminaries, wherefrom we derive that in condition (a), the sheaves $\mathcal{H}^i_A\mathcal{F}(0 \leqslant i \leqslant q)$ are locally annihilated by some powers of $\mathcal{a} = \mathcal{I}(A)$. Considering the fact just stated, the proof of the implication follows easily by induction on q.

LEMMA 4.2. *Let D be a domain in $\mathbf{C}^n$, $A \subset D$ a closed analytic subset of dimension $\leqslant d$ and $\mathcal{F} \in$ **Coh**(D). For $0 \leqslant q \leqslant n - d$, let $\mathcal{I}_q$ be the maximal ideal-sheaf which defines $S_{q+d-1}(\mathcal{F})$. Then, for any open subset $X \subset\subset D$ whose topological closure is holomorphically convex, there exists an integer k such that for any open U of X and for all $i < q$,*

$$\Gamma(U, \mathcal{I}_q)^k H^i_A(U, \mathcal{F}) = 0.$$

Proof. We first consider the case $q = n - d$. According to the hypothesis on X, there is an exact sequence of the form

$$\ldots \to \mathcal{O}^{s_q} \to \ldots \to \ldots \mathcal{O}^{s_0} \to \mathcal{F}^* \to 0$$

on a neighbourhood of $\bar{X}$, where $\mathcal{O}$ is the sheaf of germs of holomorphic functions in $\mathbf{C}^n$ and $\mathcal{F}^*$ the dual of $\mathcal{F}$. By applying the functor of passing to the duals, we obtain the following sequence of morphisms between $\mathcal{O}$-modules

(I) $$0 \xrightarrow{\varphi_{-2}} \mathcal{F} \xrightarrow{\varphi_{-1}} \mathcal{O}^{s_0} \xrightarrow{\varphi_0} \mathcal{O}^{s_1} \xrightarrow{\varphi_1} \ldots,$$

where φ_{-1} is the composite map $\mathcal{F} \to \mathcal{F}^{**} \to \mathcal{O}^{s_0}$. On $D \setminus S_{n-1}(\mathcal{F})$ the sheaf $\mathcal{F}$ is locally free; therefore the sequence (I) is exact on $\overline{X} \setminus S_{n-1}(\mathcal{F})$.

We define for $i \geqslant -1$ the sheaves

$$\mathcal{B}_i = \operatorname{Im} \varphi_{i-1},\ \mathcal{Z}_i = \operatorname{Ker} \varphi_i \text{ and } \mathcal{L}_i = \mathcal{Z}_i/\mathcal{B}_i.$$

From above, $\operatorname{Supp} \mathcal{L}_i \subset S_{n-1}(\mathcal{F})$. By Nullstellensatz we can find an integer l such that $\mathfrak{J}_q^l \mathcal{L}_i | \overline{X} = 0$. Therefore

$$\text{(II)} \qquad \Gamma(U, \mathfrak{J}_q)^l H_A^k(U, \mathcal{L}_i) = 0$$

for any open subset U of X and for all $k \geqslant 0$.

We are going to prove the following assertions by descending induction on $j \geqslant 0$:

$$(*)_j \quad \begin{cases} \text{There is an integer } l \geqslant 0 \text{ such that for any open set } U \text{ of } X, \\ \Gamma(U, \mathfrak{J}_q)^l H_A^i(U, \mathcal{Z}_j) = 0 \text{ for } i < q - j. \end{cases}$$

$$(**)_j \quad \begin{cases} \text{There is an integer } l \geqslant 0 \text{ such that for any open set } U \text{ of } X, \\ \Gamma(U, \mathfrak{J}_q)^l H_A^i(U, \mathcal{B}_j) = 0 \text{ for } i < q - j. \end{cases}$$

For $j > q$ both assertions are automatically fulfilled. From the exact sequence $0 \to \mathcal{B}_j \to \mathcal{Z}_j \to \mathcal{L}_j \to 0$ we get the exact sequence

$$H_A^{i-1}(U, \mathcal{L}_j) \to H_A^i(U, \mathcal{B}_j) \to H_A^i(U, \mathcal{Z}_j),$$

wherefrom derive, by means of (II), the implication $(*)_j \Rightarrow (**)_j$. Suppose $j \geqslant 1$. By using the exact sequence $0 \to \mathcal{Z}_{j-1} \to \mathcal{O}^{s_{j-1}} \to \mathcal{B}_j \to 0$ and the equality $H_A^i(U, \mathcal{O}) = 0$ for $i < q$ $(q = n - d)$, we get the isomorphisms $H_A^{i-1}(U, \mathcal{B}_j) \simeq$ $\simeq H_A^i(U, \mathcal{Z}_{j-1})$ for $i < q - j + 1$; hence the implication $(**)_j \Rightarrow (*)_{j-1} (j \geqslant 1)$. Thereby the assertions $(*)$ and $(**)$ are proved.

From the exact sequence $0 \to \mathcal{Z}_{-1} \to \mathcal{F} \to \mathcal{B}_0 \to 0$ we deduce the exact sequence

$$\ldots \to H_A^i(U, \mathcal{Z}_{-1}) \to H_A^i(U, \mathcal{F}) \to H_A^i(U, \mathcal{B}_0) \to \ldots$$

But $\mathcal{Z}_{-1} = \mathcal{L}_{-1}$ and the conclusion of the lemma easily follows from (II) and $(**)_0$.

The lemma being proved for $q = n - d$, we shall now study the general case and make use of descending induction on q. Let $q < n - d$. Under the assumption on X, there exists an exact sequence

$$0 \to \mathcal{G} \to \mathcal{O}^s \to \mathcal{F} \to 0$$

on a neighbourhood of $\overline{X}$. Then we get the equality $S_{q+d-1}(\mathcal{F}) = S_{(q+1)+d-1}(\mathcal{G})$ and both sets have $\mathfrak{J}_q$ as maximal ideal-sheaf. By 3.5 and 1.16 we deduce the isomorphisms

$$H^i_A(U, \mathcal{F}) \simeq H^{i+1}_A(U, \mathcal{G})$$

for any open subset U of X and for all $i < q$. Hence, for $i < q$, $H^i_A(U, \mathcal{F})$ is annihilated by some power of $\Gamma(U, \mathfrak{J}_q)$ if and only if $H^{i+1}_A(U, \mathcal{G})$ is. We conclude the proof by induction hypothesis.

LEMMA 4.3. *Let D be a domain in $\mathbb{C}^n$, $A \subset V$ two closed analytic subsets of D, $\mathcal{F} \in$ **Coh** (D), and $q \leqslant n$ an integer. Denote $S'_{k+q} = \overline{S_{k+q}(\mathcal{F}|D \setminus V)}$ and suppose* dim $A \cap S'_{k+q} \leqslant k$ *for all k. Let $\mathfrak{J}$ be the maximal ideal-sheaf of V and let X be a relatively compact open subset of D whose topological closure $\overline{X}$ is holomorphically convex. Then there exists an integer $l \geqslant 0$ such that for any open subset U of X and for all $i < q$,*

$$\Gamma(U, \mathfrak{J})^l \, H^i_A(U, \mathcal{F}) = 0.$$

Proof. For an integer $k \geqslant 0$, we set $A_k = A \cap S'_{k+q}$ and $A''_{k+1} = A_{k+1} \setminus \setminus S'_{k+q}$. We will establish by induction on k the existence of an integer $l \geqslant 0$ such that $\Gamma(U, \mathfrak{J})^l \, H^i_{A_k}(U, \mathcal{F}) = 0$ for $i < q$ and U an open subset of X and the lemma will follow since for $k = n - q$, $A_k = A$.

First of all, consider the case $k = 0$. Then by hypothesis, dim $A_0 \leqslant 0$ and our assertion follows in virtue of the previous lemma since in some neighbourhood of A_0, we have $S_{q-1}(\mathcal{F}) \subset V$; hence $\mathfrak{J} \subset \mathfrak{J}(S_{q-1}(\mathcal{F}))$.

We now pass to the general induction step. Since $S_{q+(k+1)-1}(\mathcal{F}) \cap (X \setminus \setminus S'_{q+k}) \subset V$ and $(X \setminus S'_{q+k})^- = \overline{X}$ (otherwise, V would contain on open, hence $V = D$, $\mathfrak{J} = 0$, etc.), we deduce by 4.2 that

$$\Gamma(U \setminus S'_{k+q}, \mathfrak{J})^l \, H^i_{A_{k+1}}(U \setminus S'_{q+k}, \mathcal{F}) = 0$$

for $i < q$ and for some integer $l \geqslant 0$ (which depends on X only). By the excision lemma 1.1, $H^i_{A_{k+1}}(U \setminus S'_{q+k}, \mathcal{F}) = H^i_{A''_{k+1}}(U, \mathcal{F})$. Accordingly, $\Gamma(U, \mathfrak{J})^l H^i_{A''_{k+1}}(U, \mathcal{F}) = 0$ for $i < q$. By the exact sequence

$$\ldots \to H^{i-1}_{A''_{k+1}}(U, \mathcal{F}) \to H^i_{A_k}(U, \mathcal{F}) \to H^i_{A_{k+1}}(U, \mathcal{F}) \to H^i_{A''_{k+1}}(U, \mathcal{F}) \to \ldots$$

the assertion for A_{k+1} follows from that for A_k.

Here is the proof of the theorem.

The proof of the implication $(a) \Rightarrow (b)$. We proceed by induction on q. The case $q = 0$ follows by 3.2. Assume now the implication proved for integers smaller than q and prove it for q.

The problem is local in nature, hence we can consider X a relatively compact open subset of a domain $D \subset \mathbb{C}^n$, $\overline{X}$ holomorphically convex and $\mathcal{F}$ defined on D

(by 1.13 and 1.25, the problem agrees with the immersions!). By lemma 4.3 (applied to $A = V$ and $q + 1$ instead of q), there is an integer $l \geqslant 0$ such that

$$\Gamma(U, \mathfrak{J})^l \ H^i_A(U, \mathfrak{F}) = 0$$

for all $i \leqslant q$ and any open subset U of X where $\mathfrak{J} = \mathfrak{J}(A)$. Hence $\mathfrak{J}^l \mathcal{H}^i_A \mathfrak{F} = 0$. We now notice that $\mathcal{H}^0_A(\mathfrak{F}/\mathcal{H}^0_A\mathfrak{F}) = 0$ and $\mathcal{H}^i_A(\mathfrak{F}) = \mathcal{H}^i_A(\mathfrak{F}/\mathcal{H}^0\mathfrak{F})$ for $i \geqslant 1$. Since $\mathcal{H}^0_A\mathfrak{F}$ is coherent we can assume however $\mathcal{H}^0_A\mathfrak{F} = 0$ by replacing $\mathfrak{F}$ by $\mathfrak{F}/\mathcal{H}^0_A\mathfrak{F}$. Let x be an arbitrary point of A. By 3.3 there exist a neighbourhood U of x and a section $f \in \Gamma(U, \mathfrak{J})$ such that the sequence

$$0 \to \mathfrak{F}|U \to \mathfrak{F}|U \to (\mathfrak{F}/f\mathfrak{F})|U \to 0$$

is exact where the morphism $\mathfrak{F}|U \to \mathfrak{F}|U$ is the multiplication by f^l. Since $\mathfrak{J}^l\mathcal{H}^i_A\mathfrak{F} = 0$, we get the exact sequence

$$0 \to \mathcal{H}^{i-1}_A\mathfrak{F} \to \mathcal{H}^{i-1}_A(\mathfrak{F}/f\mathfrak{F}) \to \mathcal{H}^i_A\mathfrak{F} \to 0$$

for $i \leqslant q$. Since $S_{k+q}(\mathfrak{F}/f^l\mathfrak{F}) \subset S_{q+k+1}(\mathfrak{F})$, we have

$$\dim A \cap \overline{S}_{(q-1)+k+1}((\mathfrak{F}/f\mathfrak{F})|U \setminus A) \leqslant k, \text{ for all } k.$$

By induction hypothesis, the sheaves $\mathcal{H}^{i-1}_A(\mathfrak{F}/f\mathfrak{F})$ and $\mathcal{H}^{i-1}_A\mathfrak{F}$ are coherent on U for $i \leqslant q$ and by the above exact sequence, the coherence of $\mathcal{H}^q_A\mathfrak{F}$ around x follows.

The proof of the implication (*b*) $\Rightarrow$ (*a*). If $q = 0$, then $\overline{S}_m(\mathfrak{F}|X \setminus A) \subset S_m(\mathfrak{F}/\mathcal{H}^0_A\mathfrak{F})$. Since $\mathcal{H}^0_A(\mathfrak{F}/\mathcal{H}^0_A\mathfrak{F}) = 0$, the implication will follow by the vanishing theorem.

If $q = 1$, then we have the exact sequence

$$0 \to \mathcal{H}^0_A\mathfrak{F} \to \mathfrak{F} \to \mathfrak{R}^0_A\mathfrak{F} \to \mathcal{H}^1_A\mathfrak{F} \to 0,$$

where $\mathfrak{R}^0_A\mathfrak{F}$ is the sheaf $U \mapsto \Gamma(U \setminus A, \mathfrak{F})$ defined in § 1. As $\mathcal{H}^1_A\mathfrak{F}$ is supposed coherent, $\mathfrak{R}^0_A\mathfrak{F}$ is coherent and the conclusion follows by theorem 3.6, since $\mathcal{H}^i_A(\mathfrak{R}^0_A\mathfrak{F}) = 0$ for $i = 0, 1$ and $\overline{S}_m(\mathfrak{F}|X \setminus A) \subset S_m(\mathfrak{R}^0_A\mathfrak{F})$.

The general case $q \geqslant 2$ will be proved by induction on q. Assume the implication already established for integers smaller than q and prove it for q. We have

$$\dim A \cap \overline{S}_{k+q}(\mathfrak{F}|X \setminus A) \leqslant k, \text{ for all } k.$$

On the contrary, assume that there is a point $x \in A$ such that

$$\dim_x A \cap \overline{S}_{k+q+1}(\mathfrak{F}|X \setminus A) = k + 1$$

for some k. Then there is a point $y \in A \cap \overline{S}_{k+q+1}(\mathcal{F}|X \setminus A)$ such that $y \notin \overline{S}_{k+q}(\mathcal{F}|X \setminus A)$ and

$$(*) \qquad \dim_y A \cap \overline{S}_{k+q+1}(\mathcal{F}|X \setminus A) = k + 1.$$

For a suitable neighbourhood U of y we have $U \cap \overline{S}_{k+q}(\mathcal{F}|X \setminus A) = \emptyset$. By $(*)$ and since any component of $\overline{S}_{k+q+1}(\mathcal{F}|X \setminus A)$ is not contained in A, we get the inequality $\dim_y \overline{S}_{k+q+1}(\mathcal{F}|X \setminus A) \geqslant k + 2$. By restraining eventually U, we can find a closed analytic subset B in $U \cap \overline{S}_{k+q+1}(\mathcal{F}|X \setminus A)$ such that $\dim B = k + 2$, $\dim A \cap B = k + 1$ and $B = (B \setminus A)^-$. Since $q \geqslant 2$ and by shrinking once more U, we can choose $f \in \Gamma(U, \mathcal{O})$ such that the analytic set $V(f)$ of all zeros of f contains B but does not contain any $(l + q + 1)$-dimensional irreducible component of $\overline{S}_{l+q+1}(\mathcal{F}|X \setminus A)$ for $l \geqslant k$. So we have

$$\dim V(f) \cap \overline{S}_{l+q+1}(\mathcal{F}|X \setminus A) \leqslant l + q, \text{ for } l \geqslant k.$$

Since $U \cap \overline{S}_{k+q}(\mathcal{F}|X \setminus A) = \emptyset$, we get

$$\dim V(f) \cap S_{m+1}(\mathcal{F}) \cap (U \setminus A) \leqslant m \text{ for each } m.$$

In virtue of 3.8, f_z is nonzerodivisor of $\mathcal{F}_z$ for all $z \in U \setminus A$. Then we obtain an exact sequence on $U \setminus A$

$$0 \to \mathcal{F} \to \mathcal{F} \to \mathcal{F}/f\mathcal{F} \to 0$$

where the morphism $\mathcal{F} \to \mathcal{F}$ is the multiplication by f. From the definition of the sheaves $\mathcal{R}^{\cdot}_A$ (§ 1) we derive an exact sequence on U

$$\ldots \to \mathcal{R}^i_A \mathcal{F} \to \mathcal{R}^i_A \mathcal{F} \to \mathcal{R}^i_A(\mathcal{F}/f\mathcal{F}) \to \mathcal{R}^{i+1}_A \mathcal{F} \to \ldots.$$

Since $q \geqslant 2$ and $\mathcal{R}^i_A \mathcal{F}$ are coherent for $i \leqslant q - 1$ (by hypothesis and by corollary 1.10), it follows that the sheaves $\mathcal{R}^i_A(\mathcal{F}/f\mathcal{F})$ are coherent for $i \leqslant q - 2$. Accordingly, the sheaves $\mathcal{H}^i_A(\mathcal{F}/f\mathcal{F})$ are coherent for $i \leqslant q - 1$. By induction hypothesis,

$$\dim A \cap \overline{S}_{l+q}((\mathcal{F}/f\mathcal{F})|U \setminus A) \leqslant l, \text{ for all } l.$$

Since $B \setminus A \subset S_{k+q+1}(\mathcal{F}|X \setminus A)$ and $\operatorname{prof}_z(\mathcal{F}/f\mathcal{F}) = \operatorname{prof}_z \mathcal{F} - 1$ for $z \in B \setminus A \subset V(f)$, we deduce $B \setminus A \subset S_{k+q}((\mathcal{F}/f\mathcal{F})|U \setminus A)$. Therefore, $\dim A \cap B \leqslant k$, which contradicts the very choice of B.

The implication $(b) \Rightarrow (a)$ and the finiteness theorem are proved.

Corollary 4.4. *Let X be a complex space, $A \subset X$ a closed analytic subset, and $\mathcal{F} \in \mathbf{Coh}(X)$. If for any point $x \in A$ there is a neighbourhood U such that* $\operatorname{prof}(\mathcal{F}|U \setminus A) \geqslant q + \dim_x A + 1$, *then the sheaves $\mathcal{H}^i_A \mathcal{F}$ are coherent for $i \leqslant q$.*

If $\mathcal{F}$ is a coherent analytic sheaf on a complex space X and Supp $\mathcal{F}$ contains at most a point x, then by Nullstellensatz, one gets $\dim_{\mathbb{C}} \mathcal{F}_x < \infty$. Since $\operatorname{Supp} \mathcal{H}^{\cdot}_A \mathcal{F} \subset A$, we obtain the following corollary from theorem 4.1:

Corollary 4.5. *Let X be a complex space, $x \in X$, $\mathcal{F} \in \mathbf{Coh}(X)$ and q an integer. Then the linear spaces $H^i_x(X, \mathcal{F})$ are finite-dimensional for $i \leqslant q$ if and only if there is a neighbourhood U of x such that* $\operatorname{prof}(\mathcal{F}|U \setminus \{x\}) \geqslant q + 1$.

§ 5. Absolute local cohomology

Let X be a complex space, $\mathcal{O}$ its structural sheaf and $d \geqslant 0$ an integer. For any open subset U of X denote by $\mathfrak{A}_d(U)$ the family of closed analytic subsets of U, of dimension $\leqslant$ d. Let $\mathcal{F}$ be an $\mathcal{O}$-module. Whenever $A \subset B$ lie in $\mathfrak{A}_d(U)$, $H^0_A(U, \mathcal{F}) \subset H^0_B(U, \mathcal{F})$. These inclusions are functorial and for derived functors we obtain natural maps $H^\bullet_A(U, \mathcal{F}) \to H^\bullet_B(U, \mathcal{F})$. If $U \subset V$ and $A \in \mathfrak{A}_d(V)$, then $A \cap U \in \mathfrak{A}_d(U)$. The restriction map defines natural maps $H^\bullet_A(V, \mathcal{F}) \to H^\bullet_{A \cap U}(U, \mathcal{F})$. We thus obtain the presheaves

$$U \mapsto \varinjlim \{H^\cdot_A(U, \mathcal{F}) | \mathrm{A} \in \mathfrak{A}_d\,(U)\}.$$

The associated sheaves are called the *sheaves of d-absolute local cohomology* and are denoted by $\mathcal{H}^\cdot_d\mathcal{F}$ [84]. If $\mathcal{F} \in \mathbf{Coh}(X)$, then $\mathcal{H}^0_d\mathcal{F}$ is equal to the subsheaf of $\mathcal{F}$ of all sections whose supports are of dimension $\leqslant d$.

The sheaf $\mathcal{H}^0_d\mathcal{F}$ is also denoted by $\mathcal{F}\,[d]$; it is called the *d-sheaf of Thimm* [89] and it enjoyes remarkable properties (prop. 5.3).

In the same manner, one can define the *d-absolute gap sheaves* $\mathcal{R}^\cdot_d(\mathcal{F})$, by means of the presheaves

$$U \mapsto \varinjlim \{H^\cdot(U \setminus A, \mathcal{F}) |\; A \in \mathfrak{A}_d(U)\}. \qquad [84]$$

The sheaves $\mathcal{H}^\cdot_d\mathcal{F}$ and $\mathcal{R}^\cdot_d\mathcal{F}$ have a natural structure of $\mathcal{O}$-modules. By the exactness of the inductive limit, we derive from 1.10 the exact sequence

$$0 \to \mathcal{H}^0_d\mathcal{F} \to \mathcal{F} \to \mathcal{R}^0_d\mathcal{F} \to \mathcal{H}^1_d\mathcal{F} \to 0$$

and for $i \geqslant 1$, the isomorphisms

$$\mathcal{R}^i_d\mathcal{F} \simeq \mathcal{H}^{i+1}_d\mathcal{F}.$$

The results given below will be stated in terms of the sheaves $\mathcal{H}^\cdot_d\mathcal{F}$; by the above relations one can then deduce results for the sheaves $\mathcal{R}^\cdot_d\mathcal{F}$.

THEOREM 5.1. *Let X be a complex space, $d \geqslant 0$ an integer and* $\mathcal{F} \in \mathbf{Coh}(X)$. *For an integer $q \geqslant 0$ the following conditions are equivalent*:

(*a*) $\dim S_{k+q+1}(\mathcal{F}) \leqslant k$ *for any* $k < d$.

(*b*) *For any locally closed analytic subset A of dimension* $\leqslant d$, $\mathcal{H}^i_A\mathcal{F} = 0$ *for* $i \leqslant q$.

(*c*) $\mathcal{H}^i_d\mathcal{F} = 0$ *for* $i \leqslant q$.

Proof. The implication (*a*) $\Rightarrow$ (*b*) easily follows by 3.6. Prove now (*b*) $\Rightarrow$ (*a*). Suppose on the contrary that the assertion (*a*) is not true for some $k < d$. Then we can find an open subset U and a closed analytic subset A in $S_{k+q+1}(\mathcal{F}) \cap U$ such that $\dim A = k + 1 \leqslant d$. By the assertion (*b*) and theorem 3.6, we have $k + 1 = \dim A = \dim A \cap S_{k+q+1}(\mathcal{F}) \leqslant k$, hence a contradiction.

The implication $(b) \Rightarrow (c)$ follows straightforwardly from the definitions. It only remains to prove the implication $(c) \Rightarrow (b)$. We use induction on q. The case $q = 0$ follows if we notice that $\mathcal{H}^0_A\mathcal{F}|U$ is a subsheaf of $\mathcal{H}^0_d\mathcal{F}|U$ whenever A and U are as in (b).

Let us suppose the assertion (b) is verified for indexes i, $i \leqslant p \leqslant q$ and we will actually verify it for $p + 1$. Let U be an open subset of X and $A \in \mathfrak{A}_d(U)$. Consider an arbitrary point x of U. We are going to prove that $(\mathcal{H}^{p+1}_A\mathcal{F})_x = 0$. Let V be a neighbourhood of x contained in U and let $\xi \in H^{p+1}_A(V, \mathcal{F})$. Since $(\mathcal{H}^{p+1}_d\mathcal{F})_x = 0$, there exist a neighbourhood W of x in V and $B \in \mathfrak{A}_d(W)$ such that $A \cap W \subset B$ and ξ has the image null through the composition of natural maps

$$H^{p+1}_A(V, \mathcal{F}) \to H^{p+1}_{A\cap W}(W, \mathcal{F}) \to H^{p+1}_B(W, \mathcal{F}).$$

Let $B' = B \setminus A$. Since $\mathcal{H}^i_{B'}\mathcal{F} = 0$ for $i \leqslant p$, we have $H^p_{B'}(W, \mathcal{F}) = H^p_B(W \setminus A, \mathcal{F}) = 0$. Hence the map $H^{p+1}_{A\cap W}(W, \mathcal{F}) \to H^{p+1}_B(W, \mathcal{F})$ is injective and the image of ξ in $H^{p+1}_{A\cap W}(W, \mathcal{F})$ is null. Thus the germ defined by ξ in $(\mathcal{H}^{p+1}_d\mathcal{F})_x$ is zero and the proof is completed.

Corollary 5.2. *Let X be a complex space, $d \geqslant 0$ an integer and $\mathcal{F} \in \mathbf{Coh}(X)$. The following conditions are equivalent*

(a) $\dim S_{k+1}(\mathcal{F}) \leqslant k$ for any $k < d$.

(b) $\mathcal{H}^0_A\mathcal{F} = 0$ for any locally closed analytic set A of dimension $\leqslant d$.

(c) $\mathcal{H}^0_d\mathcal{F} = 0$.

(d) $\mathcal{F}$ has no sections $\neq 0$ whose supports are of dimension $\leqslant d$.

(e) For any point $x \in X$, the $\mathcal{O}_x$-module $\mathcal{F}_x$ has no associated prime ideals of dimension $\leqslant d$.

Proof. The equivalences $(a) \Leftrightarrow (b) \Leftrightarrow (c)$ are a consequence of the theorem. The equivalence $(c) \Leftrightarrow (d)$ follows from the definition of the sheaf $\mathcal{H}^0_d\mathcal{F}$. We now prove the implication $(d) \Rightarrow (e)$. Let $\mathfrak{p}$ be an ideal of Ass $\mathcal{F}_x$. Then there is a neighbourhood U of x and a non-null section s of $\Gamma(U, \mathcal{F})$ such that $\mathfrak{p} = (0\colon s_x)$. Let A be the analytic set defined by $\mathfrak{p}$ in a neighbourhood V of x. By shrinking V we may assume $\dim A \leqslant \dim \mathcal{O}_x/\mathfrak{p}$. We have Supp s $\subset A$, hence $\dim (\mathcal{O}_x/\mathfrak{p}) > d$.

We must show only that $(e) \Rightarrow (d)$. If $s_x \in \mathcal{F}_x$ is the germ of a section s, then the annihilator sheaf $\mathcal{A}$ of s is coherent and $\mathcal{A}_x$ is contained in some prime ideal of Ass $\mathcal{F}_x$. Since Supp $s = V(\mathcal{A})$, we derive $\dim_x \operatorname{Supp} s \geqslant d + 1$.

Let us pass now to analysing the coherence of the sheaves $\mathcal{H}^{\cdot}_d$.

Proposition 5.3. *Let X be a complex space, $d \geqslant 0$ an integer and $\mathcal{F} \in \mathbf{Coh}(X)$. Then the sheaf $\mathcal{H}^0_d\mathcal{F}$ is coherent and equal to $\mathcal{H}^0_{S_d(\mathcal{F})}\mathcal{F}$. Moreover, the irreducible components of dimension d of* Supp *$\mathcal{H}^0_d\mathcal{F}$ and $S_d(\mathcal{F})$ are the same.*

Proof. Since $\dim S_d(\mathcal{F}) \leqslant d$, $\mathcal{H}^0_{S_d(\mathcal{F})}\mathcal{F} \subset \mathcal{H}^0_d\mathcal{F}$. Let x be a point of $X \setminus S_d(\mathcal{F})$ and U one of its neighbourhoods such that $U \cap S_d(\mathcal{F}) = \emptyset$. By 5.2, $\mathcal{H}^0_d\mathcal{F}|U = 0$. Hence Supp $\mathcal{H}^0_d\mathcal{F} \subset S_d(\mathcal{F})$ and $\mathcal{H}^0_d\mathcal{F}$ coincides with $\mathcal{H}^0_{S_d(\mathcal{F})}\mathcal{F}$ and in particular is coherent (3.2).

We prove now the last assertion. Since Supp $\mathcal{H}^0_d\mathcal{F} \subset S_d(\mathcal{F})$, we have to show that any irreducible component S of $S_d(\mathcal{F})$ of dimension d is an irreducible component of Supp $\mathcal{H}^0_d\mathcal{F}$. This assertion will be a consequence of the inclusion $S \subset$

$\subset \operatorname{Supp} \mathcal{H}_d^0(\mathcal{F})$. Thus, it remains to check this inclusion. Let x be arbitrary in S. Suppose on the contrary that $(\mathcal{H}_d^0\mathcal{F})_x = 0$. Since $\mathcal{H}_d^0\mathcal{F}$ is coherent, $\mathcal{H}_d^0\mathcal{F} = 0$ in some neighbourhood U of x. By applying 5.2 we deduce that $\dim(S_d(\mathcal{F}) \cap U) \leqslant \leqslant d-1$. But $\emptyset \neq S \cap U \subset S_d(\mathcal{F}) \cap U$, hence a contradiction.

COROLLARY 5.4. $\operatorname{Supp}(\mathcal{H}_d^0 \mathcal{F}/\mathcal{H}_{d-1}^0\mathcal{F})$ *is the union of all irreducible components of dimension* d *of* $S_d(\mathcal{F})$.

If M is a subset of X we say it is of dimension $\leqslant d$ if for any $x \in M$ there is a neighbourhood U of x and $A \in \mathfrak{A}_d(U)$ such that $M \cap U \subset A$.

LEMMA 5.5. *If the sheaf* $\mathcal{H}_d^i\mathcal{F}$ *is of finite type, then*

$$\dim \operatorname{Supp} \mathcal{H}_d^i\mathcal{F} \leqslant d.$$

Proof. Let $x \in X$, U one of its neighbourhoods and $s_1, \ldots, s_r \in \Gamma(U, \mathcal{H}_d^i\mathcal{F})$ generating $\mathcal{H}_d^i\mathcal{F}|U$. By shrinking U we can find $A \in \mathfrak{A}_d(U)$ and elements $\xi_1, \ldots, \xi_r \in \in H_A^i(U, \mathcal{F})$, which represent $s_1, \ldots, s_r$. It follows that $(\operatorname{Supp} \mathcal{H}_d^i\mathcal{F}) \cap U \subset A$ and the lemma is proved.

THEOREM 5.6. *Let* X *be a complex space,* $\mathcal{F} \in \mathbf{Coh}(X)$ *and* d, $q \geqslant 0$ *two integers. The following conditions are equivalent*:

(*a*) $\dim \operatorname{Supp} \mathcal{H}_{d+1}^i\mathcal{F} \leqslant d$ *for* $i \leqslant q$.

(*b*) $\dim S_{d+q+1}(\mathcal{F}) \leqslant d$.

(*c*) $\mathcal{H}_d^i\mathcal{F}$ *is coherent and equals* $\mathcal{H}_{S_{d+q+1}(\mathcal{F})}^i(\mathcal{F})$ *for* $i \leqslant q+1$.

(*c'*) $\mathcal{H}_d^i\mathcal{F}$ *is coherent for* $i \leqslant q+1$.

(*d*) $\mathcal{H}_{d+\rho}^i\mathcal{F}$ *is coherent and equals* $\mathcal{H}_d^i(\mathcal{F})$ *for* $i \leqslant q+1-\rho$.

(*d'*) $\mathcal{H}_{d+\rho}^i\mathcal{F}$ *is coherent for* $i \leqslant q+1-\rho$.

Proof. (*a*) $\Rightarrow$ (*b*). Let $S = \bigcup_{i=1}^{q} \operatorname{Supp} \mathcal{H}_{d+1}^i\mathcal{F}$. We have $\mathcal{H}_{d+1}^i\mathcal{F} = 0$ on $X \setminus S$ for $i \leqslant q$. By 5.1, $\dim(S_{d+q+1}(\mathcal{F}) \cap (X \setminus S)) \leqslant d$. Since $\dim S \leqslant d$, $\dim S_{d+q+1}(\mathcal{F}) \leqslant d$.

(*b*) $\Rightarrow$ (*c*). Let $S = S_{d+q+1}(\mathcal{F})$, hence $\dim S \leqslant d$. Since $\operatorname{prof}(\mathcal{F}|X \setminus S) \geqslant d + q + 2$, it follows by 4.4 that $\mathcal{H}_S^i\mathcal{F}$ is coherent for $i \leqslant q+1$. Let U be an arbitrary open set and $A \in \mathfrak{A}_d(U)$. By 3.9, $\mathcal{H}_A^i\mathcal{F}|U \setminus S = 0$; hence $\mathcal{H}_S^i\mathcal{F} = \mathcal{H}_{A \cup S}^i\mathcal{F}$ on U. Accordingly, $\mathcal{H}_S^i\mathcal{F} = \mathcal{H}_d^i\mathcal{F}$ for $i \leqslant q+1$.

(*c'*) $\Rightarrow$ (*b*). Let $S_i = \operatorname{Supp} \mathcal{H}_d^i\mathcal{F}$. By the previous lemma, S_i is an analytic set of dimension $\leqslant d$ for $i \leqslant q+1$. Put $S = \bigcup_{i=1}^{q+1} S_i$. Since $\mathcal{H}_d^i\mathcal{F}|X \setminus S = 0$ for $i \leqslant q+1$, by 5.1 the following inequality may result: $\dim S_{d+q+1}(\mathcal{F}|X \setminus S) \leqslant \leqslant d-1 < d$. But $\dim S \leqslant d$, hence $\dim S_{d+q+1}(\mathcal{F}) \leqslant d$.

The implication (*b*) $\Rightarrow$ (*d*) can be proved exactly as (*b*) $\Rightarrow$ (*c*), by writing $S_{d+q+1}(\mathcal{F}) = S_{(d+\rho)+(q-\rho)+1}(\mathcal{F})$.

The implications (*d*) $\Rightarrow$ (*d'*) $\Rightarrow$ (*c'*), (*d*) $\Rightarrow$ (*a*) are obvious.

COROLLARY 5.7. *The following conditions are equivalent:*
(*a*) dim Supp $\mathcal{H}^0_{d+1}(\mathcal{F}) \leqslant d$.
(*b*) $\mathcal{H}^0_{d+1}\mathcal{F} = \mathcal{H}^0_d\mathcal{F}$.
(*c*) dim $S_{d+1}(\mathcal{F}) \leqslant d$.
(*d*) $\mathcal{R}^0_d\mathcal{F}$ *is coherent.*

The proof follows by 5.3 making use of the exact sequence

$$0 \to \mathcal{H}^0_d\mathcal{F} \to \mathcal{F} \to \mathcal{R}^0_d\mathcal{F} \to \mathcal{H}^1_d\mathcal{F} \to 0.$$

APPLICATION. We are going to construct a natural complex of sheaves for any analytic space. The construction will be done in several steps.

(**a**) Let V be an open subset of a numerical space $\mathbb{C}^n$. For an integer q and an open subset U of V, we denote $\Phi^q(U) = \mathfrak{A}_{n-q}(U)$, hence $\Phi^q(U)$ is the family of all closed analytic subsets of codimension $\geqslant q$ in U.

Let $\mathcal{F}$ be an analytic sheaf on V. Denote by $\mathcal{H}^p_{\Phi^q}(\mathcal{F})$ (respectively $\mathcal{H}^p_{\Phi^q/\Phi^{q+1}}(\mathcal{F})$) the sheaf associated to the presheaf $U \mapsto H^p_{\Phi^q(U)}(U, \mathcal{F})$ (respectively $U \mapsto H^p_{\Phi^q(U)/\Phi^{q+1}(U)}$ $(U, \mathcal{F})$), p and q being arbitrary integers. Under the above notations,

$$\mathcal{H}^p_{\Phi^q}(\mathcal{F}) = \mathcal{H}^p_{n-q}\mathcal{F}.$$

By means of 1.7 we deduce for an integer p the exact sequence

$$\ldots \to \mathcal{H}^p_{\Phi^p}\mathcal{F} \to \mathcal{H}^p_{\Phi^p/\Phi^{p+1}}(\mathcal{F}) \to \mathcal{H}^{p+1}_{\Phi^{p+1}}(\mathcal{F}) \to \ldots$$

For the index $p+1$, we have the exact sequence

$$\ldots \to \mathcal{H}^{p+1}_{\Phi^{p+1}}(\mathcal{F}) \to \mathcal{H}^{+1}_{\Phi^{p+1}/\Phi^{p+2}}(\mathcal{F}) \to \mathcal{H}^{p+2}_{\Phi^{p+2}}(\mathcal{F}) \to \ldots$$

From these exact sequences we get, by composition, the morphism

$$\mathcal{H}^p_{\Phi^p/\Phi^{p+1}}(\mathcal{F}) \to \mathcal{H}^{p+1}_{\Phi^{p+1}/\Phi^{p+2}}(\mathcal{F}).$$

If we write the above exact sequences for $p = 0$, the morphism

$$\mathcal{F} \to \mathcal{H}^0_{\Phi^0/\Phi^1}(\mathcal{F})$$

is obtained. It is easy to check that in this way we get a complex

$$0 \to \mathcal{H}^0_{\Phi^0/\Phi^1}(\mathcal{F}) \to \mathcal{H}^1_{\Phi^1/\Phi^2}(\mathcal{F}) \to \mathcal{H}^2_{\Phi^2/\Phi^3}(\mathcal{F}) \to \ldots$$

which is called the *Cousin complex associated to the sheaf* $\mathcal{F}$, together with the augmentation morphism $\mathcal{F} \to \mathcal{H}^0_{\Phi^0/\Phi^1}(\mathcal{F})$.

We now consider the particular case $\mathcal{F} = \mathcal{O}$.

LEMMA 5.8. *The Cousin complex associated to the structural sheaf* $\mathcal{O}$ *is a resolution for* $\mathcal{O}$; *in other words the sequence*

$$0 \to \mathcal{O} \to \mathcal{H}^0_{\Phi^0/\Phi^1}(\mathcal{O}) \to \mathcal{H}^1_{\Phi^1/\Phi^2}(\mathcal{O}) \to \mathcal{H}^2_{\Phi^2/\Phi^3}(\mathcal{O}) \to \dots$$

is exact.

Proof. We shall prove that $\mathcal{H}^p_{\Phi^q}(\mathcal{O}) = 0$ for $q \neq p$; then the lemma easily follows from definitions only.

Since the assertion is obvious for $q = 0$, we may assume $q \geqslant 1$. Obviously, $\mathcal{H}^p_{\Phi^q}(\mathcal{O}) = \mathcal{H}^p_{n-q}\mathcal{O}$. If $p < q$, then $\mathcal{H}^p_{n-q}\mathcal{O} = 0$ by theorem 5.1. Let us consider an arbitrary point $x \in V$ and let U be a neighbourhood of x. For any $A \in \mathfrak{A}_{n-q}(U)$ we will prove that by shrinking U around x there exists $B \in \mathfrak{A}_{n-q}(U)$ such that $A \subset B$ and $H^p_B(U, \mathcal{O}) = 0$ for $p > q$. Then it will result that $(\mathcal{H}^p_{n-q}\mathcal{O})_x = 0$ for $p > q$ and therefore the conclusion. By shrinking eventually U we can further find an element C in $\mathfrak{A}_{n-q}(U)$ such that $A \subset C$ and $\dim_x C = n - q$.

In this way we can assume, for the assertion we are interested in, that $\dim_x A = n - q$. Apply the normalization lemma 1.27 for $\mathcal{O}_x$ and for the ideal $\mathfrak{I}(A)_x$. There exists a finite injective morphism $\mathbb{C}\{X_1, \dots, X_n\} \xrightarrow{f} \mathcal{O}_x$ such that $f^{-1}(\mathfrak{I}(A)_x) = (X_1, \dots, X_q)$. Passing to germs of analytic spaces, and shrinking again U, we can find an open polydisc U' in $\mathbb{C}^n$ which is centred in origin, and a finite morphism $U \xrightarrow{F} U'$ such that $F(x) = 0$, and the morphism induced by F in x between the corresponding local rings is f. Let us denote by $x_1, \dots, x_n$ the coordinates in U'. If we restrain U and U' we can assume that F is surjective (as a consequence of the finiteness and of the equality $\dim U = \dim U'$), that $F_*(\mathcal{O}_U)$ admits on U' a finite resolution with free sheaves of finite rank and moreover, the functions $x_i F (i = 1, \dots, q)$ vanish on A. We have $A \subset F^{-1}(A')$ where $A' = U' \cap \{x_1 = \dots = x_q = 0\}$. If we get $B = F^{-1}(U')$ then one will get $\dim B = n - q$ and $F(B) = A'$.

If $\mathcal{F}$ is a sheaf of abelian groups on U, then $H^0_B(U, \mathcal{F}) \simeq H^0_{A'}(U, F_*(\mathcal{F}))$. By using this and the exactness of F_* we derive canonical isomorphisms

$$H^p_B(U, \mathcal{O}) \simeq H^p_{A'}(U', F_*(\mathcal{O}_U)).$$

To conclude with, it is sufficient to show that $H^p_{A'}(U', F_*(\mathcal{O}_U)) = 0$ for $p > q$. By hypothesis on $F_*(\mathcal{O}_U)$ we have only to prove that $H^p_{A'}(U', \mathcal{O}_{U'}) = 0$ for $p > q$. But we have the natural isomorphisms $H^{p-1}(U' \setminus A', \mathcal{O}_{U'}) \simeq H^p_{A'}(U', \mathcal{O}_{U'})$ $(p > q \geqslant 1)$. If we put $U'_i = U' \cap \{x_i = 0\}$, $1 \leqslant i \leqslant q$, we get a Stein covering of $U' \setminus A'$ by q Stein open subsets and the required conclusion follows.

We shall denote by $\overline{\mathfrak{L}}^{\boldsymbol{\cdot}}_V$ the resolution of $\mathcal{O}_V$ given by the lemma, and by $\overline{\mathfrak{K}}^{\boldsymbol{\cdot}}_V$ the complex $(\overline{\mathfrak{L}}^{\boldsymbol{\cdot}}_V \otimes_{\mathcal{O}_V} \Omega_V)$ $[n]$ where Ω_V is the sheaf of germs with holomorphic forms of maximal degree, and the bracket $[n]$ means that the complex is translated to the left by n steps.

(b) Now, we study the behaviour of the previously defined complexes with respect to immersions.

LEMMA 5.9. *For any immersion* $V \xrightarrow{f} W$ *with* V, W *as open subsets in numerical spaces, there exists a natural isomorphism*

$$\bar{f}: \mathcal{K}_V^{\cdot} \simeq f^* Hom_{\mathcal{O}_W}(f_*\mathcal{O}_V, \mathcal{K}_W^{\cdot}).$$

Moreover, the correspondence $f \mapsto \bar{f}$ *agrees with the composition of immersions.*

Proof. If f is an open immersion, the construction of $\bar{f}$ is obvious. Let us pass to the case of the closed immersions and consider the following situation: the ideal-sheaf given by V in W is $t\mathcal{O}_W$ for some $t \in \Gamma(W, \mathcal{O}_W)$. Then we have the exact sequence

$$0 \to \mathcal{O}_W \xrightarrow{t} \mathcal{O}_W \to f_*(\mathcal{O}_V) \to 0.$$

We will establish an isomorphism

$$\Delta_t^q: \bar{\mathfrak{L}}_V^{q-1} \simeq f^* \, Hom_{\mathcal{O}_W}(f_*\mathcal{O}_V, \bar{\mathfrak{L}}_W^q).$$

for any integer q. The above exact sequence yields the exact sequences

$$0 \to Hom_{\mathcal{O}_W}(f_*\mathcal{O}_V, \mathcal{H}^q_{\Phi_W^q/\Phi_W^{q+1}}(\mathcal{O}_W)) \to \mathcal{H}^q_{\Phi_W^q/\Phi_W^{q+1}}(\mathcal{O}_W) \xrightarrow{t} \mathcal{H}^q_{\Phi_W^q/\Phi_W^{q+1}}(\mathcal{O}_W),$$

$$\ldots \to \mathcal{H}^{q-1}_{\Phi_W^q/\Phi_W^{q+1}}(\mathcal{O}_W) \to \mathcal{H}^{q-1}_{\Phi_W^q/\Phi_W^{q+1}}(f_*\mathcal{O}_V) \to$$

$$\to \mathcal{H}^q_{\Phi_W^q/\Phi_W^{q+1}}(\mathcal{O}_W) \xrightarrow{t} \mathcal{H}^q_{\Phi_W^q/\Phi_W^{q+1}}(\mathcal{O}_W) \to \ldots$$

Since $\mathcal{H}^p_{\Phi_W^q}(\mathcal{O}_W) = 0$ for $p \neq q$ (see the proof of lemma 5.8), it follows that $\mathcal{H}^{q-1}_{\Phi_W^q/\Phi_W^{q+1}}(\mathcal{O}_W) = 0$. Then we deduce an isomorphism

$$\mathcal{H}^{q-1}_{\Phi_W^q/\Phi_W^{q+1}}(f_*\mathcal{O}_V) \simeq Hom_{\mathcal{O}_W}(f_*\mathcal{O}_V, \bar{\mathfrak{L}}_W^q).$$

If $\mathcal{G}$ is a sheaf of abelian groups on V, then

$$\Gamma_{\Phi_V^{q-1}(f^{-1}(D) \cap V)}(f^{-1}(D) \cap V, \mathcal{G}) = \Gamma_{\Phi_W^q(D)}(D, f_*\mathcal{G})$$

for any integer q and any open subset D of W. Whence the natural isomorphisms

$$\bar{\mathfrak{L}}_V^{q-1} = \mathcal{H}^{q-1}_{\Phi_V^{q-1}/\Phi_V^q}(\mathcal{O}_V) \simeq f^*\mathcal{H}^{q-1}_{\Phi_W^q/\Phi_W^{q+1}}(f_*\mathcal{O}_V).$$

According to the above said, the construction of the isomorphism Δ_t^q becomes clear.

We then define the isomorphisms

$$\bar{f}^q : \bar{\mathcal{K}}_V^q \simeq f^* Hom_{\mathcal{O}_W}(f_* \mathcal{O}_V, \bar{\mathcal{K}}_W^q)$$

by means of $\Delta_t^\bullet$ and of the natural morphism $\Omega_W \to f_*(\Omega_V)$, which assigns the form $2\pi i \psi | V$ to any form $\psi \wedge dt$. If we replace t by another parameter t', then $\Delta_{t'} = (t/t'|V)\,\Delta_t$. It is easy to check that the difference between the morphisms $\Omega_W \to f_*(\Omega_V)$, associated to t' and t is just a multiplication by $t'/t|V$. Then the morphisms $\bar{f}^q$ are independent on the parameter t and the family $\bar{f} = (\bar{f}^q)$ is an isomorphism of complexes.

If the immersion $V \xrightarrow{f} W$ is decomposed into a finite number of immersions of the above type, then the construction of $\bar{f}$ can be made recursively, by compositions. One can verify that $\bar{f}$ does not depend on the considered decomposition. In particular, this fact allows us to construct $\bar{f}$ for an arbitrary immersion f (locally, this is of the above type; the morphisms defined locally can be glued together,...) and also to verify the functorial character of the association $f \mapsto \bar{f}$.

(c) We now pass to the general case of a complex space X. For an open subset U of X such that there is an immersion $U \xrightarrow{\varphi} V$, V open subset of a numerical space, we have

$$\bar{\mathcal{K}}_U^\bullet = \varphi^* Hom_{\mathcal{O}_V}(\varphi_* \mathcal{O}_U, \bar{\mathcal{K}}_V^\bullet),$$

where $\bar{\mathcal{K}}_V^\bullet$ is the complex previously defined. If U' is another open set enjoying the same property, the immersions obtained from φ, φ' for $U \cap U'$ can be refined by means of a third one. By the above lemma we get an isomorphism

$$\tau_{U,U'} : \bar{\mathcal{K}}_{U'}^\bullet | U \cap U' \simeq \bar{\mathcal{K}}_U^\bullet | U \cap U'.$$

If we consider a covering of X by open sets as above, then the isomorphisms τ satisfy the customary relations of compatibility (apply again 5.9) and thus, the complexes $\bar{\mathcal{K}}_U^\bullet$ can be glued together and define a complex $\bar{\mathcal{K}}_X^\bullet$.

§ 6. The separation theorem

We use the notation from the previous paragraphs and assume in addition X with countable topology. We have the exact sequence

$$\ldots \to H^{i-1}(X \setminus A, \mathcal{F}) \xrightarrow{\delta^{i-1}} H_A^i(X, \mathcal{F}) \xrightarrow{\alpha} H^i(X, \mathcal{F}) \to \ldots$$

The invariants $H^\bullet(X, \mathcal{F})$, $H^\bullet(X \setminus A, \mathcal{F})$ have natural structures of topological vector spaces, quotients of FS spaces (Fréchet-Schwartz). For $i \geqslant 1$ we endow

$H^i_A(X, \mathcal{F})$ with the finest topology such that δ^{i-1} becomes a continuous map. The map α results continuous, as $\alpha\delta^{i-1} = 0$. The invariant $H^0_A(X, \mathcal{F})$ can be endowed with the topology induced from $\Gamma(X, \mathcal{F})$ and then it is an FS space, as kernel of the map $\Gamma(X, \mathcal{F}) \to \Gamma(X \setminus A, \mathcal{F})$.

If X is Stein, then the spaces $H^i_A(X, \mathcal{F})$ and $H^{i-1}(X \setminus A, \mathcal{F})$ have the same topology for $i \geqslant 2$ and $H^1_A(X, \mathcal{F})$ has the quotient-topology $\Gamma(X, \mathcal{F}) / \Gamma(X \setminus A, \mathcal{F})$.

The main result in this paragraph is the following *separation theorem for local cohomology* of Siu and Trautmann [84]:

THEOREM 6.1. *Let X be a complex space, $A \subset X$ a closed analytic subset, q an integer and $\mathcal{F}$ a coherent analytic sheaf on X such that the sheaves $\mathcal{H}^i_A\mathcal{F}$ are coherent for $i < q$. Then for any Stein open subset Ω of X, $H^i_A(\Omega, \mathcal{F})$ is a Fréchet-Schwartz space for $i \leqslant q$.*

In order to prove this theorem we need some preparations.

LEMMA 6.2. *Let Δ be the unit polydisc in $\mathbb{C}^n$ and $A = \Delta \cap \{z_{d+1} = \ldots = z_n = 0\}$, where $0 \leqslant d \leqslant n-1$. Then $H^{n-d}_A(\Delta, \mathcal{O})$ is separated.*

Proof. Let $U_i = \Delta \cap \{z_i \neq 0\}$ for $d+1 \leqslant i \leqslant n$. Then $\mathfrak{U} = \{U_{d+1}, \ldots, U_n\}$ is a Stein covering of $\Delta \setminus A$. Consider the case $d \leqslant n-2$. Since $H^{n-d}_A(\Delta, \mathcal{O}) \simeq H^{n-d-1}(\Delta \setminus A, \mathcal{O})$, we must prove that $H^{n-d-1}(\mathfrak{U}, \mathcal{O})$ is separated. Let $U^* = \bigcap_{i=d+1}^{n} U_i$. Since $Z^{n-d-1}(\mathfrak{U}, \mathcal{O}) = \Gamma(U^*, \mathcal{O})$, every cycle $\xi \in Z^{n-d-1}(\mathfrak{U}, \mathcal{O})$ is a Laurent series

$$\xi = \sum_{\nu_{d+1}, \ldots, \nu_n = -\infty}^{\infty} a_{\nu_{d+1}\ldots\nu_n}(z_1, \ldots, z_d)\, z_{d+1}^{\nu_{d+1}} \ldots z_n^{\nu_n},$$

where the coefficients $a_{\nu_{d+1}\ldots\nu_n}$ are holomorphic functions in the unit polydisc of $\mathbb{C}^d$. It is easy to check that $\xi \in B^{n-d-1}(\mathfrak{U}, \mathcal{O})$ if and only if $a_{\nu_{d+1}\ldots\nu_n} = 0$ for $\nu_{d+1} \leqslant -1, \ldots, \nu_n \leqslant -1$. By Cauchy's integral formula, it follows that $B^{n-d-1}(\mathfrak{U}, \mathcal{O})$ is a closed subspace of $\Gamma(U^*, \mathcal{O})$ (this one endowed, as usual, with the topology of uniform convergence on compacts). Then $H^{n-d-1}(\mathfrak{U}, \mathcal{O})$, hence $H^{n-d}_A(\Delta, \mathcal{O})$ is separated. The case $d = n-1$ is similar.

We now return to the general situation $(X, A, \mathcal{F})$. For an open subset U of X denote by $N^i_A(U, \mathcal{F})$ the topological closure of zero in $H^i_A(U, \mathcal{F})$. If V is an open subset of U, then the restriction map $H^i_A(U, \mathcal{F}) \overset{\beta}{\to} H^i_A(V, \mathcal{F})$ is continuous. It results that $\beta(N^i_A(U, \mathcal{F})) \subset N^i_A(V, \mathcal{F})$. Denote by $\mathcal{N}^i_A(\mathcal{F})$ the sheaf associated to the presheaf

$U \mapsto N^i_A(U, \mathcal{F})$, $U \supset V \mapsto$ the restriction map

$N^i_A(U, \mathcal{F}) \to N^i_A(V, \mathcal{F})$ given by β.

LEMMA 6.3. *Let D be a domain in $\mathbb{C}^n$ and $A \subset D$ a closed analytic subset of dimension d. Then $\mathcal{N}^i_A(\mathcal{O}) = 0$ for $i \leqslant n-d$ and $H^{n-d}_A(D, \mathcal{O})$ is separated. Moreover if D is Stein, then $H^{n-d}_A(D, \mathcal{O})$ is just an FS space.*

Proof. Since $\mathcal{H}^i_A\mathcal{O} = 0$ for $i < n-d$, $H^{n-d}_A(U, \mathcal{O}) = \Gamma(U, \mathcal{H}^{n-d}_A\mathcal{O})$ for any open subset U of D (corollary 1.15).

For $i < n - d$ the equality $\mathcal{N}^i_A(\mathcal{O}) = 0$ is a consequence of the equality $\mathcal{H}^i_A\mathcal{O} = 0$. If $\mathcal{N}^{n-d}_A(\mathcal{O}) = 0$, then by the commutative diagram

$$\begin{array}{ccc} N^{n-d}_A(D, \mathcal{O}) & \longrightarrow & \Gamma(D, \mathcal{N}^{n-d}_A \mathcal{O}) \\ \downarrow & & \downarrow \\ H^{n-d}_A(D, \mathcal{O}) & = & \Gamma(D, \mathcal{H}^{n-d}_A \mathcal{O}) \end{array}$$

we can deduce the equality $N^{n-d}_A(D, \mathcal{O}) = 0$, whence $H^{n-d}_A(D, \mathcal{O})$ is separated.

Suppose in addition D is Stein. For $d = n$, $H^0_A(D, \mathcal{O})$ is an FS space. If $d = n - 1$, the separated space $H^1_A(D, \mathcal{O})$ is isomorphic to $\Gamma(D, \mathcal{O})/\Gamma(D \setminus A, \mathcal{O})$, hence it is FS. For $d \leqslant n - 2$, $H^{n-d}_A(D, \mathcal{O})$ is topologically isomorphic to $H^{n-d-1}(D \setminus A, \mathcal{O})$ and again it will be FS.

So the lemma is proved as soon as $\mathcal{N}^{n-d}_A(\mathcal{O}) = 0$. Let A' be the set of the singular points of A and $A'' = A \setminus A'$. Let x be a point of A''. There exists a neigbourhood U of x contained in $D \setminus A'$, which is isomorphic to the unit polydisc in $\mathbb{C}^n$ such that the analytic set $A \cap U$ corresponds to the set $\Delta \cap \cap \{z_{d+1} = \ldots = z_n = 0\}$. By the preceding lemma, $H^{n-d}_A(U, \mathcal{O})$ is separated. Therefore $\mathcal{N}^{n-d}_A \mathcal{O} | D \setminus A' = 0$. Taking into account the antecedence we derive the separation of $H^{n-d}_A(D \setminus A', \mathcal{O})$. Since $\dim A' < d$, $H^{n-d}_{A'}(D, \mathcal{O}) = 0$, we have the exact sequence

$$0 \to H^{n-d}_A(D, \mathcal{O}) \to H^{n-d}_A(D \setminus A', \mathcal{O})$$

where the second map is continuous. We thus conclude that $N^{n-d}_A(D, \mathcal{O}) = 0$. By substituting for the domain D an arbitrary open set of D, it follows $\mathcal{N}^{n-d}_A \mathcal{O} = 0$.

LEMMA 6.4. *Let D be a domain in $\mathbb{C}^n$, $A \subset D$ a closed analytic subset of dimension $\leqslant d$, and $\mathcal{F} \in$* **Coh**(D). *For $0 \leqslant q \leqslant n - d$, let $\mathfrak{I}_q$ be the maximal ideal-sheaf associated to $S_{q+d-1}(\mathcal{F})$. Then for any open set $X \subset\subset D$ such that the topological closure X^- is holomorphically convex, there exists an integer $l \geqslant 0$ so that, for any open subset U of X,*

$$\Gamma(U, \mathfrak{I}_q)^l N^q_A(U, \mathcal{F}) = 0.$$

Proof. We use the notation considered in the proof of lemma 4.2. Suppose $q = n - d$. As in lemma 4.2 we can prove by induction the following assertions

$$(*)_j \begin{cases} \text{There is an integer } l \geqslant 0 \text{ such that, for any open subset } U \text{ of } X, \\ \Gamma(U, \mathfrak{I}_q)^l \, H^{q-j}_A(U, \mathcal{Z}_j) = 0 \quad (j \geqslant 1). \end{cases}$$

$$(**)_j \begin{cases} \text{There is an integer } l \geqslant 0 \text{ such that, for any open subset } U \text{ of } X, \\ \Gamma(U, \mathfrak{I}_q)^l \, H^{q-j}_A(U, \mathcal{B}_j) = 0 \quad (j \geqslant 1). \end{cases}$$

In particular, there is an integer l such that $\Gamma(U, \mathfrak{I}_q)^l \, H^{q-1}_A(U, \mathcal{B}_1) = 0$ for any open set U of X. Then from the exact sequence

$$0 \to H^{q-1}_A(U, \mathcal{B}_1) \xrightarrow{\delta} H^q_A(U, \mathcal{Z}_0) \xrightarrow{\alpha} H^q_A(U, \mathcal{O}^{s_0})$$

and by the previous lemma, we obtain $\alpha(N^q_A(U, \mathfrak{Z}_0)) = 0$. Hence $N^q_A(U, \mathfrak{Z}_0)$ is contained in $\operatorname{Im}\delta$ and is thus annihilated by $\Gamma(U, \mathfrak{I}_q)^l$. By means of the exact sequences

$$\ldots \to H^{q-1}_A(U, \mathfrak{L}_0) \to H^q_A(U, \mathfrak{B}_0) \to H^q_A(U, \mathfrak{Z}_0) \to \ldots,$$

$$\ldots \to H^q_A(U, \mathfrak{L}_{-1}) \to H^q_A(U, \mathfrak{F}) \to H^q_A(U, \mathfrak{B}_0) \to \ldots$$

and taking into account that $\operatorname{Supp} \mathfrak{L}_i \subset S_{n-1}(\mathfrak{F})$ (which gives that some powers of $\mathfrak{I}_q = \mathfrak{I}(S_{n-1}(\mathfrak{F}))$ annihilate the sheaves $\mathfrak{L}_i$, and the associated cohomological invariants), the conclusion of lemma in the case $q = n - d$ follows easily.

Consider now the case $q < n - d$. We have an exact sequence

$$0 \to H^q_A(U, \mathfrak{F}) \to H^{q+1}_A(U, \mathfrak{G})$$

and since the coboundary map is continuous, we get $N^q_A(U, \mathfrak{F}) \subset N^{q+1}_A(U, \mathfrak{G})$. The conclusion follows from the case $q = n - d$, by descending induction on q.

LEMMA 6.5. *Let $D \subset \mathbb{C}^n$ be a domain, V a closed analytic subset of D, and $\mathfrak{F} \in \mathbf{Coh}(D)$ such that $\mathcal{H}^0_V\mathfrak{F} = 0$. Let $\mathfrak{I}$ be the maximal ideal-sheaf of V. For any holomorphically convex compact K of D there exist a neighbourhood Y and sections $f_1, \ldots, f_p \in \Gamma(Y, \mathfrak{I})$ such that $(f_i)_x$ is nonzerodivisor of $\mathfrak{F}_x$ for all $x \in Y (1 \leqslant i \leqslant p)$ and, for some integer $m \geqslant 1$, $\mathfrak{I}^m|K$ is contained in the ideal-sheaf generated on K by $f_1, \ldots, f_p$.*

Proof. There exist a neighbourhood Y of K and sections $g_1, \ldots, g_q \in \Gamma(Y, \mathfrak{I})$ such that $g_1, \ldots, g_q$ generate $\mathfrak{I}|Y$. Since $\mathcal{H}^0_V\mathfrak{F} = 0$, $\dim V \cap S_{k+1}(\mathfrak{F}) \leqslant k$ for all k. Denote by S_{k+1} the union of all irreducible components of dimension $k+1$ of $Y \cap S_{k+1}(\mathfrak{F})$. Choose a countable set $A = \{x_\nu\}$, dense in $\bigcup_k S_{k+1} \setminus V$ such that $A \cap S_{k+1}$ is dense in $S_{k+1} \setminus V$. For each x_ν, $(g_1(x_\nu), \ldots, g_q(x_\nu)) \neq 0$. Then, in the space $\mathbb{C}^q$ with coordinates $z_1, \ldots, z_q$, there exists a linear form $\sum_{i=1}^q a_i z_i (a_i \in \mathbb{C})$ such that $\sum_{i=1}^q a_i g_i(x_\nu) \neq 0$, for each ν. Let $f_1 = \sum_{i=1}^q a_i g_i \in \Gamma(Y, \mathfrak{I})$. Since $f_1(x_\nu) \neq 0$ for any ν, $\dim V(f_1) \cap S_{k+1}(\mathfrak{F}) \leqslant k$ for all k. By corollary 3.8 $(f_1)_x$ is nonzerodivisor of $\mathfrak{F}_x$ for all $x \in Y$. If $V(f_1) = V$, the lemma is over by Nullstellensatz. Otherwise, replacing A by a countable dense subset B of $(\bigcup_k S_{k+1}) \cup V(f_1) \setminus V$ whose intersection with S_{k+1} is dense in $S_{k+1} \setminus V$ and whose intersection with every irreducible component W of $V(f_1) \setminus V$ is dense in $W \setminus V$, we can find an element $f_2 \in \Gamma(Y, \mathfrak{I})$ with $f_2(x) \neq 0$ for each $x \in B$. Again, using 3.8, we obtain that $(f_2)_x$ is nonzerodivisor of $\mathfrak{F}_x$ for all $x \in Y$. We also have

$$\dim(V(f_1) \cap V(f_2) \setminus V) < \dim(V(f_1) \setminus V).$$

If $V(f_1) \cap V(f_2) = V$, then (f_1, f_2) satisfies the lemma. Otherwise, we go on and choose the elements $f_1, \ldots, f_p \in \Gamma(Y, \mathfrak{I})$ such that any $(f_i)_x$ is a nonzerodivisor of $\mathfrak{F}_x$ for $x \in Y$ and

$$\dim\left(\textstyle\bigcap_{k=1}^{i} V(f_k) \setminus V\right) < \dim\left(\textstyle\bigcap_{k=1}^{i-1} V(f_k) \setminus V\right), \quad 2 \leqslant i \leqslant p.$$

For a sufficiently large p (for instance $p \geqslant n+1$), one obtains $V = \bigcap_{k=1}^{p} V(f_k)$. By Nullstellensatz, for sufficiently large m, $\mathfrak{J}^m|K$ is contained in the sheaf generated by $f_1, \ldots, f_p$.

LEMMA 6.6. *Let $D \subset \mathbb{C}^n$ be a domain, $A \subset V$ two closed analytic subsets of D and $\mathcal{F} \in$ **Coh** (D). Suppose $q \geqslant 1$. Let S'_{k+q} be the union of all irreducible components of $S_{k+q}(\mathcal{F})$ which are not contained in V. Assume that* dim $A \cap S'_{k+q} \leqslant k$ *for all k. Let $\mathfrak{J} = \mathfrak{J}(V)$ and $X \subset\subset D$ an open subset such that the topological closure X^- is holomorphically convex.*

Under these assumptions, there is an integer $l \geqslant 0$ such that for any Stein open subset U of X

$$\Gamma(U, \mathfrak{J})^l \, N^q_A(U, \mathcal{F}) = 0.$$

Proof. Let $\mathcal{G} = \mathcal{F}/\mathcal{H}^0_V\mathcal{F}$. We have $S'_{k+q}(\mathcal{F}) = S'_{k+q}(\mathcal{G})$ ($\mathcal{F}$ and $\mathcal{G}$ are equal beyond V) and $\mathcal{H}^0_V\mathcal{G} = 0$. For some $l \geqslant 0$, $\mathfrak{J}^l\, \mathcal{H}^0_V\mathcal{F} = 0$ on a neighbourhood of $\overline{X}$, hence $\Gamma(U, \mathfrak{J})^l \, H^q_A(U, \mathcal{H}^0_V\mathcal{F}) = 0$. From the above said and the exact sequence

$$\ldots \to H^q_A(U, \mathcal{H}^0_V\mathcal{F}) \to H^q_A(U, \mathcal{F}) \to H^q_A(U, \mathcal{G}) \to \ldots$$

it follows that $\mathcal{F}$ satisfies the lemma as soon as $\mathcal{G}$ satisfies it.

Thus, we may assume that $\mathcal{H}^0_A\mathcal{F} = 0$. Like in the proof of 4.3, let $A_k = A \cap$ $\cap\, S'_{k+q}$ and $A'_{k+1} = A_{k+1} \setminus S'_{k+q}$. Since $(X \setminus S'_{q+k})^- = X^-$ (otherwise V would contain an open set, hence $V = D$, $\mathfrak{J} = 0, \ldots$) and $S_{q+k}(\mathcal{F}) \setminus S'_{q+k} \subset V$, by lemma 6.4 it follows that $\Gamma(U \setminus S'_{k+q}, \mathfrak{J})^{l_1}\, N^q_{A_{k+1}}(U \setminus S'_{k+q}, \mathcal{F}) = 0$, where l_1 is a suitable integer independent on U.

Now, consider the commutative diagram with the exact horizontal row:

$$\begin{array}{ccccccc} H^{q-1}_{A'_{k+1}}(U, \mathcal{F}) & \xrightarrow{\delta} & H^q_{A_k}(U, \mathcal{F}) & \xrightarrow{\alpha} & H^q_{A_{k+1}}(U, \mathcal{F}) & \to & H^q_{A'_{k+1}}(U, \mathcal{F}) \\ & & & & & \searrow^{\beta} \quad \swarrow^{\sim} & \\ & & & & & H^q_{A_{k+1}}(U \setminus S'_{k+q}, \mathcal{F}) & \end{array}$$

Since β is continuous, we get $\Gamma(U, \mathfrak{J})^{l_1}\, N^q_{A_{k+1}}(U, \mathcal{F}) \subset \operatorname{Im}\alpha$. Increasing if necessary the integer l_1, we may assume, by 4.3, that

$$\Gamma(U, \mathfrak{J})^{l_1}\, H^{q-1}_{A'_{k+1}}(U, \mathcal{F}) = 0.$$

Like in the proof of lemma 4.3 we will check, by induction on k, the equality

$$\Gamma(U, \mathfrak{J})^l\, N^q_{A_k}(U, \mathcal{F}) = 0$$

for l sufficiently large. For $k = 0$, dim $A_0 \leqslant 0$ and the equality follows from lemma 6.4, since $S_{q-1}(\mathcal{F}) \subset V$ in some open neighbourhood of A_0.

Suppose now there is an integer $l_2 \geqslant 0$ such that $\Gamma(U, \mathfrak{I})^{l_2} N^q_{A_k}(U, \mathfrak{F}) = 0$ for any Stein open subset U of X. We are going to prove the equality

$$\Gamma(U, \mathfrak{I})^{l_2+2l_1} N^q_{A_{k+1}}(U, \mathfrak{F}) = 0$$

for U open set. So the induction will be complete and as for sufficiently large k we have $A_k = A$, the lemma follows easily.

We apply the previous lemma for $K = X^-$ and let $Y, f_1, \ldots, f_p$, m be the entities so obtained. We may assume that $\mathfrak{I}$ coincides on X with the sheaf generated by $f_1, \ldots, f_p$, since the annihilation by suitable powers of $\Gamma(U, \mathfrak{I})$ is equivalent to the annihilation by some powers of $\Gamma(U, (f_1, \ldots, f_p)\mathcal{O})$, U open set of X. For U Stein we have the isomorphisms

$$H^1_{A_k}(U, \mathfrak{F}) \simeq \Gamma(U \setminus A_k, \mathfrak{F})/\Gamma(U, \mathfrak{F}),\ H^q_{A_k}(U, \mathfrak{F}) \simeq H^{q-1}(U \setminus A_k, \mathfrak{F}),\quad q \geqslant 2.$$

Let $\mathfrak{U}$ be a Stein covering of $U \setminus A_k$ and $\mathcal{V}$ be a Stein covering of $U \setminus A_{k+1}$ which refines the restriction of $\mathfrak{U}$ to $U \setminus A_{k+1}$. Consider the map

$$\theta: Z^{q-1}(\mathfrak{U}, \mathfrak{F}) \oplus C^{q-2}(\mathcal{V}, \mathfrak{F}) \to Z^{q-1}(\mathcal{V}, \mathfrak{F}),$$

defined by

$$\theta(\xi \oplus \eta) = \xi \mid V + \delta\eta$$

where $C^{-1}(\mathcal{V}, \mathfrak{F}) = \Gamma(U, \mathfrak{F})$ and $C^{-1}(\mathcal{V}, \mathfrak{F}) \overset{\delta}{\to} Z^0(\mathcal{V}, \mathfrak{F})$ is the restriction map $\Gamma(U, \mathfrak{F}) \to \Gamma(U \setminus A_{k+1}, \mathfrak{F})$. Denote by $\bar{B}^{q-1}(\mathcal{V}, \mathfrak{F})$ the topological closure of $B^{q-1}(\mathcal{V}, \mathfrak{F})$ in $Z^{q-1}(\mathcal{V}, \mathfrak{F})$. Since $\Gamma(U, \mathfrak{I})^{l_1} N^q_{A_{k+1}}(U, \mathfrak{F}) \subset \operatorname{Im} \alpha$, we have $g\bar{B}^{q-1}(\mathcal{V}, \mathfrak{F}) \subset \operatorname{Im} \theta$, for any $g \in \Gamma(U, \mathfrak{I})^l$.

Let $g = f_{i_1} \ldots f_{i_{l_1}}$. Then, for any $x \in X$, g_x is a nonzerodivisor of $\mathfrak{F}_x$ and hence the map

$$C^{q-1}(\mathcal{V}, \mathfrak{F}) \to C^{q-1}(\mathcal{V}, \mathfrak{F})$$

defined by multiplication by g is a strict monomorphism. Consequently $g\bar{B}^{q-1}(\mathcal{V}, \mathfrak{F})$ is a closed subspace, hence FS. By Banach theorem the map induced by θ

$$\theta^{-1}(g\,\bar{B}^{q-1}(\mathcal{V}, \mathfrak{F})) \to g\,\bar{B}^{q-1}(\mathcal{V}, \mathfrak{F})$$

is open. If $\xi \in \bar{B}^{q-1}(\mathcal{V}, \mathfrak{F})$ and $\{\xi_\nu\}$ is a sequence in $B^{q-1}(\mathcal{V}, \mathfrak{F})$ converging to ξ, then we can find a sequence $\{\zeta_\nu \oplus \eta_\nu\}$ in $\theta^{-1}(g\,\bar{B}^{q-1}(\mathcal{V}, \mathfrak{F}))$ converging to some element $\zeta \oplus \eta$, such that

$$\theta(\zeta_\nu \oplus \eta_\nu) = g\xi_\nu \in B^{q-1}(\mathcal{V}, \mathfrak{F}) \text{ and } \theta(\zeta \oplus \eta) = g\xi.$$

Since $g\xi_\nu \in B^{q-1}(\mathcal{V}, \mathcal{F})$, it follows that $h\zeta_\nu \in B^{q-1}(\mathcal{U}, \mathcal{F})$ for all $h \in \Gamma(U, \mathfrak{J})^{l_1}$, as the class in $H^q_{A_k}(U, \mathcal{F})$ defined by ζ_ν lies in the image of δ. Hence $h\zeta \in \overline{B}^{q-1}(\mathcal{U}, \mathcal{F})$ and by induction hypothesis $h'h\zeta \in B^{q-1}(\mathcal{U}, \mathcal{F})$ for any $h' \in \Gamma(U, \mathfrak{J})^{l_2}$. Therefore we have $h'hg\xi \in B^{q-1}(\mathcal{V}, \mathcal{F})$. Since $f_1 \mid U, \ldots, f_p \mid U$ generate $\Gamma(U, \mathfrak{J})$ it follows that

$$\Gamma(U, \mathfrak{J})^{l_2+2l_1} N^q_{A_{k+1}}(U, \mathcal{F}) = 0 \ (U \text{ Stein open set}).$$

The proof of lemma 6.6 is completed.

Proof of theorem 6.1. In virtue of the isomorphisms

$$H^i_A(\Omega, \mathcal{F}) \xrightarrow{\sim} \Gamma(\Omega, \mathcal{H}^i_A \mathcal{F}), \qquad i \leqslant q$$

the map $N^i_A(\Omega, \mathcal{F}) \to \Gamma(\Omega, \mathcal{N}^i_A \mathcal{F})$, $i \leqslant q$ is injective. Then it is sufficient to show that $\mathcal{N}^i_A \mathcal{F} = 0$ for $i \leqslant q$; in this case the spaces $H^i_A(\Omega, \mathcal{F})$ are consequently separated and from the exact sequences

$$\ldots \to H^{i-1}(\Omega, \mathcal{F}) \to H^{i-1}(\Omega \setminus A, \mathcal{F}) \to H^i_A(\Omega, \mathcal{F}) \to \ldots$$

they result even FS. The problem is local in nature. By a suitable embedding in a numerical space (the topologies on $H^\bullet_A(U, \mathcal{F})$ agree with such embeddings), we may assume X a relatively compact open subset of a domain $D \subset \mathbb{C}^n$, such that the compact X^- is holomorphically convex, $A \subset D$ is a closed analytic subset and $\mathcal{F} \in \mathbf{Coh}(D)$. By means of the finiteness theorem,

$$\dim A \cap \overline{S}_{k+q}(\mathcal{F} \mid D \setminus A) \leqslant k, \text{ for all } k.$$

Let $\mathfrak{J} = \mathfrak{J}(A)$. We prove by induction on q the following assertion: for any Stein open subset U of X, the space $H^i_A(U, \mathcal{F})$, $i \leqslant q$, is separated, and any $\Gamma(U, \mathcal{O})$-homomorphism $\Gamma(U, \mathcal{O})^r \to H^q_A(U, \mathcal{F})$ whose image is annihilated by some power of $\Gamma(U, \mathfrak{J})$ has closed image. In this way, the proof of the theorem will be concluded.

For $q = 0$ this assertion is trivial. Let $q \geqslant 1$, and prove the general induction step $q - 1 \mapsto q$. Let U be a Stein open subset of X. By lemmas 4.3 and 6.6, there exists an integer $l \geqslant 0$ (dependent on X) such that $\Gamma(U, \mathfrak{J})^l H^i_A(U, \mathcal{F}) = 0$ for $i < q$ and $\Gamma(U, \mathfrak{J})^l N^q_A(U, \mathcal{F}) = 0$. For the problem we are interested in, we may assume that $\mathcal{H}^0_A \mathcal{F} = 0$ when replacing $\mathcal{F}$ by $\mathcal{F}/\mathcal{H}^0_A \mathcal{F}$. In accordance with 6.5, we can find a neighbourhood Y of X^- and $g \in \Gamma(Y, \mathfrak{J})^l$ such that g_x is a nonzerodivisor of $\mathcal{F}_x$ for any $x \in Y$. We thus get an exact sequence

$$(*) \qquad H^{q-1}_A(U, \mathcal{F}) \xrightarrow{\alpha} H^{q-1}_A(U, \mathcal{F}/g\mathcal{F}) \xrightarrow{\delta} H^q_A(U, \mathcal{F}) \xrightarrow{g^*} H^q_A(U, \mathcal{F}).$$

By induction hypothesis $H^{q-1}_A(U, \mathcal{F}/g\mathcal{F})$ is an FS space. Since $\mathcal{H}^{q-1}_A \mathcal{F}$ is coherent, we have a sheaf epimorphism on X

$$\psi: \mathcal{O}^s \to \mathcal{H}^{q-1}_A \mathcal{F}.$$

Since $H_A^{q-1}(U, \mathfrak{F}) = \Gamma(U, \mathcal{H}_A^{q-1}\mathfrak{F})$, ψ induces a $\Gamma(U, \mathcal{O})$-epimorphism

$$\psi^*: \Gamma(U, \mathcal{O})^s \to H_A^{q-1}(U, \mathfrak{F}).$$

Since $\operatorname{Im} \alpha\psi^* = \operatorname{Im}\alpha$ is annihilated by $\Gamma(U, \mathfrak{J})^l$, by induction hypothesis $\operatorname{Im}\alpha$ is closed. We get $N_A^q(U, \mathfrak{F}) \subset \operatorname{Im}\delta$ because $g^*N_A^q(U, \mathfrak{F}) = 0$.

By considering Čech cohomology, as in the proof of lemma 6.6, we deduce that the map

$$\delta^{-1}(N_A^q(U, \mathfrak{F})) \to N_A^q(U, \mathfrak{F})$$

is open. Hence $N_A^q(U, \mathfrak{F})$ is isomorphic to the FS space $\delta^{-1}(N_A^q(U, \mathfrak{F}))/\operatorname{Im}\alpha$. It follows that $N_A^q(U, \mathfrak{F}) = 0$; hence $H_A^q(U, \mathfrak{F})$ will be separated. In virtue of induction hypothesis the separation of the spaces $H_A^i(U, \mathfrak{F})$ for $i < q$ follows.

Let now

$$\varphi: \Gamma(U, \mathcal{O})^r \to H_A^q(U, \mathfrak{F})$$

be a $\Gamma(U, \mathcal{O})$-morphism such that $\operatorname{Im}\varphi$ is annihilated by some power of $\Gamma(U, \mathfrak{J})$. We may assume without loss of generality that $\Gamma(U, \mathfrak{J})^l \operatorname{Im}\varphi = 0$, where l is the integer which has appeared previously. From (*) we obtain $\operatorname{Im}\varphi \subset \operatorname{Im}\delta$. There exists a $\Gamma(U, \mathcal{O})$-morphism

$$\tilde{\varphi}: \Gamma(U, \mathcal{O})^r \to H_A^{q-1}(U, \mathfrak{F}/g\mathfrak{F}),$$

such that $\varphi = \delta\tilde{\varphi}$. Define now the morphism

$$\gamma: \Gamma(U, \mathcal{O})^s \oplus \Gamma(U, \mathcal{O})^r \to H_A^{q-1}(U, \mathfrak{F}/g\mathfrak{F}),$$

by the formula

$$\gamma(a \oplus b) = \alpha\psi^*(a) + \tilde{\varphi}(b).$$

Since $\Gamma(U, \mathfrak{J})^l\, \delta(\operatorname{Im}\tilde{\varphi}) = 0$, $\operatorname{Im}\gamma$ is annihilated by some power of $\Gamma(U, \mathfrak{J})$ and, in accordance with the induction hypothesis, is closed. The surjective map $H_A^{q-1}(U, \mathfrak{F}/g\mathfrak{F}) \xrightarrow{\delta} \operatorname{Im}\delta$ of FS spaces ($\operatorname{Im}\delta = \operatorname{Ker} g^*$) is strict and $\delta(\operatorname{Im}\gamma) = \operatorname{Im}\varphi$. Since $\delta^{-1}(\operatorname{Im}\varphi) = \operatorname{Im}\gamma$ and $\operatorname{Im}\gamma$ is closed, it follows that $\operatorname{Im}\varphi$ is closed. Thereby the induction is complete and the proof of the theorem is over.

COROLLARY 6.7. *The canonical isomorphisms*

$$H_A^i(\Omega, \mathfrak{F}) \to \Gamma(\Omega, \mathcal{H}_A^i\mathfrak{F})$$

are homeomorphisms for $i < q$, where $\Gamma(\Omega, \mathcal{H}_A^i\mathfrak{F})$ is endowed with the natural FS structure given by the coherent sheaf $\mathcal{H}_A^i\mathfrak{F}$.

Proof. By the open mapping theorem, it is sufficient to show that the inverse of the isomorphism from the statement

$$\lambda^i: \Gamma(\Omega, \mathcal{H}_A^i\mathfrak{F}) \to H_A^i(\Omega, \mathfrak{F})$$

is continuous. According to the closed graph theorem, we have only to prove that λ^i has a closed graph. Let $s_\nu \to s$ in $\Gamma(\Omega, \mathcal{H}_A^i\mathfrak{F})$ and let $\xi_\nu \to \xi$ in $H_A^i(\Omega, \mathfrak{F})$ such that $\lambda^i(s_\nu) = \xi_\nu$. If $\Omega' \subset\subset \Omega$ is a Stein open set, then $\mathcal{H}_A^i\mathfrak{F} \mid \Omega'$ is generated by a finite number of sections $t_1, \ldots, t_m \in \Gamma(\Omega', \mathcal{H}^i_A\mathcal{F})$. Since $s_\nu \mid \Omega' \to s \mid \Omega'$, there are holomorphic functions $f^{(\nu)}_\mu$ on Ω' converging to some holomorphic functions f_μ on Ω' such that

$$s_\nu \mid \Omega' = \sum_{\mu=1}^{m} f^{(\nu)}_\mu t_\mu$$

(we apply the Banach theorem for the map $\Gamma(\Omega', \mathcal{O}^m) \to \Gamma(\Omega', \mathcal{H}^i_A \mathcal{F})$).

Let $\zeta_\mu \in H^i_A(\Omega', \mathcal{F})$ be the image of t_μ under the canonical isomorphism $\lambda^i_{\Omega'}: \Gamma(\Omega', \mathcal{H}^i_A\mathcal{F}) \to H^i_A(\Omega', \mathcal{F})$. Then $\xi_\nu \mid \Omega' = \lambda^i_{\Omega'}(s_\nu \mid \Omega') = \sum_{\mu=1}^{m} f^{(\nu)}_\mu \zeta_\mu$, where $\xi_\nu \mid \Omega'$ is the image of ξ_ν under the map $H^i_A(\Omega, \mathcal{F}) \to H^i_A(\Omega', \mathcal{F})$. We obtain $\xi \mid \Omega' = \sum_{\mu=1}^{m} f_\mu \zeta_\mu = \lambda^i_{\Omega'}(s \mid \Omega') = \lambda^i(s) \mid \Omega'$. Since $H^i_A(\Omega, \mathcal{F}) \simeq \Gamma(\Omega, \mathcal{H}^i_A \mathcal{F})$ for $i < q$ and Ω' is arbitrary, we conclude that $\xi = \lambda^i(s)$ and hence the graph of λ^i is closed for $i<q$.

COROLLARY 6.8. *Let X be a complex space, $A \subset X$ a closed analytic subset and $\mathcal{F} \in$ **Coh** (X). For any Stein open subset Ω of X, $H^1_A(\Omega, \mathcal{F})$ is an FS space.*

This follows by the theorem and 3.2.

Bibliographical indications

This chapter follows *Gap sheaves and extension of coherent analytic subsheaves* of Y. T. Siu and G. Trautmann.

The proof given above for theorem 2.1 is that from [86]; other proofs can be found in [15], [84]. Another proof for corollary 2.2 is in [15] as well as in [50]. Scheja's paper [71] underlines the role of the defect sets $D_m(\mathcal{F})$ in extending the analytic entities defined beyond an analytic set.

The proofs of the vanishing theorem 3.6, the finiteness theorem 4.1 and the separation theorem 5.1, which are the main results of the chapter, are those from [84]. Corollary 3.10 is proved in a different way (by using duality) in [8], [40] (cf. I. 3.3). The position of corollaries 3.12 and 3.13 in the framework of some results of Abhyankar, Oka, Rothstein, Thimm is indicated in [70]. Other proofs for the finiteness theorem 4.1 can be found in [83], [92]. A straightforward proof for corollary 4.4 is in [91]. Corollary 4.5 is proved by means of duality in [8] (cf. I. 3.3).

The results of § 5, except the given application, may be found in the book [84]. The study of relative gap sheaves, the connection with Lasker-Noether decompositions, the position of these with respect to some results of Thimm, as well as other results in this context can be found in the same book.

The complex $\overline{\mathcal{K}}^\bullet$ is built in § 5 on the analogy of the algebraic case [39] and of the dualizing complex of Ramis and Ruget [61] (cf. ch. VII). Recently, F. Fouché has shown that $\overline{\mathcal{K}}^\bullet$ is in fact a dualizing complex!

The hypothesis from the statement of the separation theorem 6.1 makes the problem be of local nature and thus, one can use the sheaves $\mathcal{H}^\bullet_A$. This hypothesis is however sufficient for the separation of the invariants $H^\bullet_A$, but not at all necessary (cf. I. 2.21).

Chapter III

Proper morphisms of complex spaces

Introduction

Let X be a compact Riemann surface and let $D = \Sigma n_i P_i$ be a divisor on X. Consider the complex vectorial space

$$L(D) = \{\varphi \in \mathcal{M}(X) \mid \varphi = 0 \text{ or } (\varphi) + D \geqslant 0\}$$

($\mathcal{M}(X)$ stands for the set of the meromorphic functions on X and (φ) is the divisor associated to φ). By a classical result, $L(D)$ is finite-dimensional.

Let now X be a compact complex manifold and V a holomorphic fiber bundle on X. Then the space of all global sections of V is also of complex finite dimension.

These two results are particular cases of the following finiteness theorem of Cartan and Serre [17]: "if X is a compact complex space and $\mathcal{F}$ a coherent analytic sheaf on X, then the complex vectorial spaces $H^q(X, \mathcal{F})$ are finite-dimensional". Our aim in this chapter is to present the extension of this result to the relative case of morphisms, and to give some applications.

If $f: X \to Y$ is a morphism of complex spaces, then the natural hypothesis of compactness required is the following: the inverse image of a compact set under f is a compact set. Such a morphism is called proper. The relativization for morphisms of the cohomology groups is given by the generalized direct images $R^q f_*$. The main result which is obtained is the following finiteness theorem of Grauert: "if $f: X \to Y$ is a proper morphism of complex spaces and $\mathcal{F} \in \mathbf{Coh}(X)$, then the sheaves $R^q f_*(\mathcal{F})$ are coherent". If the space Y is a point, then one refinds the theorem of Cartan-Serre.

Among the direct applications we recall the Remmert's projection theorem and the Stein decomposition of a morphism. The next problem is to give informations and formulae for computing the coherent sheaves $R^q f_*(\mathcal{F})$. If y is a point of Y, then the comparison theorem 3.1 gives a formula for computing the completion of the $\mathcal{O}_y$-module of finite type $R^q f_*(\mathcal{F})_y$ in the $\mathfrak{m}_y$-adic topology.

By making use of the comparison theorem one gets necessary and sufficient conditions for the sheaves $R^q f_*(\mathcal{F})$ to agree with changes of basis $Y' \to Y$ (theorem 3.4 of base change). In particular one can obtain conditions for the natural maps

$$R^q f_*(\mathcal{F})_y/\mathfrak{m}_y R^q f_*(\mathcal{F})_y \to H^q(f^{-1}(y), \mathcal{F}/\hat{\mathfrak{m}}_y\mathcal{F})$$

to be bijective, conditions for the sheaves $R^q f_*(\mathcal{F})$ to be null, locally free...

§ 4 deals with the functions

$$y \mapsto \dim H^q(f^{-1}(y), \mathcal{F}/\hat{\mathfrak{m}}_y\mathcal{F})$$

and

$$y \mapsto \sum_q (-1)^q \dim H^q(f^{-1}(y), \mathcal{F}/\hat{\mathfrak{m}}_y\mathcal{F}),$$

and establishes results of semicontinuity and continuity for them (Grauert's theorem 4.12); such results extend those obtained by Kodaira and Spencer [49].

§ 1. Preliminaries

(**a**) A continuous map between two separated topological spaces is called *proper* if the inverse image of any compact subset is compact.

LEMMA 1.1. *Any proper map between locally compact spaces is closed.*

Proof. Let $f: X \to Y$ be such a map. If $\hat{X}$ and $\hat{Y}$ are Alexandroff compactifications of X and Y, then f extends to a continuous map $\hat{f}: \hat{X} \to \hat{Y}$. Let T be a closed subset of X. If $\hat{T}$ is its closure in $\hat{X}$, then $f(T) = \hat{f}(\hat{T}) \cap Y$. Since $\hat{T}$ is compact, the lemma follows.

COROLLARY 1.2. *Let $f: X \to Y$ be a proper map. If L is a closed subset of Y and $\mathcal{V}_L$ is a fundamental system of neighbourhoods for it, then $f^{-1}(\mathcal{V}_L)$ is a fundamental system of neighbourhoods for $f^{-1}(L)$.*

Proof. Let U be an open subset which contains $f^{-1}(L)$. Then $Y \setminus f(X \setminus U)$ is an open subset containing L, hence there exists $V \in \mathcal{V}_L$ such that $V \subset Y \setminus f(X \setminus U)$. Obviously, $f^{-1}(V) \subset f^{-1}(Y \setminus f(X \setminus U)) \subset U$.

Let $f: X \to Y$ be a continuous map of topological spaces and $\mathcal{F}$ a sheaf of abelian groups on X. Denote by $R^qf_*(\mathcal{F})$ the sheaf associated to the presheaf on Y

$$V \mapsto H^q(f^{-1}(V), \mathcal{F}),$$

the restrictions being naturally defined.

The sheaves $R^qf_*(\mathcal{F})$ are called *the generalized direct images of* $\mathcal{F}$. In particular, $R^0f_*(\mathcal{F})$ equals the direct image $f_*(\mathcal{F})$. Morphisms like $R^qf_*(\mathcal{F}) \to R^qf_*(\mathcal{G})$ correspond naturally to any morphism $\mathcal{F} \to \mathcal{G}$ in **Ab** (X) and these correspondences are functorial.

If

$$0 \to \mathcal{F}' \to \mathcal{F} \to \mathcal{F}'' \to 0$$

is an exact sequence in **Ab** (X), then one obtains the exact sequence

$$0 \to f_*(\mathcal{F}') \to f_*(\mathcal{F}) \to f_*(\mathcal{F}'') \to R^1f_*(\mathcal{F}') \to R^1f_*(\mathcal{F}) \to \ \dots$$

In fact one can easily prove that $R^\bullet f_*$ are the right derived functors of the functor $f_*\colon \mathbf{Ab}(X) \to \mathbf{Ab}(Y)$.

If V is an open subset of Y and $f^V\colon f^{-1}(V) \to V$ is the restriction of f, then we obviously get isomorphisms

$$R^\bullet f_*(\mathcal{F}) \mid V \simeq R^\bullet f_*^V(\mathcal{F} \mid f^{-1}(V)).$$

LEMMA 1.3. *Let $f\colon X \to Y$ be a proper map between locally compact spaces and $\mathcal{F} \in \mathbf{Ab}(X)$. For any point $y \in Y$ and for all q,*

$$R^q f_*(\mathcal{F})_y \simeq H^q(f^{-1}(y), \mathcal{F}).$$

The proof follows from 1.2 and ([26], Ch. II, 4.11.1).

We now consider a morphism of ringed spaces

$$f\colon (X, \mathcal{O}_X) \to (Y, \mathcal{O}_Y).$$

If $\mathcal{F}$ is an $\mathcal{O}_X$-module, then the sheaves $R^q f_*(\mathcal{F})$ display naturally a structure o $\mathcal{O}_Y$-modules; in particular, if f is a morphism of complex spaces and $\mathcal{F}$ an analytic sheaf on X, then $R^q f_*(\mathcal{F})$ are analytic sheaves on Y.

Recall that $\mathcal{F}$ is called *f-flat* or *flat with respect to* Y if the $\mathcal{O}_{f(x)}$-modules $\mathcal{F}_x$ are flat for all $x \in X$ (generalities on flatness can be found in chapter V, § 1).

Suppose X, Y are paracompact, $\mathcal{F}$ is an $\mathcal{O}_X$-module and $\mathcal{M}$ an $\mathcal{O}_Y$-module. For any open subset V of Y, we get natural maps by using Čech cohomology,

$$H^q(f^{-1}(V), \mathcal{F}) \otimes_{\Gamma(V, \mathcal{O}_Y)} \Gamma(V, \mathcal{M}) \to H^q(f^{-1}(V), \mathcal{F} \otimes_{\mathcal{O}_X} f^*(\mathcal{M})).$$

Whence, by passing to associated sheaves, we derive morphisms

$$R^q f_*(\mathcal{F}) \otimes_{\mathcal{O}_Y} \mathcal{M} \to R^q f_*(\mathcal{F} \otimes_{\mathcal{O}_X} f^*(\mathcal{M})),$$

which are functorial in $\mathcal{F}$ and $\mathcal{M}$.

Consider now a commutative diagram of ringed spaces

$$\begin{array}{ccc} X & \xrightarrow{f} & Y \\ {\scriptstyle g'}\uparrow & & \uparrow{\scriptstyle g} \\ X' & \xrightarrow{f'} & Y' \end{array}$$

and an $\mathcal{O}_X$-module $\mathcal{F}$. For any open subset V of Y we have natural morphisms $H^q(f^{-1}(V), \mathcal{F}) \to H^q(g'^{-1}f^{-1}(V), g'^*(\mathcal{F})) = H^q(f'^{-1}g^{-1}(V), g'^*(\mathcal{F}))$.

From these and the morphisms $H^q(f'^{-1}\, g^{-1}(V), g'^*(\mathcal{F})) \to \Gamma(g^{-1}(V), R^q f'_*(g'^*(\mathcal{F})))$, we deduce the sheaf morphisms

$$R^q f_*(\mathcal{F}) \to g_* \,(R^q f'_*(g'^*(\mathcal{F}))),$$

hence by adjunction the morphisms

$$g^* \,(R^q f_*(\mathcal{F})) \to R^q f'_*(g'^*(\mathcal{F})).$$

It is easy to check their functorial character.

(b) Let $(Y, \mathcal{O}_Y)$ be a complex space and $\mathcal{A}$ a coherent $\mathcal{O}_Y$-algebra. In [15] one proved the existence of a complex space over Y,

$$\text{Specan } \mathcal{A} \xrightarrow{q} Y,$$

which is called the *analytic spectrum of* $\mathcal{A}$ and a morphism

$$\mathcal{A} \to q_* \,(\mathcal{O}_{\text{Specan } \mathcal{A}})$$

such that the following universal property holds: for any complex space X over Y, $X \xrightarrow{p} Y$, the natural map

$$\text{Hom}_Y (X, \text{Specan } \mathcal{A}) \to \text{Hom}_{\mathcal{O}_Y\text{-alg}}(\mathcal{A}, p_* \,(\mathcal{O}_X))$$

is bijective. One can also show that the structural morphism q is finite and that the morphism $\mathcal{A} \to q_*(\mathcal{O}_{\text{Specan } \mathcal{A}})$ is an isomorphism.

If $\mathcal{A} \to \mathcal{A}'$ is a morphism of coherent $\mathcal{O}_Y$-algebras, then we obtain a Y-morphism Specan $\mathcal{A}' \to$ Specan $\mathcal{A}$. This correspondence is functorial; moreover, if the morphism $\mathcal{A} \to \mathcal{A}'$ is an epimorphism, then the associated morphism is a closed immersion.

We also recall the property of base change: for any morphism $Y' \xrightarrow{g} Y$ there exists a canonical isomorphism

$$\text{Specan } g^*(\mathcal{A}) \simeq (\text{Specan } \mathcal{A}) \times_Y Y'.$$

(c) Let A be a ring and M an A-module. On the direct sum $A \oplus M$ one can consider the multiplication

$$(a_1, m_1) \,.\, (a_2, m_2) = (a_1 a_2, a_1 m_2 + a_2 m_1).$$

Thus we obtain a ring which is called *the Nagata ring associated to M*. $A \oplus 0$ becomes a subring of $A \oplus M$ which is isomorphic to A, and $0 \oplus M$ is an ideal (with null square) in $A \oplus M$ which is isomorphic as an $A \oplus 0$-module to M.

This construction has functorial properties and it can be easily extended to sheaves.

PROPOSITION 1.4. [22]. *Let* $(X, \mathcal{A})$ *be a complex space and* $\mathcal{M} \in \mathbf{Coh}(\mathcal{A})$. *Then the ringed space* $(X, \mathcal{A} \oplus \mathcal{M})$ *is a complex space.*

Proof. The problem is local in nature, so we can assume X an analytic subset of the open unit polydisc centred in origin P^n in a numerical space $\mathbb{C}^n$ and furthermore, we can assume $\mathcal{A} = \mathcal{O}/\mathcal{J} \mid X$, where $\mathcal{O}$ is the structural sheaf of P^n and $\mathcal{J}$ an ideal-sheaf such that $\mathrm{Supp}\,(\mathcal{O}/\mathcal{J}) = X$. We can suppose in addition that $\mathcal{M} = \mathcal{A}^k/\mathcal{R}$, where $\mathcal{R}$ is a coherent submodule of $\mathcal{A}^k$.

We first prove that the ringed space $(X, \mathcal{A} \oplus \mathcal{A}^k)$ is a complex space. To do this let $P^{n+k} = P^n \times P^k$ be the open unit polydisc centred in origin in $\mathbb{C}^{n+k}$ and denote by $x_1, \ldots, x_n; y_1, \ldots, y_k$ the coordinates. Let $\mathcal{O}'$ be the sheaf of germs of holomorphic functions in P^{n+k} and $\mathcal{J}'$ be the ideal of $\mathcal{O}'$ generated by $\mathcal{J}$ and by the functions $y_\alpha \cdot y_\beta$, $1 \leqslant \alpha, \beta \leqslant k$. Denote $X' = \mathrm{Supp}\,(\mathcal{O}'/\mathcal{J}')$ and $\mathcal{A}' = \mathcal{O}'/\mathcal{J}' \mid X'$. Obviously, $X' = X \times 0$. Define a morphism

$$(\varphi, {}^*\varphi)\colon (X, \mathcal{A} \oplus \mathcal{A}^k) \to (X', \mathcal{A}')$$

as follows: $\varphi(a) = (a, 0)$ for $a \in X$. Further on, any element $f \in \mathcal{A}'_{(a,\,0)}$ can be represented by a series in $\mathcal{O}'_{(a,\,0)}$ of the form

$$\sum_{i_1,\ldots,\,i_n=0}^{\infty} c_{i_1\cdots i_n}(x_1 - a_1)^{i_1} \ldots (x_n - a_n)^{i_n}$$

$$+\sum_{\alpha=1}^{k} \sum_{i_1,\ldots,\,i_n=0}^{\infty} c^{(\alpha)}_{i_1,\ldots,i_n}(x_1 - a_1)^{i_1} \ldots (x_n - a_n)^{i_n}\, y_\alpha.$$

Denote by g and respectively g_α, the elements of $\mathcal{A}_a$ defined by the series $\sum_{i_1,\ldots,\,i_n=0}^{\infty} c_{i_1\ldots i_n}(x_1 - a_1)^{i_1} \ldots (x_n - a_n)^{i_n} \in \mathcal{O}_a$, respectively $\sum_{i_1,\ldots,\,i_n=0}^{\infty} c^{(\alpha)}_{i_1,\ldots,i_n}$ $(x_1 - a_1)^{i_1} \ldots (x_n - a_n)^{i_n} \in \mathcal{O}_a$. Then we put ${}^*\varphi_a(f) = (g; g_1, \ldots, g_k) \in \mathcal{A}_a \oplus \mathcal{A}^k_a$. It is not difficult to check that an isomorphism

$$(\varphi, {}^*\varphi)\colon (X, \mathcal{A} \oplus \mathcal{A}^k) \to (X', \mathcal{A}')$$

of ringed spaces over the complex field is so obtained, and thus $(X, \mathcal{A} \oplus \mathcal{A}^k)$ is obviously a complex space.

The sheaf $\mathcal{R}$ determines a coherent ideal $\mathcal{R}^* = 0 \oplus \mathcal{R} \subset \mathcal{A} \oplus \mathcal{A}^k$. The set of zeros of $\mathcal{R}^*$ coincides with X and therefore $(X, \mathcal{A} \oplus \mathcal{A}^k/\mathcal{R}^*)$ is a complex space. Since

$$\mathcal{A} \oplus \mathcal{A}^k/\mathcal{R}^* \simeq \mathcal{A} \oplus (\mathcal{A}^k/\mathcal{R}) = \mathcal{A} \oplus \mathcal{M},$$

the proof is concluded.

We also remark that there exists a morphism of complex spaces

$$(X, \mathcal{A} \oplus \mathcal{M}) \to (X, \mathcal{A})$$

obtained from the identical map $X \to X$ and the inclusion $\mathcal{A} \subset \mathcal{A} \oplus \mathcal{M}$.

(d) Let $(X, \mathcal{C})$ be a ringed space in local rings. For a point $x \in X$, denote as usual the maximal ideal of $\mathcal{O}_x$ by $\mathfrak{m}_x$ and by $k(x) = \mathcal{O}_x/\mathfrak{m}_x$ the residual field in x. If $\mathcal{F}$ is an $\mathcal{O}$-module, then we set $\mathcal{F}(x) = \mathcal{F}_x/\mathfrak{m}_x \mathcal{F}_x \simeq \mathcal{F}_x \otimes_{\mathcal{O}_x} k(x)$. For a morphism $\varphi: \mathcal{F} \to \mathcal{G}$ of $\mathcal{O}$-modules, denote by $\varphi(x): \mathcal{F}(x) \to \mathcal{G}(x)$ the induced morphism. If U is an open subset of X and $f \in \Gamma(U, \mathcal{F})$, then one denotes by $f(x)$ the image of f under the composite map $\Gamma(U, \mathcal{F}) \to \mathcal{F}_x \to \mathcal{F}(x)$.

If $\mathcal{F}$ is an $\mathcal{O}$-module of finite type, then for any point $x \in X$, $\mathcal{F}(x)$ is a vectorial $k(x)$-space of finite dimension; according to the Nakayama lemma, its dimension equals the minimal number of generators of $\mathcal{F}_x$ over $\mathcal{O}_x$.

LEMMA 1.5. (i) *For any $\mathcal{O}$-module $\mathcal{F}$ of finite type, the function*

$$x \mapsto \dim_{k(x)} \mathcal{F}(x)$$

is upper semicontinuous.

(ii) *For any morphism of $\mathcal{O}$-modules $d: \mathcal{C}^p \to \mathcal{C}^q$, the function*

$$x \mapsto \mathrm{rg}_{k(x)}\, d(x)$$

is lower semicontinuous.

Proof. (*i*) Let $s_1^x, \ldots, s_n^x$ be a minimal system of generators of the $\mathcal{O}_x$-module $\mathcal{F}_x$. There exist a neighbourhood U of x and sections $s_1, \ldots, s_n \in \Gamma(U, \mathcal{F})$ such that $(s_1)_x = s_1^x, \ldots, (s_n)_x = s_n^x$. The conclusion follows if we remark that, by shrinking eventually U, the germs $(s_1)_y, \ldots, (s_n)_y$ generate the $\mathcal{O}_y$-module $\mathcal{F}_y$ for any $y \in U$.

(*ii*) The exact sequence $\mathcal{O}^p \xrightarrow{d} \mathcal{O}^q \to \mathrm{Coker}\ d \to 0$ yields the exact sequence

$$\mathcal{O}^p(x) \xrightarrow{d(x)} \mathcal{O}^q(x) \to (\mathrm{Coker}\ d)(x) \to 0$$

and apply (*i*).

We recall that a ringed space in local rings $(X, \mathcal{O})$ is said to be *reduced* if, for any open subset U of X and any section $f \in \Gamma(U, \mathcal{O})$, the relations $f(x) = 0$ for all $x \in U$, imply $f = 0$.

LEMMA 1.6. *Let $(X, \mathcal{O})$ be a reduced ringed space in local rings.*
(i) *If $\mathcal{F}$ is an $\mathcal{O}$-module of finite type such that the function*

$$x \mapsto \dim_{k(x)} \mathcal{F}(x)$$

is locally constant, then $\mathcal{F}$ is locally free.
(ii) *If $d\colon \mathcal{O}^p \to \mathcal{O}^q$ is an $\mathcal{O}$-morphism such that the function*

$$x \mapsto \mathrm{rg}_{k(x)}\, d(x)$$

is locally constant, then Coker d is locally free.

Proof. (i) Let $x \in X$ and $n = \dim_{k(x)} \mathcal{F}(x)$. There exist a neighbourhood U of x and an epimorphism $\varphi\colon \mathcal{O}^n \mid U \to \mathcal{F} \mid U$. By shrinking eventually U, we can assume $n = \dim_{k(y)} \mathcal{F}(y)$ for any point $y \in U$. We will show that φ is injective. Let V be an open subset of U and $f = (f_1, \ldots, f_n) \in \Gamma(V, \mathrm{Ker}\, \varphi)$. For any $y \in V$ we have $f_i(y) = 0$, $1 \leqslant i \leqslant n$; for otherwise the minimal number of generators of $\mathcal{F}_y$ would be smaller than n. Hence $f = 0$.

(ii) By means of the exact sequences $\mathcal{O}^p(x) \to \mathcal{O}^q(x) \to \mathrm{Coker}\, d(x) \to 0$ we deduce that the function $x \mapsto \mathrm{rg}_{k(x)}$ Coker $d(x)$ is locally constant.

PROPOSITION 1.7. *Let $(X, \mathcal{O})$ be a ringed space in local rings and $\mathcal{L}^\bullet$ a complex of free $\mathcal{O}$-modules of finite rank. Then*
(i) *For any q, the function*

$$x \mapsto \dim_{k(x)} H^q(\mathcal{L}^\bullet(x))$$

is upper semicontinuous.
(ii) *If X is reduced and if the function*

$$x \mapsto \dim_{k(x)} H^q(\mathcal{L}^\bullet(x))$$

is locally constant for a fixed integer q, then Coker d^{q-1} and Coker d^q are locally free sheaves.
(iii) *If $\mathcal{L}^\bullet$ is bounded, then the function*

$$x \mapsto \sum_q (-1)^q \dim_{k(x)} H^q(\mathcal{L}^\bullet(x))$$

is locally constant.

Proof. Denote by r_q the rank of $\mathcal{L}^q$.

(i) We have $\dim H^q(\mathcal{L}^\bullet(x)) = \dim(\mathrm{Ker}\, d^q(x)) - \dim(\mathrm{Im}\, d^{q-1}(x)) = r_q - \mathrm{rg}\, d^q(x) - \mathrm{rg}\, d^{q-1}(x)$ and apply 1.5.

(ii) We have $\dim H^q(\mathcal{L}^\bullet(x)) = r_q - \mathrm{rg}\, d^q(x) - \mathrm{rg}\, d^{q-1}(x)$. Since the functions $x \mapsto \mathrm{rg}\, d^{q-1}(x)$, $x \mapsto \mathrm{rg}\, d^q(x)$ are lower semicontinuous, we derive by the hypothesis that they are locally constant. The conclusion follows from 1.6.

(iii) Clearly, $\sum_q (-1)^q \dim H^q(\mathfrak{L}^\bullet(x)) = \sum_q (-1)^q \dim \mathfrak{L}^q(x) = \sum_q (-1)^q r^q$.

(e) Let $u: A^\bullet \to B^\bullet$ be a morphism between two complexes of abelian groups (or modules over a ring, or sheaves...). The *cone of the map u* is by definition the complex $C^\bullet(= C^\bullet(u))$ built as follows: $C^n = B^n \oplus A^{n+1}$, and the differentials $C^n \to C^{n+1}$ are given by formula $(b, a) \mapsto (d_{B^\bullet}(b) + u(a), -d_{A^\bullet}(a))$. The natural inclusions and projections define an exact sequence of complexes

$$0 \to B^\bullet \to C^\bullet \to A^\bullet[1] \to 0,$$

where $A^\bullet[1]$ is the complex obtained by translating the complex $A^\bullet$ one step to the left, together with the change in the sign of the differentials. One gets the long exact sequence

$$\ldots \to H^q(B^\bullet) \to H^q(C^\bullet) \to H^{q+1}(A^\bullet) \to H^{q+1}(B^\bullet) \to \ldots$$

It can be easily seen that the boundary morphism $H^q(A^\bullet[1]) = H^{q+1}(A^\bullet) \to H^{q+1}(B^\bullet)$ coincides with the morphism $H^{q+1}(u)$ induced by u.

We recall that the morphism u is said to be a *quasi-isomorphism* if all morphisms $H^q(u): H^q(A^\bullet) \to H^q(B^\bullet)$ are isomorphisms. The morphism u is called *n-quasi-isomorphism* if the morphisms $H^q(u): H^q(A^\bullet) \to H^q(B^\bullet)$ are isomorphisms for $q > n$ and $H^n(u): H^n(A^\bullet) \to H^n(B^\bullet)$ is an epimorphism.

From the previous assertions one obtains

LEMMA 1.8. (*i*) *Let n be an integer. The morphism u is an n-quasi-isomorphism if and only if* $H^q(C^\bullet) = 0$ *for* $q \geqslant n$.

(*ii*) *The morphism u is a quasi-isomorphism if and only if* $H^q(C^\bullet) = 0$ *for any q (i.e. if and only if the cone of u is acyclic).*

LEMMA 1.9. *Let* $A^\bullet \xrightarrow{u} B^\bullet \xrightarrow{\varphi} M^\bullet$ *be morphisms of complexes such that* φ *is a quasi-isomorphism. Under these assumptions, the canonical morphism* $C^\bullet(u) \to C^\bullet(\varphi u)$ *is a quasi-isomorphism.*

Proof. It follows from the canonical commutative diagram

$$\begin{array}{ccccccccccc}
\ldots \to & H^q(A^\bullet) & \to & H^q(B^\bullet) & \to & H^q(C^\bullet(u)) & \to & H^{q+1}(A^\bullet) & \to & H^{q+1}(B^\bullet) & \to \ldots \\
 & \downarrow \text{id} & & \downarrow \wr & & \downarrow & & \downarrow \text{id} & & \downarrow \wr & \\
\ldots \to & H^q(A^\bullet) & \to & H^q(M^\bullet) & \to & H^q(C^\bullet(\varphi u)) & \to & H^{q+1}(A^\bullet) & \to & H^{q+1}(M^\bullet) & \to \ldots
\end{array}$$

and from the lemma of the five.

COROLLARY 1.10. *For any q the sum of the maps of the diagram*

$$\begin{array}{ccc}
 & & Z^q(C^\bullet(u)) \\
 & & \downarrow \\
C^{q-1}(\varphi u) & \xrightarrow{d_{C^\bullet(\varphi u)}} & Z^q(C^\bullet(\varphi u))
\end{array}$$

is a surjective map (**Z** *means as usual the group of cycles*).

(f) Proposition 1.11. *Let A be a ring and*

$$0 \to M'_q \overset{\varphi_q}{\to} M_q \overset{\psi_q}{\to} M''_q \to 0, \qquad q \geqslant 1$$

a projective system of exact sequences of A-modules. If the modules $(M'_q)_{q\geqslant 1}$ *are artinian, then the sequence*

$$0 \to \varprojlim M'_q \to \varprojlim M_q \to \varprojlim M''_q \to 0$$

is exact.

Proof. The exactness of the sequence

$$0 \to \varprojlim M'_q \to \varprojlim M_q \to \varprojlim M''_q$$

follows from the general properties of the projective limits. It only remains to prove the surjectivity of the map $\varprojlim M_q \to \varprojlim M''_q$. By the hypothesis that all M'_q are artinian, it results that for any q there is an integer $n(q)$ such that

$$\operatorname{Im}(M'_{n(q)+i} \to M'_q) = \operatorname{Im}(M'_{n(q)} \to M'_q), \qquad i \geqslant 0.$$

We may suppose that for any $i \geqslant 0$, $q \geqslant 1$ we have

$$\operatorname{Im}(M'_{q+i} \to M'_q) = \operatorname{Im}(M'_{q+1} \to M'_q),$$

passing eventually to a subsequence.

Let now $m'' = (m''_q)_q$ be an element of $\varprojlim M''_q$. Choose $\tilde{m}_2 \in M_2$ such that $\psi_2(\tilde{m}_2) = m''_2$ and denote by m_1 the image of $\tilde{m}_2$ in M_1. Then $\psi_1(m_1) = m''_1$. Let $\alpha \in M_3$ be such that $\psi_3(\alpha) = m''_3$. There exists $\beta \in M'_2$ so that $\varphi_2(\beta) = \tilde{m}_2 - \operatorname{Im} \alpha$. Since $\operatorname{Im}(M'_3 \to M'_1) = \operatorname{Im}(M'_2 \to M'_1)$, there exists $\gamma \in M'_3$ such that the images of β and γ in M'_1 are equal. Denote $\tilde{m}_3 = \varphi_3(\gamma) + \alpha$ and let m_2 be the image of $\tilde{m}_3$ in M_2. Then one can easily check that $\psi_2(m_2) = m''_2$ and the image of m_2 in M_1 equals m_1. By the same reasoning, we will find an element $m = (m_q)_q$ of $\varprojlim M_q$ such that its image in $\varprojlim M''_q$ is just m''.

§ 2. The finiteness theorem

The main result in the paragraph is the following *finiteness theorem of Grauert* [29]:

Theorem 2.1. *Let* $X \overset{f}{\to} Y$ *be a proper morphism of complex spaces and* $\mathfrak{F}$ *a coherent analytic sheaf on X. Then the analytic sheaves* $R^q f_*(\mathfrak{F})$ *are coherent* $(q \geqslant 0)$.

The proof will follow in three steps.

I. *The construction of a free resolution.*

1. *The construction of an atlas.* Consider a complex space X, an open subset Y of a numerical space $\mathbb{C}^N$ and a proper analytic map $f: X \to Y$. Let x be a point in X. By a *relative cart in* x we mean a closed immersion $j: U \to D(r) \times V$, where U is a neighbourhood of x, V is a neighbourhood of $f(x)$ such that $f(U) \subset V$ and $D(r) = \{z \in \mathbb{C}^n; |z_k| < r\}$, $r > 0$, a polydisc in some numerical space $\mathbb{C}^n$ such that the following diagram is commutative

$$\begin{array}{ccc} U & \xrightarrow{\;j\;} & D(r) \times V \\ & {\scriptstyle f|U}\searrow \quad \swarrow{\scriptstyle \mathrm{pr}_2} & \\ & V & \end{array}$$

Let y_0 be an arbitrary point in Y. Since f is a proper map, there exist a Stein neighbourhood V_* of y_0 and a finite number of relative carts

$$j_k: U_k \to D_k(1) \times V_*, \qquad 0 \leqslant k \leqslant k_*$$

such that $\bigcup_{k=0}^{k_*} U_k = f^{-1}(V_*)$. Suppose that the unit polydisc $D_k(1)$ is taken in a numerical space $\mathbb{C}^{n(k)}$.

We shall make the following notations: for $r \leqslant 1$ and for an open subset V of V_*, we set

$$X(V) = f^{-1}(V), \; U_k(r, V) = j_k^{-1}(D_k(r) \times V) \text{ and } \mathfrak{U}(r, V) = (U_k(r, V))_{0 \leqslant k \leqslant k_*}.$$

Since f is a proper map, there exists $r_* < 1$ such that $X(V) = \bigcup_{k=0}^{k_*} U_k(r, V)$ for any r which satisfies $r_* \leqslant r \leqslant 1$ and for any open subset $V \subset V_*$. If V is a Stein open set, then $\mathfrak{U}(r, V)$ is a Stein covering of $X(V)$, therefore by the Leray theorem, $H^{\bullet}(X(V), \mathcal{F}) \simeq H^{\bullet}(C^{\bullet}(\mathfrak{U}(r, V), \mathcal{F}))$, $r_* \leqslant r \leqslant 1$. In this formula denote by $C^{\bullet}(\mathfrak{U}(r, V), \mathcal{F})$ the Čech complex of alternate cochains of the covering $\mathfrak{U}(r, V)$ with values in the sheaf $\mathcal{F}$.

2. *Link systems of sheaves* ("Verbundene Garbensysteme"). We use the above notations. Let

$$\Delta_n = \{(k_0, \ldots, k_n) \mid 0 \leqslant k_0 < \ldots < k_n \leqslant k_*\} \text{ and } \Delta = \bigcup_{n \geqslant 0} \Delta_n.$$

If $\alpha = (k_0, \ldots, k_n) \in \Delta$ and $\beta = (l_0, \ldots, l_m) \in \Delta$ we connive to write

$$\alpha \subset \beta \Leftrightarrow \{k_0, \ldots, k_n\} \subset \{l_0, \ldots, l_m\}.$$

We further define

$$U_\alpha(r, V) = \bigcap_{\nu=0}^{n} U_{k_\nu}(r, V) \quad \text{and} \quad D_\alpha(r) = \prod_{\nu=0}^{n} D_{k_\nu}(r).$$

The fibred product of the maps j_{k_ν} determines a closed immersion

$$j_\alpha: U_\alpha(r, V) \to D_\alpha(r) \times V.$$

If $\alpha \subset \beta$ we denote by

$$\pi_{\alpha\beta}: D_\beta(r) \times V \to D_\alpha(r) \times V$$

the canonical projection. The diagram

$$\begin{array}{ccc} U_\beta(r, V) & \to & U_\alpha(r, V) \\ j_\beta \downarrow & & j_\alpha \downarrow \\ D_\beta(r) \times V & \to & D_\alpha(r) \times V, \end{array}$$

where the map $U_\beta(r, V) \to U_\alpha(r, V)$ is the canonical inclusion, is commutative.

By a *link system of sheaves over* $(D_\alpha(r) \times V, \pi_{\alpha\beta})$ we mean:

(i) a family $(\mathcal{G}_\alpha)_{\alpha\in\Delta}$ of analytic sheaves $\mathcal{G}_\alpha$ over $D_\alpha(r) \times V$;

(ii) a family $(\varphi_{\beta\alpha})_{\alpha\subset\beta}$ of morphisms of analytic sheaves

$$\varphi_{\beta\alpha}: \mathcal{G}_\alpha \to (\pi_{\alpha\beta})_* \mathcal{G}_\beta$$

such that $\varphi_{\alpha\alpha} = id$ and $\varphi_{\gamma\alpha} = ((\pi_{\alpha\beta})_* \varphi_{\gamma\beta})\varphi_{\beta\alpha}$ for $\alpha \subset \beta \subset \gamma$.

If $(\mathcal{G}_\alpha, \varphi_{\beta\alpha})$ and $(\mathcal{G}'_\alpha, \varphi'_{\beta\alpha})$ are two such systems, then a morphism between them

$$\theta: (\mathcal{G}_\alpha, \varphi_{\beta\alpha}) \to (\mathcal{G}'_\alpha, \varphi'_{\beta\alpha})$$

is by definition a family $(\theta_\alpha)_{\alpha\in\Delta}$ of sheaf morphisms $\theta_\alpha: \mathcal{G}_\alpha \to \mathcal{G}'_\alpha$, which are compatible with the morphisms $\varphi_{\beta\alpha}, \varphi'_{\beta\alpha}$.

One can easily prove that one gets an abelian category, which is called the *category of the link systems of sheaves over* $(D_\alpha(r) \times V, \pi_{\alpha\beta})$.

If $\mathcal{F}$ is an analytic sheaf on X, then it defines naturally a link system of sheaves $j_*(\mathcal{F})$, by $(j_*\mathcal{F})_\alpha = (j_\alpha)_*(\mathcal{F})$. A morphism $\varphi: \mathcal{F} \to \mathcal{F}'$ of analytic sheaves on X induces in a natural way a morphism $j_*(\varphi): j_*(\mathcal{F}) \to j_*(\mathcal{F}')$. One can easily see that the correspondences are functorial.

3. *The Čech complex associated to a link system of sheaves.* Let $\mathcal{G} = (\mathcal{G}_\alpha, \varphi_{\beta\alpha})$ be a link system of sheaves over $(D_\alpha(r) \times V, \pi_{\alpha\beta})$. For any $n \geqslant 0$ we define

$$C^n(r; V; \mathcal{G}) = \prod_{\alpha\in\Delta_n} \Gamma(D_\alpha(r) \times V, \mathcal{G}_\alpha).$$

$C^n(r, V; \mathcal{G})$ is a $\Gamma(V, \mathcal{O}_V)$-module. Define the differentials

$$\delta: C^n(r, V; \mathcal{G}) \to C^{n+1}(r, V; \mathcal{G})$$

by the formulae: for $\xi = (\xi_\alpha)_{\alpha \in \Delta_n} \in C^n(r, V; \mathcal{G})$, $(\delta\xi)_\beta = \sum_{i=0}^{n+1} (-1)^i \varphi_{\beta\beta_i}(\xi_{\beta_i})$, where $\beta = (l_0, \ldots, l_{n+1}) \in \Delta_{n+1}$ and $\beta_i = (l_0, \ldots, l_{i-1}, l_{i+1}, \ldots, l_{n+1})$. We denote by $C^n(r, \mathcal{G})$ the sheaf $V \mapsto C^n(r, V; \mathcal{G})$, V open subset of V_*. $C^n(r, \mathcal{G})$ is an $\mathcal{O}_{V_*}$-module. If $\mathcal{F}$ is an analytic sheaf on X, one gets an isomorphism of complexes

$$C^\bullet(\mathfrak{U}(r, V), \mathcal{F}) \xrightarrow{\sim} C^\bullet(r, V; j_*(\mathcal{F})).$$

4. *The construction of a free resolution.* A link system of sheaves $\mathcal{R} = (\mathcal{R}_\alpha, \psi_{\beta\alpha})$ is called a *free system of finite rank* if all $\mathcal{O}_{D_\alpha(r) \times V}$-modules $\mathcal{R}_\alpha$ are free of finite rank.

LEMMA 2.2. *Let $\mathcal{G} = (\mathcal{G}_\alpha, \varphi_{\beta\alpha})$ be a link system of sheaves and, for any $\alpha \in \Delta$, let $\varepsilon_\alpha : \mathcal{S}_\alpha \to \mathcal{G}_\alpha$ be a morphism of analytic sheaves $\mathcal{S}_\alpha$ being a free $\mathcal{O}_{D_\alpha(r) \times V}$-module of finite rank. Then there exist a free system of finite rank $\mathcal{R} = (\mathcal{R}_\alpha, \psi_{\beta\alpha})$ and a morphism $\theta : \mathcal{R} \to \mathcal{G}$ such that* $\operatorname{Im} \theta_\alpha \supset \operatorname{Im} \varepsilon_\alpha$ *for all* $\alpha \in \Delta$.

Proof. For every $\gamma \in \Delta$ we define a link system $\mathcal{R}^\gamma = (\mathcal{R}^\gamma_\alpha, \varphi^\gamma_{\beta\alpha})$ as follows:

$$\mathcal{R}^\gamma_\alpha = \begin{cases} 0 & \text{if } \gamma \not\subset \alpha. \\ \pi^*_{\gamma\alpha}(\mathcal{S}_\gamma) & \text{if } \gamma \subset \alpha. \end{cases}$$

If $\gamma \subset \alpha \subset \beta$, then $\varphi^\gamma_{\beta\alpha}: \pi^*_{\gamma\alpha}(\mathcal{S}_\gamma) \to (\pi_{\alpha\beta})_* \pi^*_{\gamma\beta}(\mathcal{S}_\gamma)$ is the morphism associated by adjunction to the identical map $(\pi_{\gamma\beta})^*(\mathcal{S}_\gamma) \simeq \pi^*_{\alpha\beta}\pi^*_{\gamma\alpha}(\mathcal{S}_\gamma) \to \pi^*_{\gamma\beta}(\mathcal{S}_\gamma)$. If $\gamma \not\subset \alpha$, then we put $\varphi^\gamma_{\beta\alpha} = 0$. The morphism ε_γ induces naturally a morphism $\theta^\gamma: \mathcal{R}^\gamma \to \mathcal{G}$. We denote $\mathcal{R} = \bigoplus_{\gamma \in \Delta} \mathcal{R}^\gamma$ and define $\theta: \mathcal{R} \to \mathcal{G}$ as being the sum of the morphisms θ^γ. Obviously, $\operatorname{Im} \theta_\alpha \supset \operatorname{Im} \varepsilon_\alpha$ for any $\alpha \in \Delta$.

LEMMA 2.3. *Let $\mathcal{F} \in$ **Coh** (X). For any relatively compact Stein open subset $V' \subset V_*$ and for any r', $r_* \leqslant r' < 1$, there exists a resolution*

$$\ldots \to \mathcal{R}^{k+1} \to \mathcal{R}^k \to \ldots \to \mathcal{R}^1 \to \mathcal{R}^0 \to j_*(\mathcal{F}) \to 0$$

over $(D_\alpha(r') \times V', \pi_{\alpha\beta})$, where the systems $\mathcal{R}^k$ are free of finite rank.

(Such a resolution will be briefly called a *free resolution*).

The proof will derive from the following assertion which will be proved by induction on k: "there exist a number r, $r' < r < 1$, a Stein open set V such that $V' \subset\subset V \subset\subset V_*$, and a resolution by free systems of finite rank over $(D_\alpha(r) \times V, \pi_{\alpha\beta})$, $\mathcal{R}^k \to \mathcal{R}^{k-1} \to \ldots \to \mathcal{R}^0 \to j_*(\mathcal{F}) \to 0$". For any r, $r' < r < 1$ and whenever V is a Stein open set, $V' \subset\subset V \subset\subset V_*$, there are surjective morphisms $\varepsilon_\alpha: \mathcal{S}_\alpha \to j_*(\mathcal{F})_\alpha$, $\mathcal{S}_\alpha$ being free $\mathcal{O}_{D_\alpha(r) \times V}$-modules of finite rank. By applying the preceding lemma we find a free system $\mathcal{R}^0$ over $(D_\alpha(r) \times V, \pi_{\alpha\beta})$ and a surjective morphism $\mathcal{R}^0 \to j_*(\mathcal{F})$. For the general induction step $k \mapsto k+1$ one proceeds analogously, by considering the system of sheaves $\mathcal{Z}^k = \text{Ker}\,(\mathcal{R}^k \to \mathcal{R}^{k-1})$ instead of $j_*(\mathcal{F})$.

5. *The calculation of the direct images.* Let $\mathcal{F} \in \mathbf{Coh}(X)$ and r_{**} be a number so that $r_* < r_{**} < 1$. By lemma 2.3, if we denote V' by V_* too, we can assume that there is a free resolution over $(D_\alpha(r_{**}) \times V_*, \pi_{\alpha\beta})$

$$\ldots \to \mathcal{R}^k \to \mathcal{R}^{k-1} \to \ldots \to \mathcal{R}^1 \to \mathcal{R}^0 \to j_*(\mathcal{F}) \to 0.$$

For any open set $V \subset V_*$ and for all r, $r_* \leqslant r \leqslant r_{**}$, we consider the double complex $(C^l(r, V; \mathcal{R}^k))_{l,k}$. We shall denote by $C^\bullet(r, V)$ the associated simple complex, that is

$$C^n(r, V) = \prod_{l-k=n} C^l(r, V; \mathcal{R}^k).$$

The $\mathcal{O}_{V_*}$-module $V \mapsto C^n(r, V)$, V open subset of V_*, will be denoted by $C^n(r)$. $C^\bullet(r)$ is a complex of $\mathcal{O}_{V_*}$-modules. Since $C^l(r, V; \mathcal{R}^k) = 0$ for $l > k_*$, it follows that $C^n(r) = 0$ for $n > k_*$. The morphisms $C^n(r, V) \to C^n(r, V; \mathcal{R}^0) \to C^n(r, V; j_*(\mathcal{F}))$ define the morphisms of complexes

$$C^\bullet(r, V) \to C^\bullet(r, V; j_*(\mathcal{F})),\ \ C^\bullet(r) \to C^\bullet(r, j_*(\mathcal{F})).$$

If $r' \leqslant r$, then we have canonical restriction maps

$$C^\bullet(r, V) \to C^\bullet(r', V)$$

which induce morphisms of $\mathcal{O}_{V_*}$-modules $C^\bullet(r) \to C^\bullet(r')$.

LEMMA 2.4. *For any r, $r_* \leqslant r \leqslant r_{**}$ and whenever V is a Stein open subset of V_*, the canonical morphism of complexes $C^\bullet(r, V) \to C^\bullet(r, V; j_*(\mathcal{F}))$ is a quasi-isomorphism. In particular, the $\mathcal{O}_{V*}$-morphism of complexes $C^\bullet(r) \to C^\bullet(r, j_*(\mathcal{F}))$ is a quasi-isomorphism.*

Proof. Since every $D_\alpha(r) \times V$ is Stein, it results by theorem B that $H^q(C^l(r, V; \mathcal{R}^\bullet)) = 0$ for $q \geqslant 1$ and $H^0(C^l(r, V; \mathcal{R}^\bullet)) \simeq C^l(r, V; j_*(\mathcal{F}))$. Then the assertion of the lemma follows from the properties of the spectral sequences associated to the bicomplex $(C^l(r, V; \mathcal{R}^k))_{l,k}$.

COROLLARY 2.5. *For any r, $r_* \leqslant r \leqslant r_{**}$, and for any Stein open subset $V \subset V_*$, the canonical morphisms $H^n(C^\bullet(r, V)) \to H^n(X(V), \mathcal{F})$ are isomorphisms for all n. In particular, $H^n(C^\bullet(r)) \simeq R^n f_*(\mathcal{F})|V_*$ for any n.*

COROLLARY 2.6. *Whatever the numbers r and r', $r_* \leqslant r' \leqslant r \leqslant r_{**}$, and the Stein open subset V of V_*, the restriction map $C^\bullet(r, V) \to C^\bullet(r', V)$ is a quasi-isomorphism. In particular, the restriction $\mathcal{O}_{V_*}$-morphism $C^\bullet(r) \to C^\bullet(r')$ is a quasi-isomorphism.*

II. *The induction scheme.*

1. *The statement of the induction assertions.* We will state two lemmas to be proved by induction.

Let $C^{\bullet}(r)$ be the complex of $\mathcal{O}_{V_*}$-modules built above, $r_* \leqslant r \leqslant r_{**}$ and Q_* be a compact subset of V_*.

LEMMA $A(n)$. *There exist a Stein open set V_n such that $Q_* \subset V_n \subset V_*$, a number r_n so that $r_* < r_n \leqslant r_{**}$, a complex $\mathcal{L}^{\bullet}$ of free $\mathcal{O}_{U_n}$-modules of finite rank*

$$\ldots \to 0 \to \mathcal{L}^n \xrightarrow{\alpha} \mathcal{L}^{n+1} \to \ldots \xrightarrow{\alpha} \mathcal{L}^{k_*} \to 0,$$

and a morphism of complexes $\mathcal{L}^{\bullet} \xrightarrow{\sigma} C^{\bullet}(r_n)$ where, for any Stein open subset V of V_n, the morphism $\mathcal{L}^{\bullet}(V) \xrightarrow{\sigma(V)} C^{\bullet}(r_n, V)$ is an n-quasi-isomorphism. In particular, σ is an n-quasi-isomorphism.

In accordance with corollary 2.6, for any r satisfying $r_* \leqslant r \leqslant r_n$, the composite map $\mathcal{L}^{\bullet} \xrightarrow{\sigma} C^{\bullet}(r_n) \to C^{\bullet}(r)$ denoted by σ too, still verifies the assertion of lemma $A(n)$.

Again under the hypothesis of lemma $A(n)$, we are going to formulate the second lemma.

Denote by $K^{\bullet}(r)$ the cone of the morphism $\mathcal{L}^{\bullet} \to C^{\bullet}(r)$, hence $K^m(r) = C^m(r) \oplus \mathcal{L}^{m+1}$ for $m \geqslant n-1$ and $K^m(r) = C^m(r)$ for $m < n-1$. The differentials of $K^{\bullet}(r)$ will be denoted by δ.

For any open subset $V \subset V_n$, let $K^m(r, V) = \Gamma(V, K^m(r))$. If V is a Stein open set, then the hypothesis of quasi-isomorphism from $A(n)$ implies the exactness of the sequence

$$K^{n-1}(r, V) \xrightarrow{\delta} K^n(r, V) \xrightarrow{\delta} K^{n+1}(r, V) \to \ldots$$

Let $Z^{n-1}(r) = \operatorname{Ker}(K^{n-1}(r) \xrightarrow{\delta} K^n(r))$ and

$$Z^{n-1}(r, V) = \operatorname{Ker}(K^{n-1}(r, V) \xrightarrow{\delta} K^n(r, V)).$$

LEMMA $B(n-1)$ (the projection of cycles). *For any Stein open set $V' \subset\subset V_n$ and for any pair of real numbers r, r' such that $r_* \leqslant r' < r \leqslant r_n$, there exists over V' a continuous morphism of $\mathcal{O}_{V'}$-modules $\tau : K^{n-1}(r) \to Z^{n-1}(r')$ such that the following diagram is commutative*

$$\begin{array}{ccc} K^{n-1}(r) & \xrightarrow{\tau} & Z^{n-1}(r') \\ & \nwarrow \quad \nearrow \text{restr.} & \\ & Z^{n-1}(r) & \end{array}$$

The significance of the word "continuous" in the above statement will be made clear in the followings.

The above stated lemmas will be proved by induction in section III, following the scheme:

0. For $n > k_*$, $A(n)$ and $B(n)$ are clearly true.

I. $A(n)$ and $B(n) \Rightarrow B(n-1)$.

II. $A(n)$ and $B(n-1) \Rightarrow A(n-1)$.

2. *Preparation for induction.* Let V be an open subset of V_* and $D(r)$ be the polydisc of radius r centred in origin which lies in the space $\mathbb{C}^m$. Let $t_1, \ldots, t_m$ be the coordinates in $\mathbb{C}^m$. Every element $f \in \Gamma(D(r) \times V, \mathcal{O}_{\mathbb{C}^m \times V})$ can be expanded in convergent series $f = \sum_\nu a_\nu t^\nu$ with $a_\nu \in \Gamma(V, \mathcal{O}_V)$, $\nu = (\nu_1, \ldots, \nu_m) \in \mathbb{N}^m$ and $t^\nu = t_1^{\nu_1} \ldots t_m^{\nu_m}$. For a compact $Q \subset V$ and for any positive real number $\rho < r$ define

$$\|f\|_{\rho Q} = \sum \|a_\nu\|_Q \, \rho^{|\nu|},$$

where $\|a_\nu\|_Q = \sup_{y \in Q} |a_\nu(y)|$ and $|\nu| = \nu_1 + \ldots + \nu_m$. $\| \ \|_{\rho Q}$ is a seminorm on $\Gamma(D(r) \times V, \mathcal{O}_{\mathbb{C}^m \times V})$. The family of seminorms $\| \ \|_{\rho Q}$, Q compact in V and $\rho < r$, defines the usual Fréchet topology of $\Gamma(D(r) \times V, \mathcal{O}_{\mathbb{C}^m \times V})$.

We will also consider $\| \ \|_{\rho Q}$ for $\rho = r$ or $Q = V$, but in this case one obtains only a pseudonorm. For $\rho' \leqslant \rho$ and $Q' \subset Q$, we have

$$\| \ \|_{\rho' Q'} \leqslant \| \ \|_{\rho Q}.$$

If $a \in \Gamma(V, \mathcal{O}_V)$ and $f \in \Gamma(D(r) \times V, \mathcal{O}_{\mathbb{C}^m \times V})$, then $\|af\|_{\rho Q} \leqslant \|a\|_Q \, \|f\|_{\rho Q}$.

LEMMA 2.7. *If* $0 < r' < r'' < r$, *then the family* $(t/r'')^\nu$, $\nu \in \mathbb{N}^m$, *has the following properties:*

(i) *every element* $f \in \Gamma(D(r) \times V, \mathcal{O}_{\mathbb{C}^m \times V})$ *can be uniquely written* $f = \sum_\nu a_\nu (t/r'')^\nu$ *such that* $\|a_\nu\|_Q \leqslant \|f\|_{r''Q}$ *for any compact* $Q \subset V$.

(ii) $\sum_\nu \|(t/r'')^\nu\|_{r'V} < \infty$.

The lemma follows from definitions.

We now consider a finite number of polydiscs $D_k(r) \subset \mathbb{C}^{m(k)}$ and denote

$$K(r, V) = \prod_k \Gamma(D_k(r) \times V, \mathcal{O}_{\mathbb{C}^{m(k)} \times V}).$$

For $f = (f_k) \in K(r, V)$ we define

$$\|f\|_{\rho Q} = \max_k \|f_k\|_{\rho Q}.$$

From the previous lemma one easily deduces the following

LEMMA 2.8. *Let* $0 < r' < r'' < r$. *There is a countable family* $(e_i)_{i \in I}$ *of elements of* $K(r, V)$ *enjoying the following properties*:

(i) *For any open subset* $V' \subset V$, *every element* $f \in K(r, V')$ *can be uniquely expanded in convergent series* $f = \sum_i a_i e_i$ *with* $a_i \in \Gamma(V', \mathcal{O}_V)$ *and* $\|a_i\|_Q \leqslant \|f\|_{r''Q}$ *where* Q *is an arbitrary compact in* V'.

(ii) $\sum_i \|e_i\|_{r'V} < \infty$.

The $\Gamma(V, \mathcal{O}_V)$-modules $C^l(r, V)$ and $K^l(r, V)$ already defined in section I are of the type $K(r, V)$; they are even topological $\Gamma(V, \mathcal{O}_V)$-modules. We also remark that the morphisms σ in lemma $A(n)$ and the differentials δ are continuous.

Under the hypothesis of lemma $A(n+1)$ we give another form of lemma $B(n)$.

LEMMA $B^*(n)$. *For any Stein open set $V' \subset\subset V_{n+1}$ and for any pair of real numbers r, r' enjoying the property $r_* \leqslant r' < r \leqslant r_{n+1}$, there exists a continuous morphism of $\mathcal{O}_V$-modules over V', $\tau: K^n(r) \to Z^n(r')$, such that the diagram*

$$\begin{array}{ccc} K^n(r) & \xrightarrow{\ \tau\ } & Z^n(r') \\ & \nwarrow \quad \nearrow {\scriptstyle \text{restr.}} & \\ & Z^n(r) & \end{array}$$

is commutative and moreover, the following assertion is fulfilled:

There are a countable family $(e_i)_{i\in I}$ of elements of $K^n(r, V')$ and a number $\tilde{r}$, $r' < \tilde{r} < r$, such that

(i) For any open subset $V'' \subset V'$, every element $f \in K^n(r, V'')$ can be uniquely expanded in convergent series $f = \sum_i a_i e_i$ with $a_i \in \Gamma(V'', \mathcal{O}_{V'})$ and $\|a_i\|_Q \leqslant \|f\|_{\tilde{r}Q}$ Q compact of V''.

(ii) $\sum_i \| \tau e_i \|_{r'V'} < \infty$.

PROPOSITION 2.9. *$A(n+1)$ and $B(n) \Rightarrow B^*(n)$.*

Proof. Choose a Stein open set $\tilde{V}$ such that $V' \subset\subset \tilde{V} \subset\subset V_{n+1}$ and real numbers $\tilde{r}, \rho, \rho'$ enjoying the property $r' < \rho' < \rho < \tilde{r} < r$. By $B(n)$, there exists a projection of cycles $\tilde{\tau}: K^n(\rho) \to Z^n(\rho')$ over $\tilde{V}$. Consider the diagram

$$\begin{array}{ccccccc} K^n(r) & \xrightarrow{\ \beta\ } & K^n(\rho) & \xrightarrow{\ \tilde{\tau}\ } & Z^n(\rho') & \xrightarrow{\ \beta'\ } & Z^n(r') \\ \uparrow & & \uparrow & \nearrow & & & \\ Z^n(r) & \longrightarrow & Z^n(\rho) & & & & \end{array}$$

where β and β' are the restriction maps. We will show that the morphism $\tau = \beta' \tilde{\tau} \beta$ has the properties required by $B^*(n)$ over V'. Indeed, by making use of lemma 2.8, we find a family $(e_i)_{i\in I}$ of elements of $K^n(r, \tilde{V})$ so that $\sum \|\beta e_i\|_{\rho\tilde{V}} < \infty$ such that the condition $i)$ from $B^*(n)$ holds. Since $\tilde{\tau}: K^n(\rho, \tilde{V}) \to Z^n(\rho', \tilde{V})$ is continuous, there is a constant M such that $\|\tilde{\tau} g\|_{r'V'} \leqslant M \|g\|_{\rho\tilde{V}}$ for any $g \in K^n(\rho, \tilde{V})$. We get $\|\beta' \tilde{\tau} \beta e_i\|_{r'V'} = \| \tilde{\tau} \beta e_i\|_{r'V'} \leqslant M \|\beta e_i\|_{\rho\tilde{V}}$, hence $\sum_{i\in I} \| \tau e_i\|_{r'V'} < \infty$.

III. *The proof of the induction assertions and of the theorem.*

1. *$A(n)$ and $B(n) \Rightarrow B(n-1)$*. Let r, r' be real numbers such that $r_* \leqslant r' < r \leqslant r_n$ and V' be a relatively compact Stein open subset of V_n. Choose a real

number r'', $r' < r'' < r$ and a Stein open set V'', $V' \subset\subset V'' \subset\subset V_n$. Let $\tau: K^n(r) \to Z^n(r'')$ and $(e_i)_{i\in I}$ be the entities obtained by applying lemma $B^*(n)$, τ being a morphism of $\mathcal{O}_{V''}$-modules and e_i elements of $K^n(r, V'')$. We have $\sum_i \|\tau e_i\|_{r''V''} < \infty$. According to lemma $A(n)$, the linear continuous map $K^{n-1}(r'', V'') \xrightarrow{\delta} Z^n(r'', V'')$ is surjective. By applying the Banach theorem we find a constant M and some elements $\xi_i \in K^{n-1}(r'', V'')$ such that $\delta\xi_i = \tau e_i$ and $\|\xi_i\|_{r'V'} \leqslant \leqslant M\|\tau e_i\|_{r''V''}$. We get $\sum_i \|\xi_i\|_{r'V'} < \infty$. The correspondence $\sum_i a_i e_i \mapsto \sum_i a_i \xi_i$ determines a continuous $\mathcal{O}_{V'}$-morphism

$$h: K^n(r) \to K^{n-1}(r'),$$

which makes the diagram commutative

$$\begin{array}{ccc} K^n(r) & \longleftarrow & Z^n(r) \\ {\scriptstyle h}\downarrow & & \downarrow \\ K^{n-1}(r') & \xrightarrow{\delta} & Z^n(r'). \end{array}$$

Consider the diagram

$$\begin{array}{ccc} K^{n-1}(r) & \longrightarrow & K^n(r) \\ {\scriptstyle \beta}\downarrow & \overset{h}{\swarrow} & \downarrow \\ K^{n-1}(r') & \xrightarrow{\delta} & K^n(r'). \end{array}$$

The morphism $\tau: K^{n-1}(r) \to Z^{n-1}(r')$, $\tau = \beta - h\delta$ verifies $B(n-1)$.

2. *$A(n)$ and $B(n-1) \Rightarrow A(n-1)$.*

(a) Let V_{n-1} be a Stein open set such that $Q_* \subset V_{n-1} \subset\subset V_n$ and let r_{n-1} be a real number, $r_* < r_{n-1} < r_n$. By lemma $A(n)$, for any ρ, $r_{n-1} \leqslant \rho \leqslant r_n$, we have the diagram

$$\begin{array}{ccccccccc} & & & & \mathfrak{L}^n & \to & \mathfrak{L}^{n+1} & \to & \dots \\ & & & & {\scriptstyle \sigma^n}\downarrow & & \downarrow & & \\ \dots \to C^{n-2}(\rho) & \to & C^{n-1}(\rho) & \xrightarrow{\partial^{n-1}} & C^n(\rho) & \to & C^{n+1}(\rho) & \to & \dots, \end{array}$$

which induces for any Stein open set $V \subset V_n$ an epimorphism $\Gamma(V, \operatorname{Ker} \alpha^n) \to \to H^n(C^\bullet(\rho, V))$. Over V_{n-1}, we have to find a free sheaf of finite rank $\mathfrak{L}^{n-1}$ and morphisms α^{n-1} and σ^{n-1}

$$\begin{array}{ccc} \mathfrak{L}^{n-1} & \xrightarrow{\alpha^{n-1}} & \mathfrak{L}^n \\ {\scriptstyle \sigma^{n-1}}\downarrow & & \\ C^{n-1}(r_{n-1}) & & \end{array}$$

which enjoy the following properties:

i) $\alpha^n \alpha^{n-1} = 0$, $\sigma^n \alpha^{n-1} = \delta^{n-1}\sigma^{n-1}$,

ii) for any Stein open set $V \subset V_{n-1}$, the induced morphism $\Gamma(V, \mathrm{Ker}\, \alpha^n/\mathrm{Im}\alpha^{n-1}) \to \to H^n(C^\bullet(r_{n-1}, V))$ is an isomorphism and $\Gamma(V, \mathrm{Ker}\, \alpha^{n-1}) \to H^{n-1}(C^\bullet(r_{n-1}, V))$ is an epimorphism.

Since $Z^{n-1}(r_{n-1}) = \{(\eta, s) \in C^{n-1}(r_{n-1}) \oplus \mathfrak{L}^n | \alpha^n s = 0, \ \sigma^n s = -\partial^{n-1}\eta\}$, it is sufficient to construct $\mathfrak{L}^{n-1}$ and a morphism

$$\omega: \mathfrak{L}^{n-1} \to Z^{n-1}(r_{n-1})$$

such that, for any Stein open set $V \subset V_{n-1}$, the sum of the morphisms in the diagram

$$\begin{array}{ccc} & & \mathfrak{L}^{n-1}(V) \\ & & \downarrow \omega \\ C^{n-2}(r_{n-1}, V) & \xrightarrow{\delta} & Z^{n-1}(r_{n-1}, V) \end{array}$$

is a surjective morphism.

(**b**) Let $r', r_{n-1} < r' < r_n$. For any Stein open set $V \subset V_n$ the restriction map $C^\bullet(r_n, V) \to C^\bullet(r', V)$ is a quasi-isomorphism (corollary 2.6). Thereby, it results that the sum of the maps in the diagram

$$\begin{array}{ccc} & & Z^{n-1}(r_n, V) \\ & & \downarrow \\ C^{n-2}(r', V) & \to & Z^{n-1}(r', V) \end{array}$$

is surjective (1.10).

(**c**) Consider a Stein open set V', $V_{n-1} \subset\subset V' \subset\subset V_n$ and a real number r, $r' < r < r_n$. By means of lemma $B^*(n-1)$ we get a projection of cycles $\tau: K^{n-1}(r) \to Z^{n-1}(r')$ over V', a family $(e_i)_{i \in I}$ of elements $e_i \in K^{n-1}(r, V')$ and a real number $\tilde{r}$, $r' < \tilde{r} < r$ such that the property *i*) of $B^*(n-1)$ holds and $\sum_i \|\tau e_i\|_{r'V'} < \infty$. Since

$$\mathrm{Im}\,(K^{n-1}(r_n) \xrightarrow{\beta} K^{n-1}(r) \xrightarrow{\tau} Z^{n-1}(r')) \supset \mathrm{Im}\,(Z^{n-1}(r_n) \xrightarrow{\text{restr.}} Z^{n-1}(r')),$$

it results that the sum of the maps of the diagram

$$\begin{array}{ccc} & & K^{n-1}(r_n, V') \\ & & \downarrow \tilde{\tau} = \tau\beta \\ C^{n-2}(r', V') & \to & Z^{n-1}(r', V') \end{array}$$

is surjective.

According to the Banach theorem, there exist a constant M and elements $\xi_i \in K^{n-1}(r_n, V')$, $\eta_i \in C^{n-2}(r', V')$ $(i \in I)$ such that $\bar{\tau}\xi_i + \partial\eta_i = \tau e_i$ and $\max(\|\xi_i\|_{rV_{n-1}}, \|\eta_i\|_{r_{n-1}V_{n-1}}) \leqslant M\|\tau e_i\|_{r'V'}$. We then derive that $\sum_i \|\xi_i\|_{rV_{n-1}} < \infty$ and $\sum_i \|\eta_i\|_{r_{n-1}V_{n-1}} = M_1 < \infty$. Then there exists a finite set $J \subset I$ such that

$$\sum_{i \in I \setminus J} \|\xi_i\|_{rV_{n-1}} \leqslant \frac{1}{2}.$$

Put $\mathfrak{L}^{n-1} = \mathcal{O}^J_{V_{n-1}}$ and let $\omega\colon \mathfrak{L}^{n-1} \to Z^{n-1}(r_{n-1})$ be the morphism which maps the canonical generators $(g_i)_{i \in J}$ of $\mathfrak{L}^{n-1}$ onto the elements $(\beta'\bar{\tau}\xi_i)_{i \in J}$. By $\beta'\colon Z^{n-1}(r') \to Z^{n-1}(r_{n-1})$ we have denoted the restriction map.

(d) We are going to prove the following assertion: for any open set $V \subset V_{n-1}$ and for any element $f \in K^{n-1}(r, V)$, there are elements $f_1 \in K^{n-1}(r, V)$, $g \in \Gamma(V, \mathfrak{L}^{n-1})$ and $\eta \in C^{n-2}(r_{n-1}, V)$ such that $\beta'\tau(f) = \omega(g) + \delta\eta + \beta'\tau(f_1)$ and

$$\|f_1\|_{rQ} \leqslant \frac{1}{2}\|f\|_{\bar{r}Q}, \|g\|_Q \leqslant \|f\|_{\bar{r}Q}, \|\eta\|_{r_{n-1}Q} \leqslant M_1\|f\|_{\bar{r}Q},$$

Q being an arbitrary compact of V. In order to do this, let f be expanded in convergent series $f = \sum_i a_i e_i$, where $a_i \in \Gamma(V, \mathcal{O}_V)$, $\|a_i\|_Q \leqslant \|f\|_{\bar{r}Q}$ for any compact $Q \subset V$. If we set $f_1 = \sum_{i \in I \setminus J} a_i\xi_i$, $g = \sum_{i \in J} a_i g_i$ and $\eta = \sum_{i \in I} a_i\eta_i$, then we have

$$\|f_1\|_{rQ} \leqslant \sum_{i \in I \setminus J} \|a_i\|_Q\|\xi_i\|_{rQ} \leqslant \|f\|_{\bar{r}Q} \sum_{i \in I \setminus J} \|\xi_i\|_{rQ} \leqslant \frac{1}{2}\|f\|_{\bar{r}Q},$$

$$\|g\|_Q = \max_{i \in J} \|a_i\|_Q \leqslant \|f\|_{\bar{r}Q}, \|\eta\|_{r_{n-1}Q} \leqslant \sum_{i \in I} \|a_i\|_Q\|\eta_i\|_{r_{n-1}Q} \leqslant M_1\|f\|_{\bar{r}Q}$$

and the assertion is proved.

(e) We now check that ω is the morphism we need. Let V be a Stein open subset of V_{n-1} and $f \in K^{n-1}(r, V)$. By iterating the assertion proved in **(d)**, we find elements $g \in \Gamma(V, \mathfrak{L}^{n-1})$ and $\eta \in C^{n-2}(r_{n-1}, V)$ so that

$$(*) \qquad \beta'\tau(f) = \omega(g) + \partial\eta.$$

Since the restriction map $C^\bullet(r, V) \to C^\bullet(r_{n-1}, V)$ is a quasi-isomorphism, the sum of the maps in the diagram

$$\begin{array}{ccc} & & Z^{n-1}(r, V) \\ & & \downarrow \\ C^{n-2}(r_{n-1}, V) & \xrightarrow{\delta} & Z^{n-1}(r_{n-1}, V) \end{array}$$

is a surjective map (1.10). By using this fact, the equalities (*) and the equalities $\beta'\tau(f) = \text{restr.}\ (f)$ for $f \in Z^{n-1}(r, V)$, it follows that the sum of the maps in the diagram

$$\begin{array}{ccc} & & \Gamma(V, \mathfrak{L}^{n-1}) \\ & & \downarrow \omega \\ C^{n-2}(r_{n-1}, V) & \xrightarrow{\delta} & Z^{n-1}(r_{n-1}, V) \end{array}$$

is surjective. In this way the lemma $A(n-1)$ is proved.

3. *Proof of the theorem.* We may assume that Y is an open subset of a numerical space $\mathbb{C}^N$. Let $y_0 \in Y$. By lemma $A(-1)$ and corollary 2.5, there exists a complex $\mathfrak{L}^\bullet$ of free $\mathcal{O}_V$-modules of finite rank in a neighbourhood V of y_0

$$0 \to \mathfrak{L}^{-1} \to \mathfrak{L}^0 \to \mathfrak{L}^1 \to \ldots \to \mathfrak{L}^{k_*} \to 0$$

such that $H^q(\mathfrak{L}^\bullet) \simeq R^q f_*(\mathfrak{F})|V$ for any $q \in \mathbb{N}$. Accordingly, every sheaf $R^q f_*(\mathfrak{F})$ is coherent and the theorem is completely proved.

We end this paragraph giving some applications of theorem 2.1. The first one is the following *Cartan-Serre finiteness theorem* [17].

THEOREM 2.10. *Let X be a compact analytic space and $\mathfrak{F}$ a coherent analytic sheaf on X. Then the complex vectorial spaces $H^n(X, \mathfrak{F})$ are finite-dimensional for any $n \geqslant 0$.*

Proof. Consider the complex space $e = (*, \mathbb{C})$ which is reduced to one point and has as stalk the complex field. The unique morphism $f\colon X \to e$ is proper and since $R^\bullet f_*(\mathfrak{F}) = H^\bullet(X, \mathfrak{F})$, the assertion follows obviously from Grauert's theorem.

The next application is the *projection theorem of Remmert* [65]:

THEOREM 2.11. *The image of a closed analytic set by a proper morphism is a closed analytic set.*

Proof. It is sufficient to show that $f(X)$ is a closed analytic subset of Y, for any proper morphism $f : X \to Y$. Since the support of any coherent analytic sheaf is an analytic set and since we have $f(X) = \text{Supp}\,(f_*(\mathcal{O}_X))$, the conclusion results from 2.1.

The last application deals with Stein decomposition.

Let $f : X \to Y$ be a proper morphism of complex spaces. Any connected component of a fiber $f^{-1}(y)$, $y \in Y$, is called *a level set* of f. We shall denote their set by Y' and let $p : Y' \to Y$ be the corresponding natural map.

If we assign to any point $x \in X$ the connected component of $f^{-1}(f(x))$ which contains x, then one gets a map $f' : X \to Y'$ such that $f = pf'$. The map f' is surjective and we can endow Y' with the quotient topology. Then f' will be a proper map and p is a finite continuous map. We further consider on Y' the structure of ringed space, where $f'_*(\mathcal{O}_X)$ stands for the structural sheaf.

The topological decomposition $X \to Y' \to Y$ of f naturally determines a decomposition of ringed spaces and the next theorem shows that it is even a decomposition of complex spaces, which is called *the Stein decomposition of the morphism f* [14], [85].

THEOREM 2.12. *The ringed space* $(Y', f'_*(\mathcal{O}_X))$ *is a complex space.*

Proof. By 2.1 the $\mathcal{O}_X$-algebra $f_*(\mathcal{O}_X)$ is coherent. Denote $Z = \operatorname{Specan}(f_*(\mathcal{O}_X))$ and let $q : Z \to Y$ be the structural morphism. The natural morphism $f^*f_*(\mathcal{O}_X) \to \to \mathcal{O}_X$ gives rise to a morphism $g : X \to Z$ such that $f = qg$.

We are going to prove that for any $y \in Y$, the connected components of $f^{-1}(y)$ are in $1-1$ correspondence with the points of $q^{-1}(y)$ and that within this correspondence, for a point $x \in X$, $g(x)$ is carried into the connected component of $f^{-1}(f(x))$, where x belongs. In this case a $1-1$ correspondence θ between Z and Y' which agrees with g, f', p, q is obtained. The map g is proper and surjective, hence Z has the quotient topology from X. We derive that θ is a topological isomorphism. Since $\mathcal{O}_Y \simeq q_*(\mathcal{O}_Z)$, it will result that the image of the morphism $\mathcal{O}_Z \to g_*(\mathcal{O}_X)$ under the functor q_* is an isomorphism; therefore $\mathcal{O}_Z \to g_*(\mathcal{O}_X)$ is an isomorphism and θ can be extended to an isomorphism of ringed spaces

$$(Y', f'_*(\mathcal{O}_X)) \simeq \operatorname{Specan} f_*(\mathcal{O}_X).$$

We now prove that the fibres of g are nonempty and connected; if so, the assignments $z \in q^{-1}(y) \mapsto g^{-1}(z)$, $y \in Y$, will establish the above stated bijective correspondences and the proof of the theorem will be concluded.

It remains to prove the assertion concerning the fibers of g. Suppose $f : X \to Y$ is a proper morphism such that the morphism $\mathcal{O}_Y \to f_*(\mathcal{O}_X)$ is an isomorphism. The map f is then surjective. We now consider an arbitrary point $y \in Y$ and claim that $f^{-1}(y)$ is a connected set. For otherwise, let $f^{-1}(y) = T_1 \cup T_2$ where T_1, T_2 are disjoint closed proper subsets of $f^{-1}(y)$. Then there are neighbourhoods U_i of T_i, $i = 1, 2$ such that $U_1 \cap U_2 = \emptyset$. By shrinking eventually this neighbourhood, we can assume that $U = U_1 \cup U_2$ has the form $f^{-1}(V)$, where V is a neighbourhood of y. The section of $\mathcal{O}_X$ which equals the unity on U_1 and zero on U_2 determines a section φ of $\mathcal{O}_Y$ on V; therefore $\varphi(y) = 0$, $\varphi(y) = 1$, a contradiction.

COROLLARY 2.13. *Let* $f : X \to Y$ *be a proper morphism of complex spaces and* X' *be the set of all points of* X *isolated in fiber (it is an open subset). Then the restriction of* f *to* X' *can be decomposed into an open immersion and a finite morphism; more precisely, if* $X \xrightarrow{f'} Y' \xrightarrow{p} Y$ *is the Stein decomposition of* f*, then there exists an open set* $V' \subset Y'$ *such that* $X' = f'^{-1}(V')$ *and* $f'|X' : X' \to V'$ *is an isomorphism.*

Proof. For a point $x \in X$, $f'^{-1}f'(x)$ is the connected component of $f^{-1}f(x)$ to which x belongs, hence it is an open subset of $f^{-1}f(x)$. Thus x is isolated in $f^{-1}f(x)$ if and only if it is isolated in $f'^{-1}f'(x)$. Let x be a point in X'. Since the fibers of f' are connected, $f'^{-1}f'(x) = \{x\}$. Hence the inverse image under f' of a fundamental system of neighbourhoods of $y' = f'(x)$ is a fundamental system of neighbourhoods of x. This fact and the equality $\mathcal{O}_{Y'} = f'_*(\mathcal{O}_X)$ imply that the canonical morphism $\mathcal{O}_{Y', y'} \to \mathcal{O}_{X, x}$ is an isomorphism. Therefore there exists a neighbourhood $V'_{y'}$ of y' so that $f'^{-1}(V'_{y'}) \subset X'$ and $f'|f'^{-1}(V'_{y'}): f'^{-1}(V'_{y'}) \to V'_{y'}$ is an isomorphism. The set $V' = \bigcup_{x \in X} V'_{f'(x)}$ satisfies the assertion of the corollary.

§ 3. The comparison and the base change theorems

Let $f : X \to Y$ be a morphism of complex spaces and $y \in Y$. We denote by $\mathfrak{m}_y$ both the maximal ideal of $\mathcal{O}_{Y,y}$ and the natural ideal-sheaf given by this; by $\hat{\mathfrak{m}}_y$ we mean the ideal-sheaf of $\mathcal{O}_X$ generated by the inverse image of $\mathfrak{m}_y$. Consider an analytic sheaf $\mathcal{F}$ on X and $q \geqslant 0$ an integer. Define

$$(R^q f_*(\mathcal{F})_y)^\wedge = \varprojlim_k (R^q f_*(\mathcal{F})_y / \mathfrak{m}_y^k (R^q f_*(\mathcal{F})_y)).$$

If $R^q f_*(\mathcal{F})$ is $\mathcal{O}_Y$-coherent, then this module is the completion of the $\mathcal{O}_Y$-module of finite type $R^q f_*(\mathcal{F})_y$ in the $\mathfrak{m}_y$-adic topology.

We now define a natural morphism

$$\varphi^q : (R^q f_*(\mathcal{F})_y)^\wedge \to \varprojlim_k H^q(f^{-1}(y), \mathcal{F}/\hat{\mathfrak{m}}_y^k \mathcal{F})$$

as follows. For any integer $k \geqslant 0$, the exact sequence

$$0 \to \hat{\mathfrak{m}}_y^k \mathcal{F} \to \mathcal{F} \to \mathcal{F}/\hat{\mathfrak{m}}_y^k \mathcal{F} \to 0$$

yields the exact sequence

$$\ldots \to R^q f_*(\hat{\mathfrak{m}}_y^k \mathcal{F}) \to R^q f_*(\mathcal{F}) \to R^q f_*(\mathcal{F}/\hat{\mathfrak{m}}_y^k \mathcal{F}) \to \ldots$$

The relations $\mathfrak{m}_y^k R^q f_*(\mathcal{F})_y \subset \operatorname{Im}(R^q f_*(\hat{\mathfrak{m}}_y^k \mathcal{F})_y \to R^q f_*(\mathcal{F})_y)$ and $H^q(f^{-1}(y), \mathcal{F}/\hat{\mathfrak{m}}_y^k \mathcal{F}) = R^q f_*(\mathcal{F}/\hat{\mathfrak{m}}_y^k \mathcal{F})_y$ give rise to a natural morphism

$$\varphi_k^q : R^q f_*(\mathcal{F})_y / \mathfrak{m}_y^k (R^q f_*(\mathcal{F})_y) \to H^q(f^{-1}(y), \mathcal{F}/\hat{\mathfrak{m}}_y^k \mathcal{F}).$$

This morphism can be deduced also from the morphisms

$$R^q f_*(\mathcal{F}) \otimes \mathcal{M} \to R^q f_*(\mathcal{F} \otimes f^*(\mathcal{M}))$$

constructed in § 1, **(a)** by taking $\mathcal{M} = \mathcal{O}_Y / \mathfrak{m}_y^k \mathcal{O}_Y$.

The family $(\varphi_k^q)_k$ is compatible with the projective systems and we thus obtain the desired morphism φ^q.

The following theorem, which is called *the Grauert comparison theorem* (Vergleichssatz), is the main result of the paragraph.

THEOREM 3.1. *Let $f : X \to Y$ be a proper morphism of complex spaces, y a point in Y, $q \geqslant 0$ an integer, and $\mathcal{F}$ a coherent analytic sheaf on X. Under these assumptions:*

(*i*) *there is a function $F : \mathbb{N} \to \mathbb{N}$ such that $\lim_{k\to\infty} F(k) = \infty$ and for any $k \geqslant 0$,*

$$\operatorname{Im}(R^q f_*(\hat{\mathfrak{m}}_y^k \mathcal{F})_y \to R^q f_*(\mathcal{F})_y) \subset \mathfrak{m}_y^{F(k)}(R^q f_*(\mathcal{F})_y);$$

(*ii*) *the natural morphism defined above*

$$\varphi^q : (R^q f_*(\mathcal{F})_y)^\wedge \to \varprojlim_k H^q(f^{-1}(y), \mathcal{F}/\hat{\mathfrak{m}}_y^k \mathcal{F})$$

is an isomorphism.

First of all, we recall the following fact of algbera. Consider a complex space X, an $\mathcal{O}_X$-module $\mathcal{F}$, a section $f \in \Gamma(X, \mathcal{O}_X)$ and a point $x \in X$. We say $\mathcal{F}$ *is without f-torsion in x* if the canonical morphism

$$\mathcal{F}_x \to \mathcal{F}_x, \qquad \varphi \mapsto f_x \varphi$$

is injective.

LEMMA 3.2. *If $\mathcal{F} \in \mathbf{Coh}(X)$, then for any compact K of X there exists an integer $d \geqslant 0$ such that the sheaf $f^d \mathcal{F}$ is without f-torsion in all points of K.*

Proof. Let $x \in X$. The ascending sequence of $\mathcal{O}_x$-submodules of $\mathcal{F}_x$

$$(0 : f_x) \subset (0 : f_x^2) \subset \ldots \subset \mathcal{F}_x$$

is stationary and let d be such that $(0 : f_x^d) = (0 : f_x^{d+1}) = \ldots$ Then $f^d \mathcal{F}$ will result without f-torsion in x. The morphism $f^d \mathcal{F} \to f^d \mathcal{F}$, $\varphi \mapsto f\varphi$ is injective in x, hence it remains injective in a neighbourhood of x. In its points, $f^d \mathcal{F}$ will be without f-torsion, etc...

The smallest d enjoing the property required in this lemma will be denoted by $d(f, \mathcal{F}; K)$.

Proof of the first assertion of the theorem. We may assume that Y is an open subset of a numerical space $\mathbb{C}^m$ and that y is the origin of this space. Let $t_1, \ldots, t_m$ be the coordinate functions on $\mathbb{C}^m$.

If $\alpha \in \mathbb{N}^m$, then by $\mathfrak{m}^\alpha$ we mean the ideal-sheaf $\sum_{i=1}^{m} t_i^{\alpha_i} \mathcal{O}_Y$ and by $\hat{\mathfrak{m}}_\alpha$ the ideal-sheaf on X generated by the inverse image of $\mathfrak{m}^\alpha$. We will prove the following assertion: for any $\alpha \in \mathbb{N}^m$ there is $\beta \in \mathbb{N}^m$ such that

$$\operatorname{Im}(R^q f_*(\hat{\mathfrak{m}}^\beta \mathcal{F})_0 \to R^q f_*(\mathcal{F})_0) \subset \mathfrak{m}^\alpha(R^q f_*(\mathcal{F})_0).$$

For an integer $k \geqslant 0$ we denote by $F(k)$ the largest integer $\geqslant 0$ where

$$\operatorname{Im}(R^q f_*(\hat{\mathfrak{m}}_y^k \mathcal{F})_y \to R^q f_*(\mathcal{F})_y) \subset \mathfrak{m}_y^{F(k)}(R^q f_*(\mathcal{F})_y)$$

(if the left term vanishes we agree to set $F(k) = k$).

For any $\gamma \in \mathbb{N}^m$,

$$\operatorname{Im}(R^q f_*(\hat{\mathfrak{m}}^\gamma \mathscr{F})_0 \to R^q f_*(\mathscr{F})_0) \subset \operatorname{Im}(R^q f_*(\hat{\mathfrak{m}}^{|\gamma|} \mathscr{F})_0 \to R^q f_*(\mathscr{F})_0),$$

where $|\gamma| = \gamma_1 + \ldots + \gamma_m$, because $\hat{\mathfrak{m}}^\gamma \mathscr{F} \subset \hat{\mathfrak{m}}^{|\gamma|} \mathscr{F}$. Then it follows that $\lim_{k\to\infty} F(k) = \infty$ and part (*i*) of the theorem is so proved. It remains to prove only the above stated assertion: let $\alpha \in \mathbb{N}^m$. Denote

$$\mathscr{F}^{[0]} = 0, \quad \mathscr{F}_{[0]} = \mathscr{F}/\mathscr{F}^{[0]} = \mathscr{F}.$$

For $1 \leqslant r \leqslant m$ we define recursively the entities

$$\mathscr{F}^{[r]} = \sum_{i=1}^{r} t_i^{\beta_i} \mathscr{F}, \quad \mathscr{F}_{[r]} = \mathscr{F}/\mathscr{F}^{[r]},$$

$$d'_r = d(t_r, \mathscr{F}_{[r-1]}; f^{-1}(0)), \quad d_r = d(t_r, R^{q+1} f_* \mathscr{F}^{[r-1]}; 0)$$

and $\beta_r = \alpha_r + d'_r + d_r$. Obviously, $\mathscr{F}_{[m]} = \mathscr{F}/\hat{\mathfrak{m}}^\beta \mathscr{F}$.

If V is a neighbourhood of the origin in $\mathbf{C}^m$, which is small enough, then we get the exact sequences of $f^{-1}(V)$

$$0 \to t_r^{d'_r} \mathscr{F}_{[r-1]} \xrightarrow{t_r^{\alpha_r + d_r}} \mathscr{F}_{[r-1]} \xrightarrow{\tau_r^{r-1}} \mathscr{F}_{[r]} \to 0 \quad (1 \leqslant r \leqslant m).$$

We also have the exact sequences on X

$$0 \to \mathscr{F}^{[r-1]} \to \mathscr{F} \xrightarrow{\tau_{r-1}} \mathscr{F}_{[r-1]} \to 0$$

(by τ_r^{r-1}, τ_{r-1} we have denoted the canonical morphisms . . .). One then derives the exact sequences

$$(*) \qquad H^q(f^{-1}(V), t_r^{d'_r} \mathscr{F}_{[r-1]}) \xrightarrow{t_r^{\alpha_r + d_r}} H^q(f^{-1}(V), \mathscr{F}_{[r-1]}) \xrightarrow{\tau_r^{r-1}} H^q(f^{-1}(V), \mathscr{F}_{[r]}),$$

$$(**) \qquad R^q f_*(\mathscr{F})_0 \xrightarrow{\tau_{r-1}} R^q f_*(\mathscr{F}_{[r-1]})_0 \xrightarrow{\delta} R^{q+1} f_*(\mathscr{F}^{[r-1]})_0.$$

It is enough to prove the following fact: if $s \in R^q f_*(\mathscr{F})_0$ satisfies $\tau_r(s) = 0$ $(1 \leqslant r \leqslant m)$, then there exists $s' \in R^q f_*(\mathscr{F})_0$ such that $\tau_{r-1}(s - t_r^{\alpha_r} s') = 0$. We thus suppose $s \in R^q f_*(\mathscr{F})_0$ enjoys the property $\tau_r(s) = 0$. Since $\tau_r^{r-1} \tau_{r-1} = \tau_r$, we derive that $\tau_{r-1}(s) \in \operatorname{Ker} \tau_r^{r-1}$. In virtue of $(*)$, there exists $s_1 \in R^q f_*(\mathscr{F}_{[r-1]})_0$ so that $t_r^{\alpha_r + d_r} s_1 = \tau_{r-1}(s)$. From $(**)$ it follows that $t_r^{\alpha_r + d_r} \delta(s_1) = \delta(t_r^{\alpha_r + d_r} s_1) = \delta(\tau_{r-1}(s)) = 0$. By the definition of d_r we deduce $\delta(t_r^{d_r} s_1) = 0$; hence there exists an element $s' \in R^q f_*(\mathscr{F})_0$ such that $\tau_{r-1}(s') = t_r^{d_r} s_1$. Hence $\tau_{r-1}(s - t_r^{\alpha_r} s') = 0$.

Proof of the second assertion of the theorem. We first check the *injectivity* of φ^q. Let $s = (s_k)_{k\geqslant 0} \in \varprojlim_k (R^q f_*(\mathcal{F})_y/\mathfrak{m}_y^k(R^q f_*(\mathcal{F})_y)$ be such that $\varphi^q(s) = 0$. Let F be the function from (i); if we eventually replace F by the function $k \mapsto \min(k, F(k))$, we can assume $F(k) \leqslant k$.

For any $k \in \mathbb{N}$, there exists $l \in \mathbb{N}$ such that $F(l) \geqslant k$. From the commutative diagram

$$\begin{array}{ccccc} R^qf_*(\hat{\mathfrak{m}}_y^l\mathcal{F})_y & \rightarrow & R^qf_*(\mathcal{F})_y & \xrightarrow{\eta_l} & R^qf_*(\mathcal{F}/\hat{\mathfrak{m}}_y^l\mathcal{F})_y \\ & & \searrow{\scriptstyle\varepsilon_l} & \nearrow{\scriptstyle\varphi_l^q} & \\ & & & R^qf_*(\mathcal{F})_y/\hat{\mathfrak{m}}_y^l(R^qf_*(\mathcal{F})_y) & \end{array}$$

where ε_l, η_l are the natural morphisms...), one can find an element $s_l' \in R^qf_*(\mathcal{F})_y$ such that $\varepsilon_l(s_l') = s_l$. Since $\varphi_l^q(s_l) = 0$, we get $\eta_l(s_l') = 0$. According to (i), $\varepsilon_{F(l)}(s_l') = 0$. Since $\varepsilon_{F(l)}(s_l') = s_{F(l)}$ and $F(l) \geqslant k$, it follows that $s_k = 0$. As k is an arbitrary integer, we derive that $s = 0$.

We now check the *surjectivity* of φ^q. Let $s = (s_k)_{k\in\mathbb{N}} \in \varprojlim_k (R^qf_*(\mathcal{F}/\hat{\mathfrak{m}}_y^k\mathcal{F})_y)$ Let us fix an integer $l \geqslant 0$. For $k \geqslant l$ we have the commutative diagram

$$\begin{array}{ccccc} R^qf_*(\mathcal{F}) & \xrightarrow{\eta_k} & R^qf_*(\mathcal{F}/\hat{\mathfrak{m}}_y^k\mathcal{F}) & \xrightarrow{\delta_k} & R^{q+1}f_*(\hat{\mathfrak{m}}_y^k\mathcal{F}) \\ \| & & \downarrow{\scriptstyle\eta_{l,k}} & & \downarrow{\scriptstyle\eta^{l,k}} \\ R^qf_*(\mathcal{F}) & \xrightarrow{\eta_l} & R^qf_*(\mathcal{F}/\hat{\mathfrak{m}}_y^l\mathcal{F}) & \xrightarrow{\delta_l} & R^{q+1}f_*(\hat{\mathfrak{m}}_y^l\mathcal{F}) \end{array}$$

$\eta_{l,k}$ and $\eta^{l,k}$ are the morphisms given by the natural maps $\mathcal{F}/\hat{\mathfrak{m}}_y^k\mathcal{F} \to \mathcal{F}/\hat{\mathfrak{m}}_y^l\mathcal{F}$, $\hat{\mathfrak{m}}_y^k\mathcal{F} \to \hat{\mathfrak{m}}_y^l\mathcal{F}$).

We now apply the assertion (i) for the sheaf $\hat{\mathfrak{m}}_y^l\mathcal{F}$. Then there exists a function $F : \mathbb{N} \to \mathbb{N}$ such that $\lim_{k\to\infty} F(k) = \infty$ and for $k \geqslant l$,

$$\operatorname{Im}(\eta^{l,k})_y \subset \mathfrak{m}_y^{F(k-l)}(R^{q+1}f_*(\hat{\mathfrak{m}}_y^l\mathcal{F})_y).$$

From the above diagram we have $\delta_l(s_l) \in \mathfrak{m}_y^{F(k-l)}(R^{q+1}f_*(\hat{\mathfrak{m}}_y^l\mathcal{F})_y)$ for $k \geqslant l$. In accordance with Krull's separation theorem we deduce that $\delta_l(s_l) = 0$. Therefore there exists an element $s_l' \in R^qf_*(\mathcal{F})_y$ such that $\eta_l(s_l') = s_l$. The image of s_l' in $R^qf_*(\mathcal{F})_y/\mathfrak{m}_y^l(R^qf_*(\mathcal{F})_y)$ has s_l as image by φ_l^q. The kernels of the maps φ_l^q are artinian (they are $\mathcal{O}_y/\mathfrak{m}_y^l$-modules of finite type) and, as in the proof of 1.11, we can find s'' such that $\varphi^q(s'') = s$. This completes the proof.

COROLLARY 3.3. *Under the assumptions of the theorem, the canonical morphism*

$$(f_*(\mathcal{F})_y)^\wedge \xrightarrow{\varphi^0} \varprojlim_k \Gamma(f^{-1}(y),\ \mathcal{F}/\hat{\mathfrak{m}}_y^k\mathcal{F})$$

is an isomorphism.

We will give now applications of the comparison theorem. Let $f : X \to Y$ be a morphism of complex spaces and $y \in Y$. The ringed space $(X_y = f^{-1}(y), \mathcal{O}_X/\hat{\mathfrak{m}}_y\mathcal{O}_X|X_y)$ is an analytic space, which is isomorphic to $X \times_Y (y, \mathcal{O}_y/\mathfrak{m}_y)$; it is called *the analytic fiber over* y. If $\mathcal{F} \in \mathbf{Coh}(X)$, then we denote $\mathcal{F}_y = \mathcal{F}/\hat{\mathfrak{m}}_y\mathcal{F}$. $\mathcal{F}_y$ is a coherent $\mathcal{O}_X/\hat{\mathfrak{m}}_y\mathcal{O}_X$-module and is called the *analytic fiber of* $\mathcal{F}$ *over* y.

Let M be an $\mathcal{O}_y$-module of finite type. In a neighbourhood V of y we may define a coherent analytic sheaf $\mathcal{M}$ such that $\mathcal{M}_y = M$.

If $f^V : f^{-1}(V) \to V$ is the restriction of f then we will denote

$$R^q f_*(\mathcal{F} \otimes_{\mathcal{O}_Y} M)_y = R^q f_*^V(\mathcal{F} \otimes_{\mathcal{O}_Y} \mathcal{M})_y$$

(for convenience we put $\mathcal{F} \otimes_{\mathcal{O}_Y} \mathcal{M}$ instead of $\mathcal{F} \otimes_{\mathcal{O}_X} f^*(\mathcal{M})$ and we will use this notation in what follows). $R^q f_*(\mathcal{F} \otimes_{\mathcal{O}_Y} M)_y$ has naturally a structure of an $\mathcal{O}_y$-module and it does not depend on $\mathcal{M}$; moreover, if f is a proper map, then it will be of finite type over $\mathcal{O}_y$. The canonical morphism

$$R^q f_*^V(\mathcal{F}) \otimes_{\mathcal{O}_Y} \mathcal{M} \to R^q f_*^V (\mathcal{F} \otimes_{\mathcal{O}_Y} \mathcal{M})$$

induces a canonical morphism

$$R^q f_*(\mathcal{F})_y \otimes_{\mathcal{O}_y} M \to R^q f_*^V(\mathcal{F} \otimes_{\mathcal{O}_Y} M)_y.$$

If M is of the form $\mathcal{O}_y/\mathfrak{m}_y^k$, then

$$R^q f_*(\mathcal{F} \otimes_{\mathcal{O}_Y} M)_y = R^q f_*(\mathcal{F}/\hat{\mathfrak{m}}_y^k\mathcal{F})_y = H^q(f^{-1}(y), \mathcal{F}/\hat{\mathfrak{m}}_y^k\mathcal{F}),$$

and the map $R^q f_*(\mathcal{F})_y \otimes_{\mathcal{O}} M \to R^q f_*(\mathcal{F} \otimes_{\mathcal{O}_Y} M)_y$ equals φ_k^q.

We also remark that all the above associations are functorial in M.

Now we prove, by applying 3.1, the following *theorem of base change:*

THEOREM 3.4. *Let $f : X \to Y$ be a proper morphism of complex spaces, $\mathcal{F}$ a coherent analytic sheaf on X which is flat with respect to f, y a point in Y, and q an integer. The following assertions are equivalent:*

(*a*) *The functor $M \mapsto R^q f_*(\mathcal{F} \otimes_{\mathcal{O}_Y} M)_y$ is right exact.*

(*a'*) *The functor $M \mapsto R^{q+1} f_*(\mathcal{F} \otimes_{\mathcal{O}_Y} M)_y$ is left exact.*

(*a''*) *The canonical morphisms $R^q f_*(\mathcal{F})_y \otimes_{\mathcal{O}_y} M \to R^q f_*(\mathcal{F} \otimes_{\mathcal{O}_Y} M)_y$ are isomorphisms, M being an arbitrary $\mathcal{O}_y$-module of finite type.*

(*b*) *The canonical map* $R^q f_*(\mathcal{F})_y \to R^q f_*(\mathcal{F}/\hat{\mathfrak{m}}_y\mathcal{F})_y$ *is surjective.*

(*c*) *The canonical maps* $R^q f_*(\mathcal{F}/\hat{\mathfrak{m}}_y^{k+1}\mathcal{F})_y \to R^q f_*(\mathcal{F}/\hat{\mathfrak{m}}_y\mathcal{F})_y$ *are surjective,* $k \geqslant 0$ *being an arbitrary integer.*

(*d*) *For any base change* $g : Y' \to Y$ *and for any* $y' \in Y'$ *such that* $g(y') = y$, *and taking into account the morphism* $f' : X' = X \times_Y Y' \to Y'$ *deduced from* f *and the sheaf* $\mathcal{F}'$ *on* X' *which is the inverse image of* $\mathcal{F}$ *on* X', *the canonical morphism* $g^*(R^q f_*(\mathcal{F})) \to R^q f'_*(\mathcal{F}')$ *is an isomorphism in a neighbourhood of* y'.

Proof. Let $0 \to M' \to M \to M'' \to 0$ be an exact sequence of $\mathcal{O}_y$-modules of finite type. In a neighbourhood V of y there exists an exact sequence of coherent analytic sheaves $0 \to \mathcal{M}' \to \mathcal{M} \to \mathcal{M}'' \to 0$, which induces the given sequence in y. Since $\mathcal{F}$ is f-flat we get the exact sequence of $f^{-1}(V)$

$$0 \to \mathcal{F} \otimes_{\mathcal{O}_Y} \mathcal{M}' \to \mathcal{F} \otimes_{\mathcal{O}_Y} \mathcal{M} \to \mathcal{F} \otimes_{\mathcal{O}_Y} \mathcal{M}'' \to 0.$$

Thus one obtains the exact sequence

$$\dots \to R^q f_*(\mathcal{F} \otimes_{\mathcal{O}_Y} M)_y \to R^q f_*(\mathcal{F} \otimes_{\mathcal{O}_Y} M'')_y \to R^{q+1} f_*(\mathcal{F} \otimes_{\mathcal{O}_Y} M')_y \to \dots$$

Consequently, the equivalence $(a) \Leftrightarrow (a')$ becomes obvious. The implication $(a'') \Rightarrow (a)$is also clear.

In order to prove that $(a) \Rightarrow (a'')$ we consider an exact sequence of the form $\mathcal{O}_y^{n_1} \to \mathcal{O}_y^{n_0} \to M \to 0$; a suitable commutative diagram and the fact that the morphism from (a'') is obviously an isomorphism for a free $\mathcal{O}_y$-module of finite rank lead immediately to the conclusion.

The implication $(a'') \Rightarrow (b)$ follows by taking $M = \mathcal{O}_y/\mathfrak{m}_y$ and the implication $(b) \Rightarrow (c)$ results from the factorization

$$R^q f_*(\mathcal{F})_y \to R^q f_*(\mathcal{F}/\hat{\mathfrak{m}}_y^{k+1}\mathcal{F})_y \to R^q f_*(\mathcal{F}/\hat{\mathfrak{m}}_y\mathcal{F})_y.$$

We prove that $(c) \Rightarrow (b)$. It is sufficient to show that we get a surjection if we pass to $\mathfrak{m}_y$-adic completion. According to the comparison theorem,

$$(R^q f_*(\mathcal{F})_y^{\wedge} \simeq \varprojlim_k (R^q f_*(\mathcal{F}/\hat{\mathfrak{m}}_y^{k+1}\mathcal{F})_y).$$

Since $R^q f_*(\mathcal{F}/\hat{\mathfrak{m}}_y\mathcal{F})_y = H^q(X_y, \mathcal{F}_y)$ is a vectorial space of finite dimension over $\mathcal{O}_y/\mathfrak{m}_y \simeq \mathbf{C}$, it follows that

$$R^q f_*(\mathcal{F}/\hat{\mathfrak{m}}_y\mathcal{F})_y^{\wedge} \simeq R^q f_*(\mathcal{F}/\hat{\mathfrak{m}}_y\mathcal{F})_y.$$

The conclusion results from 1.11 because the kernels of the maps from (*c*) are modules of finite type over the artinian rings $\mathcal{O}_y/\mathfrak{m}_y^{k+1}$, hence they are artinian modules.

We prove the implication (*b*) ⇒ (*a*). We must show that for any epimorphism $M \to M''$ of $\mathcal{O}_y$-modules of finite type, the map

$$R^q f_*(\mathcal{F} \otimes_{\mathcal{O}_Y} M)_y \to R^q f_*(\mathcal{F} \otimes_{\mathcal{O}_Y} M'')$$

is surjective. By making use of the commutative diagram

$$\begin{array}{ccc} R^q f_*(\mathcal{F})_y \otimes_{\mathcal{O}_y} M & \to & R^q f_*(\mathcal{F})_y \otimes_{\mathcal{O}_y} M'' \to 0 \\ \downarrow & & \downarrow \\ R^q f_*(\mathcal{F} \otimes_{\mathcal{O}_Y} M)_y & \to & R^q f_*(\mathcal{F} \otimes_{\mathcal{O}_Y} M'')_y \end{array}$$

the assertion (*a*) is proved as soon as the following assertion is verified:

(∗) "For any module of finite type M, the map

$R^q f_*(\mathcal{F})_y \otimes_{\mathcal{O}_y} M \to R^q f_*(\mathcal{F} \otimes_{\mathcal{O}_Y} M)_y$ is surjective".

We prove this fact. It is enough to verify that the map from the assertion (∗) becomes surjective if we pass to $\mathfrak{m}_y$-adic completion. We have $(R^q f_*(\mathcal{F})_y \otimes_{\mathcal{O}_y} M)^\wedge = \varprojlim_k (R^q f_*(\mathcal{F})_y \otimes_{\mathcal{O}_y} M \otimes_{\mathcal{O}_y} \mathcal{O}_y/\mathfrak{m}_y^{k+1}) = \varprojlim_k (R^q f_*(\mathcal{F})_y \otimes_{\mathcal{O}_y} M/\mathfrak{m}_y^{k+1} M)$. By 3.1 we also have $(R^q f_*(\mathcal{F} \otimes_{\mathcal{O}_Y} M)_y)^\wedge$

$$\simeq \varprojlim_k (R^q f_*(\mathcal{F} \otimes_{\mathcal{O}_Y} M/\hat{\mathfrak{m}}_y^{k+1}(\mathcal{F} \otimes_{\mathcal{O}_Y} M))_y) = \varprojlim_k (R^q f_*(\mathcal{F} \otimes_{\mathcal{O}_Y} M/\mathfrak{m}_y^{k+1} M)_y).$$

By applying 1.11 we have to prove that the maps

$$R^q f_*(\mathcal{F})_y \otimes_{\mathcal{O}_y} M/\mathfrak{m}_y^{k+1} M \to R^q f_*(\mathcal{F} \otimes_{\mathcal{O}_Y} M/\mathfrak{m}_y^{k+1} M)_y$$

are surjective. In other words, it remains to check (∗) in the particular case when M is annihilated by some power of $\mathfrak{m}_y$. We will show by descending induction that the maps

$$R^q f_*(\mathcal{F})_y \otimes_{\mathcal{O}_y} (\mathfrak{m}_y^k M) \to R^q f_*(\mathcal{F} \otimes_{\mathcal{O}_Y} \mathfrak{m}_y^k M)$$

are surjective ($k \geqslant 0$ integer). For $k = 0$ we obtain our very assertion. For sufficiently large k, $\mathfrak{m}_y^k M = 0$ and the conclusion is patent. For the general induction step we consider the exact sequence

$$0 \to \mathfrak{m}_y^{k+1} M \to \mathfrak{m}_y^k M \to \mathfrak{m}_y^k M/\mathfrak{m}_y^{k+1} M \to 0.$$

We get the exact commutative diagram:

$$\begin{array}{ccccc} R^q f_*(\mathcal{F})_y \otimes_{\mathcal{O}_y} (\mathfrak{m}_y^{k+1} M) & \to & R^q f_*(\mathcal{F})_y \otimes_{\mathcal{O}_y} (\mathfrak{m}_y^k M) & \to & R^q f_*(\mathcal{F})_y \otimes_{\mathcal{O}_y} (\mathfrak{m}_y^k M/\mathfrak{m}_y^{k+1} M) \to 0 \\ \downarrow & & \downarrow & & \downarrow \\ R^q f_*(\mathcal{F} \otimes_{\mathcal{O}_Y} \mathfrak{m}_y^{k+1} M)_y & \to & R^q f_*(\mathcal{F} \otimes_{\mathcal{O}_Y} \mathfrak{m}_y^k M)_y & \to & R^q f_*(\mathcal{F} \otimes_{\mathcal{O}_Y} (\mathfrak{m}_y^k M/\mathfrak{m}_y^{k+1} M))_y. \end{array}$$

The module $\mathfrak{m}_y^k M/\mathfrak{m}_y^{k+1}M$ is a vectorial space of finite dimension over $\mathcal{O}_y/\mathfrak{m}_y \simeq \mathbb{C}$, hence by using (b) and additivity the third vertical arrow is surjective. Thereby the induction step $k+1 \mapsto k$ is achieved.

The proof of the implication $(a'') \Rightarrow (d)$. We will prove that the maps

$$(**) \qquad g^*(R^q f_*(\mathcal{F}))_{y'}/\mathfrak{m}_{y'}^{k+1}(g^*(R^q f_*(\mathcal{F}))_{y'}) \to R^q f'_*(\mathcal{F}'/\hat{\mathfrak{m}}_{y'}^{k+1}\mathcal{F}'))_{y'},$$

which are deduced from the composition of natural maps

$$g^*(R^q f_*(\mathcal{F})) \to R^q f'_*(\mathcal{F}') \to R^q f'_*(\mathcal{F}'/\hat{\mathfrak{m}}_{y'}^{k+1}\mathcal{F}'),$$

are isomorphisms. By 3.1, we get that the maps

$$g^*(R^q f_*(\mathcal{F}))^{\wedge}_{y'} \to (R^q f'_*(\mathcal{F}')_{y'})^{\wedge}$$

are bijective.

The assertion (d) will then follow from the properties of the functor of completion. It is convenient to set

$$Y^{(k)} = (y, \mathcal{O}_y/\mathfrak{m}_y^{k+1}), \qquad Y'^{(k)} = (y', \mathcal{O}_{y'}/\mathfrak{m}_{y'}^{k+1}),$$

$$X^{(k)} = (f^{-1}(y), \mathcal{O}_X/\hat{\mathfrak{m}}_y^{k+1}\mathcal{O}_X|f^{-1}(y)), \; X'^{(k)} = (f'^{-1}(y'), \mathcal{O}_{X'}/\hat{\mathfrak{m}}_{y'}^{k+1}\mathcal{O}_{X'}|f'^{-1}(y')).$$

We denote by $f^{(k)}$, $g^{(k)}$, $f'^{(k)}$, respectively, the morphisms which correspond to f g, f' and consider the diagram

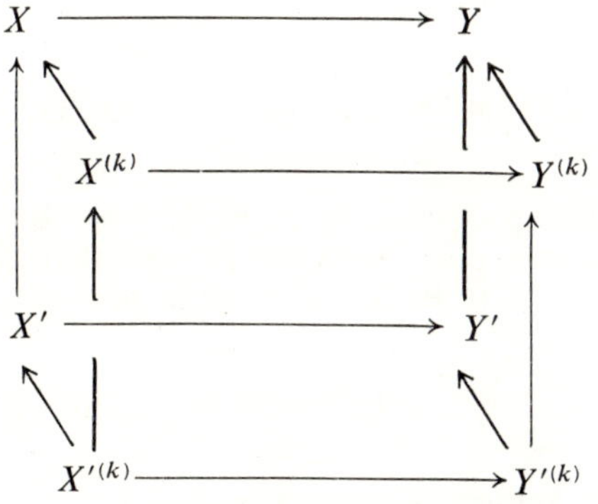

By means of (a'') applied to modules M of type $\mathcal{O}_y/\mathfrak{m}_y^{k+1}$, we can easily deduce that the source module in $(**)$ is canonically isomorphic to $g^{(k)*}(R^q f_*^{(k)}(\mathcal{F}^{(k)}))$ and the target module is canonically isomorphic to $R^q f_*'^{(k)}(\mathcal{F}'^{(k)})$, where $\mathcal{F}^{(k)}$ is the restriction at $X^{(k)}$ of $\mathcal{F}/\hat{\mathfrak{m}}_y^{k+1}\mathcal{F}$ and $\mathcal{F}'^{(k)}$ is its inverse image under the morphism $X'^{(k)} \to X^{(k)}$. $\mathcal{F}^{(k)}$ is flat over $Y^{(k)}$ and the hypothesis (a'') can be easily checked for $(X^{(k)}, Y^{(k)}, f^{(k)}, \mathcal{F}^{(k)}, y, q)$.

To prove the bijectivity of the maps$(**)$, it is then sufficient to prove the assertion (d) in case when Y and Y' are reduced to a point and their structural sheaves are artinian rings. In this case, (d) easily follows by applying (a'') for the module $M = \mathcal{O}_{y'}$ (which is an $\mathcal{O}_y$-module of finite type!).

It remains to prove *the implication* $(d) \Rightarrow (a'')$. Let M be an $\mathcal{O}_y$-module of finite type. If we substitute eventually Y for a neighbourhood of y, we can assume that there exists a coherent analytic sheaf $\mathcal{M}$ on Y such that $\mathcal{M}_y = M$. Consider the analytic space $Y' = (Y, \mathcal{O}_Y \oplus \mathcal{M})$ where $\mathcal{O}_Y \oplus \mathcal{M}$ is the Nagata sum (1.4) and let $g : Y' \to Y$ be the canonical morphism. In accordance with (d) we have a canonical isomorphism in a neighbourhood of y

$$R^q f_*(\mathcal{F}) \otimes_{\mathcal{O}_Y} (\mathcal{O}_Y \oplus \mathcal{M}) \cong R^q f_*(\mathcal{F} \otimes_{\mathcal{O}_Y} (\mathcal{O}_Y \oplus \mathcal{M})).$$

Then it will result that the morphism

$$R^q f_*(\mathcal{F}) \otimes_{\mathcal{O}_Y} \mathcal{M} \to R^q f_*(\mathcal{F} \otimes_{\mathcal{O}_Y} \mathcal{M})$$

is an isomorphism and passing to stalks in y, one gets the isomorphism stated in (a'').

The proof of the change base theorem is thus over.

We now consider a case when this theorem applies.

COROLLARY 3.5. *To the conditions of the theorem we suppose in addition* $H^q(X_y, \mathcal{F}_y) = 0$. *Then for any module* M *of finite type over* $\mathcal{O}_y$, $R^q f_*(\mathcal{F} \otimes_{\mathcal{O}_Y} M)_y = 0$; *in particular,* $R^q f_*(\mathcal{F})$ *vanishes in a neighbourhood of* y. *As a consequence, the equivalent conditions from 3.4 are fulfilled for the integer* $q - 1$; *in particular, for any* $k \geqslant 0$, *the canonical maps*

$$R^{q-1} f_*(\mathcal{F})_y / \mathfrak{m}_y^k (R^{q-1} f_*(\mathcal{F})_y) \to H^{q-1}(f^{-1}(y), \mathcal{F}/\hat{\mathfrak{m}}_y^k \mathcal{F})$$

are bijective.

Proof. Let M be an $\mathcal{O}_y$-module of finite type. By shrinking Y around y, we may assume the existence of $\mathcal{M} \in \mathbf{Coh}(Y)$ such that $\mathcal{M}_y = M$. We must show that $R^q f_*(\mathcal{F} \otimes_{\mathcal{O}_X} f^*(\mathcal{M}))_y = 0$. By 3.1 we have to prove that for any integer $k \geqslant 0$,

$$H^q(X_y, \mathcal{F} \otimes_{\mathcal{O}_X} f^*(\mathcal{M}) / \mathfrak{m}_y^k (\mathcal{F} \otimes_{\mathcal{O}_X} f^*(\mathcal{M}))) = 0.$$

We shall prove this by induction on k. For $k = 1$,

$$\mathcal{F} \otimes_{\mathcal{O}_X} f^*(\mathcal{M}) / \hat{\mathfrak{m}}_y (\mathcal{F} \otimes_{\mathcal{O}_X} f^*(\mathcal{M})) \simeq \mathcal{F}_y \otimes_{\mathbf{C}} \mathcal{M}_y / \mathfrak{m}_y \mathcal{M}_y$$

(we identify $\mathcal{O}_y/\mathfrak{m}_y$ with $\mathbf{C}$) and the conclusion follows by the hypothesis already made and by additivity.

In order to verify the general induction step, by using a suitable exact sequence, we have to prove that

$$H^q(X_y, \hat{\mathfrak{m}}_y^k (\mathcal{F} \otimes_{\mathcal{O}_X} f^*(\mathcal{M})) / \hat{\mathfrak{m}}_y^{k+1} (\mathcal{F} \otimes_{\mathcal{O}_X} f^*(\mathcal{M}))) = 0.$$

Reasoning on stalks and using the flatness of $\mathcal{F}$ over Y, one can easily deduce the isomorphism

$$\hat{\mathfrak{m}}_y^k(\mathcal{F} \otimes_{\mathcal{O}_X} f^*(\mathcal{M}))/\hat{\mathfrak{m}}_y^{k+1}(\mathcal{F} \otimes_{\mathcal{O}_X} f^*(\mathcal{M})) \simeq \mathcal{F}_y \otimes_{\mathbb{C}} (\mathfrak{m}_y^k\mathcal{M}/\mathfrak{m}_y^{k+1}\mathcal{M})$$

and again the conclusion follows by additivity from the hypothesis.

Remark. In the course of the proof the complex field has been identified with the constant sheaf $\mathbb{C}$ which was defined on X_y and the vectorial spaces $\mathcal{M}_y/\mathfrak{m}_y\mathcal{M}_y$, $\mathfrak{m}_y^k\mathcal{M}_y/\mathfrak{m}_y^{k+1}\mathcal{M}_y$ were identified with the constant sheaves (of $\mathbb{C}$-modules) defined on X_y by them. The structure of $\mathbb{C}$-module of $\mathcal{F}/\hat{\mathfrak{m}}_y\mathcal{F}$ is given by the inclusion $\mathbb{C} \to \mathcal{O}_X/\hat{\mathfrak{m}}_y\mathcal{O}_X$ and coincides with that deduced naturally from the identification $\mathcal{O}_y/\mathfrak{m}_y \simeq \mathbb{C}$.

A consequence of 3.5 is the following:

COROLLARY 3.6. *Let $f : X \to Y$ be a flat proper morphism of complex spaces and let $y \in Y$ be such that $H^1(X_y, \mathcal{O}_X/\hat{\mathfrak{m}}_y\mathcal{O}_X) = 0$. If $\mathcal{F}$ and $\mathcal{G}$ are two invertible sheaves on X such that the sheaves $\mathcal{F}_y$ and $\mathcal{G}_y$ are isomorphic, then there exists a neigbourhood V of y so that $\mathcal{F}|f^{-1}(V) \simeq \mathcal{G}|f^{-1}(V)$ and this isomorphism induces the given one $\mathcal{F}_y \simeq \mathcal{G}_y$.*

Proof. Consider the sheaf $\mathcal{H} = Hom_{\mathcal{O}}(\mathcal{F}, \mathcal{G}) \simeq \mathcal{F}^{-1} \otimes_{\mathcal{O}} \mathcal{G}$. We have $\mathcal{H}_y \simeq \simeq \mathcal{F}_y^{-1} \otimes_{\mathcal{O}_X/\hat{\mathfrak{m}}_y\mathcal{O}_X} \mathcal{G}_y \simeq \mathcal{O}_X/\hat{\mathfrak{m}}_y\mathcal{O}_X$, hence $H^1(X_y, \mathcal{H}_y) = 0$. According to 3.5, the canonical map

$$f_*(\mathcal{H})_y/\mathfrak{m}_y(f_*(\mathcal{H})_y) \to \Gamma(X_y, \mathcal{H}_y)$$

is bijective.

Suppose $\xi_y : \mathcal{F}_y \to \mathcal{G}_y$ is an isomorphism; it defines an element in $\Gamma(X_y, \mathcal{H}_y)$. Then there exists on some neighbourhood V of y a morphism $\xi : \mathcal{F}|f^{-1}(V) \to \to \mathcal{G}|f^{-1}(V)$ which induces the isomorphism ξ_y. Analogously, by shrinking eventually V, we deduce a morphism $\eta: \mathcal{G}|f^{-1}(V) \to \mathcal{F}|f^{-1}(V)$, which induces the isomorphism η_y, which is the inverse of ξ_y.

Let x be an arbitrary point of the fiber $f^{-1}(y)$. By identifying $\mathcal{F}_x$ with $\mathcal{O}_x$, the morphism $\eta_x\xi_x : \mathcal{F}_x \to \mathcal{F}_x$ corresponds to an element α of $\mathcal{O}_x$. The image of this in $\mathcal{O}_x/\mathfrak{m}_y\mathcal{O}_x$ corresponds to the morphism $(\eta_y)_x(\xi_y)_x =$ identity, hence it is just the unity. It will then result that α is invertible, hence $\eta_x\xi_x$ is an isomorphism. Similarly, we can deduce that $\xi_x\eta_x$ is an isomorphism. In this way, the composite maps $\xi\eta$ and $\eta\xi$ induce isomorphisms in all points of $f^{-1}(y)$. The same property holds on a neighbourhood of $f^{-1}(y)$; since f is proper, we may assume this neighbourhood of the form $f^{-1}(V)$.

Another consequence of the base theorem is the following *exactness criterion*:

COROLLARY 3.7. *Under the assumptions of theorem 3.4, the following assertions are equivalent:*

(*a*) *The functor $M \mapsto R^qf_*(\mathcal{F} \otimes_{\mathcal{O}_Y} M)_y$ is exact.*

(*b*) *The maps*

$$R^q f_*(\mathfrak{F})_y \to R^q f_*(\mathfrak{F}/\hat{\mathfrak{m}}_y \mathfrak{F})_y, \quad R^{q-1} f_*(\mathfrak{F})_y \to R^{q-1} f_*(\mathfrak{F}/\hat{\mathfrak{m}}_y \mathfrak{F})_y$$

are surjective.

(*c*) *For any integer* $k \geqslant 0$ *the maps*

$$R^q f_*(\mathfrak{F}/\hat{\mathfrak{m}}_y^{k+1}\mathfrak{F})_y \to R^q f_*(\mathfrak{F}/\hat{\mathfrak{m}}_y \mathfrak{F})_y, R^{q-1} f_*(\mathfrak{F}/\hat{\mathfrak{m}}_y^{k+1}\mathfrak{F})_y \to R^{q-1} f_*(\mathfrak{F}/\hat{\mathfrak{m}}_y \mathfrak{F})_y$$

are surjective.

(*d*) *The map*

$$R^q f_*(\mathfrak{F})_y \to R^q f_*(\mathfrak{F}/\hat{\mathfrak{m}}_y \mathfrak{F})_y$$

is surjective and $R^q f_*(\mathfrak{F})_y$ *is a free* $\mathcal{O}_y$*-module.*

The proof follows straightforwardly from 3.4 (recall that whenever N is an $\mathcal{O}_y$-module of finite type and the functor $M \mapsto N \otimes_{\mathcal{O}_y} M$ is exact, N will be free!).

If the equivalent conditions of this criterion are fulfilled, one says that f is *cohomologically flat in dimension q in the point y*. Here is a case when this criterion can be applied.

COROLLARY 3.8. *Under the assumptions of theorem 3.4, we suppose in addition that* $H^q(X_y, \mathfrak{F}_y) = H^{q-2}(X_y, \mathfrak{F}_y) = 0$. *Then* $R^{q-1} f_*(\mathfrak{F})$ *is free in a neighbourhood of y and the canonical maps*

$$R^{q-1} f_*(\mathfrak{F})_y / \mathfrak{m}_y^k (R^{q-1} f_*(\mathfrak{F}))_y \to R^{q-1} f_*(\mathfrak{F}/\hat{\mathfrak{m}}_y^k \mathfrak{F})_y$$

are bijective, $k \geqslant 0$ *being an arbitrary integer.*

The proof is immediate. In particular one obtains:

COROLLARY 3.9. *Let* $f : X \to Y$ *be a proper morphism of complex spaces,* $\mathfrak{F}$ *a coherent analytic sheaf on X flat with respect to f and* $y \in Y$. *If* $H^1(X_y, \mathfrak{F}_y) = 0$, *then* $f_*(\mathfrak{F})$ *is free in a neighbourhood of y and the maps*

$$f_*(\mathfrak{F})_y / \mathfrak{m}_y^k (f_*(\mathfrak{F})_y) \to H^0(f^{-1}(y), \mathfrak{F}/\hat{\mathfrak{m}}_y^k \mathfrak{F})$$

are bijective.

COROLLARY 3.10. *Under the assumptions of theorem 3.4, the following assertions are equivalent:*

(*a*) $\mathfrak{F}$ *is cohomologically flat in y, whatever dimension* $p, p \geqslant q$.

(*b*) $R^p f_*(\mathfrak{F})_y$ *is a free* $\mathcal{O}_y$*-module for any* $p \geqslant q$.

Proof. The implication (*a*) $\Rightarrow$ (*b*) follows obviously from the exactness criterion. For the inverse implication one proceeds by descending induction on p, and considers the fact that $R^p f_*(\mathfrak{F})_y = 0$ for p sufficiently large.

COROLLARY 3.11. *Under the assumptions of 3.4, the following assertions are equivalent:*

(*a*) $R^p f_*(\mathfrak{F})_y = 0$ *for any* $p \geqslant q$.

(*b*) $H^p(X_y, \mathfrak{F}_y) = 0$ *for any* $p \geqslant q$.

The proof derives by means of 3.7 and 3.10.

Next we consider an example.

PROPOSITION 3.12. *Let $f : X \to Y$ be a flat proper morphism of complex spaces and $y \in Y$. Suppose that the analytic fiber X_y is a reduced space. Then $\mathcal{O}_X$ is cohomologically flat in dimension zero in y; in particular $f_*(\mathcal{O}_X)$ is free around y and*

$$f_*(\mathcal{O}_X)_y/\mathfrak{m}_y f_*(\mathcal{O}_X)_y \xrightarrow{\sim} H^0(X_y, \mathcal{O}_X/\hat{\mathfrak{m}}_y\mathcal{O}_X).$$

Proof. By 3.7 it is sufficient to show that the maps

$$f_*(\mathcal{O}_X/\hat{\mathfrak{m}}_y^{k+1}\mathcal{O}_X)_y = \Gamma(X, \mathcal{O}_X/\hat{\mathfrak{m}}_y^{k+1}\mathcal{O}_X) \to \Gamma(X, \mathcal{O}_X/\hat{\mathfrak{m}}_y\mathcal{O}_X)$$

are surjective for any $k \geqslant 0$. By the change base $Y' = (y, \mathcal{O}_y/\mathfrak{m}_y^{k+1}) \to Y$, it is enough to prove the surjectivity of the map

$$\Gamma(X, \mathcal{O}_X) \to \Gamma(f^{-1}(y), \mathcal{O}_X/\hat{\mathfrak{m}}_y\mathcal{O}_X)$$

under the hypothesis of the statement, where Y is supposed in addition reduced to a point. Since $f^{-1}(y) = X$, we may assume X connected. In virtue of the assumptions previously made, the structural morphism $\mathbb{C} \to \Gamma(X_y, \mathcal{O}_X/\hat{\mathfrak{m}}_y\mathcal{O}_X)$ is an isomorphism and the conclusion follows from the composition of natural maps

$$\mathbb{C} \to \Gamma(Y, \mathcal{O}_Y) \to \Gamma(X, \mathcal{O}_X) \to \Gamma(X_y, \mathcal{O}_X/\hat{\mathfrak{m}}_y\mathcal{O}_X).$$

COROLLARY 3.13. *Under the assumptions of the proposition, if we require X_y connected, then the canonical morphism $\mathcal{O}_Y \to f_*(\mathcal{O}_X)$ is an isomorphism in a neighbourhood of y.*

§ 4. The semicontinuity and continuity theorems. The invariance of Euler-Poincaré characteristic

Let $f : X \to Y$ be a proper morphism of complex spaces, $\mathcal{F} \in \mathbf{Coh}(X)$ and $\mathcal{M} \in \mathbf{Coh}(Y)$. We denote, as in the previous paragraph, $\mathcal{F} \otimes_{\mathcal{O}_X} f^*(\mathcal{M})$ by $\mathcal{F} \otimes_{\mathcal{O}_Y} \mathcal{M}$. If V is an open subset of Y and $\mathcal{M} \in \mathbf{Coh}(V)$, then we set for convenience $\mathcal{F} \otimes_{\mathcal{O}_Y} \mathcal{M}$ (resp. $R^{\cdot}f_*(\mathcal{F} \otimes_{\mathcal{O}_Y} \mathcal{M})$) instead of $\mathcal{F} \,|\, f^{-1}(V) \otimes_{\mathcal{O}_V} \mathcal{M}$ (resp. $R^{\cdot}f_*^V(\mathcal{F} \,|\, f^{-1}(V) \otimes_{\mathcal{O}_Y} \mathcal{M})$), where $f^V : f^{-1}(V) \to V$ is the restriction of f.

The sheaf $\mathcal{F}$ is said to be *transversal on* $\mathcal{M}$ [47] if for any $x \in f^{-1}(V)$,

$$\mathrm{Tor}_i^{\mathcal{O}_y}(\mathcal{F}_x, \mathcal{M}_y) = 0, \quad \text{for } i \geqslant 1, \quad y = f(x).$$

If $\mathcal{F}$ is flat with respect to Y, then it is transversal on any $\mathcal{M}$.

The first important result of the paragraph is the following theorem of Kiehl-Verdier and Schneider ([47] and [72]):

THEOREM 4.1. *Let* $X \xrightarrow{f} Y$ *be a proper morphism of complex spaces,* $\mathcal{F}$ *a coherent analytic sheaf on* X*, and* $y_0 \in Y$*. Then there exist a neighbourhood* V *of* y_0 *and a complex* $\mathcal{L}^{\cdot}$ *of free sheaves of finite rank on* V*, which is bounded above, and which satisfies the following assertion:*

"For any coherent analytic sheaf $\mathcal{M}$ *defined on an open subset of* Y *such that* $\mathcal{F}$ *is transversal on it, there is an isomorphism*

$$R^{\cdot}f_*(\mathcal{F} \otimes_{\mathcal{O}_Y} \mathcal{M}) \simeq H^{\cdot}(\mathcal{L}^{\cdot} \otimes_{\mathcal{O}_Y} \mathcal{M})$$

which is functorial in $\mathcal{M}$ *and compatible with the restrictions and with the short exact sequences in the argument* $\mathcal{M}$ *(hence it is a functorial isomorphism of* ∂*-functors!)".*

Moreover, if $\mathcal{F}$ *is flat with respect to* Y*, then the complex* $\mathcal{L}^{\cdot}$ *can be supposed bounded.*

The proof needs some preparations. We shall use the notations and the facts from § 2.

(a) Let $V_*, r_*, r_{**}, j_k, \ldots$ be the entities from the proof of the finiteness theorem. Let $\mathcal{R}$ be a link system of sheaves. If $\mathcal{M} \in \mathbf{Coh}(Y)$, then we denote by $\mathcal{R} \otimes_{\mathcal{O}_Y} \mathcal{M}$ the link system of sheaves, of components $\mathcal{R}_\alpha \otimes_{\mathcal{O}_Y} \mathcal{M} (= \mathcal{R}_\alpha \otimes_{\mathcal{O}_{D_\alpha(r_{**}) \times V_*}} p_\alpha^*(\mathcal{M})$, where $p_\alpha : D_\alpha(r_{**}) \times V_* \to V_*$ are the canonical projections).

In fact we will consider coherent sheaves $\mathcal{M}$ on open subsets on Y; in this case we adopt the abuse of notation which has been recalled at the beginning of the paragraph.

If $\mathcal{R} \to \mathcal{R}'$ is a morphism of link systems of sheaves, then one naturally obtains a morphism $\mathcal{R} \otimes_{\mathcal{O}_Y} \mathcal{M} \to \mathcal{R}' \otimes_{\mathcal{O}_Y} \mathcal{M}$. On the other hand, if $\mathcal{M} \to \mathcal{N}$ is a morphism of sheaves, then one naturally gets a morphism of link systems $\mathcal{R} \otimes_{\mathcal{O}_Y} \mathcal{M} \to \mathcal{R}' \otimes_{\mathcal{O}_Y} \mathcal{N}$. One can easily check the functorial character of these correspondences.

If $0 \to \mathcal{M}' \to \mathcal{M} \to \mathcal{M}'' \to 0$ is an exact sequence of coherent analytic sheaves (defined on an open subset of Y), then the sequence

$$0 \to \mathcal{R} \otimes_{\mathcal{O}_Y} \mathcal{M}' \to \mathcal{R} \otimes_{\mathcal{O}_Y} \mathcal{M} \to \mathcal{R} \otimes_{\mathcal{O}_Y} \mathcal{M}'' \to 0$$

is exact for any free system $\mathcal{R}$; this fact follows easily from the flatness of the projections $D_\alpha \times V_* \to V_*$.

One can also check without difficulty the existence of a canonical isomorphism

$$j_*(\mathcal{F}) \otimes_{\mathcal{O}_Y} \mathcal{M} \xrightarrow{\sim} j_*(\mathcal{F} \otimes_{\mathcal{O}_Y} \mathcal{M}).$$

(b) Let $\mathcal{R}^{\cdot}$ be the resolution of $j_*(\mathcal{F})$ built in § 2. Thus we have the exact sequence of systems of sheaves

$$\ldots \to \mathcal{R}^2 \to \mathcal{R}^1 \to \mathcal{R}^0 \to j_*(\mathcal{F}) \to 0.$$

Let $\mathcal{M}$ be a coherent analytic sheaf defined on an open subset of Y. By **(a)** one gets a complex $\mathcal{R}^{\bullet} \otimes_{\mathcal{O}_Y} \mathcal{M}$ and an augmentation morphism

$$\mathcal{R}^{\bullet} \otimes_{\mathcal{O}_Y} \mathcal{M} \to j_*(\mathcal{F} \otimes_{\mathcal{O}_Y} \mathcal{M}).$$

LEMMA 4.2. *If $\mathcal{F}$ is transversal on $\mathcal{M}$, then $\mathcal{R}^{\bullet} \otimes_{\mathcal{O}_Y} \mathcal{M}$ is a resolution for $j_*(\mathcal{F} \otimes_{\mathcal{O}_Y} \mathcal{M})$.*

Proof. We have to show that for any simplex α, the sequence

$$\ldots \to \mathcal{R}^2_\alpha \otimes_{\mathcal{O}_Y} \mathcal{M} \to \mathcal{R}^1_\alpha \otimes_{\mathcal{O}_Y} \mathcal{M} \to \mathcal{R}^0_\alpha \otimes_{\mathcal{O}_Y} \mathcal{M} \to j_\alpha(\mathcal{F}) \otimes_{\mathcal{O}_Y} \mathcal{M} \to 0$$

is exact. This fact follows easily by dividing the exact sequence

$$\ldots \to \mathcal{R}^2_\alpha \to \mathcal{R}^1_\alpha \to \mathcal{R}^0_\alpha \to j_\alpha(\mathcal{F}) \to 0$$

into short exact sequences; the reasoning goes from the right to the left and by using the hypothesis of transversality and the flatness of the projection $D_\alpha \times V_* \to V_*$.

(c) Recall that for any open subset $V \subset V_*$ and for any r, $r_* \leqslant r \leqslant r_{**}$ we denote by $C^{\bullet}(r, V; \mathcal{R}^{\bullet})$ the Čech complex of components

$$C^l(r, V; \mathcal{R}^k) = \prod_{\alpha \in \Delta_l} \Gamma(D_\alpha(r) \times V, \mathcal{R}^k_\alpha).$$

The simple associated complex was denoted by $C^{\bullet}(r, V)$. The assignement $V \mapsto$ $\mapsto C^{\bullet}(r, V)$, together with the restriction maps that have been deduced canonically, give rise to a complex of sheaves $C^{\bullet}(r)$ on V_*.

Let now $\mathcal{M}$ be a coherent analytic sheaf defined on an open subset of V_*. We denote by $C^{\bullet}(r, \mathcal{M})$ the complex of sheaves defined as above, replacing $\mathcal{R}^{\bullet}$ by $\mathcal{R}^{\bullet} \otimes_{\mathcal{O}_Y} \mathcal{M}$. Therefore, if $V \subset V_*$ is an open subset contained in the open set where $\mathcal{M}$ is defined, then

$$\Gamma(V, C^n(r, \mathcal{M})) = \prod_{l-k=n} C^l(r, V; \mathcal{R}^k \otimes_{\mathcal{O}_Y} \mathcal{M}),$$

where $C^l(r, V; \mathcal{R}^k \otimes_{\mathcal{O}_Y} \mathcal{M}) = \prod_{\alpha \in \Delta_l} \Gamma(D_\alpha(r) \times V, \mathcal{R}^k_\alpha \otimes_{\mathcal{O}_Y} \mathcal{M})$.

LEMMA 4.3. *There exists a canonical isomorphism of complexes*

$$C^{\bullet}(r) \otimes_{\mathcal{O}_Y} \mathcal{M} \to C^{\bullet}(r, \mathcal{M}),$$

functorial in $\mathcal{M}$.

Proof. The lemma easily results from the definitions if we use the following fact: if Y is a complex space, D an open polydisc, $X = D \times Y$, $p : X \to Y$ the projection, $\mathcal{M} \in \mathbf{Coh}(Y)$ and $\mathcal{R}$ a free sheaf of finite rank on X, then the natural morphism

$$p_*(\mathcal{R}) \otimes_{\mathcal{O}_Y} \mathcal{M} \to p_*(\mathcal{R} \otimes_{\mathcal{O}_X} p^*(\mathcal{M}))$$

is an isomorphism.

We prove this fact; we can assume that $\mathcal{R} = \mathcal{O}_X$, hence we must prove that the morphism

$$p_*(\mathcal{O}_X) \otimes_{\mathcal{O}_Y} \mathcal{M} \to p_* p^*(\mathcal{M})$$

is an isomorphism. If $\mathcal{M}$ is a free $\mathcal{O}_Y$-module of finite rank this is obvious. For the general case, the problem being local on Y, we may assume the existence of an exact sequence of the form

$$\mathcal{O}_Y^m \to \mathcal{O}_Y^n \to \mathcal{M} \to 0.$$

Then the conclusion follows from a suitable diagram, by means of the fact that functor p_* is exact on coherent sheaves.

The augmentation morphism

$$\mathcal{R}^0 \otimes_{\mathcal{O}_Y} \mathcal{M} \to j_*(\mathcal{F} \otimes_{\mathcal{O}_Y} \mathcal{M})$$

induces naturally an augmentation morphism

$$C^\bullet(r, \mathcal{M}) \to C^\bullet(r, j_*(\mathcal{F} \otimes_{\mathcal{O}_Y} \mathcal{M})),$$

where the target complex is the Čech complex with respect to the system $j_*(\mathcal{F} \otimes_{\mathcal{O}_Y} \mathcal{M})$.

If V is as above, then we have:

$$\Gamma(V, C^n(r, j_*(\mathcal{F} \otimes_{\mathcal{O}_Y} \mathcal{M}))) = \prod_{\alpha \in \Delta_n} \Gamma(D_\alpha(r) \times V, j_*(\mathcal{F} \otimes_{\mathcal{O}_Y} \mathcal{M})) =$$

$$= \prod_{\alpha \in \Delta_n} \Gamma(U_\alpha \cap f^{-1}(V), \mathcal{F} \otimes_{\mathcal{O}_Y} \mathcal{M}).$$

Hence $\Gamma(V, C^\bullet(r, j_*(\mathcal{F} \otimes_{\mathcal{O}_Y} \mathcal{M}))) = C^\bullet(r, V; j_*(\mathcal{F} \otimes_{\mathcal{O}_Y} \mathcal{M}))$ coincides with the Čech complex associated to the covering $\mathcal{U}(r, V)$ and to the sheaf $\mathcal{F} \otimes_{\mathcal{O}_Y} \mathcal{M}$.

LEMMA 4.4. *If $\mathcal{F}$ is transversal on $\mathcal{M}$, then for any r, $r_* \leqslant r \leqslant r_{**}$, we have natural isomorphisms*

$$R^n f_*(\mathcal{F} \otimes_{\mathcal{O}_Y} \mathcal{M}) \simeq H^n(C^\bullet(r, \mathcal{M})).$$

In particular the restriction maps

$$C^\bullet(r, \mathcal{M}) \to C^\bullet(r', \mathcal{M})$$

are quasi-isomorphisms (even on Stein open subsets) for all real numbers $r, r_* \leqslant r' \leqslant r$.

Proof. Let $V \subset V_*$ be a Stein open subset contained in the open set where $\mathcal{M}$ is defined. Since the sheaves $\mathcal{R}_\alpha^k \otimes_{\mathcal{O}_Y} \mathcal{M}$ are coherent, exactly as in 2.4 one deduces that the augmentation maps

$$\Gamma(V, C^\bullet(r, \mathcal{M})) \to C^\bullet(r, V; j_*(\mathcal{F} \otimes_{\mathcal{O}_Y} \mathcal{M}))$$

are quasi-isomorphisms; we thus get isomorphisms

$$H^n(\Gamma(V, C^\bullet(r, \mathcal{M}))) \xrightarrow{\sim} H^n(f^{-1}(V), \mathcal{F} \otimes_{\mathcal{O}_Y} \mathcal{M}).$$

All these agree with the restrictions given by inclusions $V' \subset V$; in this way, we derive the isomorphisms stated in the lemma.

(d) We use the conditions and the notation from above and prove that for any integer n the following holds.

LEMMA $C(n)$. *Let V_n, r_n, $\mathcal{L}^\bullet$, $\mathcal{L}^\bullet \to C^\bullet(r_n)$ be the entities which verify lemma $A(n)$, which was built in the proof of the finiteness theorem. Then for any Stein neighbourhood V_n' of y_0, $V_n' \subset\subset V_n$ and for any real number ρ_n, $r_* < \rho_n < r_n$, the following assertion is true:*

"For any coherent analytic sheaf $\mathcal{M}$ defined on an open subset of V_n' such that $\mathcal{F}$ is transversal on it, the induced morphism

$$\mathcal{L}^\bullet \otimes_{\mathcal{O}_Y} \mathcal{M} \to C^\bullet(\rho_n, \mathcal{M}) = C^\bullet(\rho_n) \otimes_{\mathcal{O}_Y} \mathcal{M}$$

s an n-quasi-isomorphism".

Proof. Recall that for any r, $r_* \leqslant r \leqslant r_{**}$, we denoted by $K^\bullet(r)$ the cone of the map $\mathcal{L}^\bullet \to C^\bullet(r)$. Let V_n', ρ_n and $\mathcal{M}$ be as in the statement. Similarly, we denote by $K^\bullet(r, \mathcal{M})$ the cone of the map $\mathcal{L}^\bullet \otimes_{\mathcal{O}_Y} \mathcal{M} \to C^\bullet(r, \mathcal{M})$. From lemma 4.3 one easily derives a natural isomorphism

$$K^\bullet(r) \otimes_{\mathcal{O}_Y} \mathcal{M} \xrightarrow{\sim} K^\bullet(r, \mathcal{M}).$$

We shall prove the exactness of the complex $K^\bullet(\rho_n, \mathcal{M})$ in any dimension $i \geqslant n$ and the lemma will be concluded.

The problem is of local nature, so we can suppose that there is an epimorphism $\mathcal{O}^q \xrightarrow{\theta} \mathcal{M} \to 0$ on the domain where $\mathcal{M}$ is defined. Let i be an integer, $i \geqslant n$. Denote by ∂ the boundary and by β the maps induced naturally by restrictions. We agree to use also θ for different maps naturally induced by θ.

By the very way of constructing the complex $\mathcal{L}^\bullet$, there exists a continuous $\mathcal{O}_{V_n}$-morphism of sheaves

$$h : K^{i+1}(r_n) \to K^i(\rho_n)$$

such that the diagram

$$\begin{array}{ccc} K^{i+1}(r_n) & \hookleftarrow & Z^{i+1}(K^\bullet(r_n)) \\ {\scriptstyle h}\downarrow & & \downarrow{\scriptstyle \beta} \\ K^i(\rho_n) & \xrightarrow{\partial} & Z^{i+1}(K^\bullet(\rho_n)) \end{array}$$

is commutative. Then we obtain canonically maps of $\mathcal{O}_{V_n}$-modules, which are also denoted by h, $K^{i+1}(r_n, \mathcal{O}_Y^q) \to K^i(\rho_n, \mathcal{O}_Y^q)$, $K^{i+1}(r_n, \mathcal{M}) \to K^i(\rho_n, \mathcal{M})$, such that the diagram

$$\begin{array}{ccc} K^{i+1}(r_n, \mathcal{C}_Y^q) & \xrightarrow{\theta} & K^{i+1}(r_n, \mathfrak{M}) \\ {\scriptstyle h}\downarrow & & \downarrow{\scriptstyle h} \\ K^i(\rho_n, \mathcal{C}_Y^q) & \xrightarrow{\theta} & K^i(\rho_n, \mathfrak{M}) \end{array}$$

is commutative. Moreover, the maps θ result to be epimorphisms at the level of sections over Stein open sets.

The commutativity of the diagram $(*)$ implies by additivity the commutativity of the diagram

$$(**) \qquad \begin{array}{ccc} K^{i+1}(r_n, \mathcal{O}_Y^q) & \hookleftarrow & Z^{i+1}(K^\bullet(r_n, \mathcal{O}_Y^q)) \\ {\scriptstyle h}\downarrow & & \downarrow{\scriptstyle \beta} \\ K^i(\rho_n, \mathcal{O}_Y^q) & \xrightarrow{\partial} & Z^{i+1}(K^\bullet(\rho_n, \mathcal{O}_Y^q)). \end{array}$$

By applying lemmas 4.4, 1.9, it results that the map $K^\bullet(r_n, \mathfrak{M}) \to K^\bullet(\rho_n, \mathfrak{M})$ is a quasi-isomorphism. We now prove the exactness of the sequence

$$K^{i-1}(\rho_n, \mathfrak{M}) \to K^i(\rho_n, \mathfrak{M}) \to K^{i+1}(\rho_n, \mathfrak{M}).$$

Let $\varphi \in Z^i(K^\bullet(\rho_n, \mathfrak{M}))$ (we omit to indicate the domain of definition which is always supposed to be a Stein open set). Choose $\psi \in Z^i(K^\bullet(r_n, \mathfrak{M}))$ such that $\varphi = \beta(\psi) +$ + coboundary. Let $\psi' \in K^i(r_n, \mathcal{O}_Y^q)$ such that $\theta(\psi') = \psi$. As follows from the diagram $(**)$, the element $\beta(\psi') - h(\partial\psi')$ of $K^i(\rho_n, \mathcal{C}_Y^q)$ is just a cocycle. Since the sequence

$$K^{i-1}(\rho_n, \mathcal{O}_Y^q) \to K^i(\rho_n, \mathcal{C}_Y^q) \to K^{i+1}(\rho_n, \mathcal{C}_Y^q)$$

is exact, there exists $\varphi' \in K^{i-1}(\rho_n, \mathcal{O}_Y^q)$ such that $\partial\varphi' = \beta(\psi') - h(\partial\psi')$. We then get $\partial(\theta\varphi') = \theta(\partial\varphi') = \theta(\beta(\psi')) - \theta(h(\partial\psi')) = \beta(\theta(\psi')) - h(\theta(\partial\psi')) = \beta(\psi) -$ $- h(\partial(\theta\psi')) = \beta(\psi) - h(\partial\psi) = \beta(\psi)$. Therefore φ is a coboundary and the lemma $C(n)$ is completely proved.

(e) *The proof of the first assertion of the theorem* follows from lemmas $C(-1)$ and 4.4. The functoriality and the compatibility with the short exact sequences follow easily from the functoriality of the isomorphisms given by 4.4 and from the exact commutative diagrams of the type

$$\begin{array}{ccc} \mathfrak{L}^\bullet \otimes \mathfrak{M} & \longrightarrow & \mathfrak{L}^\bullet \otimes \mathfrak{N} \\ \downarrow & & \downarrow \\ C^\bullet(\rho_n, \mathfrak{M}) & \to & C^\bullet(\rho_n, \mathfrak{N}) \end{array} \qquad \begin{array}{ccccccccc} 0 & \to & \mathfrak{L}^\bullet \otimes \mathfrak{M}' & \to & \mathfrak{L}^\bullet \otimes \mathfrak{M} & \to & \mathfrak{L}^\bullet \otimes \mathfrak{M}'' & \to & 0 \\ & & \downarrow & & \downarrow & & \downarrow & & \\ 0 & \to & C^\bullet(\rho_n, \mathfrak{M}') & \to & C^\bullet(\rho_n, \mathfrak{M}) & \to & C^\bullet(\rho_n, \mathfrak{M}'') & \to & 0 \end{array}$$

In order to prove the second assertion we need some other preparations. We use the preceding notations.

(f) LEMMA 4.5. *Let A be a noetherian local ring and M an A-module of finite type which has a resolution*

$$0 \to L^{n_0} \to \ldots \to L^1 \to L^0 \to M \to 0$$

by free modules. Under these assumptions, if

$$\ldots \to R^1 \to R^0 \to M \to 0$$

is a resolution of M by free modules of finite rank, then $N = \operatorname{Ker}(R^{n_0-1} \to R^{n_0-2})$ is a free module.

The proof is an elementary fact of homological algebra [16]: from the first resolution we deduce that $\operatorname{Ext}^i_A(M, *) = 0$ for $i > n_0$ and from the second one, $\operatorname{Ext}^i_A(N, *) \simeq \operatorname{Ext}^{n_0+i}_A(M, *)$, hence N is a projective module, therefore it is free.

PROPOSITION 4.6 [20]. *Let Y be a complex space, let U be an open subset of a numerical space, and $\mathcal{F} \in \mathbf{Coh}\,(U \times Y)$, flat with respect to Y. Under these assumptions, for any point $(x_0, y_0) \in U \times Y$ there is on a neighbourhood of (x_0, y_0) a finite resolution of $\mathcal{F}$ by free sheaves of finite rank.*

Proof. Let (x_0, y_0) be an arbitrary point of $U \times Y$. Consider the closed immersion $U \overset{j}{\to} U \times Y$, $x \mapsto (x, y_0)$. Denote by $\mathcal{F}(y_0)$ the sheaf $j^*(\mathcal{F})$. If $p : U \times Y \to Y$ is the projection, then $j_*(\mathcal{F}(y_0))$ is the sheaf $\mathcal{F}/\hat{\mathfrak{m}}_{y_0}\mathcal{F}$. By syszggies theorem there exists on a neighbourhood of x_0 a finite resolution $\mathcal{L}^0_\bullet$ for $\mathcal{F}(y_0)$ by free sheaves of finite rank. We set $\mathcal{L}^0_{-1} = \mathcal{F}(y_0)$. If $\mathcal{L}^0_i$ is of the form $\mathcal{O}^{r_i}_U$, then we set $\mathcal{L}_i = \mathcal{O}^{r_i}_{U \times Y}$, $i \geqslant 0$. Denote also $\mathcal{L}_{-1} = \mathcal{F}$ and $\mathcal{K}^0_i = \operatorname{Ker}(\mathcal{L}^0_i \xrightarrow{d^0_i} \mathcal{L}^0_{i-1})$. We construct in a neighbourhood of (x_0, y_0), morphisms $d_i : \mathcal{L}_i \to \mathcal{L}_{i-1}$ such that $j^*(d_i) = d^0_i$ and show that the sheaves $\mathcal{K}_i = \operatorname{Ker} d_i$ are Y-flat and moreover $\mathcal{K}_i(y_0)\,(= j^*(\mathcal{K}_i)) = \mathcal{K}^0_i$. To do this we proceed by induction on i. Suppose d_i already constructed and the properties relative to $\mathcal{K}_i$ verified. One can construct $d_{i+1} : \mathcal{L}_{i+1} \to \mathcal{L}_i$ such that the diagram

$$\begin{array}{ccc} \mathcal{L}_{i+1} & \xrightarrow{d_{i+1}} & \mathcal{K}_i \\ \downarrow & & \downarrow \\ j_*(\mathcal{L}^0_{i+1}) & \xrightarrow{j_*(d^0_{i+1})} & j_*(\mathcal{K}^0_i) \end{array}$$

is commutative. By Nakayama lemma we shall then deduce that $\operatorname{Im} d_{i+1} = \mathcal{K}_i$ in the point (x_0, y_0), hence also in a neighbourhood of this point. The exact sequence

$$0 \to \mathcal{K}_{i+1} \to \mathcal{L}_{i+1} \to \mathcal{K}_i \to 0$$

where $\mathcal{K}_i$ and $\mathcal{L}_{i+1}$ are Y-flat shows that the sheaf $\mathcal{K}_{i+1}$ is Y-flat and that $\mathcal{K}_{i+1}(y_0) = \mathcal{K}^0_{i+1}$. The beginning of the recurrence is similar when the hypothesis of flatness concerning $\mathcal{F}$ is used. One thus obtains the resolution required in the proposition (remark, by construction, that $\operatorname{Im} d_{i+1} = \mathcal{K}_i$).

(g) *The proof of the last assertion of the theorem.* Since Δ is finite, then by lemma 4.5 and proposition 4.6 we may suppose that the resolution $\mathcal{R}^\bullet$ of $j_*(\mathcal{F})$ is bounded. Accordingly, the complexes $C^\bullet(r)$ are bounded and let n_0 be such that $C^n(r) = 0$ for $n < n_0$.

By the proof of theorem 2.1, there exist a Stein neighbourhood V' of y_0, a number $r, r_* < r \leqslant r_{**}$, a bounded complex

$$\mathcal{L}^\bullet : \ldots \to 0 \to \mathcal{L}^{n_0-1} \to \mathcal{L}^{n_0} \to \mathcal{L}^{n_0+1} \to \ldots$$

of free $\mathcal{O}_{V'}$-modules of finite rank and an $(n_0 - 1)$-quasi-isomorphism $\sigma : \mathcal{L}^\bullet \to C^\bullet(r)$. We set $\mathcal{N} = \operatorname{Ker}(\mathcal{L}^{n_0-1} \to \mathcal{L}^{n_0})$ and let $K^\bullet(r)$ be the cone of σ. We have $K^{n_0-2}(r) = \mathcal{L}^{n_0-1}$ and $K^{n_0-1}(r) = \mathcal{L}^{n_0}$, hence $Z^{n_0-2}(K^\bullet(r)) = \mathcal{N}$.

By lemma $B(n_0 - 2)$, there is a morphism $\tau : \mathcal{L}^{n_0-1} \to \mathcal{N}$ such that $\tau\nu = \mathrm{id}$ where $\nu : \mathcal{N} \to \mathcal{L}^{n_0-1}$ is the inclusion. Thus the sheaf $\mathcal{N}$ results to be locally free and by shrinking eventually V' we may suppose it free. Consider the complex

$$\hat{\mathcal{L}}^\bullet : \ldots \to 0 \to \mathcal{N} \to \mathcal{L}^{n_0-1} \to \mathcal{L}^{n_0} \to \ldots$$

The morphism σ extends to a quasi-isomorphism $\hat{\sigma} : \hat{\mathcal{L}}^\bullet \to C^\bullet(r)$. The complex $\hat{\mathcal{L}}^\bullet$ and the morphism $\hat{\sigma}$ satisfy lemmas $A(n)$ and $B(n)$. The proof of the theorem is concluded by applying again lemma $C(n_0 - 3)$ and lemma 4.4.

Remark. Theorem 4.1 allows us to give a new proof for the comparison theorem for sheaves of the type $\mathcal{F} \otimes_{\mathcal{O}_Y} \mathcal{M}$, $\mathcal{F}$ flat with respect to Y (EGA III, 7.4.8). In this way the change base theorem is also a consequence of 4.1.

We confine ourselves to prove the following *complement to the change base theorem.*

THEOREM 4.7. *The equivalent conditions of 3.4 are also equivalent to the following ones:*

(e) If $\mathcal{L}^\bullet$ is a complex of sheaves defined in a neighbourhood of y and satisfying the assertions from 4.1, then the sheaf $Z'^{q+1}(\mathcal{L}^\bullet) = \operatorname{Coker}(\mathcal{L}^q \to \mathcal{L}^{q+1})$ *is free around y.*

(f) One can find a complex $\mathcal{L}'^\bullet$ in a neighbourhood of y, which verifies the assertions from 4.1 and such that the differential $\mathcal{L}'^q \to \mathcal{L}'^{q+1}$ is null.

(g) There exists an $\mathcal{O}_y$-module of finite type $\mathcal{T}_y$ (which is unique, up to isomorphism) such that the functorial isomorphism

$$R^{q+1}f_*(\mathcal{F} \otimes_{\mathcal{O}_Y} M)_y \simeq \operatorname{Hom}_{\mathcal{O}_y}(\mathcal{T}_y, M)$$

holds, M being a module of finite type over $\mathcal{O}_y$.

Proof. $(a') \Rightarrow (e)$. It is sufficient to prove that $Z'^{q+1}(\mathcal{L}^\bullet)_y = Z'^{q+1}(\mathcal{L}^\bullet_y)$ is a flat $\mathcal{O}_y$-module. Let $0 \to M' \to M$ be a monomorphism of $\mathcal{O}_y$-modules of finite type. From the right exactness of the tensor product one can easily deduce the exact commutative diagram

$$\begin{array}{ccccccc} 0 \to R^{q+1}f_*(\mathcal{F} \otimes_{\mathcal{O}_Y} M')_y & \to & Z'^{q+1}(\mathcal{L}^\bullet)_y \otimes_{\mathcal{O}_y} M' & \to & \mathcal{L}_y^{q+2} \otimes_{\mathcal{O}_y} M' \\ \downarrow & & \downarrow & & \downarrow \\ 0 \to R^{q+1}f_*(\mathcal{F} \otimes_{\mathcal{O}_Y} M)_y & \to & Z'^{q+1}(\mathcal{L}^\bullet)_y \otimes_{\mathcal{O}_y} M & \to & \mathcal{L}_y^{q+2} \otimes_{\mathcal{O}_y} M, \end{array}$$

whence the conclusion.

$(e) \Rightarrow (f)$. We have the exact sequences

$$0 \to Z^q(\mathcal{L}^\bullet) \to \mathcal{L}^q \to B^{q+1}(\mathcal{L}^\bullet) \to 0, \qquad 0 \to B^{q+1}(\mathcal{L}^\bullet) \to \mathcal{L}^{q+1} \to Z'^{q+1}(\mathcal{L}^\bullet) \to 0$$

where $B^{q+1}(\mathcal{L}^\bullet) = \operatorname{Im}(\mathcal{L}^q \to \mathcal{L}^{q+1})$ and $Z^q(\mathcal{L}^\bullet) = \operatorname{Ker}(\mathcal{L}^q \to \mathcal{L}^{q+1})$. Then one gets that $Z^q(\mathcal{L}^\bullet)_y$ and $B^{q+1}(\mathcal{L}_y^\bullet)$ are free $\mathcal{O}_y$-modules. By shrinking eventually the neighbourhood V from the statement of theorem 4.1, we may assume that $Z'^{q+1}(\mathcal{L}^\bullet)$, $Z^q(\mathcal{L}^\bullet)$ and $B^{q+1}(\mathcal{L}^\bullet)$ are free sheaves.

Consider the complex $\mathcal{L}'^\bullet$ of components $\mathcal{L}'^i = \mathcal{L}^i$ for $i \neq q$, $i \neq q+1$ and $\mathcal{L}'^q = Z^q(\mathcal{L}^\bullet)$, $\mathcal{L}'^{q+1} = Z'^{q+1}(\mathcal{L}^\bullet)$. The differentials d'^i are equal to d^i for $i \neq q-1, q, q+1$ and $d'^{q-1} : \mathcal{L}^{q-1} \to Z^q$, $d'^q : Z^q \to Z'^{q+1}$, $d'^{q+1} : Z'^{q+1} \to \mathcal{L}^{q+2}$ are deduced naturally from the differentials of $\mathcal{L}^\bullet$. Obviously $d'^q = 0$. It remains to check that $\mathcal{L}'^\bullet$ satisfies the assertions of the theorem. Let $\mathcal{M}$ be a coherent analytic sheaf defined on an open subset of V. One can deduce without difficulty the equalities

$$\operatorname{Ker}(\mathcal{L}^{q-1} \otimes \mathcal{M} \to \mathcal{L}^q \otimes \mathcal{M}) = \operatorname{Ker}(\mathcal{L}^{q-1} \otimes \mathcal{M} \to Z^q \otimes \mathcal{M}),$$

$$\operatorname{Im}(\mathcal{L}^{q+1} \otimes \mathcal{M} \to \mathcal{L}^{q+2} \otimes \mathcal{M}) = \operatorname{Im}(Z'^{q+1} \otimes \mathcal{M} \to \mathcal{L}^{q+2} \otimes \mathcal{M})$$

and isomorphisms

$$H^q(\mathcal{L}^\bullet \otimes \mathcal{M}) = Z^q(\mathcal{L}^\bullet) \otimes \mathcal{M} / \operatorname{Im}(\mathcal{L}^{q-1} \otimes \mathcal{M} \to Z^q(\mathcal{L}^\bullet) \otimes \mathcal{M}),$$

$$H^{q+1}(\mathcal{L}^\bullet \otimes \mathcal{M}) \simeq \operatorname{Ker}(Z'^{q+1} \otimes \mathcal{M} \to \mathcal{L}^{q+2} \otimes \mathcal{M}).$$

One thus obtains an isomorphism

$$H^\bullet(\mathcal{L}^\bullet \otimes \mathcal{M}) \simeq H^\bullet(\mathcal{L}'^\bullet \otimes \mathcal{M}).$$

One verifies canonically that this isomorphism is functorial in $\mathcal{M}$ and agrees with the short exact sequences. Thereby the assertion of 4.1 is verified.

$(f) \Rightarrow (g)$. Let V be a neighbourhood where $\mathcal{L}'^\bullet$ is defined. Denote by "$\vee$" the passing to dual. Let $\mathcal{T}_V$ be the cokernel of the map $\check{\mathcal{L}}'^{q+2} \to \check{\mathcal{L}}'^{q+1}$ and $\mathcal{T}_y$ be its stalk in y. $\mathcal{T}_y$ is an $\mathcal{O}_y$-module of finite type. Let M be an $\mathcal{O}_y$-module of finite type. We have the exact sequence

$$0 \to \mathrm{Hom}_{\mathcal{O}_y}(\mathcal{F}_y, M) \to \mathrm{Hom}_{\mathcal{O}_y}(\check{\mathcal{L}}_y'^{q+1}, M) \to \mathrm{Hom}_{\mathcal{O}_y}(\check{\mathcal{L}}_y'^{q+2}, M).$$

On the other hand

$$R^{q+1}f_*(\mathcal{F} \otimes M)_y \simeq (H^{q+1}(\mathcal{L}'^{\bullet} \otimes \mathcal{M}))_y \simeq H^{q+1}(\mathcal{L}_y'^{\bullet} \otimes_{\mathcal{O}_y} M)$$
$$= \mathrm{Ker}\,(\mathcal{L}_y'^{q+1} \otimes M \to \mathcal{L}_y'^{q+2} \otimes M),$$

where $\mathcal{M}$ is a coherent analytic sheaf defined in some neighbourhood of y and $\mathcal{M}_y = M$. The isomorphism from (*g*) follows by the natural identification

$$\mathcal{L}_y'^{\bullet} \otimes_{\mathcal{O}_y} M \xrightarrow{\sim} \mathrm{Hom}_{\mathcal{O}_y}(\check{\mathcal{L}}_y', M).$$

Its functoriality can be proved in a canonical way. The uniqueness from (*g*) follows from functoriality in a well-known way.

Since the implication (*g*) ⇒ (*a*′) is clear, the proof of the theorem is complete.

COROLLARY 4.8. *Let* $X \xrightarrow{f} Y$ *be a proper morphism of complex spaces,* $\mathcal{F}$ *a coherent analytic sheaf on* X *which is flat with respect to* Y, *and* q *an integer. If the change base theorem is true in a point* $y \in Y$, *it remains true in all points of a neighbourhood of* y.

The proof arises straightforwardly from (*e*).

In particular, we get

COROLLARY 4.9. *Under the assumptions of the previous corollary, if* $\mathcal{F}$ *is cohomologically flat in dimension* q *in a point* $y \in Y$, *then the same property holds in a neighbourhood of* y.

We actually consider *the change base theorem in the global form.*

We shall write $\mathcal{M}$ for a coherent analytic sheaf defined on an open subset of Y.

THEOREM 4.10. *Let* $f : X \to Y$ *be a proper morphism of complex spaces,* $\mathcal{F}$ *a coherent analytic sheaf on* X *which is flat with respect to* Y, *and* q *an integer. Then the following assertions are equivalent:*

(*a*) *The functor* $\mathcal{M} \mapsto R^q f_*(\mathcal{F} \otimes_{\mathcal{O}_Y} \mathcal{M})$ *is right exact.*

(*a*′) *The functor* $\mathcal{M} \mapsto R^{q+1} f_*(\mathcal{F} \otimes_{\mathcal{O}_Y} \mathcal{M})$ *is left exact.*

(*a*″) *The canonical morphisms*

$$R^q f_*(\mathcal{F}) \otimes_{\mathcal{O}_Y} \mathcal{M} \to R^q f_*(\mathcal{F} \otimes_{\mathcal{O}_Y} \mathcal{M})$$

are isomorphisms.

(*a*‴) *The functor* $\mathcal{M} \mapsto R^q f_*(\mathcal{F} \otimes_{\mathcal{O}_Y} \mathcal{M})$ *is isomorphic to a functor of the form* $\mathcal{M} \mapsto \mathcal{N} \otimes_{\mathcal{O}_Y} \mathcal{M}$, $\mathcal{N} \in \mathbf{Coh}\,(Y)$ ($\mathcal{N}$ *will result isomosphic to* $R^q f_*(\mathcal{F})$).

(*b*) *The canonical morphisms*

$$R^q f_*(\mathcal{F}) \to R^q f_*(\mathcal{F}_y)$$

are epimorphisms for any $y \in Y$.

(*c*) *The canonical morphisms*

$$R^q f_*(\mathcal{F}/\hat{\mathfrak{m}}_y^{k+1}\mathcal{F}) \to R^q f_*(\mathcal{F}/\hat{\mathfrak{m}}_y\mathcal{F})$$

are epimorphisms for any $y \in Y$ *and for any* $k \geqslant 0$.

(*d*) *For any change base* $g : Y' \to Y$, *if one considers the morphism* $f' : X' = X \times_Y Y' \to Y'$ *deduced from* f *and the sheaf* $\mathcal{F}'$ *on* X', *the inverse image of* $\mathcal{F}$, *then the canonical morphism*

$$g^*(R^q f_*(\mathcal{F})) \to R^q f'_*(\mathcal{F}')$$

is an isomorphism.

(*e*) *There exists* $\mathcal{T} \in \mathbf{Coh}(Y)$ *(unique up to isomorphism) such that the following functorial (with respect to* $\mathcal{M}$*) isomorphism*

$$R^{q+1} f_*(\mathcal{F} \otimes_{\mathcal{O}_Y} \mathcal{M}) \simeq Hom_{\mathcal{O}_Y}(\mathcal{T}, \mathcal{M})$$

holds.

Proof. All the implications, but for the construction of $\mathcal{T}$, follow straightforwardly from the previous considerations. In order to construct $\mathcal{T}$, we take again the proof of the implication (*f*) $\Rightarrow$ (*g*) from 4.7. The sheaf $\mathcal{T}_V$ (defined on V) enjoies the following property: for any coherent analytic sheaf $\mathcal{M}$ defined on an open subset of V there exists an isomorphism $R^{q+1} f_*(\mathcal{F} \otimes_{\mathcal{O}_Y} \mathcal{M}) \simeq Hom_{\mathcal{O}_Y}(\mathcal{T}_V, \mathcal{M})$ and moreover, such isomorphisms are functorial in $\mathcal{M}$ and agree with the restrictions. If $(V_1, \mathcal{T}_{V_1})$ and $(V_2, \mathcal{T}_{V_2})$ are two pairs which verify the above properties, then there is an isomorphism

$$\xi_{V_1}^{V_2} : \mathcal{T}_{V_2} | V_1 \cap V_2 \xrightarrow{\sim} \mathcal{T}_{V_1} | V_1 \cap V_2 .$$

One easily checks the conditions of gluing together the $\mathcal{T}_V$'s and in this way one gets the required sheaf $\mathcal{T}$.

COROLLARY 4.11. *Let* $f : X \to Y$ *be a proper morphism of complex spaces and* $\mathcal{F}$ *a coherent* $\mathcal{O}_X$*-module which is flat with respect to* Y. *Then there exists a coherent* $\mathcal{O}_Y$*-module* $\mathcal{T}$ *(unique up to isomorphism) and an isomorphism functorial in* $\mathcal{M}$

$$f_*(\mathcal{F} \otimes_{\mathcal{O}_Y} \mathcal{M}) \xrightarrow{\sim} Hom_{\mathcal{O}_Y}(\mathcal{T}, \mathcal{M}).$$

The proof follows by means of (*a'*) $\Leftrightarrow$ (*e*) when $q = -1$.

We shall say that $\mathcal{F}$ is *cohomologically flat in dimension q over Y* if the functor $\mathcal{M} \mapsto R^q f_*(\mathcal{F} \otimes_{\mathcal{O}_Y} \mathcal{M})$ is exact. This means that $\mathcal{F}$ is cohomologically flat in dimension q in any point of Y. By 3.4, 3.7, 4.7 and 4.10 we obtain criteria of cohomological flatness. In particular, it will result that $\mathcal{F}$ is cohomologically flat in dimension q over Y if and only if the maps

$$R^i f_*(\mathcal{F})_y/\mathfrak{m}_y R^i f_*(\mathcal{F})_y \to R^i f_*(\mathcal{F}/\hat{\mathfrak{m}}_y\mathcal{F})_y = H^i(X_y, \mathcal{F}_y)$$

are bijective for any $y \in Y$ and $i = q,\ q-1$, if and only if $R^q f_*(\mathcal{F})$ is locally free and the maps

$$R^q f_*(\mathcal{F})_y/\mathfrak{m}_y R^q f_*(\mathcal{F})_y \to H^q(X_y, \mathcal{F}_y)$$

are bijective for any $y \in Y$.

The main result of the paragraph is the following theorem of Grauert:

THEOREM 4.12. *Let $f : X \to Y$ be a proper morphism of complex spaces, $\mathcal{F}$ a coherent analytic sheaf on X which is flat with respect to Y. Then:*

(*i*) (*Semicontinuity theorem*). *For any fixed integer q, the function*

$$y \mapsto \dim H^q(X_y, \mathcal{F}_y)$$

is upper semicontinuous.

(*ii*) (*Continuity theorem*). *Consider an integer q. If $\mathcal{F}$ is cohomologically flat in dimension q over Y, then the function*

$$y \mapsto \dim H^q(X_y, \mathcal{F}_y)$$

is locally constant. Conversely, if this function is locally constant and Y is a reduced space, then $\mathcal{F}$ is cohomologically flat in dimension q over Y; in particular, the sheaf $R^q f_*(\mathcal{F})$ is locally free.

(*iii*) (*Invariance of Euler-Poincaré characteristic*). *The function*

$$y \mapsto \sum_q (-1)^q \dim H^q(X_y, \mathcal{F}_y)$$

is locally constant.

Proof. We apply theorem 4.1 and consider sheaves $\mathcal{M}$ of the form $\mathcal{O}_Y/\mathfrak{m}_y\mathcal{O}_Y$. The assertions of the theorem being of local nature on Y, we may assume that there exists a bounded complex $\mathcal{L}^\bullet$ of free $\mathcal{O}_Y$-modules of finite rank such that

$$H^\bullet(\mathcal{L}^\bullet) \simeq R^\bullet f_*(\mathcal{F}) \text{ and } H^\bullet(\mathcal{L}^\bullet/\mathfrak{m}_y\mathcal{L}^\bullet) \simeq H^\bullet(X_y, \mathcal{F}_y)$$

for any point $y \in Y$.

The assertions (*i*) and (*iii*) follow straightforwardly from 1.7.

(*ii*) If $\mathcal{F}$ is cohomologically flat in dimension q over Y, then $R^q f_*(\mathcal{F})$ is locally free and

$$R^q f_*(\mathcal{F})_y / \mathfrak{m}_y (R^q f_*(\mathcal{F})_y) \simeq H^q(X_y, \mathcal{F}_y).$$

Accordingly, the function

$$y \mapsto \dim H^q(X_y, \mathcal{F}_y)$$

is locally constant.

We now prove the converse. By 1.7 (*ii*), Coker $(\mathcal{L}^{q-1} \xrightarrow{d^{q-1}} \mathcal{L}^q)$ and Coker $(\mathcal{L}^q \xrightarrow{d^q} \mathcal{L}^{q+1})$ are locally free. From 4.10 it will then result that $\mathcal{F}$ is cohomologically flat in dimension q over Y.

Remark. The continuity theorem is also called the *exactness criterion of Grauert* (EGA, III). The "clasical" statement (Y reduced, $y \mapsto \dim H^q(X_y, \mathcal{F}_y)$ locally constant $\Rightarrow R^q f_*(\mathcal{F})$ locally free) can be directly obtained from the exact sequence

$$0 \to H^q(\mathcal{L}^\bullet) \xrightarrow{\sim} R^q f_*(\mathcal{F}) \to \text{Coker } d^{q-1} \to \mathcal{L}^{q+1} \to \text{Coker } d^q \to 0.$$

Bibliographical indications

The proof of Grauert's finiteness theorem is taken from Forster and Knorr's paper [23] and it is reproduced here without any modification! Another proof besides that "classical" one ([29], [48], [58]) can be found in the paper of Kiehl and Verdier [47]. For a more general context one can consult the work [42] of Houzel and [21].

The original proof of Cartan-Serre finiteness theorem, of the Remmert projection theorem, respectively, can be found in [17] and [65]. The Stein factorization theorem is presented following the algebraic model [(37], Ch. III, 4.3). The original proof (for manifolds) is given in [85]. The paper [14] of Cartan and [95] present a general theorem of quotient spaces (in whose proof the main argument is again the finiteness theorem).

The proof of the comparison 3.1 is that of the paper [48] of Knorr. In Chapter VI (according to [6]) a more general and more precise statement will be proved; in particular some supplementary informations about the projective systems $(H^q(X_y, \mathcal{F}/\hat{\mathfrak{m}}_y^k\mathcal{F}))_{k\geq 0}$ will be obatined.

The base change theorem as well as the other facts from § 3, are either taken from [15] or they are transpositions from the algebraic case ([36], [37]).

Theorem 4.1 is proved for the case of sheaves of the form $\mathcal{M} = \mathcal{O}_Y/\mathfrak{m}_y\mathcal{O}_Y$ (sufficient for theorem 4.12) in Schneider's paper [72] and it was stated in the general form in ([47], Bemerkung 4.4.1). The proof given here follows [72]. We would also like to mention that this theorem is analogous to theorem 6.10.5 from EGA, Ch. III, [37].

As soon as theorem 4.1 is proved, the proof of the other facts from § 4 can be made without any difficulty following the algebraic model (EGA III, SGA 6). For another proof of theorem 4.12 one can consult [66].

The case of the analytic families of compact complex manifolds is treated in the paper of Kodaira and Spencer [49].

Chapter IV

Projective morphisms of complex spaces

Introduction

In his paper *Faisceaux algébriques cohérents* [76], Serre proved the following theorem:

"Let $\mathbb{P}^r$ be the projective space of dimension r over an algebraically closed field and $\mathcal{F}$ be a coherent algebraic sheaf on $\mathbb{P}^r$. Then there exists an integer m_0 such that:

(*A*) the sheaves $\mathcal{F}(m) = \mathcal{F} \otimes \mathcal{O}(m)$ are generated by their global sections for any $m \geqslant m_0$.

(*B*) $$H^q(\mathbb{P}^r, \mathcal{F}(m)) = 0 \text{ for any } q \geqslant 1 \text{ and } m \geqslant m_0.\text{"}$$

The case of coherent analytic sheaves over the complex projective space is analysed in [77]. The authors' aim in this chapter is to present the extension, due to Grauert and Remmert [32], of this result to projective morphisms of complex spaces.

The main result is theorem 2.1, which shows that the projective morphisms "verify at $+\infty$ theorems *A* and *B*". As a particular case we find again the absolute case of the projective spaces. As an application, we obtain by means of the above stated theorem of Serre, the comparison theorem 2.6 between coherent algebraic sheaves and coherent analytic sheaves on a projective manifold [77].

Following Grothendieck [36], the behaviour at $-\infty$ of the projective morphisms is analysed in § 3. The result so obtained (theorem 3.1) corresponds to the characterization of the dimension and the depth of a coherent analytic sheaf on a Stein space and shows that the projective morphisms (under suitable hypothesis of flatness) "behave even at $-\infty$ like the affine morphisms". Some results from the next chapter are used in this paragraph.

Two criteria for ampleness are proved in § 4. The first one is a converse of theorem 2.1. The second one gives conditions for a morphism $X \xrightarrow{f} Y$ to be projective around a point $y \in Y$ as soon as the fiber X_y is a projective manifold.

The first paragraph contains as usual preliminaries: here we recall the notion of projective fiber bundle, ample sheaf, projective morphism...

§ 1. Preliminaries

(a) We consider the projective space $\mathbb{P}^r$ over the complex field as a complex manifold. Let $\mathcal{O}$ be its structural sheaf, that is the sheaf of germs of holomorphic functions.

Denote by $t_0, \ldots, t_r$ a system of homogeneous coordinates in $\mathbb{P}^r$. By $\mathcal{U} = (U_0, \ldots, U_r)$ we mean the natural covering of $\mathbb{P}^r$ with affine spaces given by these coordinates, hence $U_i =$ the set of the points such that $t_i \neq 0$, $0 \leqslant i \leqslant r$.

For an $\mathcal{O}$-module $\mathcal{F}$ we shall denote by $\mathcal{F}_i$ the restriction of $\mathcal{F}$ to U_i, $0 \leqslant i \leqslant r$. Let m be an integer. The multiplication by t_j^m/t_i^m defines an isomorphism on $U_i \cap U_j$ between $\mathcal{F}_i$ and $\mathcal{F}_j$. The sheaf obtained by gluing together the sheaves $\mathcal{F}_i$ via these isomorphisms is denoted by $\mathcal{F}(m)$. If $\mathcal{F}$ is coherent, then the sheaves $\mathcal{F}(m)$ are clearly coherent.

We have canonical isomorphisms

$$\mathcal{O}(m + m') \simeq \mathcal{O}(m) \otimes_{\mathcal{O}} \mathcal{O}(m'),$$

$$\mathcal{F}(0) \simeq \mathcal{F},\ \mathcal{F}(m) \simeq \mathcal{F} \otimes_{\mathcal{O}} \mathcal{O}(m),\ \mathcal{F}(m + m') \simeq \mathcal{F}(m)(m')$$

and also,

$$Hom_{\mathcal{O}}(\mathcal{F}, \mathcal{G}(m)) \simeq Hom_{\mathcal{O}}(\mathcal{F}(-m), \mathcal{G}) \simeq Hom_{\mathcal{O}}(\mathcal{F}, \mathcal{G})(m),$$

$\mathcal{G}$ being another $\mathcal{O}$-module. In particular, $\mathcal{O}(-m)$ is isomorphic to the dual of $\mathcal{O}(m)$.

Recall that if $(X, \mathcal{O})$ is a ringed space and $\mathcal{L}$ an invertible $\mathcal{O}$-module, then one sets

$$\mathcal{L}^{\otimes m} = \begin{cases} \overbrace{\mathcal{L} \otimes_{\mathcal{O}} \ldots \otimes_{\mathcal{O}} \mathcal{L}}^{m \text{ times}} & \text{if } m > 0 \\ \mathcal{O} & \text{if } m = 0 \\ Hom_{\mathcal{O}}(\mathcal{L}^{\otimes -m}, \mathcal{O}) & \text{if } m < 0. \end{cases}$$

From the above said one derives that $\mathcal{O}(m) \simeq \mathcal{O}(1)^{\otimes m}$, hence $\mathcal{F}(m) \simeq \mathcal{F} \otimes_{\mathcal{O}} \mathcal{O}(1)^{\otimes m}$.

Let t be an homogeneous polynomial of degree k in the coordinates $t_0, \ldots, t_r$. If $s = (s_i)_i$ is a section of $\mathcal{F}(m)$, then the system $((t/t_i^k)s_i)_i$ is a section of $\mathcal{F}(m + k)$. We thus obtain natural morphisms $\mathcal{F}(m) \to \mathcal{F}(m + k)$; these will be called *the morphisms obtained by multiplication with t;* the image of a section s will be denoted by ts.

One can easily verify the existence of an isomorphism

$$\mathcal{O}(-r - 1) \simeq \Omega$$

where Ω is the sheaf of germs of differential forms of the type $(r, 0)$ with analytic coefficients.

We shall use the following

LEMMA 1.1. *$H^q(\mathbb{P}^r, \mathcal{O}) = 0$ for any $q \geqslant 1$.*

Proof [24]. We use the isomorphisms $H^{\bullet}(\mathbb{P}^r, \mathcal{O}) \simeq H^{\bullet}(\mathfrak{U}, \mathcal{O})$. If $q > r$ then $H^q(\mathbb{P}^r, \mathcal{O}) = 0$. Let now $\pi: \mathbb{C}^{r+1} \setminus 0 \to \mathbb{P}^r$ be the natural morphism which corresponds to the homogeneous coordinates $t_0, \ldots, t_r$. Denote $\mathfrak{U}' = \pi^{-1}(\mathfrak{U})$ and let $\mathcal{O}'$ be the structural sheaf of $\mathbb{C}^{r+1}$. Let q be an integer, $1 \leqslant q \leqslant r-1$. By lemma II. 3.4, $H^q(\mathfrak{U}', \mathcal{O}') = 0$. Consider a cocycle $\xi = (\xi_{i_0 \ldots i_q}) \in C^q(\mathfrak{U}, \mathcal{O})$. By composition with π we get a cocycle $\xi' = (\pi\, \xi_{i_0 \ldots i_q}) \in C^q(\mathfrak{U}', \mathcal{O}')$, hence there exists $\eta' \in C^{q-1}(\mathfrak{U}', \mathcal{O}')$, so that $\delta\eta' = \xi'$. Every holomorphic function $\eta'_{i_0 \ldots i_{q-1}}$ can be uniquely written as a sum of the homogeneous components of given degree by expansion in Laurent series. Denote by $\eta_{i_0 \ldots i_{q-1}}$ the component of zero degree of $\eta'_{i_0 \ldots i_{q-1}}$. Since the holomorphic functions on an open subset U of $\mathbb{P}^r$ correspond to the homogeneous holomorphic functions of zero degree on $\pi^{-1}(U)$, $\eta = (\eta_{i_0 \ldots i_{q-1}})$ defines an element of $C^{q-1}(\mathfrak{U}, \mathcal{O})$ such that $\delta\eta = \xi$.

It remains to check that $H^r(\mathbb{P}^r, \mathcal{O}) = 0$. We first remark that any cocycle $\xi' \in C^r(\mathfrak{U}', \mathcal{O}') = \mathcal{O}'(U'_0 \cap \ldots \cap U'_r)$ can be expanded in a Laurent series

$$\sum_{\alpha_0, \ldots, \alpha_r = -\infty}^{\infty} c_{\alpha} t_0^{\alpha_0} \ldots t_r^{\alpha_r}, \quad c_{\alpha} \in \mathbb{C}.$$

Moreover, $\xi' \in B^r(\mathfrak{U}', \mathcal{O}')$ if and only if $a_\alpha = 0$ as soon as $\alpha_0 \leqslant -1, \ldots, \alpha_r \leqslant -1$. Consequently, ξ' is cohomologic with a holomorphic function that admits a Laurent expansion of the form

$$\sum_{\alpha_0 \leqslant -1, \ldots, \alpha_r \leqslant -1} a_\alpha t_0^{\alpha_0} \ldots t_r^{c_r}.$$

The homogeneous component of zero degree is null within such an expansion so that the desired conclusion follows easily as above.

(b) We now consider the projective space $\mathbb{P}^r$ as an algebraic manifold and let $\mathcal{O}$ be the structural sheaf, that is the sheaf of germs of rational functions. In order to prove the comparison theorem 2.6 we shall use the following result of Serre [76]:

"Let $\mathcal{F}$ be a coherent algebraic sheaf on $\mathbb{P}^r$. Then there exists an integer $m_0 = m_0(\mathcal{F})$ such that

(A) The sheaves $\mathcal{F}(m)$ are generated by the global sections for $m \geqslant m_0$.

(B) $H^q(\mathbb{P}^r, \mathcal{F}(m)) = 0$ for $q \geqslant 1$ and $m \geqslant m_0$."

The following fact will be useful too:

"The cohomology groups $H^q(\mathbb{P}^r, \mathcal{O})$ vanish for any $q \geqslant 1$."

(c) Let us fix a complex space Y and a sheaf $\mathcal{S} \in \mathbf{Coh}(Y)$. Let $X \xrightarrow{f} Y$ be a complex space over Y. Consider the set of all pairs $(\mathcal{L}, \varphi)$, where $\mathcal{L}$ is an invertible $\mathcal{O}_X$-module and $\varphi : f^*(\mathcal{S}) \to \mathcal{L}$ is a surjective $\mathcal{O}_X$-morphism. Two such pairs,

$(\mathcal{L}, \varphi)$ and $(\mathcal{L}', \varphi')$, will be called equivalent if there is an $\mathcal{O}_X$-isomorphism $\tau : \mathcal{L} \to \mathcal{L}'$ such that $\varphi' = \tau\varphi$. If $X' \xrightarrow{f'} Y$ is another complex space over Y and $g: X' \to X$ a Y-morphism $(f' = fg)$, then any pair $(\mathcal{L}, \varphi)$ on X defines a pair $(g^*(\mathcal{L}), g^*(\varphi))$ on X' (for convenience, we identify $g^*f^*(\mathcal{E})$ with $f'^*(\mathcal{E})$).

As can be easily seen, one obtains a functor from the category of complex spaces over Y with values into the category of sets. In [15] the author proves the existence of a complex space denoted $P(\mathcal{E})$, which is called *the projective fiber bundle associated to* $\mathcal{E}$, together with a morphism $p : P(\mathcal{E}) \to Y$, which is called *the structural morphism* and a pair $(\mathcal{O}(1), \varepsilon)$ on $P(\mathcal{E})$, which represents this functor. So for any complex space $X \xrightarrow{f} Y$ over Y, there exists a $1-1$ correspondence between the Y-morphisms $X \to P(\mathcal{E})$ and the equivalence classes of pairs $(\mathcal{L}, \varphi)$ on X defined above; if $r : X \to P(\mathcal{E})$ is a Y-morphism, then the associated pair is $(\mathcal{L} = r^*(\mathcal{O}(1)), \varphi = r^*(\varepsilon))$. The system $(P(\mathcal{E}), p, \mathcal{O}(1), \varepsilon)$ is uniquely determined (up to isomorphism) by this property.

If $\mathcal{E} = \mathcal{O}_Y^{r+1}$, then one can actually conclude that $P(\mathcal{E})$ coincides (modulo an isomorphism) with $Y \times \mathbb{P}^r$, the structural morphism p coincides with the projection, $\mathcal{O}(1)$ with the inverse image of $\mathcal{O}_{\mathbb{P}^r}(1)$ by the projection $Y \times \mathbb{P}^r \to \mathbb{P}^r$ and ε is deduced via this projection from the morphism $\mathcal{O}_{\mathbb{P}^r}^{r+1} \to \mathcal{O}_{\mathbb{P}^r}(1)$ which is defined by means of formula $(s_0, \ldots, s_r) \mapsto \sum_{i=1}^{r} t_i s_i$. In this case one denotes by $\mathcal{O}(m)$ the inverse image of $\mathcal{O}_{\mathbb{P}^r}(m)$ by the projection $Y \times \mathbb{P}^r \to \mathbb{P}^r$; for an $\mathcal{O}_{Y \times \mathbb{P}^r}$-module $\mathcal{F}$, we denote $\mathcal{F}(m) = \mathcal{F} \otimes_{\mathcal{O}} \mathcal{O}(m)$ (these sheaves can also be easily obtained by patching just as in **(a)**). Properties similar to those of the absolute case hold here.

We also recall the following two properties of the projective fiber bundles.

If $\mathcal{F} \to \mathcal{E}$ is a surjective morphism in **Coh** (Y), then one obtains a closed immersion $P(\mathcal{E}) \to P(\mathcal{F})$, which agrees with the structural morphisms. In particular, we derive that, locally on Y, any projective fiber bundle $P(\mathcal{E})$ can be embedded in a space of the form $Y \times \mathbb{P}^r$, and the embedding is compatible with the structural morphisms.

The other property is the following: if $(Y, \mathcal{E})$ is as above and $\varphi: Y' \to Y$ is a morphism of complex spaces, then there exists a natural isomorphism

$$P(\mathcal{E}) \times_Y Y' \to P(\varphi^*(\mathcal{E})).$$

In particular, for any point $y \in Y$, the analytic fiber $P(\mathcal{E})_y$ (which is isomorphic to the fiber product of X and $(y, \mathcal{O}_y/\mathfrak{m}_y)$ over Y) can be identified naturally with the projective space associated to the vectorial space $\mathcal{E}_y/\mathfrak{m}_y\mathcal{E}_y$.

(d) Let $X \xrightarrow{f} Y$ be a proper morphism of complex spaces and $\mathcal{L}$ an invertible $\mathcal{O}_X$-module. We assume that the canonical morphism $f^*f_*(\mathcal{L}) \to \mathcal{L}$ is surjective. By **(c)**, it defines a morphism

$$X \to P(f_*(\mathcal{L})),$$

which agrees with f and with the structural morphism $P(f_*(\mathcal{L})) \to Y$.

$\mathcal{L}$ is said to be *very ample with respect to Y (or f)* if the morphism $X \to P(f_*(\mathcal{L}))$ is a closed immersion. The sheaf $\mathcal{L}$ is called *ample with respect to Y* if for any point $y \in Y$ there exists a neighbourhood V and an integer $n \geqslant 1$ such that the sheaf $\mathcal{L}^{\otimes n} | f^{-1}(V)$ is very ample with respect to V.

A morphism $f : X \to Y$ of complex spaces is called *projective* if there exists an ample sheaf with respect to Y on X.

In particular we consider the case when Y is the final object $(*, \mathbb{C})$. An invertible sheaf $\mathcal{L}$ on a compact analytic space X is called *very ample* if the global sections $\Gamma(X, \mathcal{L})$ generate the stalks $\mathcal{L}_x$, $x \in X$, and the canonical morphism $X \to P(\Gamma(X, \mathcal{L}))$ is a closed immersion. $\mathcal{L}$ is said to be *ample* if some of its power $\mathcal{L}^{\otimes n}$ $(n \geqslant 1)$ is very ample. These definitions agree with those given above.

(e) Recall some facts in local algebra.

LEMMA 1.2. *Let $A \to B$ be a local morphism of noetherian local rings and M a B-module of finite type, which is flat over A and such that $M/\mathfrak{m}_A M$ is a free $B/\mathfrak{m}_A B$-module. Under these assumptions M is a free B-module.*

Proof. Choose $x_1, \ldots, x_r \in M$ such that their classes in $M/\mathfrak{m}_A M$ form a basis over $B/\mathfrak{m}_A B$. By Nakayama lemma, the morphism $B^r \to M$ defined by these elements is surjective and let N be its kernel. Hence we have the exact sequence of B-modules $0 \to N \to B^r \to M \to 0$ and by tensoring $\otimes_A A/\mathfrak{m}_A$ we obtain an exact sequence

$$0 \to N/\mathfrak{m}_A N \to (B/\mathfrak{m}_A B)^r \to M/\mathfrak{m}_A M \to 0.$$

It results that $N/\mathfrak{m}_A N = 0$. Since $\mathfrak{m}_A N = (\mathfrak{m}_A B)N$, we deduce again by Nakayama lemma $N = 0$, therefore M is B-free.

PROPOSITION 1.3. *Let $A \to B$ be a local morphism of noetherian local rings such that B is a flat A-module and $B/\mathfrak{m}_A B$ is a regular ring of dimension r. Let M be an A-flat B-module of finite type and N an arbitrary B-module. Under these assumptions*

(*a*) $\operatorname{Ext}^i_B(M, N) = 0$ *for $i > r$.*

(*b*) *If moreover the modules N and $\operatorname{Ext}^i_B(M, N)$ are A-flat, then*

$$\operatorname{Ext}^{\bullet}_B(M, N)/\mathfrak{m}_A \operatorname{Ext}^{\bullet}_B(M, N) \simeq \operatorname{Ext}^{\bullet}_{B/\mathfrak{m}_A B}(M/\mathfrak{m}_A M, N/\mathfrak{m}_A N).$$

Proof. (*a*) Let us consider an exact sequence of B-modules

$$L_{r-1} \to \ldots \to L_0 \to M \to 0,$$

where L_i are free modules of finite type and let P be the kernel of the map $L_{r-1} \to L_{r-2}$. P is a B-module of finite type which is flat over A (one can see this by splitting the above exact sequence into short exact sequences). By tensoring $\otimes_A A/\mathfrak{m}_A$ one then obtains the exact sequence

$$0 \to P/\mathfrak{m}_A P \to L_{r-1}/\mathfrak{m}_A L_{r-1} \to \ldots \to L_0/\mathfrak{m}_A L_0 \to M/\mathfrak{m}_A M \to 0.$$

By the hypothesis on $B/\mathfrak{m}_A B$ one derives that $P/\mathfrak{m}_A P$ is free as a $B/\mathfrak{m}_A B$-module. In virtue of the previous lemma, P is B-free. The conclusion follows by calculating $\mathrm{Ext}_B^{\cdot}(M, N)$, taking into account the exact sequence

$$(*) \qquad 0 \to L_r \to L_{r-1} \to \ldots \to L_0 \to M \to 0,$$

where we have put L_r instead of P.

(*b*) Denote by $L.$ the resolution of M given by $(*)$. By tensoring $\otimes_A A/\mathfrak{m}_A$, one gets a resolution $L_\bullet/\mathfrak{m}_A L_\bullet$ of $M/\mathfrak{m}_A M$. Then $\mathrm{Ext}^{\bullet}_{B/\mathfrak{m}_A B}(M/\mathfrak{m}_A M, N/\mathfrak{m}_A N)$ are the cohomology groups of the complex $\mathrm{Hom}_{B/\mathfrak{m}_A B}(L_\bullet/\mathfrak{m}_A L_\bullet, N/\mathfrak{m}_A N)$. Consider the complex $K^\bullet = \mathrm{Hom}_B(L_\bullet, N)$. Its components are flat A-modules and the associated homology groups enjoy the same property since they are isomorphic to $\mathrm{Ext}_B^{\cdot}(M, N)$. Whence (by taking into account the short exact sequences given by cocycles, coboundaries and cohomology...), some isomorphisms

$$H^i(K^\bullet)/\mathfrak{m}_A H^i(K^\bullet) \simeq H^i(K^\bullet/\mathfrak{m}_A K^\bullet).$$

Since $K^\bullet/\mathfrak{m}_A K^\bullet \simeq \mathrm{Hom}_{B/\mathfrak{m}_A B}(L_\bullet/\mathfrak{m}_A L_\bullet, N/\mathfrak{m}_A N)$, the conclusion follows.

We shall also use the following

LEMMA 1.4. *Let $A \to B$ be a local morphism of noetherian local rings such that the morphism induced between completions $\hat{A} \to \hat{B}$ is flat. Then $A \to B$ is flat.*

Proof. The morphisms $A \to \hat{A}$ and $B \to \hat{B}$ are flat. Let now $0 \to M \to N$ be a monomorphism of A-modules of finite type and $P = \mathrm{Ker}\,(M \otimes_A B \to N \otimes_A B)$. Then one obtains the exact sequence

$$0 \to P \otimes_B \hat{B} \to (M \otimes_A B) \otimes_B \hat{B} \to (N \otimes_A B) \otimes_B \hat{B}.$$

But $(M \otimes_A B) \otimes_B \hat{B} \simeq (M \otimes_A \hat{A}) \otimes_{\hat{A}} \hat{B}$, $(N \otimes_A B) \otimes_B \hat{B} \simeq (N \otimes_A \hat{A}) \otimes_{\hat{A}} \hat{B}$. By hypothesis we have $P \otimes_B \hat{B} = 0$, therefore $P = 0$ according to Krull's theorem.

§ 2. The behaviour at $+\infty$ of the sheaves $\mathcal{F}(m)$

The main result in this paragraph is the following theorem of Grauert and Remmert [32]:

THEOREM 2.1. *Let $f : X \to Y$ be a projective morphism of complex spaces and $\mathcal{L}$ be an invertible sheaf on X, which is ample with respect to f. For any $\mathcal{O}_X$-module $\mathcal{F}$, denote $\mathcal{F}(m) = \mathcal{F} \otimes_{\mathcal{O}_X} \mathcal{L}^{\otimes m}$, m arbitrary integer.*

Then for any coherent analytic sheaf $\mathcal{F}$ on X and for any compact K of Y there exists an integer $m_0 = m_0(K, \mathcal{F})$ such that:

(A) The canonical morphism

$$f^* f_*(\mathcal{F}(m)) \to \mathcal{F}(m)$$

is surjective in all points of K for any $m \geqslant m_0$;

(B) $R^q f_*(\mathcal{F}(m))$ vanishes on K for all $q \geqslant 1$, $m \geqslant m_0$.

Proof. We first remark that if the theorem is still true when $\mathcal{L}$ is replaced by $\mathcal{L}^{\otimes d}$, $d \geqslant 1$, then it is true in the stated form. In order to prove this, let $\mathcal{F} \in \mathbf{Coh}\ X$. For any integer $m \geqslant 1$ we can write $\mathcal{F}(m) = (\mathcal{F} \otimes \mathcal{L}^{\otimes r}) \otimes \mathcal{L}^{\otimes hd}$ where h, r are integers, $h > 0$, $0 \leqslant r < d$; the conclusion follows for the coherent sheaves $\mathcal{F} \otimes \mathcal{L}^{\otimes r}$, $0 \leqslant r < d$, which is nothing else than our hypothesis.

Since the theorem is local on Y we may assume that $\mathcal{L}$ is very ample with respect to f. Hence we have a natural commutative diagram

$$\begin{array}{ccc} X & \longrightarrow & Y \\ & \searrow \quad \nearrow & \\ & P(f_*(\mathcal{L})) & \end{array}$$

where $X \to P(f_*(\mathcal{L}))$ is a closed immersion and $P(f_*(\mathcal{L})) \to Y$ is the structural morphism. Again, if we take into account the fact that the assertions of the theorem are local on Y, we can assume that there exists a surjective morphism $\mathcal{O}_Y^{r+1} \to f_*(\mathcal{L}) \to$ $\to 0$. We thus obtain a closed immersion $P(f_*(\mathcal{L})) \to P(\mathcal{O}_Y^{r+1}) = Y \times \mathbb{P}^r$, which is compatible with the structural morphisms. Thus we find a closed immersion i: $X \to Y \times \mathbb{P}^r$ such that the diagram

$$\begin{array}{ccc} X & \xrightarrow{f} & Y \\ & {\scriptstyle i}\searrow \quad \swarrow{\scriptstyle r} & \\ & Y \times \mathbb{P}^r & \end{array}$$

is comutative; moreover, we have an isomorphism $i^*(\mathcal{O}_{Y \times \mathbb{P}^r}(1)) \simeq \mathcal{L}$. For any $\mathcal{O}_X$-module $\mathcal{F}$, we get $i_*(\mathcal{F}(m)) \simeq (i_*(\mathcal{F}))(m)$.

Taking all these into account we can easily reduce the proof of the theorem to the case when X *is of the form* $Y \times \mathbb{P}^r, f$ *is the canonical projection and* $\mathcal{L}$ *is the sheaf* $\mathcal{O}_X(1)$ (the inverse image under the projection $Y \times \mathbb{P}^r \to \mathbb{P}^r$ of the sheaf $\mathcal{O}_{\mathbb{P}^r}(1)$). We shall use in this case induction on r. If $r = 0$ then the assertions from the statement are obviously fulfilled for any integer m. We suppose now that the theorem is true for $r - 1$ and prove it for r. Let $\mathcal{F} \in \mathbf{Coh}\ (X)$.

The existence of some m_0 *which verifies* (A). Denote by $t_0, \ldots, t_r$ the homogeneous coordinates in $\mathbb{P}^r$. Let x be a point in X and m an integer such that the morphism $f^*f_*(\mathcal{F}(m))_x \to \mathcal{F}(m)_x$ is surjective. We show that this fact holds for any integer m', $m' \geqslant m$ too. Let $x = (y, z)$, $y \in Y$, $z \in \mathbb{P}^r$ and let U_k be one of the canonical affine carts of $\mathbb{P}^r$ which contains z. Consider the morphism

$$\theta: \mathcal{O}_{\mathbb{P}^r}(m) \to \mathcal{O}_{\mathbb{P}^r}(m)$$

defined by multiplication by $t_k^{m'-m}$(for any cart U_i this means the multiplication by $(t_k/t_i)^{m'-m}$). This morphism is an isomorphism on U_k. Thus one derives canonically a morphism of $\mathcal{O}_X$-modules $\mathcal{F}(m) \to \mathcal{F}(m')$, which is an isomorphism on $Y \times U_k$, particularly in the point x. The conclusion follows immediately. We also remark that as soon as for an integer m the morphism $f^*f_*(\mathcal{F}(m))_x \to \mathcal{F}(m)_x$ is surjective in x, the same property holds in virtue of the coherence in a neighbourhood of x.

These two remarks reduce the proof of the existence of the integer m_0 to the proof of the following assertion:

(*) For every $x \in X$ there exists an integer m which depends on x and $\mathcal{F}$, such that the morphism

$$f^*f_*(\mathcal{F}(m)) \to \mathcal{F}(m)$$

is surjective in x.

Let $x = (y, z)$, $y \in Y$ and $z \in \mathbb{P}^r$. Consider a hyperplane E which passes through z, whose equation is $t = 0$. Denote by $\mathcal{O}_E$ the structural sheaf of E, which is extended trivially in $\mathbb{P}^r \setminus E$. $Y \times E$ is a closed subspace of $Y \times \mathbb{P}^r$ and denote by $\mathcal{O}_{Y \times E}$ its structural sheaf trivially extended out of $Y \times E$. We have the exact sequence

$$0 \to \mathcal{O}_{\mathbb{P}^r}(-1) \to \mathcal{O}_{\mathbb{P}^r} \to \mathcal{O}_E \to 0$$

where the morphism $\mathcal{O}_{\mathbb{P}^r}(-1) \to \mathcal{O}_{\mathbb{P}^r}$ is given by multiplying by t. Taking into account the projection $Y \times \mathbb{P}^r \to \mathbb{P}^r$ one obtains an exact sequence

$$0 \to \mathcal{O}_X(-1) \to \mathcal{O}_X \to \mathcal{O}_{Y \times E} \to 0.$$

By tensoring $\otimes_{\mathcal{O}_X} \mathcal{F}$ one gets the exact sequence

$$0 \to \mathrm{Tor}_1^{\mathcal{O}_X}(\mathcal{F}, \mathcal{O}_{Y \times E}) \to \mathcal{F}(-1) \to \mathcal{F} \to \mathcal{F} \otimes_{\mathcal{O}_X} \mathcal{O}_{Y \times E} \to 0.$$

For convenience we denote

$$\mathcal{B} = \mathcal{F} \otimes_{\mathcal{O}_X} \mathcal{O}_{Y \times E} \text{ and } \mathcal{C} = \mathrm{Tor}_1^{\mathcal{O}_X}(\mathcal{F}, \mathcal{O}_{Y \times E}).$$

$\mathcal{B}$ and $\mathcal{C}$ are coherent $\mathcal{O}_X$-modules and for any integer m we have an exact sequence

$$0 \to \mathcal{C}(m) \to \mathcal{F}(m-1) \to \mathcal{F}(m) \to \mathcal{B}(m) \to 0.$$

Denote $\mathcal{Q}_m = \mathrm{Ker}\,(\mathcal{F}(m) \to \mathcal{B}(m))$. We thus obtain the exact sequences

$$0 \to \mathcal{C}(m) \to \mathcal{F}(m-1) \to \mathcal{Q}_m \to 0, \; 0 \to \mathcal{Q}_m \to \mathcal{F}(m) \to \mathcal{B}(m) \to 0.$$

By applying the functor f_* we derive the exact sequences

$$R^1f_*(\mathcal{F}(m-1)) \to R^1f_*(\mathcal{Q}_m) \to R^2f_*(\mathcal{C}(m)),$$

$$R^1f_*(\mathcal{Q}_m) \to R^1f_*(\mathcal{F}(m)) \to R^1f_*(\mathcal{B}(m)).$$

If $\mathfrak{J}(Y \times E)$ is the maximal ideal-sheaf associated to the subspace $Y \times E \subset Y \times \mathbb{P}^r$, then

$$\mathfrak{J}(Y \times E)\mathfrak{B} = 0 \text{ and } \mathfrak{J}(Y \times E)\mathfrak{C} = 0.$$

We realize that $\mathfrak{B}$ and $\mathfrak{C}$ can be regarded as $\mathcal{O}_{Y \times E}$-modules, even as coherent $\mathcal{O}_{Y \times E}$-modules. By the induction hypothesis applied to the morphism $Y \times E \to Y$ and to these sheaves, there exists an integer m_0 such that $R^2f_*(\mathfrak{C}(m))$ and $R^1f_*(\mathfrak{B}(m))$ vanish in a neighbourhood of y whenever $m \geqslant m_0$. Then the maps

$$R^1f_*(\mathfrak{F}(m-1)) \to R^1f_*(\mathfrak{Q}_m),\ R^1f_*(\mathfrak{Q}_m) \to R^1f_*(\mathfrak{F}(m))$$

are surjective in y for any $m \geqslant m_0$. We deduce that for these integers m the composite maps

$$R^1f_*(\mathfrak{F}(m-1))_y \to R^1f_*(\mathfrak{F}(m))_y$$

are surjective. The sheaf $R^1f_*(\mathfrak{F}(m_0 - 1))$ is $\mathcal{O}_Y$-coherent by the finiteness theorem. By the noetherianity of the $\mathcal{O}_y$-module $R^1f_*(\mathfrak{F}(m_0-1))_y$ we deduce the existence of an integer $m_1 \geqslant m_0$ such that the maps $R^1f_*(\mathfrak{F}(m-1))_y \to R^1f_*(\mathfrak{F}(m))_y$ are bijective for all $m \geqslant m_1$. Then it follows that the maps $R^1f_*(\mathfrak{Q}_m) \to R^1f_*(\mathfrak{F}(m))$ are bijective in y for m $\geqslant m_1$. From the exact sequence

$$f_*(\mathfrak{F}(m)) \to f_*(\mathfrak{B}(m)) \to R^1f_*(\mathfrak{Q}_m) \to R^1f_*(\mathfrak{F}(m))$$

we deduce that the maps $f_*(\mathfrak{F}(m))_y \to f_*(\mathfrak{B}(m))_y$ are surjective for any $m \geqslant m_1$. As a consequence, the maps $f^*f_*(\mathfrak{F}(m))_x \to f^*f_*(\mathfrak{B}(m))_x$ are surjective for any $m \geqslant m_1$. By induction hypothesis there is an integer m_2 such that the map $f^*f_*(\mathfrak{B}(m))_x \to$ $\to \mathfrak{B}(m)_x$ is surjective for any $m \geqslant m_2$. Therefore, for m sufficiently large, the maps

$$f^*f_*(\mathfrak{F}(m))_x \to f^*f_*(\mathfrak{B}(m))_x \text{ and } f^*f_*(\mathfrak{B}(m))_x \to \mathfrak{B}(m)_x$$

are surjective. We have

$$\mathfrak{B}(m)_x = \mathfrak{F}(m)_x / \mathfrak{J}_x(Y \times E)\, \mathfrak{F}(m)_x.$$

The assertion (*) will be concluded from these facts by means of the commutative diagram

$$\begin{array}{ccc} f^*f_*(\mathfrak{F}(m))_x & \to & f^*f_*(\mathfrak{B}(m))_x \\ \downarrow & & \downarrow \\ \mathfrak{F}(m)_x & \longrightarrow & \mathfrak{B}(m)_x \end{array}$$

and by Nakayama lemma.

The existence of some m_0 which verifies (B). We proceed by descending induction on q. For q sufficiently large the functor $R^q f_*$ is null. We suppose now that for an integer $q \geqslant 2$ and for any triplet $(Y, K, \mathcal{F})$ there is an integer m_0 which verifies (B) and we are going to prove the corresponding assertion for the integer $q-1$. Let $(Y, K, \mathcal{F})$ be as in the statement of the theorem (recall that we are considering the case $X = Y \times \mathbb{P}^r$, $\mathcal{L} = \mathcal{O}(1)$ and f projection). Let $y \in Y$ be an arbitrary point. It is enough to find a neighbourhood V of y such that $R^{q-1}f_*(\mathcal{F}(m))$ vanishes on V for m sufficiently large. Let V be a relatively compact Stein neighbourhood of y. Consider an integer m_0 such that the morphism

$$f^*f_*(\mathcal{F}(m_0)) \to \mathcal{F}(m_0)$$

is surjective on $f^{-1}(V)$. There exists a surjective morphism $\mathcal{O}_Y^p \to f_*(\mathcal{F}(m_0))$ on V. We then deduce via f^* a surjective morphism $\mathcal{O}_X^p \to f^*f_*(\mathcal{F}(m_0))$ on $f^{-1}(V)$ and by composition, a surjective morphism $\mathcal{O}_X^p \to \mathcal{F}(m_0)$ on $f^{-1}(V)$. Let $\mathcal{O}_X^p(-m_0) \to \mathcal{F}$ be the morphism obtained from this and let $\mathcal{G}$ be its kernel, $\mathcal{G} \in \mathbf{Coh}\,(f^{-1}(V))$. For any integer m we have an exact sequence on $f^{-1}(V)$.

$$0 \to \mathcal{G}(m) \to \mathcal{O}_X^p(m - m_0) \to \mathcal{F}(m) \to 0.$$

Denote by f too the morphism $f^{-1}(V) \to V$ deduced from f. One obtains the exact sequence on V

$$R^{q-1}f_*(\mathcal{O}_X^p(m - m_0)) \to R^{q-1}f_*(\mathcal{F}(m)) \to R^q f_*(\mathcal{G}(m)).$$

By the induction hypothesis and by shrinking eventually V around y, we get $R^q f_*(\mathcal{G}(m)) = 0$ for m sufficiently large.

The proof of the theorem will be over by applying the following

LEMMA 2.2. *Let Y be a complex space, $X = Y \times \mathbb{P}^r$ and $f\colon X \to Y$ the projection. Then*

$$R^q f_*(\mathcal{O}_X(m)) = 0$$

for any $m \geqslant 0$, $q \geqslant 1$.

Proof. If V is a Stein open subset sufficiently small in Y, then $H^q(V \times \mathbb{P}^r, \mathcal{O}_X) = 0$ for $q \geqslant 1$; this fact can be proved either exactly as in lemma 1.1 or by using this lemma and a simple Künneth formula. Thus we get $R^q f_*(\mathcal{O}_X) = 0$.

We shall prove the assertion of the lemma by induction on m. The case $m = 0$ is already proved. For the general induction step we use again induction on r. If E is an arbitrary hyperplane of $\mathbb{P}^r$ then we have already pointed out the exact sequence

$$0 \to \mathcal{O}_X(-1) \to \mathcal{O}_X \to \mathcal{O}_{Y \times E} \to 0.$$

The conclusion easily follows by means of the exact sequences

$$0 \to \mathcal{O}_X(m-1) \to \mathcal{O}_X(m) \to \mathcal{O}_{Y \times E}(m) \to 0.$$

Remark. The theorem can be applied in the particular case when X is the r-dimensional projective space over Y, $\mathcal{L} = \mathcal{O}_X(1)$ and f is the projection. However, the theorem follows easily from this case, as one could realize from the proof.

COROLLARY 2.3. *Under the assumptions of the theorem, let* $\mathcal{F} \to \mathcal{G} \to \mathcal{H}$ *be an exact sequence of coherent* $\mathcal{O}_X$*-modules. Then, for any compact* K *of* Y, *there is an integer* m_0 *such that the sequence*

$$f_*(\mathcal{F}(m)) \to f_*(\mathcal{G}(m)) \to f_*(\mathcal{H}(m))$$

is exact on K *whatever* $m \geqslant m_0$.

Proof. Let $\mathcal{F}'$, $\mathcal{G}'$, $\mathcal{G}''$ be the kernel, image and cokernel, respectively, of the morphism $\mathcal{F} \to \mathcal{G}$. Then $\mathcal{G}'$ is the kernel and $\mathcal{G}''$ is the image of the morphism $\mathcal{G} \to \mathcal{H}$; let $\mathcal{H}''$ be the cokernel of the latter morphism. All these are coherent $\mathcal{O}_X$-modules. Since the functor $\mathcal{F} \mapsto \mathcal{F}(m)$ is exact, it is sufficient to prove that for m sufficiently large, each of the sequences

$$0 \to f_*(\mathcal{F}'(m)) \to f_*(\mathcal{F}(m)) \to f_*(\mathcal{G}'(m)) \to 0,$$

$$0 \to f_*(\mathcal{G}'(m)) \to f_*(\mathcal{G}(m)) \to f_*(\mathcal{G}''(m)) \to 0,$$

$$0 \to f_*(\mathcal{G}''(m)) \to f_*(\mathcal{H}(m)) \to f_*(\mathcal{H}''(m)) \to 0,$$

is exact on K. In other words, we may assume that the sequence

$$0 \to \mathcal{F} \to \mathcal{G} \to \mathcal{H} \to 0$$

is exact. To conclude with, we apply the theorem and the exact sequences:

$$0 \to f_*(\mathcal{F}(m)) \to f_*(\mathcal{G}(m)) \to f_*(\mathcal{H}(m)) \to R^1f_*(\mathcal{F}(m)) \to \dots .$$

Consider Y as a complex space reduced to a point, whose stalk is the complex field and $X = \mathbb{P}^r$; then one gets *the theorems A and B on the projective space of Serre* [77]:

THEOREM 2.4. *Let* $\mathbb{P}^r$ *be the complex projective space of dimension* r *and* $\mathcal{O}$ *be the sheaf of germs of holomorphic functions on it. For any coherent* $\mathcal{O}$*-module* $\mathcal{F}$ *there exists an integer* $m_0 = m_0(\mathcal{F})$ *such that*

(A) The sheaves $\mathcal{F}(m) = \mathcal{F} \otimes_{\mathcal{O}} \mathcal{O}(m)$ *are generated by the global sections for* $m \geqslant m_0$;

(B) $H^q(\mathbb{P}^r, \mathcal{F}(m)) = 0$ for $q \geqslant 1$ and $m \geqslant m_0$.

COROLLARY 2.5. *Any coherent analytic sheaf on* $\mathbb{P}^r$ *admits a finite resolution by locally free sheaves of finite rank.*

Proof. Considering the regularity of the local rings of $\mathbb{P}^r$, it is enough to show that any sheaf $\mathcal{F} \in$ **Coh** $(\mathbb{P}^r)$ is the quotient of a locally free sheaf of finite rank. For m sufficiently large we deduce from the theorem an epimorphism of the form $\mathcal{O}^n_{\mathbb{P}^r} \to \mathcal{F}(m) \to 0$, hence an epimorphism

$$\mathcal{O}^n_{\mathbb{P}^r}(-m) \to \mathcal{F} \to 0$$

and the conclusion follows.

We conclude this paragraph with the proof of *the comparison theorem of Serre* [77].

Denote by X the projective space $\mathbb{P}^r$ considered as algebraic manifold and by X^h the same space considered as complex manifold. By $\mathcal{O}$ we mean the structural sheaf of X and by $\mathcal{O}^h$ the structural sheaf of X^h. Since any Zariski open set is open in the usual topology and since any rational function is holomorphic, one naturally obtains a morphism of ringed space $X^h \to X$.

For any point $x \in X$ the morphism $\mathcal{O}_x \to \mathcal{O}^h_x$ induces an isomorphism by passing to completions (the $\mathbb{C}$-algebras $\hat{\mathcal{O}}_x$ and $\hat{\mathcal{O}}^h_x$ are in fact isomorphic to the algebra $\mathbb{C}[[z_1, \ldots, z_r]]$ of formal series in r indeterminates). By applying lemma 1.4 it follows that the morphism $X^h \to X$ is flat.

If $\mathcal{F}$ is a coherent algebraic sheaf on X, then its inverse image under the morphism $X^h \to X$ is a coherent analytic sheaf on X^h, denoted by $\mathcal{F}^h$. In this way one defines an exact functor from the category $\mathbf{Coh}(X)$ to the category $\mathbf{Coh}(X^h)$. We have $\mathcal{O}(m)^h = \mathcal{O}^h(m)$, m being an arbitrary integer.

Let $\mathcal{F} \in \mathbf{Coh}(X)$, $\mathfrak{U}$ a finite affine covering of X and $\mathfrak{U}^h$ its inverse image in X^h. Since any Zariski affine open set is a Stein open set, $\mathfrak{U}^h$ turns out to be a Stein covering. We naturally deduce a morphism between the associated Čech complexes $C^\bullet(\mathfrak{U}, \mathcal{F}) \to C^\bullet(\mathfrak{U}^h, \mathcal{F}^h)$, hence a morphism between the cohomology groups

$$H^\bullet(X, \mathcal{F}) \to H^\bullet(X^h, \mathcal{F}^h)$$

(the existence of this morphism is in fact, a general property of the ringed spaces).

It is easy to check the functoriality of these morphisms and their compatibility with the short exact sequences.

THEOREM 2.6. *The functor $\mathcal{F} \mapsto \mathcal{F}^h$ from the category of coherent algebraic sheaves on the projective space $X = \mathbb{P}^r$ to the category of coherent analytic sheaves is an equivalence of categories.*

Moreover, the canonical morphisms

$$H^q(X, \mathcal{F}) \to H^q(X^h, \mathcal{F}^h)$$

are isomorphisms, q being an arbitrary integer.

Proof. We first prove the last assertion. If $\mathcal{F} = \mathcal{O}$, then $H^0(X, \mathcal{O}) = H^0(X^h, \mathcal{O}^h) = \mathbb{C}$ and $H^q(X, \mathcal{O})$, $H^q(X^h, \mathcal{O}^h)$ vanish for $q \geqslant 1$ (lemma 1.1 and § 1. *b*).

Next, consider the case of the sheaves $\mathcal{O}(m)$. We proceed by induction on r. As the case $r = 0$ is obvious, we prove the general induction step. Let E be a hyperplane of the homogeneous equation $t = 0$. We have the exact sequence

$$0 \to \mathcal{O}(-1) \to \mathcal{O} \to \mathcal{O}_E \to 0$$

where the morphism $\mathcal{O}(-1) \to \mathcal{O}$ is obtained by multiplication by t and $\mathcal{O}_E$ is the structural sheaf of $E = \mathbb{P}^{r-1}$ which is trivially extended on $\mathbb{P}^r \setminus E$. For any integer m we get the exact sequence

$$0 \to \mathcal{O}(m-1) \to \mathcal{O}(m) \to \mathcal{O}_E(m) \to 0.$$

Similarly we get exact sequences

$$0 \to \mathcal{O}^h(m-1) \to \mathcal{O}^h(m) \to \mathcal{O}_{E^h}(m) \to 0.$$

We have a commutative diagram

$$\begin{array}{ccccccccc} \ldots \to & H^q(X, \mathcal{O}(m-1)) & \to & H^q(X, \mathcal{O}(m)) & \to & H^q(E, \mathcal{O}_E(m)) & \to & H^{q+1}(X, \mathcal{O}(m-1)) & \to \ldots \\ & \downarrow & & \downarrow & & \downarrow & & \downarrow & \\ \ldots \to & H^q(X^h, \mathcal{O}(m-1)^h) & \to & H^q(X^h, \mathcal{O}(m)^h) & \to & H^q(E^h, \mathcal{O}_E(m)^h) & \to & H^{q+1}(X^h, \mathcal{O}(m-1)^h) & \to \ldots \end{array}$$

According to the induction hypothesis the morphisms $H^q(E, \mathcal{O}_E(m)) \to H^q(E^h, \mathcal{O}_E(m)^h)$ are isomorphisms. By the lemma of the five the morphisms in the statement are isomorphisms for $\mathcal{F} = \mathcal{O}(m)$ if and only if they are isomorphisms for $\mathcal{F} = \mathcal{O}(m-1)$. Since $m = 0$, this fact has been already verified; the required conclusion follows.

We now consider the case of arbitrary coherent algebraic sheaves and proceed by descending induction on q. For q sufficiently large $H^q(X, \mathcal{F})$ and $H^q(X^h, \mathcal{F}^h)$ are null (by using Čech cohomology!), hence we have to prove the general induction step only. By § 1 (**b**) there exists an exact sequence $0 \to \mathcal{R} \to \mathcal{L} \to \mathcal{F} \to 0$ where $\mathcal{L}$ is a direct sum of sheaves of the form $\mathcal{O}(m)$. We have the commutative diagram

$$\begin{array}{ccccccccccc} \ldots \to & H^q(X, \mathcal{R}) & \to & H^q(X, \mathcal{L}) & \to & H^q(X, \mathcal{F}) & \to & H^{q+1}(X, \mathcal{R}) & \to & H^{q+1}(X, \mathcal{L}) & \to \ldots \\ & \downarrow{\scriptstyle \varepsilon_1} & & \downarrow{\scriptstyle \varepsilon_2} & & \downarrow{\scriptstyle \varepsilon_3} & & \downarrow{\scriptstyle \varepsilon_4} & & \downarrow{\scriptstyle \varepsilon_5} & \\ \ldots \to & H^q(X^h, \mathcal{R}^h) & \to & H^q(X^h, \mathcal{L}^h) & \to & H^q(X^h, \mathcal{F}^h) & \to & H^{q+1}(X^h, \mathcal{R}^h) & \to & H^{q+1}(X^h, \mathcal{L}^h) & \to \ldots \end{array}$$

The morphism ε_2 is an isomorphism; by the induction hypothesis the morphisms ε_4 and ε_5 are also isomorphisms. The lemma of the five shows that ε_3 is surjective. This fact holds for any $\mathcal{F} \in$ **Coh** (X), hence for $\mathcal{R}$ too, therefore ε_1 is surjective. By using again the lemma of the five we deduce that ε_3 is bijective and the proof of the second assertion of the theorem is completed.

We now prove the first assertion of the statement. Let $\mathcal{F}, \mathcal{G} \in$ **Coh** (X). There exists a natural morphism

$$\mathcal{H}om_{\mathcal{O}}(\mathcal{F}, \mathcal{G})^h \to \mathcal{H}om_{\mathcal{O}^h}(\mathcal{F}^h, \mathcal{G}^h).$$

We will show that it is an isomorphism. The problem is local on X, hence we may assume that there exists an exact sequence $\mathcal{O}_X^p \to \mathcal{O}_X^q \to \mathcal{F} \to 0$. The conclusion follows without difficulty from the exactness of the functor "h". By applying the last assertion of the theorem, the morphism

$$\mathrm{Hom}_{\mathcal{O}}(\mathcal{F}, \mathcal{G}) = H^0(X, \mathcal{H}om_{\mathcal{O}}(\mathcal{F}, \mathcal{G})) \to H^0(X^h, \mathcal{H}om_{\mathcal{O}}(\mathcal{F}, \mathcal{G})^h)$$

$$\simeq H^0(X^h, \mathcal{H}om_{\mathcal{O}^h}(\mathcal{F}^h, \mathcal{G}^h)) = \mathrm{Hom}_{\mathcal{O}^h}(\mathcal{F}^h, \mathcal{G}^h)$$

is bijective.

Thus, in order to complete the proof of the theorem, it remains to show that for any $\mathcal{F}' \in \mathbf{Coh}\ (X^h)$ there exists (and hence is unique) a sheaf $\mathcal{F} \in \mathbf{Coh}\ (X)$ such that $\mathcal{F}' \simeq \mathcal{F}^h$. By 2.4, for m sufficiently large, $\mathcal{F}'(m)$ is isomorphic to a quotient of a sheaf by the form $(\mathcal{O}^h)^n$ and hence $\mathcal{F}'$ is isomorphic to a quotient of a sheaf of the form $\mathcal{O}^h(-m)^n$. If we denote by $\mathcal{L}_0$ the sheaf $\mathcal{O}(-m)^n$, then we obtain an exact sequence $0 \to \mathcal{G}' \to \mathcal{L}_0^h \to \mathcal{F}' \to 0$. By iterating the above reasoning we find an exact sequence

$$\mathcal{L}_1^h \xrightarrow{\varphi'} \mathcal{L}_0^h \to \mathcal{F}' \to 0$$

where the sheaves $\mathcal{L}_0$ and $\mathcal{L}_1$ are isomorphic to some finite direct sums of sheaves of the type $\mathcal{O}(m)$. By the antecedence, there is $\varphi \in \mathrm{Hom}_{\mathcal{O}}(\mathcal{L}_1, \mathcal{L}_0)$ such that $\varphi^h = \varphi'$. If we denote by $\mathcal{F}$ the cokernel of φ, then we deduce that $\mathcal{F}' \simeq \mathcal{F}^h$ and the theorem is proved.

Remark. The results from 2.4, 2.5, 2.6 could be indeed stated for an arbitrary projective manifold X (in 2.5 X is assumed nonsingular). The proofs easily follow from the above considered case by embeddings $X \to \mathbb{P}^r$.

§ 3. The behaviour at $-\infty$ of the sheaves $\mathcal{F}(m)$

Let $f: X \to Y$ be an arbitrary morphism of complex spaces, $\mathcal{F}$ an $\mathcal{O}_X$-module and y a point of Y. Denote as usual by X_y the fiber $f^{-1}(y)$ endowed with the restriction of the sheaf $\mathcal{O}_{X_y} = \mathcal{O}_X/\hat{\mathfrak{m}}_y\mathcal{O}_X$ to it; by $\mathcal{F}_y$ we mean the sheaf $\mathcal{F}/\hat{\mathfrak{m}}_y\mathcal{F}$ ($\mathfrak{m}_y$ is the maximal ideal of $\mathcal{O}_y$ and "$\wedge$" stands for the sheaf generated by it in $\mathcal{O}_X$). $\mathcal{F}_y$ is an $\mathcal{O}_{X_y}$-module and denote by prof $\mathcal{F}_y$ its depth,

$$\mathrm{prof}\ \mathcal{F}_y = \inf_{x, f(x)=y} \mathrm{prof}_{\mathcal{O}_{X_y,x}} (\mathcal{F}_y)_x = \inf_{x, f(x)=y} \mathrm{prof}_{\mathcal{O}_x/\mathfrak{m}_y\mathcal{O}_x} (\mathcal{F}_x/\mathfrak{m}_y\mathcal{F}_x).$$

Analogously,

$$\dim \mathcal{F}_y = \sup_{x, f(x)=y} \dim_{\mathcal{O}_{X_y,x}} (\mathcal{F}_y)_x = \sup_{x, f(x)=y} \dim_{\mathcal{O}_x/\mathfrak{m}_y\mathcal{O}_x} (\mathcal{F}_x/\mathfrak{m}_y\mathcal{F}_x).$$

For convenience we introduce the notation $\operatorname{prof}_Y \mathcal{F} = \inf_{y\in Y} \operatorname{prof} \mathcal{F}_y$, *the depth of* $\mathcal{F}$ *with respect to* Y, $\dim_Y \mathcal{F} = \sup_{y\in Y} \dim \mathcal{F}_y$, *the dimension of* $\mathcal{F}$ *with respect to* Y.

We have the inequality

$$\operatorname{prof}_Y \mathcal{F} \leqslant \dim_Y \mathcal{F} (\text{if } \mathcal{F} \neq 0).$$

The main result in the paragraph is the following theorem of Grothendieck:

THEOREM 3.1. *Let* $f\colon X \to Y$ *be a projective morphism,* $\mathcal{L}$ *be an invertible* $\mathcal{O}_X$-*module very ample with respect to* Y, $\mathcal{F}$ *a coherent* $\mathcal{O}_X$-*module flat over* Y, *and* $q \geqslant 0$ *an integer.*

(*a*) *If* $\operatorname{prof}_Y \mathcal{F} \geqslant q+1$, *then for any compact* K *of* Y *there exists an integer* $m_0 = m_0(K, \mathcal{F})$ *such that* $R^i f_*(\mathcal{F}(-m)) \mid K = 0$ *for any* $i \leqslant q$ *and* $m \geqslant m_0$.

(*b*) $\dim_Y \mathcal{F} \leqslant q$ *if and only if for any compact* K *of* Y, *there exists an integer* $m_0 = m_0(K, \mathcal{F})$ *so that* $R^i f_*(\mathcal{F}(-m)) \mid K = 0$ *for any* $i > q$ *and* $m \geqslant m_0$.

Proof. The assertions of the theorem are of local nature on Y, hence we may assume the existence of a surjective morphism of sheaves $\mathcal{O}_Y^{r+1} \to f_*(\mathcal{L})$. This induces a closed immersion $X \to X' = Y \times \mathbb{P}^r$ such that $\mathcal{L}$ is induced by $\mathcal{O}_{X'}(1)$ and, in addition, the diagram

$$\begin{array}{ccc} X & \longrightarrow & Y \\ & \searrow \quad \nearrow & f' = \mathrm{pr} \\ & X' & \end{array}$$

is commutative.

If $\mathcal{F}'$ is the image of $\mathcal{F}$ by the immersion $X \to X'$, then $\mathcal{F}'$ is a coherent $\mathcal{O}_{X'}$-module which is flat over Y. We have

$$R^{\bullet} f'_*(\mathcal{F}') \simeq R^{\bullet} f_*(\mathcal{F}), \ \operatorname{prof}_Y \mathcal{F} = \operatorname{prof}_Y \mathcal{F}', \ \dim_Y \mathcal{F} = \dim_Y \mathcal{F}'.$$

Therefore, in order to prove the theorem we may *assume* X *of the form* $Y \times \mathbb{P}^r$, $\mathcal{L} = \mathcal{O}_X(1)$ *and* f *the projection.*

For an analytic subset Z of Y (locally closed) and for an integer $j \geqslant 0$ we denote by X_Z the product $Z \times \mathbb{P}^r$ $(= Z \times_Y X)$, by $\mathcal{F}_Z$ the inverse image of $\mathcal{F}$ under the morphism $X_Z \to X$ and by $\mathcal{E}^j(Z)$ the sheaf $\mathcal{E}xt^j_{\mathcal{O}_{X_Z}}(\mathcal{F}_Z, \mathcal{O}_{X_Z}(-r-1))$. $\mathcal{F}_Z$ is flat over Z. $\mathcal{E}^j(Z)$ is a coherent $\mathcal{O}_{X_Z}$-module which is equal to zero if j does not belong to the interval $[0, r]$, as it can be seen by passing to stalks and applying proposition 1.3. In particular, for the sets $Z = \{y\}$, $y \in Y$ we get a family of coherent analytic sheaves $\mathcal{E}^j(y) = \mathcal{E}xt^j_{\mathcal{O}_{X_y}}(\mathcal{F}_y, \mathcal{O}_{X_y}(-r-1))$ on the fibers X_y, $y \in Y$. We are going to prove the following assertion:

(*) *For any compact* K *of* Y *there exists an integer* m_0 *such that* $H^i(X_y, \mathcal{E}^j(y)(m)) = 0$ *for any* $i \geqslant 1$, $j \geqslant 0$, $m \geqslant m_0$, $y \in K$.

It is enough to confine the problem to semianalytic Stein compacts ([25] or Chapter V, § 1). Let K be such a compact. The ring $\mathcal{O}(K)$ of the germs of holomorphic functions on K is noetherian ([25] or Chapter V, § 3).

Let $\alpha \in \operatorname{Spec} \mathcal{O}(K)$; we denote by Z_α the germ of analytic set defined by α around K, and we designate so any representative of this germ. Let $X_\alpha = X_{Z_\alpha}$ and $\mathcal{E}^j(\alpha) = \mathcal{E}^j(Z_\alpha)$. Apply V. 4.3 for the morphism $X_\alpha \to Z_\alpha$ and the sheaf $\mathcal{E}^j(\alpha)$: there exists a dense open subset V_α of Z_α such that $Z_\alpha \setminus V_\alpha$ is an analytic set and $\mathcal{E}^j(\alpha)$ is flat over V_α. Since the sheaves $\mathcal{E}^j(\alpha)$ are null, except a finite number of indices j, one may assume the open set V_α "good" for all indexes j. The analytic set $Z_\alpha \setminus V_\alpha$ yields a closed set in $\overline{\{\alpha\}} \in \operatorname{Spec} \mathcal{O}(K)$ and let D_α be its complement in $\overline{\{\alpha\}}$: it is a nonempty open subset of $\overline{\{\alpha\}}$ (otherwise, by restricting eventually the representative Z_α around K, one finds $V_\alpha = \emptyset$, a contradiction), so $\alpha \in D_\alpha$.

We are going to prove, by noetherian induction, that there exists a finite partition of $\operatorname{Spec} \mathcal{O}(K)$ by locally closed sets of the form D_α. Consider the set of all closed subsets of $\operatorname{Spec} \mathcal{O}(K)$ which do not enjoy this property. We shall prove that it is empty and the desired conclusion will follow. On the contrary, suppose that this set is nonempty: since $\mathcal{O}(K)$ is noetherian, this set has a minimal element T. The minimality implies that T is irreducible, hence $T = \overline{\{\alpha\}}$, $\alpha \in \operatorname{Spec} \mathcal{O}(K)$. Let D_α be the open subset of $\overline{\{\alpha\}} = T$ constructed above; the minimality of T again and the equality $T = D_\alpha \cup (T \setminus D_\alpha)$ lead us to a contradiction. Thus let $(D_\alpha)_\alpha$ be such a finite partition of $\operatorname{Spec} \mathcal{O}(K)$. By the above constructions we derive that the corresponding sets V_α cover the compact K (if $x \in K$ and $m_x \in D_\alpha$, then $x \in V_\alpha$!). Apply theorem 2.1 to the morphisms $f_\alpha\colon X_\alpha \to Z_\alpha$ deduced from f and to the sheaves $\mathcal{E}^j(\alpha)$: there exists (eventually by restricting the sets Z_α around K) an integer m_0 such that $R^i f_\alpha (\mathcal{E}^j(\alpha))(m) = 0$ for all $i \geqslant 1$, $j \geqslant 0$, $m \geqslant m_0$ and for any index α of the above partition. Since the sheaves $\mathcal{E}^j(\alpha)(m)$ are flat over V_α then by III. 3.11,

$$H^i(X_y, \mathcal{E}^j(\alpha)(m)_y) = 0 \text{ for } i \geqslant 1,\ j \geqslant 0,\ m \geqslant m_0,\ y \in V_\alpha.$$

By virtue of proposition 1.3 we derive isomorphisms

$$\mathcal{E}^j(\alpha)(m)_y \simeq \mathcal{E}^j(y)(m),$$

for any $j \geqslant 0$, m arbitrary integer, $y \in V_\alpha$ and the assertion (*) is thus proved.

From (*) we now deduce the following assertion:

(**) *for any compact K of Y there exists an integer m_0 such that*

$$H^i(X_y, \mathcal{F}_y(-m))' \simeq \Gamma(X_y, \mathcal{E}xt^{r-i}_{\mathcal{O}_{X_y}}(\mathcal{F}_y, \Omega_{X_y})(m))$$

for any $m \geqslant m_0$, $y \in K$ and $i \geqslant 0$ (the accent means the algebraic dual over $\mathbb{C}$ and Ω, as usual means the sheaf of germs of holomorphic forms in the maximal degree). Let K be an arbitrary compact of Y and let m_0 be an integer which verifies (*). For any integer m and for any point $y \in Y$ there is a spectral sequence which converges to $\mathrm{Ext}^{\bullet}_{\mathcal{O}_{X_y}}(X_y; \mathcal{F}_y, \Omega_{X_y}(m))$, such that

$$E_2^{i,j}(y, m) = H^j(X_y, \mathit{Ext}^i_{\mathcal{O}_{X_y}}(\mathcal{F}_y, \Omega_{X_y}(m))).$$

We have

$$\mathit{Ext}^{\bullet}_{\mathcal{O}_{X_y}}(\mathcal{F}_y, \Omega_{X_y}(m)) \simeq \mathit{Ext}^{\bullet}_{\mathcal{O}_{X_y}}(\mathcal{F}_y, \Omega_{X_y})(m).$$

For the projective space $\mathbb{P}^r$, $\mathcal{O}_{\mathbb{P}^r}(-r-1) \simeq \Omega_{\mathbb{P}^r}$. The assertion $(*)$ shows that the spectral sequences $E^{i,j}(y, m)$ degenerate, hence

$$\mathrm{Ext}^{\bullet}_{\mathcal{O}_{X_y}}(X_y; \mathcal{F}_y, \Omega_{X_y})(m) \simeq \Gamma(X_y, \mathit{Ext}^{\bullet}_{\mathcal{O}_{X_y}}(\mathcal{F}_y, \Omega_{X_y})(m))$$

for any $y \in K$ and $m \geqslant m_0$.

The assertion $(**)$ will follow then from the duality theorem on the projective space (VII, § 4) and from the isomorphisms

$$\mathit{Ext}^{\bullet}_{\mathcal{O}_{X_y}}(X_y; \mathcal{F}_y(-m), \Omega_{X_y}) \simeq \mathit{Ext}^{\bullet}_{\mathcal{O}_{X_y}}(X_y; \mathcal{F}_y, \Omega_{X_y}(m)).$$

We prove now the assertions of the theorem.

(*a*) Let K be a compact of Y and m_0 an integer which verifies $(**)$. Let y be a point of K. We have prof $\mathcal{F}_y \geqslant q+1$. By passing to stalks and applying I. 1.15, it follows that

$$\mathit{Ext}^{r-i}_{\mathcal{O}_{X_y}}(\mathcal{F}_y, \Omega_{X_y})(m) = 0$$

for any $i \leqslant q$ and whatever m. From $(**)$ we derive that

$$H^i(X_y, \mathcal{F}_y(-m)) = 0 \text{ for any } i \leqslant q,\ m \geqslant m_0 \text{ and } y \in K.$$

By applying III. 3.5 it follows that

$$R^i f_*(\mathcal{F}(-m)) \mid K = 0 \text{ for any } i \leqslant q \text{ and } m \geqslant m_0.$$

(*b*) The direct implication can be proved in the same manner. We prove the converse. Let $y \in K$. By hypothesis, the sheaves $R^i f_*(\mathcal{F}(-m))$ are null in a neighbourhood of y, for $i > q$ and m large enough. According to III. 3.11, $H^i(X_y, \mathcal{F}_y(-m)) = 0$ for $i > q$ and m large enough. For m sufficiently large the sheaves $\mathit{Ext}^{\bullet}_{\mathcal{O}_{X_y}}(\mathcal{F}_y, \Omega_{X_y})(m)$ are generated by the global sections. By applying the assertion $(**)$ (to the compact $K = \{y\}$, a case when the use of $(*)$ is superfluous!) we deduce that $Ext^i_{\mathcal{O}_{X_y}}(\mathcal{F}_y, \Omega_{X_y}) = 0$ for $i < r - q$. By I.1.17, $\dim \mathcal{F}_y \leqslant q$. Accordingly, $\dim_Y \mathcal{F} \leqslant q$ and the theorem is proved.

Remarks. (i) The proof of (*b*) can be made without the use of the assertion (∗). For the converse implication, we have already remarked this fact. For the direct implication it is enough to check that $R^i f_*(\mathcal{F}) = 0$ for $i > q$, for any $\mathcal{F} \in \mathbf{Coh}\,(X)$ which is flat over Y and such that $\dim_Y \mathcal{F} \leqslant q$. For any $y \in Y$, $H^i(X_y, \mathcal{F}_y) = 0$, whenever $i > q \geqslant \dim \mathcal{F}_y$ (apply [64]); the conclusion will follow by III. 3.11.

(*ii*) In the course of the proof we have applied the duality theorem on the projective space $\mathbb{P}^r$ for coherent analytic sheaves. By using 2.5 this duality can be deduced from the classical case [75] of the locally free sheaves of finite rank.

COROLLARY 3.2. *Let $f\colon X \to Y$ be a flat proper morphism such that the fibres X_y are perfect spaces of dimension r, let $\mathcal{L}$ be an invertible $\mathcal{O}_X$-module very ample with respect to Y and $\mathcal{F} \in$ **Coh** (X) locally free.*

Under these assumptions, for any compact K of Y, there exists an integer $m_0 = m_0(K, \mathcal{F})$ such that

$$R^i f_*(\mathcal{F}(-m)) \mid K = 0 \text{ for any } i \neq r \text{ and } m \geqslant m_0.$$

Proof. For any point $y \in Y$ the sheaf $\mathcal{F}_y$ is locally free of finite rank over X_y, hence $\operatorname{prof} \mathcal{F}_y = \dim \mathcal{F}_y = r$. As a consequence $\operatorname{prof}_Y \mathcal{F} = \dim_Y \mathcal{F} = r$ and apply the theorem.

The corollary can be applied in the following situation: Y an arbitrary complex space, $X = Y \times \mathbb{P}^r$, f the projection and $\mathcal{L} = \mathcal{O}_X(1)$.

We now restate theorem 3.1 in the absolute case.

COROLLARY 3.3. *Let X be a projective manifold, $\mathcal{L}$ a very ample sheaf on it, $\mathcal{L} \in$ **Coh** (X), and q be an integer. Denote $\mathcal{F}(m) = \mathcal{F} \otimes_{\mathcal{O}_X} \mathcal{L}^{\otimes m}$, m arbitrary integer.*

(*a*) $\operatorname{prof} \mathcal{F} \geqslant q + 1$ *if and only only if $H^i(X, \mathcal{F}(-m)) = 0$ for $i \leqslant q$ and m large enough;*

(*b*) $\dim \mathcal{F} \leqslant q$ *if and only if $H^i(X, \mathcal{F}(-m)) = 0$ for $i > q$ and m large enough.*

Proof. We have to prove the inverse implication from (*a*) only. By a suitable embedding we reduce the problem to the case $X = \mathbb{P}^r$ and $\mathcal{L} = \mathcal{O}_{\mathbb{P}^r}(1)$. The conclusion follows from the assertion (∗∗) (which can be easily established in the absolute case), exactly as in the case of assertion (*b*) from 3.1.

COROLLARY 3.4. *Let X be a projective manifold, $\mathcal{L}$ a very ample sheaf on it and $\mathcal{F} \in$ **Coh** (X). Then $\mathcal{F}$ is Cohen-Macauley (prof $\mathcal{F}$ = dim $\mathcal{F}$) if and only if there exists an integer q such that*

$$H^i(X, \mathcal{F}(-m)) = 0 \text{ for } i \neq q \text{ and } m \text{ large enough.}$$

To conclude this paragraph, we give the following result due to Grauert.

THEOREM 3.5. *Let $f\colon X \to Y$ be a projective morphism of complex spaces, $\mathcal{L}$ an $\mathcal{O}_X$-module very ample with respect to Y, t a global section of $\mathcal{L}$, $\mathcal{J}$ the maximal ideal-sheaf associated to the subspace of all zeros of t, $\mathcal{F} \in$ **Coh**(X), and q an integer. Suppose in addition that $\mathcal{F}$ is flat over Y, the section t is $\mathcal{F}$-regular and $\operatorname{prof}_Y \mathcal{F} \geqslant q + 1$.*

Under these assumptions, the canonical morphism

$$R^i f_*(\mathfrak{F}) \to \varprojlim_m R^i f_*(\mathfrak{F}/\mathfrak{J}^m \mathfrak{F})$$

is an isomorphism for $i < q$ *and a monomorphism for* $i = q$.

(Recall that a point x is said to be a zero for t if $t_x \in \mathfrak{m}_x \mathfrak{L}_x$).

Proof. The problem is local on Y and so we may replace $\mathfrak{J}$ by any coherent sheaf of ideals having the same zeros. If we replace $\mathfrak{L}$ by some of its tensor power and section t by some of its power, we may assume that $\mathfrak{L}$ is very ample with respect to Y. Moreover, by 3.1 we can suppose that there exists an integer m_0 such that

$$R^i f_*(\mathfrak{F}(-m)) = 0 \text{ for } i \leqslant q \text{ and } m \geqslant m_0.$$

In order to prove the theorem it is sufficient to check that if moreover Y is Stein, the canonical morphism

$$H^i(X, \mathfrak{F}) \to \varprojlim_m H^i(X, \mathfrak{F}/\mathfrak{J}^m \mathfrak{F})$$

is an isomorphism for $i < q$ and a monomorphism for $i = q$.

We have isomorphisms $H^i(X, \mathfrak{F}(-m)) \simeq \Gamma(X, R^i f_*(\mathfrak{F}(-m))) = 0$ for $i \leqslant q$ and $m \geqslant m_0$. On the other hand, the multiplication by t^m considered as morphism from $\mathfrak{F}(-m)$ to $\mathfrak{F}$ yields an exact sequence

$$0 \to \mathfrak{F}(-m) \to \mathfrak{F} \to \mathfrak{F}/\mathfrak{J}^m \mathfrak{F} \to 0 \quad (m \geqslant 0).$$

From the associated cohomology exact sequence, one derives that the canonical morphism $H^i(X, \mathfrak{F}) \to H^i(X, \mathfrak{F}/\mathfrak{J}^m \mathfrak{F})$ is bijective for $i < q$ and injective for $i = q$. Thereby the proof is completed.

§ 4. Two criteria for ampleness

The first criterion stands for a converse of theorem 2.1.

THEOREM 4.1. *Let* $f\colon X \to Y$ *be a proper morphism of complex spaces and let* $\mathfrak{L}$ *be an invertible sheaf on* X *which enjoys the following property: for any compact* K *of* Y *and for any* $\mathfrak{F} \in \mathbf{Coh}(X)$, *there is an integer* $n_0 = n_0(K, \mathfrak{F})$ *such that*

$$R^q f_*(\mathfrak{F} \otimes \mathfrak{L}^{\otimes n}) | K = 0 \text{ for } q \geqslant 1,\ n \geqslant n_0.$$

Under these assumptions, $\mathfrak{L}$ *is very ample with respect to* Y.

Proof. The problem is local on Y. Let us fix a point $y_0 \in Y$. Take $x \in X_{y_0} = f^{-1}(y_0)$. We first prove the following assertion:

($*$) For any invertible sheaf $\mathfrak{L}$ on X, as soon as the canonical morphism $f^*f_*(\mathfrak{L}) \to \mathfrak{L}$ is surjective in x, the same property holds for the morphisms $f^*f_*(\mathfrak{L}^{\otimes n}) \to \mathfrak{L}^{\otimes n}$, $n \geqslant 1$ arbitrary integer.

Indeed, the hypothesis shows that any element of $\mathfrak{L}_x$ is a linear combination of germs in x of elements from $f_*(\mathfrak{L})_{y_0} = \Gamma(X_{y_0}, \mathfrak{L})$ with coefficients in $\mathcal{O}_x$; then any element of $(\mathfrak{L}^{\otimes n})_x \simeq (\mathfrak{L}_x)^{\otimes n}$ is a linear combination of elements of the form $s_x^1 \cdot \ldots \cdot s_x^n$, $s_i \in \Gamma(X_{y_0}, \mathfrak{L})$ with coefficients in $\mathcal{O}_x$; the conclusion follows.

By $\mathfrak{m}_x$ we denote for convenience the maximal ideal of $\mathcal{O}_x$ as well as the coherent ideal-sheaf on X defined by it. From the exact sequence

$$0 \to \mathfrak{m}_x \to \mathcal{O}_X \to \mathcal{O}_X/\mathfrak{m}_x\mathcal{O}_X \to 0$$

we get the exact sequences

$$0 \to \mathfrak{m}_x \otimes \mathfrak{L}^{\otimes n} \to \mathfrak{L}^{\otimes n} \to \mathfrak{L}^{\otimes n}/\mathfrak{m}_x \mathfrak{L}^{\otimes n} \to 0.$$

For n sufficiently large, $R^1f_*(\mathfrak{m}_x \otimes \mathfrak{L}^n)_{y_0} = 0$, hence the morphisms

$$f_*(\mathfrak{L}^{\otimes n})_{y_0} \to f_*(\mathfrak{L}^{\otimes n}/\mathfrak{m}_x \mathfrak{L}^{\otimes n})_{y_0}$$

are surjective. The sheaves $\mathfrak{L}^{\otimes n}/\mathfrak{m}_x \mathfrak{L}^{\otimes n}$ are concentrated in the point x where their stalks are equal to $\mathfrak{L}_x^{\otimes n}/\mathfrak{m}_x \mathfrak{L}_x^{\otimes n}$. In this way the morphisms $\Gamma(X_{y_0}, \mathfrak{L}^{\otimes n}) \to \mathfrak{L}_x^{\otimes n}/\mathfrak{m}_x \mathfrak{L}_x^{\otimes n}$ are surjective for n sufficiently large. By Nakayamma lemma, the morphisms $f^*f_*(\mathfrak{L}^{\otimes n})_x \to \mathfrak{L}_x^{\otimes n}$ result surjective.

So there exist a neighbourhood U_x of x and an integer n_x such that the morphism $f^*f_*(\mathfrak{L}^{\otimes n_x}) \to \mathfrak{L}^{\oplus n_x}$ is surjective on U_x. Since the fiber X_{y_0} is compact, we find a finite covering $(U_i)_{1 \leqslant i \leqslant k}$ of it and some integers n_i such that the morphisms $f^*f_*(\mathfrak{L}^{\otimes n_i}) \to \mathfrak{L}^{\otimes n_i}$ are surjective on U_i. If we set $n = n_1 \ldots n_k$, then the morphism $f^*f_*(\mathfrak{L}^{\otimes n}) \to \mathfrak{L}^{\otimes n}$ will be surjective on $\bigcup_i U_i$, hence it will be surjective on an open subset of the form $f^{-1}(V)$, V neighbourhood of y_0. If we replace $\mathfrak{L}$ by $\mathfrak{L}^{\otimes n}$ and Y by V, then we may assume that the canonical morphism $f^*f_*(\mathfrak{L}) \to \mathfrak{L}$ is surjective. By ($*$) the same property holds for the morphisms

$$f^*f_*(\mathfrak{L}^{\otimes n}) \to \mathfrak{L}^{\otimes n},$$

$n \geqslant 1$ being an arbitrary integer.

Denote by

$$\theta : X \to P(f_*(\mathfrak{L}))$$

the morphism over Y which is deduced from the epimorphism $f^*(f_*(\mathfrak{L})) \to \mathfrak{L}$. This morphism becomes explicit by passing to fibers. The fiber in y_0 of the structural morphism $P(f_*(\mathfrak{L})) \to Y$ can be identified with $P(f_*(\mathfrak{L})_{y_0}/\mathfrak{m}_{y_0}f_*(\mathfrak{L})_{y_0})$. We have a natural commutative diagram

$$\begin{array}{ccccc} X_{y_0} & \hookrightarrow & X & & \\ \downarrow & & \downarrow & \searrow & \\ & & & & Y \\ & & & \nearrow & \\ P(f_*(\mathfrak{L})_{y_0}/\mathfrak{m}_{y_0}f_*(\mathfrak{L})_{y_0}) & \to & P(f_*(\mathfrak{L})) & & \end{array}$$

Let $s^0, \ldots, s^r \in f_*(\mathfrak{L})_{y_0} = \Gamma(X_{y_0}, \mathfrak{L})$ be such that their images in the vectorial $\mathcal{O}_{y_0}/\mathfrak{m}_{y_0}$-space $f_*(\mathfrak{L})_{y_0}/\mathfrak{m}_{y_0}f_*(\mathfrak{L})_{y_0}$ form a basis. If x is a point of X_{y_0}, then $\theta(x)$ regarded as point in $P(f_*(\mathfrak{L})_{y_0}/\mathfrak{m}_{y_0}f_*(\mathfrak{L})_{y_0})$ has the homogeneous coordinates $s^0(x), \ldots, s^r(x)$ (as usual, for a section s of $\mathfrak{L}$, $s(x)$ is its image in $\mathfrak{L}_x/\mathfrak{m}_x\mathfrak{L}_x \simeq \mathbb{C}$). If $\dfrac{P(s^0, \ldots, s^r)}{Q(s^0, \ldots, s^r)}$ is a rational function defined around $\theta(x)$, P and Q being homogeneous polynomials of the same degree such that $Q(s^0, \ldots, s^r)(\theta(x)) \neq 0$, then the germ of holomorphic function defined by it around the point $x \in X_y$ through θ is just $\dfrac{P(\sigma(s_x^0), \ldots, \sigma(s_x^r))}{Q(\sigma(s_x^0), \ldots, \sigma(s_x^r))}$, where σ is an isomorphism $\mathfrak{L}_x \simeq \mathcal{O}_x$ (we remark that the element $Q(\sigma(s_x^0), \ldots, \sigma(s_x^r))$ is invertible).

We will denote by $\theta^{\otimes n}$ the morphism $X \to P(f_*(\mathfrak{L}^{\otimes n}))$ and show that for n large enough, by shrinking eventually Y around y_0, $\theta^{\otimes n}$ is a closed immersion. In this way, $\mathfrak{L}^{\otimes n}$ will be very ample with respect to Y and the theorem will be proved.

We first prove that for n sufficiently large $\theta^{\otimes n}$ is a local immersion in every point. In order to do this we will use the characterization of local immersions by means of Zariski tangent spaces. Consider a point $x \in X_{y_0}$. We prove the following assertion:

(**) For any invertible sheaf $\mathfrak{L}$ such that the morphism $f^*f_*(\mathfrak{L}) \to \mathfrak{L}$ is surjective, if moreover the induced morphism $X \to P(f_*(\mathfrak{L}))$ is a local immersion in x, then the morphisms $X \to P(f_*(\mathfrak{L}^{\otimes n}))$ enjoy the same property, $n \geqslant 1$ being an arbitrary integer.

Again, consider sections $s^0, \ldots, s^r \in \Gamma(X_{y_0}, \mathfrak{L})$ such that their images in $f_*(\mathfrak{L})_{y_0}/\mathfrak{m}_{y_0}f_*(\mathfrak{L})_{y_0}$ form a basis. The morphism $X_{y_0} \to P(f_*(\mathfrak{L})_{y_0}/\mathfrak{m}_{y_0}f_*(\mathfrak{L})_{y_0})$ is an immersion in x. Consider a section $s \in \Gamma(X_y, L)$ such that $s(x) \neq 0$ and let $s^{n-1} = \overbrace{s \otimes \ldots \otimes s}^{n-1 \text{ times}}$; the elements $s^{n-1} \otimes s^0, \ldots, s^{n-1} \otimes s^r$ lie in $\Gamma(X_{y_0}, \mathfrak{L}^{\otimes n})$ and the morphism $X_{y_0} \to \mathbb{P}^r$ induced by their images in $\Gamma(X_{y_0}, \mathfrak{L}^{\otimes n}/\mathfrak{m}_{y_0}\mathfrak{L}^{\otimes n})$ coincides, in a neighbourhood of x in X_{y_0}, with the morphism induced by θ. One then derives that $\theta^{\otimes n}$ is immersion in the point x and the assertion (**) is verified.

From the exact sequence $0 \to \mathfrak{m}_x^2 \to \mathcal{O}_X \to \mathcal{O}_X/\mathfrak{m}_x^2 \to 0$ we deduce the exact sequences

$$0 \to \mathfrak{m}_x^2 \otimes \mathfrak{L}^{\otimes n} \to \mathfrak{L}^{\otimes n} \to \mathfrak{L}^{\otimes n}/\mathfrak{m}_x^2\mathfrak{L}^{\otimes n} \to 0.$$

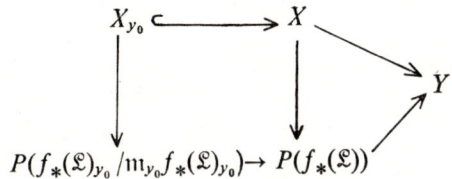

If n is large enough, then the morphism

$$f_*(\mathfrak{L}^{\otimes n})_{y_0} \to f_*(\mathfrak{L}^{\otimes n}/\mathfrak{m}_x^2\mathfrak{L}^{\otimes n})_{y_0}$$

is an epimorphism. The sheaf $\mathfrak{L}^{\otimes n}/\mathfrak{m}_x^2\mathfrak{L}^{\otimes n}$ is concentrated in x and its stalk in this point coincides with $\mathfrak{L}_x^{\otimes n}/\mathfrak{m}_x^2\mathfrak{L}_x^{\otimes n}$. We thus realize that the morphisms $\Gamma(X_{y_0}, \mathfrak{L}^{\otimes n}) \to \mathfrak{L}_x^{\otimes n}/\mathfrak{m}_x^2\mathfrak{L}_x^{\otimes n}$ are surjective. Then it will result that the morphisms induced by $\theta^{\otimes n}$ between Zariski tangent spaces in the point x and $\theta^{\otimes n}(x)$ are surjective. Accordingly, for an integer n as above, the morphism $\theta^{\otimes n}$ is a local immersion in the point x ([15]).

We actually find open sets $U_1, \ldots, U_k$ which cover X_{y_0} and integers $n_1', \ldots, n_k'$ such that the morphisms

$$\theta^{\otimes n_i'} : X \to P(f_*(\mathfrak{L}^{\otimes n_i'}))$$

are immersions on U_i. Let $n' = n_1' \ldots n_k'$. By (**) the morphism $\theta^{\otimes n'} : X \to$ $\to P(f_*(\mathfrak{L}^{\otimes n'}))$ is a local immersion in every point of $\bigcup_i U_i$. By shrinking eventually Y around y_0, we may assume that $X = \bigcup_i U_i$. We will show that for n large enough, $\theta^{\otimes n}$ is an injective map. Let x, x' be points in X such that $f(x) = f(x')$. We are going to prove the following assertion:

(***) For any invertible sheaf $\mathfrak{L}$ on X such that the morphism $f^*f_*(\mathfrak{L}) \to \mathfrak{L}$ is surjective, provided that the map $\theta : X \to P(f_*(\mathfrak{L}))$ verifies the condition $\theta(x) \neq \theta(x')$, then the same property holds also for the maps $\theta^{\otimes n}$ for any $n \geqslant 1$.

Let $y = f(x) = f(x')$ and $s^0, \ldots, s^r \in \Gamma(X_y, \mathfrak{L})$ be as in the proof of (**). By hypothesis the points $(s^0(x), \ldots, s^r(x))$, $(s^0(x'), \ldots, s^r(x'))$ differ from each other in the projective space $\mathbb{P}^r$. Let $s \in \Gamma(X_y, \mathfrak{L})$ be such that $s(x) \neq 0$, $s(x') \neq 0$. The elements $s^{n-1} \otimes s^0, \ldots, s^{n-1} \otimes s^r$ are in $\Gamma(X_y, \mathfrak{L}^{\otimes n})$ and separate the points x and x'. One then derives that $\theta^{\otimes n}(x) \neq \theta^{\otimes n}(x')$ and (***) is proved.

Let K be a compact neighbourhood of y_0. Denote $U = \bigcup_{i=1}^k U_i \times_Y U_i$. Consider a point $(x, x') \in (X \times_Y X) \setminus U$ such that $y = f(x) = f(x') \in K$. We have an exact sequence

$$0 \to \mathfrak{m}_x \cap \mathfrak{m}_{x'} \to \mathcal{O}_X \to \mathcal{O}_X/\mathfrak{m}_x \oplus \mathcal{O}_X/\mathfrak{m}_{x'} \to 0.$$

As in the previous reasonings, one derives that for n large enough the canonical morphism

$$\Gamma(X_y, \mathfrak{L}^{\otimes n}) \to \mathfrak{L}_x^{\otimes n}/\mathfrak{m}_x\mathfrak{L}_x^{\otimes n} \oplus \mathfrak{L}_{x'}^{\otimes n}/\mathfrak{m}_{x'}\mathfrak{L}_{x'}^{\otimes n}$$

is surjective. Then one can easily check that $\theta^{\otimes n}(x) \neq \theta^{\otimes n}(x')$. Since $\theta^{\otimes n}$ is a proper map, it follows that there exists a neighbourhood W in $X \times_Y X$ for (x, x') such that $\theta^{\otimes n}(\tilde{x}) \neq \theta^{\otimes n}(\tilde{x}')$ for any $(\tilde{x}, \tilde{x}') \in W$.

Thus there is a finite number of open sets W_i which cover $((X \times_Y X) \setminus U) \cap \cap (f \times_Y f)^{-1}(K)$ and some integers n_i'' such that the maps $\theta^{\otimes n_i''}$ separate the points of W_i. Let n'' be the product of these integers. The map $\theta^{\otimes n''}$ separates the points of $((X \times_Y X) \setminus U) \cap (f \times_Y f)^{-1}(K)$. By shrinking eventually Y around y_0, we may assume that the map $\theta^{\otimes n''}$ separates the points of $(X \times_Y X) \setminus U$. Let $n = n'n''$. We show that the map $\theta^{\otimes n}$ is injective. Let $x \neq x'$ such that $f(x) = f(x')$ ($\theta^{\otimes n}$ being a morphism over Y, it is sufficient to confine to such pairs of points). If x and x' belong to an open set U_i, then the conclusion follows because n_i' divides n. If $(x, x') \in (X \times_Y X) \setminus U$, then $\theta^{\otimes n}(x) \neq \theta^{\otimes n}(x')$ as n'' divides n.

Therefore the morphism $\theta^{\otimes n}$ is injective, proper and local immersion in any point of X. Consequently, $\theta^{\otimes n}$ is a closed immersion and this completes the proof.

The second amplitude criterion is a consequence of the results established in Chapter III.

THEOREM 4.2. *Let $f : X \to Y$ be a flat proper morphism of complex spaces, let $\mathfrak{L}$ be an invertible sheaf on X, and y a point in Y. Assume that $\mathfrak{L}_y$ is a very ample sheaf on X_y and $H^1(X_y, \mathfrak{L}_y) = 0$.*

Then there exists a neighbourhood V of y such that $\mathfrak{L}|f^{-1}(V)$ is very ample with respect to V; moreover, the sheaf $f_*(\mathfrak{L})$ is free in a neighbourhood of y.

Proof. We set $\mathfrak{E} = f_*(\mathfrak{L})$. By III. 3.9, $\mathfrak{E}$ is free in a neighbourhood of y and $\mathfrak{E}_y/\mathfrak{m}_y\mathfrak{E}_y = \Gamma(X_y, \mathfrak{L}_y)$. Since the sheaf $\mathfrak{L}_y = \mathfrak{L}/\hat{\mathfrak{m}}_y\mathfrak{L}$ is very ample, hence the elements of $\Gamma(X_y, \mathfrak{L}_y)$ generate the stalks $(\mathfrak{L}_y)_x = \mathfrak{L}_x/\mathfrak{m}_y\mathfrak{L}_x$, $x \in f^{-1}(y)$. Hereof, by Nakayama lemma one easily derives that the natural morphism of sheaves $f^*f_*(\mathfrak{L}) = f^*(\mathfrak{E}) \to \mathfrak{L}$ is surjective in the points of the fiber $f^{-1}(y)$. Then it will be surjective in a neighbourhood of $f^{-1}(y)$.

By shrinking eventually Y around y we can assume that the morphism $f^*(\mathfrak{E}) \to \mathfrak{L}$ is surjective and let $i : X \to P(\mathfrak{E}) = P$ be the morphism over Y given by it. The morphism induced by i on the analytic fiber over y is just the morphism

$$X_y \to P_y = P(\mathfrak{E}(y)), \quad \mathfrak{E}(y) = \mathfrak{E}_y/\mathfrak{m}_y\mathfrak{E}_y = H^0(X_y, \mathfrak{L}_y),$$

defined by the very ample sheaf $\mathfrak{L}_y$ on X_y, whence it is a closed immersion.

The conclusion of the theorem will result from the following:

PROPOSITION 4.3. *Let Z be a complex space, X and Y two complex spaces over Z such that the structural morphisms g: $X \to Z$, h: $Y \to Z$ are proper and let $f : X \to Y$ be a Z-morphism. Let $z \in Z$ and $f_z : X_z \to Y_z$ be the map induced by f.*

(i) If f_z is a finite morphism (viz., a closed immersion), then there exists a neighbourhood U of z such that the morphism $g^{-1}(U) \to h^{-1}(U)$, which is the restriction of f, is a finite morphism (viz., a closed immersion).

(ii) Suppose in addition g flat. If f_z is an isomorphism, then there exists a neighbourhood U of z such that the morphism $g^{-1}(U) \to h^{-1}(U)$, the restriction of f, is an isomorphism.

Proof. In both cases it is sufficient to prove that for any $y \in h^{-1}(z)$ there exists a neighbourhood V_y of y such that the restriction of f, $f^{-1}(V_y) \to V_y$ is a finite morphism (respectively closed immersion, isomorphism). Indeed the restriction of f, $f^{-1}(V) \to V$, results clearly to be a finite morphism (closed immersion, isomorphism, respectively), where V is the union of the neighbourhoods V_y; since h is proper, there exists a neighbourhood U of z such that $h^{-1}(U) \subset V$ and the proposition is proved.

We remark that f is also proper.

(*i*) Let y be an arbitrary point in the fiber $h^{-1}(z)$. We first analyse the case of the finite morphisms. Consider a point $x \in f^{-1}(y)$. The morphism $\mathcal{O}_y/\mathfrak{m}_z\mathcal{O}_y \to \mathcal{O}_x/\mathfrak{m}_z\mathcal{O}_x$ is a finite morphism of rings. Then $\mathcal{O}_x/\mathfrak{m}_z\mathcal{O}_x$, hence $\mathcal{O}_x/\mathfrak{m}_y\mathcal{O}_x$ is an $\mathcal{O}_y$-module of finite type. According to Weierstrass preparation theorem ([15], [33]), $\mathcal{O}_x$ will be an $\mathcal{O}_y$-module of finite type, hence f is finite in x.

Then one can deduce easily the existence of a neighbourhood V_y of y such that the morphism $f^{-1}(V_y) \to V_y$ is finite in every point. Since this restriction of f is even proper, it is finite.

We now deal with the case of closed immersions. By the previous considerations, we may assume that f is a finite morphism. Thus one gets $X = \text{Specan}(\mathcal{A})$ where $\mathcal{A}$ is a coherent $\mathcal{O}_Y$-algebra. Let $u : \mathcal{O}_Y \to \mathcal{A}$ be the structural morphism. We are going to prove that there exists a neighbourhood V_y of y such that the restriction $u|V_y$ is surjective; in virtue of the properties of the functor Specan it follows that the morphism induced by f, $f^{-1}(V_y) \to V_y$ is a closed immersion. It is enough to check that the morphism $u_y : \mathcal{O}_y \to \mathcal{A}_y$ is surjective. By hypothesis, the morphism $\mathcal{O}_y/\mathfrak{m}_z\mathcal{O}_y \to \mathcal{A}_y/\mathfrak{m}_z\mathcal{A}_y$ deduced from u_y is surjective and the conclusion follows from Nakayama lemma.

(*ii*) By the same reasoning as above, all we need is to prove that u_y is bijective, as soon as the morphism $\mathcal{O}_y/\mathfrak{m}_z\mathcal{O}_y \to \mathcal{A}_y/\mathfrak{m}_z\mathcal{A}_y$, induced by it is bijective and that $\mathcal{A}_y$ is a flat $\mathcal{O}_z$-module ($\mathcal{A}_y$ is isomorphic to the direct sum of the $\mathcal{O}_z$-modules $\mathcal{O}_x$, $f(x) = y$). The morphism u_y is surjective and let $K = \text{Ker } u_y$. The exact sequence

$$0 \to K \to \mathcal{O}_y \to \mathcal{A}_y \to 0$$

yields the exact sequence

$$0 \to K/\mathfrak{m}_zK \to \mathcal{O}_y/\mathfrak{m}_z\mathcal{O}_y \to \mathcal{A}_y/\mathfrak{m}_z\mathcal{A}_y \to 0.$$

Then $K/\mathfrak{m}_zK = 0$, so much the more $K/\mathfrak{m}_yK = 0$. In virtue of Nakayama lemma, $K = 0$ and the proof of the proposition is over.

Recall that a morphism of complex spaces $f : X \to Y$ is said *smooth* (or *simple*) if, for any point $x \in X$, there exist some neighbourhoods U_x (V_y respectively) of x ($y = f(x)$, respectively) and an open set D_x in a numerical space such that $f(U_x) \subset V_y$, $U_x \simeq V_y \times D_x$ and $f|U_x : U_x \to V_y$ is obtained by means of the projection $V_y \times D_x \to V_y$. In particular, the morphism f results to be flat. Moreover, if f is proper and the fibers are Riemann surfaces of genus g, then X is called *curve of genus g over Y*. In this case the sheaf $\Omega^1_{X/Y}$ of relative differentials is an invertible $\mathcal{O}_X$-module [15].

COROLLARY 4.4. *Let X be a "curve of genus g over Y" and let $\Omega = \Omega^1_{X/Y}$ the sheaf of relative differentials.*

(*a*) *If $g = 0$ then $\Omega^{\otimes -n}$ is very ample with respect to Y for any $n \geqslant 1$.*

(*b*) *If $g = 1$ and s is the marked section of X over Y which corresponds to an $\mathcal{O}_X$-ideal $\mathfrak{J}$ on X, then $\mathfrak{J}^{\otimes -n}$ is very ample with respect to Y for any $n \geqslant 3$.*

(*c*) *If $g = 2$, then $\Omega^{\otimes n}$ is very ample with respect to Y for $n \geqslant 3$.*

(*d*) *If $g \geqslant 3$, then $\Omega^{\otimes n}$ is very ample with respect to Y for $n \geqslant 2$.*

Proof. Apply the theorem, the conditions to be checked being consequences of the Riemann-Roch theorem [78].

Corollary 3.2. can be applied if X is a "curve of genus g over Y": by the preceding corollary we can obtain very ample sheaves $\mathfrak{L}$ by means of $\Omega^1_{X/Y}$. For instance we have:

COROLLARY 4.5: *Let $f: X \to Y$ be a "curve of genus 0 over Y", $\Omega = \Omega^1_{X/Y}$ the sheaf of relative holomorphic differential forms and $\mathfrak{F} \in$ **Coh** (X), locally free.*

Then, locally on Y, $f_*(\mathfrak{F} \otimes \Omega^{\otimes m}) = 0$ for m large enough.

Indeed, $\mathfrak{L} = \Omega^{\otimes -1}$ is very ample with respect to Y and the conclusion follows from 3.2.

Bibliographical indications

The former proof of the Grauert-Remmert theorem in the case $X = Y \times \mathbb{P}^r$, $\mathfrak{L} = \mathcal{O}_X(1)$, $f =$ projection, can be found at [32]; at the same time, the authors proved the coherence of the sheaves $R^q f_*(\mathfrak{F})$. The proof given here, which uses the finiteness theorem for proper morphisms, is the transposition to the relative case of the proof given by Serre [77] in the absolute case.

Theorem 2.4 is essentially equivalent (*modulo* its algebraic analogous [76]) to the comparison theorem 2.6; for this reason, its extension 2.1 is also called the comparison theorem.

A systematical treating of the comparison of algebraic and analytic properties can be found in ([36], SGA 1, Exp. XII).

Theorem 3.1 was proved by Grothendieck in the algebraic case ([36], SGA 2, Exp. XII); the analytic case was considered in [7].

Theorems 4.1 and 4.2 are taken from [30], [57] and [15], respectively.

Chapter V

Flat morphisms of complex spaces

Introduction

Let $f : X \to Y$ be a morphism of complex spaces and $\mathcal{F}$ a coherent analytic sheaf on X. For any point $y \in Y$, the fiber $X_y = (f^{-1}(y), \mathcal{O}_X/\hat{\mathfrak{m}}_y\mathcal{O}_X|f^{-1}(y))$ is a complex space. The sheaf $\mathcal{F}_y = \mathcal{F}/\hat{\mathfrak{m}}_y\mathcal{F}$ defines a coherent analytic sheaf on X_y. One thus obtains an "analytic" family $(X_y)_{y\in Y}$ of complex spaces and a family $(\mathcal{F}_y)_{y\in Y}$ of coherent sheaves.

The key point of the flatness consists in the fact that under the hypothesis that f is flat a "continuous" behaviour of the properties of the above families is provided, as well as some connections between the properties of the source X, the target space Y and the fibers X_y. Assertions 2.4—2.6, 2.8—2.11 stand for examples in this respect; a systematical treating of these problems in the algebraic case can be found in ([37], Ch. IV).

The hypothesis of the flatness of a morphism also implies the remarkable fact that it is open (theorem 2.12); theorem 2.13 stands as a converse result.

The main result in the chapter asserts that the set of the points where $\mathcal{F}$ is not flat over Y is a closed analytic subset (theorem 4.5). In its proof one makes essential use of the following theorem: the ring of germs of analytic functions defined on a Stein semianalytic compact is noetherian (theorem 3.1).

The flat morphisms can be interpreted as the extension of the notion of analytic family of complex manifolds to the singular case.

All the results of this chapter were first proved in algebraic geometry and were conjectured for the analytic case by A. Grothendieck [15].

§ 1. Preliminaries

(a) (cf. [82]) Let (X, d_X) be a metric space. For $x \in X$ and A, C subsets of X, denote $d_X(x, A) = \inf_{y\in A} d_X(x, y)$ and $d_X(C, A) = \inf_{y\in C} d_X(y, A)$. If $\varepsilon > 0$ and $x \in X$ we set $B_X(x; \varepsilon) = \{y \in X | d_X(x, y) < \varepsilon\}$. If $\varepsilon > 0$ and $x, y \in X$, then a finite sequence of points $x_0, \ldots, x_m$ of X is called an *ε-chain linking x with y* whenever $x_0 = x$, $x_m = y$ and $d_X(x_i; x_{i+1}) < \varepsilon$ for $0 \leqslant i < m$.

PROPOSITION 1.1. *If x and y are points of a compact subset K of a metric space X, then x and y lie in the same connected component of K if and only if, for any $\varepsilon > 0$, x can be linked with y by an ε-chain in K.*

Proof. We may assume $K = X$. Let C be the connected component of X which contains x. Denote by T_ε the set of the points $z \in X$ which can be linked with x by an ε-chain. Let $T = \bigcap_{\varepsilon>0} T_\varepsilon$. We will prove the equality $C = T$ and the conclusion will follow. The sets T_ε are simultaneously open and closed, hence $C \subset T$. For the other inclusion, it is sufficient to prove that the closed set T is connected. Suppose, on the contrary, that $T = M \cup N$, where M and N are closed disjoint nonempty subsets which are distinct of T. Let $d = d(M, N)$, $d > 0$ and $P = \left\{ z \in X \mid d(z, T) \geqslant \frac{d}{3} \right\}$. Suppose $x_0 \in M$. For any natural number n there exists a $\frac{1}{n}$-chain linking x_0 with a point of N. For any integer $n > 3/d$ there exists a point x_n of this chain which lies in P. Any accumulation point of the sequence x_n lies in $P \cap T$ and we thus obtain a contradiction.

LEMMA 1.2. *Let $f : X \to Y$ be an open proper map between metric spaces. Then for any compact connected subset K of Y, $f^{-1}(K)$ has only a finite number of connected components.*

Proof. Let $K \subset Y$ be a compact connected set. We may assume $K = Y$. Since f is open and proper, $f(X)$ is simultaneously open and closed in Y; therefore $f(X) = Y$. We first prove the following assertion:

(*) For any $\varepsilon > 0$, there is $\delta > 0$ such that for any $x \in X$,

$$f(B_X(x; \varepsilon)) \supset B_Y(f(x); \delta).$$

Suppose, on the contrary, that this is not true. Then there exist $\varepsilon > 0$, a descending sequence of positive numbers $\{\delta_n\}$ converging to 0 and a sequence $\{x_n\}$ in X such that $f(B_X(x_n; \varepsilon)) \not\supset B_Y(f(x_n); \delta_n)$. Let $y_n \in B_Y(f(x_n); \delta_n) \setminus f(B_X(x_n; \varepsilon))$. Since X and Y are compact spaces, we can assume that there are $x^* \in X$, $y^* \in Y$ such that $x_n \to x^*$, $y_n \to y^*$ (passing eventually to subsequences). We have $f(x^*) = y^*$, as $d_Y(f(x_n), y_n) < \delta_n$ and $\delta_n \to 0$. Since $B_X(x^*; \varepsilon/2)$ and $f(B_X(x^*; \varepsilon/2))$ are neighbourhoods of x^* and y^*, there exists m_0 so that $x_m \in B_X(x^*; \varepsilon/2)$ and $y_m \in f(B_X(x^*; \varepsilon/2))$ for any $m \geqslant m_0$. Therefore, for $m \geqslant m_0$, $y_m \in f(B_X(x^*; \varepsilon/2)) \subset f(B_X(x_m; \varepsilon))$, a contradiction. In this way the assertion (*) is proved.

Let us fix a point $y \in Y$ and let $f^{-1}(y) = \{x_1, \ldots, x_k\}$. Consider an arbitrary point $x \in X$. We shall prove now the assertion:

(**) For every natural number n there exists j_n, $1 \leqslant j_n \leqslant k$ such that x can be linked with x_{j_n} by a $1/n$-chain in X.

Fix a natural number n. By (*) there exists $\delta > 0$ such that $f(B_X(z; 1/n)) \supset B_Y(f(z); \delta)$ for any $z \in X$. Since Y is compact and connected, there exists a δ-chain $y_0, \ldots, y_n$ in Y which links $f(x)$ with y (1.1). By induction on i, $0 \leqslant i \leqslant m$, we can find points $z_i \in X$ such that $z_0 = x$, $f(z_i) = y_i$ and $d_X(z_{i-1}, z_i) < 1/n$, $1 \leqslant i \leqslant m$. Since $f(z_m) = y_m = y$, there exists j_n, $1 \leqslant j_n \leqslant k$, such that $z_m = x_{j_n}$. So the assertion (**) is proved too.

There exists j^*, $1 \leqslant j^* \leqslant k$, such that $j^* = j_n$ for infinitely many natural numbers n. Consequently, for any $\varepsilon > 0$, x can be linked with x_{j^*} by an ε-chain in X. According to proposition 1.1, x lies in the same connected component of X as x_{j^*}.

In this way, we have proved that X has only a finite number of connected components.

LEMMA 1.3. *Let X be a compact metric space and A a compact connected component of an open subset U of X. Then A is a connected component of X.*

Proof. Let $\tilde{A}$ be the connected component of X containing A. It is enough to check that $\tilde{A} \subset U$. Suppose the contrary. Let us fix $y_0 \in \tilde{A} \setminus U$ and $x_0 \in A$. We have $\varepsilon = d(A, X \setminus U) > 0$ and let $C = \{x \in X \mid \varepsilon/3 \leqslant d(x, A) \leqslant 2\varepsilon/3\}$, $D = \{x \in X \mid d(x, A) \leqslant 2\varepsilon/3\}$. We need the following assertion:

$(*)$ If n is a natural number $> 3/\varepsilon$, then x_0 can be linked with an element z_n of C by a $1/n$-chain in D.

Let $n > 3/\varepsilon$. The point x_0 can be linked with y_0 by a $1/n$-chain $x_0, \ldots, x_m$ in X. Let i be the smallest integer such that $x_i \notin D$. We have $0 < i \leqslant m$. Since $d(x_{i-1}, x_i) < 1/n < \varepsilon/3$, it follows that $x_{i-1} \in C$. Let $z_n = x_{i-1}$. Then $x_0, \ldots, x_{i-1}$ is a $1/n$-chain in D linking x_0 with z_n and $(*)$ is thus proved.

Since C is a compact set, by passing eventually to a subsequence, we may assume that the sequence z_n tends to a point z_0. For any $\delta > 0$, x_0 can be linked with z_0 by a δ-chain in D. According to 1.1, z_0 belongs to the connected component of D which contains x_0. As $D \subset U$ and A is the connected component of U containing x_0, $z_0 \in A$. This fact contradicts the inequality $d(z_0, A) \geqslant \varepsilon/3$ and the lemma follows.

(b) We would like to call attention to several facts concerning flat modules.

"Let $A \to B$ be a morphism of noetherian rings, I an ideal of A such that IB is contained in the Jacobson radical of B (the intersection of all maximal ideals) and M a B-module of finite type. Then M is a flat A-module if and only if M/IM is a flat A/I-module and $\mathrm{Tor}_1^A(M, A/I) = 0$" ([10], Ch. III, § 5, th. 1 and prop. 2).

In particular, the following result holds:

"Let A and B be noetherian local rings, k the residual field of A, $A \to B$ a local morphism and M a B-module of finite type. Then M is A-flat if and only if $\mathrm{Tor}_1^A(M, k) = 0$".

COROLLARY 1.4. *Let $A \to B$ be a local morphism of noetherian local rings, t a nonzerodivisor of A belonging to the maximal ideal, and M a B-module of finite type. If t is M-regular and M/tM is A/tA-flat, then M is A-flat.*

Proof. We have to verify that $\mathrm{Tor}_1^A(M, A/tA) = 0$. This fact can be easily checked by tensoring $\otimes_A M$ the exact sequence

$$0 \to A \xrightarrow{t} A \to A/tA \to 0.$$

COROLLARY 1.5. ([37], Ch. $\mathrm{O_{III}}$, 10.2.4). *Let $A \to B$ be a local morphism of noetherian local rings, let k be the residual field of A and let M and N be two*

B-modules of finite type, where N is flat over A. Let $u: M \to N$ be a B-morphism. The following conditions are equivalent:

(*a*) *u is injective and Coker u is A-flat.*

(*b*) *$u \otimes 1 : M \otimes_A k \to N \otimes_A k$ is injective.*

Proof. The implication (*a*) ⇒ (*b*) follows by tensoring $\otimes_A k$ the exact sequence

$$0 \to M \xrightarrow{u} N \to \text{Coker}\, u \to 0.$$

(*b*) ⇒ (*a*). Let $P = \text{Im}\, u$, $Q = \text{Coker}\, u$ and $R = \text{Ker}\, u$. The composition $M \otimes_A k \to P \otimes_A k \to N \otimes_A k$ is injective and the map $M \otimes_A k \to P \otimes_A k$ is surjective. Therefore the map $M \otimes_A k \to P \otimes_A k$ is bijective and the map $P \otimes_A k \to N \otimes_A k$ is injective. The exact sequence $0 \to P \to N \to Q \to 0$ yields the exact sequence

$$0 = \text{Tor}_1^A(N, k) \to \text{Tor}_1^A(Q, k) \to P \otimes_A k \to N \otimes_A k \to Q \otimes_A k \to 0.$$

It will then result that $\text{Tor}_1^A(Q, k) = 0$, hence by the result recalled above, $Q = \text{Coker}\, u$ is A-flat. From the exact sequence $0 \to P \to N \to Q \to 0$ it follows that P is A-flat. The exact sequence $0 \to R \to M \to P \to 0$ yields the exact sequence

$$0 \to R \otimes_A k \to M \otimes_A k \to P \otimes_A k \to 0.$$

Since the map $M \otimes_A k \to P \otimes_A k$ is bijective, $R \otimes_A k = 0$. By Nakayama lemma we get $R = 0$, hence u is injective. The corollary is proved.

Let A be a ring (commutative and unitary), I an ideal of it and M an A-module. Denote by $\hat{M}$ the completion of M in the I-adic topology, $\hat{M} = \varprojlim_k (M/I^k M)$. $\hat{M}$ is an $\hat{A}$-module and the correspondence $M \mapsto \hat{M}$ is functorial. Recall the following result:

"If A is a noetherian local ring, M an A-module of finite type and I an ideal included in the maximal ideal of A, then $\hat{M}$ is canonically isomorphic to $M \otimes_A \hat{A}$ and the natural map $M \to \hat{M}$ is injective. Moreover, the functor $M \mapsto \hat{M}$ from the category of A-modules of finite type to the category of $\hat{A}$-modules of finite type is exact and faithfull ($\hat{M} = 0 \Rightarrow M = 0$)" ([10], Ch. III, § 3).

Corollary 1.6. ([10], Ch. III, § 5, n° 4, prop. 4). *Let $A \to B$ be a local morphism of noetherian local rings, $\mathfrak{m}_A$ and $\mathfrak{m}_B$ the maximal ideals of A and B and M a B-module of finite type. Then M is A-flat if and only if $\hat{M}$ (the completion in the $\mathfrak{m}_B$-adic topology) is $\hat{A}$ (the completion in the $\mathfrak{m}_A$-adic topology)-flat.*

Proof. By our above remark, it is sufficient to prove that $\text{Tor}_1^{\hat{A}}(\hat{A}/\mathfrak{m}_{\hat{A}}, \hat{M}) \simeq \text{Tor}_1^A(A/\mathfrak{m}_A, M)^\wedge$ ($\mathfrak{m}_{\hat{A}} = \mathfrak{m}_A \hat{A}$ is the maximal ideal of $\hat{A}$). Let $L_\bullet$ be a resolution for $A/\mathfrak{m}_A$ by free A-modules of finite type. $\mathrm{Tor}^A_\bullet(A/\mathfrak{m}_A, M)$ are the homology groups of the complex $L_\bullet \otimes_A M$. Since the completion is an exact functor on the modules of finite type, $\mathrm{Tor}^A_\bullet(A/\mathfrak{m}_A, M)^\wedge$ can be canonically identified with the homology groups of the complex $(L_\bullet \otimes_A M) \otimes_B \hat{B}$. This complex is canonically isomorphic to the complex $\hat{L}_\bullet \otimes_{\hat{A}} \hat{M}$. Since $\hat{L}_\bullet$ is a resolution for $\hat{A}/\mathfrak{m}_{\hat{A}}$ by free $\hat{A}$-modules, the desired conclusion follows.

PROPOSITION 1.7. ([37], Ch. 0_{III}, 10.2.6). *Let A and B be two nowtherian local rings, $A \to B$ a local morphism, I a proper ideal of B and M a B-module of finite type. If moreover $M_n = M/I^{n+1} M$ is a flat A-module for any $n \geqslant 0$, then M is A-flat.*

Proof. We must show that for any injective morphism $u : N' \to N$ of A-modules of finite type, the morphism $v = 1 \otimes u : M \otimes_A N' \to M \otimes_A N$ is injective. $M \otimes_A N'$ and $M \otimes_A N$ are B-modules of finite type, hence the morphisms $M \otimes_A N' \to \to (M \otimes_A N')^\wedge$, $M \otimes_A N \to (M \otimes_A N)^\wedge$ are injective, where the completions are considered with respect to the I-adic topology.

Consequently, it is enough to prove that the morphism $\hat{v} : (M \otimes_A N')^\wedge \to \to (M \otimes_A N)^\wedge$ is injective. We have $v = \varprojlim v_n$, where v_n is the morphism $1 \otimes u : M_n \otimes_A N' \to M_n \otimes_A N$. By hypothesis M_n is A-flat, hence v_n is injective for any n; the functor $\varprojlim$ being left exact, the conclusion follows.

COROLLARY 1.8. ([37], Ch. 0_{III}, 10.2.7). *Let $A \to B$ be a local morphism of noetherian local rings, f an element of the maximal ideal of B and M a B-module of finite type. Suppose f is M-regular and M/fM is a flat A-module. Under these assumptions, M is a flat A-module.*

Proof. Denote $M^n = f^{n+1}M$, $n \geqslant 0$. Since f is M-regular, the modules M^n/M^{n+1} are isomorphic to M/fM, hence A-flat. By using an exact sequence of the form

$$0 \to M^n/M^{n+1} \to M/M^{n+1} \to M/M^n \to 0$$

and by induction on n, one derives that the modules $M_n = M/M^n = M/f^{n+1}M$ are A-flat. The conclusion then follows from 1.7.

PROPOSITION 1.9. ([37], Ch. 0_{IV}, 15.1.16). *Let $A \to B$ be a local morphism of noetherian local rings, k the residual field of A and M a B-module of finite type. Let $(f_i)_{1 \leqslant i \leqslant r}$ be a sequence of elements of the maximal ideal of B and $(g_i)_{1 \leqslant i \leqslant r}$ its image in $B \otimes_A k$. The following conditions are equivalent:*

(*a*) *The sequence $(g_i)_{1 \leqslant i \leqslant r}$ is $M \otimes_A k$-regular and M is A-flat.*

(*b*) *The sequence $(f_i)_{1 \leqslant i \leqslant r}$ is M-regular and $M \Big/ \sum_{i=1}^{r} f_i M$ is A-flat.*

Proof. We use induction on r. For $r = 1$ the equivalence is true in virtue of 1.5 and 1.8. We now assume the assertion already proved for $r - 1$ and prove the equivalence of the statement for r ($r \geqslant 2$).

$(a) \Rightarrow (b)$. By induction hypothesis, the sequence $(f_i)_{1 \leqslant i \leqslant r-1}$ is M-regular and $M \Big/ \sum_{i=1}^{r-1} f_i M$ is A-flat. We have a canonical isomorphism

$$(M \otimes_A k) \Big/ \sum_{i=1}^{r-1} g_i (M \otimes_A k) \simeq \left(M \Big/ \sum_{i=1}^{r-1} f_i M \right) \otimes_A k.$$

Consequently, the element g_r is $\left(M \Big/ \sum_{i=1}^{r-1} f_i M \right) \otimes_A k$-regular. According to the implication $(a) \Rightarrow (b)$ for $r = 1$, f_r is $M \Big/ \sum_{i=1}^{r-1} f_i M$-regular and $\left(M \Big/ \sum_{i=1}^{r-1} f_i M \right) \Big/ f_r \left(M \Big/ \sum_{i=1}^{r-1} f_i M \right) \simeq M \Big/ \sum_{i=1}^{r} f_i M$ is A-flat. The conclusion follows.

The other implication can be proved similarly.

(c) LEMMA 1.10. *Let A be an integral ring, f a non-null element of A and M an A-module such that the module of quotients M_f is A_f-free. Then for any $g \neq 0$ in A, M_{fg} is A_{fg}-free.*

Proof. If $\bar{g}$ is the image of g in A_f, then $M_{fg} \simeq (M_f)_{\bar{g}}$. The conclusion follows then from the fact that passing to quotients commutes with the direct sums.

LEMMA 1.11. *Let A be an integral ring, f and g non-null elements of A and $0 \to M \to N \to P \to 0$ an exact sequence of A-modules. If M_f is A_f-free and P_g is A_g-free, then N_{fg} is A_{fg}-free.*

Proof. By the previous lemma it results that M_{fg} and P_{fg} are free A_{fg}-modules. The sequence $0 \to M_{fg} \to N_{fg} \to P_{fg} \to 0$ splits and the conclusion follows.

Recall a fundamental result on the algebras of finite type over a field ([10], Ch. V, § 3, n°=1, Th. 1).

"(E. Noether's lemma of normalization). Let A be an algebra of finite type over a field k. Then there exists a finite number of elements $t_1, \ldots, t_m$ in A, which are algebraically independent over k and such that A is integral over $k[t_1, \ldots, t_m]$".

Consequently one obtains the following result:

"Let A be an integral domain and B an A-algebra of finite type such that the structural morphism $A \to B$ is injective. Then there exist an element $g \neq 0$ in A and elements $t_1, \ldots, t_m$ in B, which are algebraically independent over A and such that the ring of quotients B_g is integral over the ring $A_g[t_1, \ldots, t_m]$" ([10], Ch. V, § 3, n°=1, cor. 1, th. 1).

We also recall the following fact:

"If B is an A-algebra of finite type, integral over A, then B is an A-module of finite type. If moreover the structural morphism $A \to B$ is injective, then $\dim A = \dim B$" ([79], Ch. III).

LEMMA 1.12. *Let A be a noetherian integral domain, B an A-algebra of finite type and M a B-module of finite type. Then there exists $f \neq 0$ in A such that M_f is a free A_f-module.*

Proof. Denote by K the field of the quotients of A; then $B\otimes_A K$ is an algebra of finite type over K and $M\otimes_A K$ is a $B\otimes_A K$-module of finite type. We use induction on the integer $n = \dim_{B\otimes_A K}(M\otimes_A K)$. Consider first of all the case when $M\otimes_A K = 0$. Let $(m_i)_{1\leqslant i\leqslant r}$ be a system of generators of M over B. There exists an element $f \neq 0$ in A such that $fm_i = 0$ $(i = 1,\ldots, r)$; accordingly, $M_f = 0$ and the lemma is concluded in this case.

So we may assume $n > 0$ and the lemma proved for integers $< n$. Let $M = M_1 \supset M_2 \supset \ldots \supset M_q = 0$ be a composition series such that the B-modules $N_i = M_i/M_{i+1}$ are isomorphic to B-modules of the form $B/\mathfrak{p}_i$, $\mathfrak{p}_i$ being prime ideals of B. If the lemma is true for every N_i, then it will be true for M too (1.10, 1.11).

In this way, if we replace B by a ring of the form $B/\mathfrak{p}$ ($\mathfrak{p}$ being a prime ideal of B), we may suppose $M = B$ and B an integral domain. Let $A' \simeq A/\mathfrak{p}$ the image of A in B, $\mathfrak{p}$ being the kernel of $A \to B$. Consider the open set $\operatorname{Spec} A \setminus V(\mathfrak{p})$. If this set is empty, then the morphism $A \to B$ is injective. Otherwise, there exists $f \neq 0$ such that $D(f) \subset \operatorname{Spec} A \setminus V(\mathfrak{p})$; accordingly, $(A/\mathfrak{p})_f = 0$, hence the desired assertion is fulfilled for the morphism $A \to B$ as soon as it holds for the morphism $A' \to B$. We may thus assume that the structural morphism $A \to B$ is injective. Then there exist an element $g \neq 0$ in A and elements t_i in B, $1 \leqslant i \leqslant m$, which are algebraically independent over A and such that B_g is integral over $A_g[t_1,\ldots, t_m]$. We may further replace A by A_g and B by B_g and in this way we can assume the ring B integral over the polynomial ring $C = A[t_1,\ldots, t_m]$. The extension $C\otimes_A K = K[t_1,\ldots, t_m] \to B\otimes_A K$ is integral (and obviously injective), therefore $m = n$.

B is a C-module of finite type without torsion, hence there exists an exact sequence of C-modules of the form

$$0 \to C^r \to B \to M' \to 0,$$

where M' is a torsion C-module (one considers elements $y_1,\ldots y_r$ in B, whose images in $B\otimes_C L$ form a basis over the field L of the quotients of C; the morphism $C^r \to B$ defined by them is injective and its cokernel is a torsion C-module). $M'\otimes_A K$ is $C\otimes_A K$-module of finite type, hence

$$\dim_{C\otimes_A K}(M'\otimes_A K) < \dim (C\otimes_A K) = n.$$

By the induction hypothesis, there exists $f \neq 0$ such that M'_f is A_f-free. Since $(C^r)_f \simeq A_f[t_1,\ldots, t_m]^r$ is A_f-free, it follows that B_f is a free A_f-module and the lemma is proved.

LEMMA 1.13. ([37], Ch. 0_{III}, 9.2.6). *Let X be a noetherian topological space* (*any nonempty family of closed subsets contains a minimal element*) *and E a subset of X. Then E is open if and only if, for any irreducible closed subset Y of X which intersects E, $E \cap Y$ contains a nonempty open subset of Y.*

Proof. The condition is obviously necessary. We shall prove now it is also sufficient. Consider the set of all closed subspaces (hence noetherian) of X which do not verify the lemma. We will show that this is void and the proof will be over. Suppose, on the contrary, that it is non-void, hence it has a minimal element. Then we obviously come to a contradiction as soon as we are able to prove the lemma under the supplementary hypothesis that it holds for any closed subspace X' of X, $X' \neq X$. If the space X is of the form $X = X' \cup X''$ where X' and X'' are closed proper subsets of X, then the lemma holds for X, since it is true for X', X''.

Thus we may assume that X is an irreducible space. Let E be like in the statement. There exists a nonempty open subset U of X, $U \subset E$. In virtue of the supplementary hypothesis, $E \cap (X \setminus U)$ is open in $X \setminus U$. Then it will easily result that $E = U \cup (X \setminus \overline{(X \setminus E)})$ and this concludes the proof.

Let $A \xrightarrow{\varphi} B$ be a ring morphism, $\mathfrak{q} \in \operatorname{Spec} B$ and M an A-module. We say that M *is A-flat* (or *φ-flat*) in the ideal $\mathfrak{q}$ if $M_\mathfrak{q}$ is a flat $A_{\varphi^{-1}(\mathfrak{q})}$-module. By the next lemma this is equivalent to the fact that $M_\mathfrak{q}$ is A-flat.

LEMMA 1.14. *Let A be a ring and let $S \subset A$ be a multiplicative system. If M is a flat A-module, then M_S is flat over the rings A and A_S. If M is an A_S-module, then M is A-flat if and only if it is A_S-flat.*

The proof immediately follows from the properties of the modules of quotients.

THEOREM 1.15. ([37], Ch. IV, 11.1.1). *Let $A \to B$ be a morphism of noetherian rings such that B is an A-algebra of finite type and M a B-module of finite type. Then the set of the prime ideals of B in which M is A-flat is an open subset of Spec B.*

Proof. By applying lemma 1.13 (since the set from the statement cuts an irreducible closed subset $V(\mathfrak{q})$ of Spec B if and only if M is A-flat in $\mathfrak{q}$), it is sufficient to prove the assertion:

(∗) "Let $\mathfrak{q}$ be a prime ideal of B in which M is A-flat and $\mathfrak{p}$ is its inverse image in A. Then there exists an element $g \in B \setminus \mathfrak{q}$ such that for any prime ideal $\mathfrak{q}' \supset \mathfrak{q}$ of B enjoying the property that $g \notin \mathfrak{q}'$, M is A-flat in $\mathfrak{q}'$."

Let us fix $\mathfrak{q}$ such that $M_\mathfrak{q}$ is A-flat and let $\mathfrak{p}$ be its inverse image in A. Let $\mathfrak{q}' \supset \mathfrak{q}$ be a prime ideal in B. Consider $B_{\mathfrak{q}'}$ as an A-algebra; we obviously have $\mathfrak{p}B_{\mathfrak{q}'} \subset \mathfrak{q}'B_{\mathfrak{q}'}$. We now apply the criterion of flatness recalled at the beginning of section **(b)**: $M_{\mathfrak{q}'}$ is A-flat if and only if $M_{\mathfrak{q}'}/\mathfrak{p}M_{\mathfrak{q}'}$ is a flat $A/\mathfrak{p}$-module and $\operatorname{Tor}_1^A(M_{\mathfrak{q}'}, A/\mathfrak{p}) = 0$. There are natural identifications

$$M_{\mathfrak{q}'}/\mathfrak{p}M_{\mathfrak{q}'} \simeq (M/\mathfrak{p}M)_{\mathfrak{q}'} \text{ and } \operatorname{Tor}_1^A(M_{\mathfrak{q}'}, A/\mathfrak{p}) \simeq (\operatorname{Tor}_1^A(M, A/\mathfrak{p}))_{\mathfrak{q}'}$$

(for the second one, the functor Tor could be calculated by means of a projective resolution of $A/\mathfrak{p}$). Let now $g \in B \setminus \mathfrak{q}$ be such that $g \notin \mathfrak{q}'$. There are also natural isomorphisms

$$(M/\mathfrak{p}M)_{\mathfrak{q}'} \simeq ((M/\mathfrak{p}M)_g)_{\mathfrak{q}'B_\mathfrak{q}} \text{ and } (\operatorname{Tor}_1^A(M, A/\mathfrak{p}))_{\mathfrak{q}'} \simeq (\operatorname{Tor}_1^A(M, A/\mathfrak{p})_g)_{\mathfrak{q}'B_\mathfrak{q}}$$

where $M/\mathfrak{p}M$ and $\operatorname{Tor}_1^A(M, A/\mathfrak{p})$ are considered as B-modules.

Since the passing to quotients is an exact functor, in order to prove (∗) it is enough to verify the assertion:

(∗∗) "Under the conditions from (∗), there exists an element $g \in B \setminus \mathfrak{q}$ such that $(M/\mathfrak{p}M)_g$ is a flat $A/\mathfrak{p}$-module and $\operatorname{Tor}_1^A(M, A/\mathfrak{p})_g = 0$."

We prove this assertion. Apply lemma 1.12 to the integral domain $A/\mathfrak{p}$, $A/\mathfrak{p}$-algebra of finite type $B/\mathfrak{p}B$ and the $B/\mathfrak{p}B$-module of finite type $M/\mathfrak{p}M$: there exists an element $h \in A \setminus \mathfrak{p}$ such that if $\bar{h}$ is its image in $A/\mathfrak{p}$, then $(M/\mathfrak{p}M)_{\bar{h}}$ is a free $(A/\mathfrak{p})_{\bar{h}}$-module, hence a flat $A/\mathfrak{p}$-module. On the other hand, since $M_\mathfrak{q}$ is a flat $A_\mathfrak{p}$-module (consequently A-flat), one gets

$$(\mathrm{Tor}_1^A(M, A/\mathfrak{p}))_\mathfrak{q} \simeq \mathrm{Tor}_1^A(M_\mathfrak{q}, A/\mathfrak{p}) = 0.$$

Since A and B are noetherian, $\mathrm{Tor}_1^A(M, A/\mathfrak{p})$ is a B-module of finite type, hence there exists $g' \in B \setminus \mathfrak{q}$ such that

$$g'(\mathrm{Tor}_1^A(M, A/\mathfrak{p})) = 0, \text{ therefore } \mathrm{Tor}_1^A(M,A/\mathfrak{p})_{g'} = 0.$$

The element $g = hg'$ satisfies (**). Indeed, $(M/\mathfrak{p}M)_g = (M/\mathfrak{p}M)_{\bar{g}} \simeq ((M/\mathfrak{p}M)_{\bar{h}})_{\bar{g}}$ will be $A/\mathfrak{p}$-flat ($\bar{g}$ and $\bar{h}$ are the images of g and h in $B/\mathfrak{p}B$) and $\mathrm{Tor}_1^A(M, A/\mathfrak{p})_g = (\mathrm{Tor}_1^A(M, A/\mathfrak{p})_h)_{g'} = 0$. The theorem is proved.

The following variant, due to Kiehl [46], will be useful:

THEOREM 1.16. *Let $A \to B$ be a morphism of noetherian rings, let I be an ideal in B such that B/I is an A-algebra of finite type and M a B-module of finite type. Then the set of the prime ideals of $V(I)$ in which M is A-flat is open in $V(I)$.*

The proof can be made similarly by using the followings instead of 1.12:

LEMMA 1.17. *Under the conditions of the statement of the theorem, suppose in addition that A is an integral domain. Then there exists an element $f \neq 0$ in A such that M_f is A-flat in all prime ideals of $V(I_f)$.*

Proof. The ring

$$\mathrm{gr}_I(B) = B/I \oplus I/I^2 \oplus \dots$$

is naturally a B-algebra. By the hypothesis on I, $\mathrm{gr}_I(B)$ is an A-algebra of finite type

$$\mathrm{gr}_I(M) = M/IM \oplus IM/I^2M \oplus \dots$$

is $\mathrm{gr}_I(B)$-module of finite type. There exists an element $f \neq 0$ in A such that $(\mathrm{gr}_I(M))_f \simeq \mathrm{gr}_{I_f}(M_f)$ is A_f-free, hence A-flat (1.12).

For any $\mathfrak{p} \in V(I_f)$ we obtain that the $\mathrm{gr}_{I\mathfrak{p}}(B_\mathfrak{p})$-module $\mathrm{gr}_{I\mathfrak{p}}(M_\mathfrak{p}) \simeq (\mathrm{gr}_{I_f}(M_f))_\mathfrak{p}$ is A-flat (we identify the ideals of Spec B_f with the corresponding ideals of Spec B). In this way, all modules $M_\mathfrak{p}/I_\mathfrak{p}M_\mathfrak{p}$, $I_\mathfrak{p}M_\mathfrak{p}/I_\mathfrak{p}^2M_\mathfrak{p}, \dots$ are A-flat; it then results that for any n, $M_\mathfrak{p}/I_\mathfrak{p}^nM_\mathfrak{p}$ is A-flat, hence $A_{\mathfrak{p}\cap A}$-flat. By (1.7), the module $M_\mathfrak{p}(\simeq (M_f)_\mathfrak{p})$ is $A_{\mathfrak{p}\cap A}$-flat.

Remark. Under the assumptions of the theorem, there exists however an ideal $\mathfrak{a}$ in B such that

$$V(\mathfrak{a}) \cap V(I) = \{\mathfrak{p} \in V(I) | M_\mathfrak{p} \text{ is not } A\text{-flat}\}.$$

We will use 1.16 just in this form.

(d) Let $f : X \to Y$ be a morphism of complex spaces and let $\mathcal{F}$ be an $\mathcal{O}_X$-module. Consider a point x of X. By means of the natural morphism $\mathcal{O}_{Y,f(x)} \to \mathcal{O}_{X,x}$, $\mathcal{F}_x$ has a structure of $\mathcal{O}_{Y,f(x)}$-module. The sheaf $\mathcal{F}$ is called *f-flat* (or *Y-flat*) *in the point* x if $\mathcal{F}_x$ is a flat $\mathcal{O}_{Y,f(x)}$-module. We shall say that $\mathcal{F}$ is *f-flat* (or *Y-flat*) if it is f-flat in any point x of X. In this case, if

$$0 \to \mathcal{G}' \to \mathcal{G} \to \mathcal{G}'' \to 0$$

is an exact sequence of $\mathcal{O}_Y$-modules, then the sequence

$$0 \to f^*(\mathcal{G}') \otimes_{\mathcal{O}_X} \mathcal{F} \to f^*(\mathcal{G}) \otimes_{\mathcal{O}_X} \mathcal{F} \to f^*(\mathcal{G}'') \otimes_{\mathcal{O}_X} \mathcal{F} \to 0$$

is exact.

The morphism f itself is called *flat* if the structural sheaf $\mathcal{O}_X$ is f-flat. Again, if $0 \to \mathcal{G}' \to \mathcal{G} \to \mathcal{G}'' \to 0$ is an exact sequence as above and if f is flat, then one obtains the exact sequence

$$0 \to f^*(\mathcal{G}') \to f^*(\mathcal{G}) \to f^*(\mathcal{G}'') \to 0.$$

Recall the following properties:

"Let $f : X \to Y$ be a morphism of complex spaces, $\mathcal{F} \in \mathbf{Coh}\,(X)$, $Y' \to Y$ another morphism of complex spaces, $f' : X' = X \times_Y Y' \to Y'$ the morphism deduced from f by base change, $\mathcal{F}'$ the inverse image of $\mathcal{F}$ on X', x' a point of X' and x, y, y' its images in X, Y, Y'. Under these conditions, if $\mathcal{F}$ is f-flat in x, then $\mathcal{F}'$ is f'-flat in x'. Conversely, if the morphism $Y' \to Y$ is flat in y' and $\mathcal{F}'$ is f'-flat in x', then $\mathcal{F}$ is f-flat in x ([15], Exp. 13, prop. 2.4 and Cor. 2.5)."

As a consequence, one obtains:

"If X, Y, X', Y', f, f' are as above and f is flat, then f' is flat. Conversely, if $Y' \to Y$ and f' are flat, then f is flat".

"If Y and Z are complex spaces, then the projection morphism $X = Y \times Z \to Y$ is flat" (indeed, if e is the final complex space, then the unique morphism $Z \to e$ is flat and $Y \times Z = Y \times_e Z$; another proof of this fact can be found in [20], th. 2, p. 60).

(e) PROPOSITION 1.18. *Let $A \to B$ be a flat local morphism of local rings. If M is an A-module and $M \otimes_A B = 0$, then $M = 0$.*

Proof. If M' is a submodule of M, then the map $M' \otimes_A B \to M \otimes_A B$ is injective. It will then result that we can suppose M of the form $A/\mathfrak{a}$, where $\mathfrak{a}$ is an ideal of A. Since $A/\mathfrak{a} \otimes_A B \simeq B/\mathfrak{a}B = 0$, it follows that $B = \mathfrak{a}B$. Thus $\mathfrak{a} = A$ (hence $M = 0$), otherwise, $\mathfrak{a} \subset \mathfrak{m}_A$ and $\mathfrak{a}B \subset \mathfrak{m}_B$.

PROPOSITION 1.19. *Let $A \overset{\varphi}{\to} B$ be a flat local morphism of local rings. Then the following properties hold:*

i) The morphism φ is injective.

ii) If $\mathfrak{a}$ *is an ideal of* A, *then* $\mathfrak{a}B \cap A = \mathfrak{a}$ *(one identifies* A *with its image under* φ*).*

iii) If moreover B *is noetherian, then the natural map* $\operatorname{Spec} B \to \operatorname{Spec} A$ *is surjective; more precisely, if* $\mathfrak{p} \in \operatorname{Spec} A$ *and* $\mathfrak{q} \in \operatorname{Ass}(B/\mathfrak{p}B)$, *then* $\mathfrak{q} \cap A = \mathfrak{p}$.

Proof. (*i*) We have $\operatorname{Ker} \varphi \otimes_A B \simeq (\operatorname{Ker}\varphi)B = 0$ and the conclusion follows from proposition 1.18.

(*ii*) The morphism $A/\mathfrak{a} \to B/\mathfrak{a}B \simeq A/\mathfrak{a} \otimes_A B$ is a flat local morphism of local rings. It results to be injective, therefore $\mathfrak{a} = \mathfrak{a}B \cap A$.

(*iii*) Let $\mathfrak{p}$, $\mathfrak{q}$ be as in the statement. If we replace $A \to B$ by $A/\mathfrak{p} \to B/\mathfrak{q}$ we may suppose $\mathfrak{p} = 0$ (hence A is an integral domain). Suppose, on the contrary, that there exists a non-null element $x \in \mathfrak{q} \cap A$. The homotety defined by it in A is injective. By the flatness hypothesis, the homotety defined by x in B is also injective. So x is nonzerodivisor in B, and this contradicts the hypothesis $x \in \mathfrak{q} \in \operatorname{Ass} B$.

Lemma 1.20. *Let* $A \to B$ *be a local morphism of noetherian local rings and let* M *be a* B*-module of finite type, which is flat over* A. *Suppose in addition that* $\operatorname{Supp} M = \operatorname{Spec} B$. *If* $\mathfrak{p} \in \operatorname{Spec} A$, *then any prime ideal* $\mathfrak{q}$ *of* B *which contains* $\mathfrak{p}B$ *and is minimal with respect to this property verifies the relation* $\varphi^{-1}(\mathfrak{q}) = \mathfrak{p}$.

Proof. We proceed as in 1.19 (*iii*). If we replace φ by $A/\mathfrak{p} \to B/\mathfrak{p}B$ and M by $M/\mathfrak{p}M \simeq M \otimes_A A/\mathfrak{p}$, we may assume $\mathfrak{p} = 0$, hence $\mathfrak{q}$ is a minimal prime ideal of B. Suppose, on the contrary, that $\varphi^{-1}(\mathfrak{q}) \neq 0$. Let $x \neq 0$ be an element of A such that $\varphi(x) \in \mathfrak{q}$. The multiplication by x in A being injective, it results that the multiplication by $\varphi(x)$ in M is also injective. The hypothesis $\operatorname{Supp} M = \operatorname{Spec} B$ implies $\mathfrak{q} \in \operatorname{Ass} M$, hence $\varphi(x) \in \mathfrak{q}$ is a zerodivizor of M, a contradiction.

Proposition 1.21. ([37], Ch. IV, 6.1.2, 6.3.1). *Let* $A \xrightarrow{\varphi} B$ *be a local morphism of noetherian local rings,* $\mathfrak{m}$ *the maximal ideal of* A, $k = A/\mathfrak{m}$ *its residual field and* $M \neq 0$ *a* B*-module of finite type which is flat over* A. *Under these conditions*

$$(i) \qquad \dim_B M = \dim A + \dim_{B\otimes_A k}(M \otimes_A k).$$

In particular, if φ *is flat, then* $\dim B = \dim A + \dim (B \otimes_A k)$.

$$(ii) \qquad \operatorname{prof}_B M = \operatorname{prof} A + \operatorname{prof}_{B\otimes_A k}(M \otimes_A k).$$

In particular, if φ *is flat, then* $\operatorname{prof} B = \operatorname{prof} A + \operatorname{prof} (B \otimes_A k)$.

Proof. (*i*) We may replace B by $B/\operatorname{Ann} M$ and we can thus suppose that $\operatorname{Supp} M = \operatorname{Spec} B$. By Nakayama lemma, $\operatorname{Supp}(M \otimes_A k) = \operatorname{Spec}(B \otimes_A k)$. So we have to prove the equality $\dim B = \dim A + \dim (B \otimes_A k)$ under the assumption of the existence of a B-module of finite type M, which is flat over A and such that $\operatorname{Supp} M = \operatorname{Spec} B$. We proceed by induction on $n = \dim A$.

If $n = 0$, then $\mathfrak{m}$ (hence $\mathfrak{m}B$ too) is nilpotent and the conclusion is clear.

We now suppose $n \geqslant 1$ and the assertion proved for $n - 1$. Let $\mathfrak{q}_i$, $1 \leqslant i \leqslant s$, be the minimal prime ideals of B and $\mathfrak{p}'_j$, $1 \leqslant j \leqslant t$, the minimal prime ideals of A. For any i, the ideal $\mathfrak{p}_i = \varphi^{-1}(\mathfrak{q}_i)$ is different from $\mathfrak{m}$. Indeed, if we would have

$\mathfrak{p}_i = \mathfrak{m}$ for an index i, then there would be prime ideal $\mathfrak{p}$ in A, $\mathfrak{p} \subset \mathfrak{p}_i$ and $\mathfrak{p} \neq \mathfrak{p}_i$; we easily obtain a contradiction by using 1.20. Consider an element $x \in \mathfrak{m}$ which does not lie in any ideal $\mathfrak{p}_i$ or $\mathfrak{p}'_j$ ([10], Ch. II, § 1, n $\overset{\circ}{=}$ 1, prop. 2).

Denote $A' = A/xA$, $B' = B/xB$ and $M' = M/xM$. Then

$$\dim A' = \dim A - 1, \ \dim B' = \dim B - 1, \ \operatorname{Supp} M' = \operatorname{Spec} B'$$

and M' is A'-flat. Thus we may apply the induction hypothesis and get the equality

$$\dim B' = \dim A' + \dim (B' \otimes_{A'} k).$$

On the other hand, $B' \otimes_{A'} k = B \otimes_A k = B/\mathfrak{m}B$.

(*ii*) By Nakayama lemma the module $M \otimes_A k$ is non-null, hence the integer $n = \operatorname{prof} A + \operatorname{prof}_{B \otimes_A k}(M \otimes_A k)$ is finite. We proceed by induction on n. We first consider the case $n = 0$. It follows that $\operatorname{prof} A = \operatorname{prof}_{B \otimes_A k}(M \otimes_A k) = 0$, hence $\mathfrak{m}_A \in \operatorname{Ass} A$, $\mathfrak{m}_B/\mathfrak{m}_A B \in \operatorname{Ass}\ (M/\mathfrak{m}_A M)$. There exists an injective A-morphism $A/\mathfrak{m}_A \to A$. By tensoring $\otimes_A M$ we get an injective B-morphism $M/\mathfrak{m}_A M \to M$, therefore $\operatorname{Ass}_B(M/\mathfrak{m}_A M) \subset \operatorname{Ass} M$. Accordingly, $\mathfrak{m}_B \in \operatorname{Ass}_B M$, hence $\operatorname{prof}_B M = 0$ and the case $n = 0$ is thus treated.

We suppose now $n > 0$ and distinguish two cases:

a) Assume $\operatorname{prof} A > 0$. Let $x \in \mathfrak{m}_A$ be an A-regular element. Denote $A' = A/xA$, $B' = B/xB$ and $M' = M/xM$. Since $x \in \mathfrak{m}_A$, $B' \otimes_{A'} k = B \otimes_A k$ and $M' \otimes_{A'} k = M \otimes_A k$. By the flatness hypothesis on M one derives that M' is A'-flat and x is M-regular. Then

$$\operatorname{prof} A' = \operatorname{prof} A - 1, \ \operatorname{prof}_{B'} M' = \operatorname{prof}_B M - 1 \text{ and}$$

$$\operatorname{prof}_{B' \otimes_{A'} k}(M' \otimes_{A'} k) = \operatorname{prof}_{B \otimes_A k}(M \otimes_A k).$$

Thereby, the general induction step is proved in the case *a*).

b) Assume $\operatorname{prof}_{B \otimes_A k}(M \otimes_A k) > 0$. Let $y \in \mathfrak{m}_B$ be a regular element with respect to $M/\mathfrak{m}_A M = M \otimes_A k$. By 1.9, y is M-regular and $M' = M/yM$ is A-flat. We have

$$M' \otimes_{A'} k \simeq (M/\mathfrak{m}_B M)/y(M/\mathfrak{m}_B M), \text{ hence } \operatorname{prof}_{B \otimes_{A'} k}(M' \otimes_A k) = \operatorname{prof}_{B \otimes_A k}(M \otimes_A k) - 1.$$

In this case the proof of the general induction step is obvious too.

COROLLARY 1.22. *Let $A \to B$ be a flat local morphism of noetherian local rings. For any ideal $\mathfrak{a}$ of $A (\mathfrak{a} \neq A)$ the following equality*

$$\operatorname{ht} \mathfrak{a} = \operatorname{ht} (\mathfrak{a} B)$$

holds.

Proof. By definition, $\text{ht}\,\mathfrak{a} = \inf_{\mathfrak{p}\supset\mathfrak{a}} \text{ht}\,\mathfrak{p}$ and $\text{ht}\,(\mathfrak{a}B) = \inf_{\mathfrak{q}\supset\mathfrak{a}B} \text{ht}\,\mathfrak{q}$, $\mathfrak{p}$ and $\mathfrak{q}$ being prime ideals in A, B, respectively. The conclusion follows straightforwardly from the following two assertions:

(*) $\mathfrak{q} \in \text{Spec}\, B \Rightarrow \text{ht}\,(\mathfrak{q} \cap A) \leqslant \text{ht}\,\mathfrak{q}$;

(**) $\mathfrak{p} \in \text{Spec}\, A$, $\mathfrak{q}$ prime ideal in B containing $\mathfrak{p}B$ and minimal with respect to this property $\Rightarrow \text{ht}\,\mathfrak{p} = \text{ht}\,\mathfrak{q}$.

In order to prove (*), consider the flat local morphism of noetherian local rings $A_{\mathfrak{q}\cap A} \to B_{\mathfrak{q}}$. We have $\text{ht}\,(\mathfrak{q} \cap A) = \dim A_{\mathfrak{q}\cap A}$ and $\text{ht}\,\mathfrak{q} = \dim B_{\mathfrak{q}}$ and the conclusion follows from the proposition.

As for the assertion (**), we remark moreover that $\mathfrak{p} = \mathfrak{q} \cap A$ (1.19) and that $\dim B_{\mathfrak{q}}/\mathfrak{p}B_{\mathfrak{q}} = 0$, since $\text{Spec}\,(B_{\mathfrak{q}}/\mathfrak{p}B_{\mathfrak{q}})$ consists of the prime ideal determined by $\mathfrak{q}$ only.

COROLLARY 1.23. *Under the assumptions of the proposition,*

$$\text{coprof}_B M = \text{coprof}\, A + \text{coprof}_{B\otimes_A k}(M \otimes_A k).$$

In particular, if φ *is flat, then*

$$\text{coprof}\, B = \text{coprof}\, A + \text{coprof}\,(B \otimes_k A).$$

The proof follows from the equality coprof = dim$-$prof.

COROLLARY 1.24. *Under the assumptions of the proposition, M is a Cohen-Macauley B-module if and only if A is a Cohen-Macauley ring and $M \otimes_A k$ is a Cohen-Macauley $B \otimes_A k$-module. In particular, if φ is flat, then B is Cohen-Macauley if and only if the rings A and $B \otimes_A k$ are so.*

We will also use the following results.

PROPOSITION 1.25. *Let $A \to B$ be a local morphism of noetherian local rings and M a B-module of finite type. If B and M are A-flat, then*

$$\text{dh}_B M = \text{dh}_{B/\mathfrak{m}_A B}(M/\mathfrak{m}_A M).$$

Proof. If $L^\bullet \to M$ is a resolution of M by free B-modules, then the flatness hypothesis shows that $L^\bullet/\mathfrak{m}_A L^\bullet$ is a resolution of $M/\mathfrak{m}_A M$ by free $B/\mathfrak{m}_A B$-modules. Thus $\text{dh}_{B/\mathfrak{m}_A B}(M/\mathfrak{m}_A M) \leqslant \text{dh}_B M$.

We now prove the other inequality. If $n = \text{dh}_{B/\mathfrak{m}_A B}(M/\mathfrak{m}_A M) = \infty$, this is clear. Suppose $n < \infty$ and let

$$L^{n-1} \to \ldots \to L^0 \to M \to 0$$

be an exact sequence of B-modules, L^i being free B-modules of finite rank. Denote $N = \text{Ker}\,(L^{n-1} \to L^{n-2})$. By splitting the exact sequence into short exact sequences from right to left, we derive that N is A-flat. By tensoring $\otimes_A A/\mathfrak{m}_A$, we get an exact sequence

$$0 \to N/\mathfrak{m}_A N \to L^{n-1}/\mathfrak{m}_A L^{n-1} \to \ldots\, L^0/\mathfrak{m}_A L^0 \to M/\mathfrak{m}_A M \to 0.$$

It will then result that $N/\mathfrak{m}_A N$ is $B/\mathfrak{m}_A B$-free, hence N will be B-free (IV.1.2) and the proof is completed.

PROPOSITION 1.26 ([31], 2.4). *Let $A \to B$ be a flat local morphism of noetherian local rings. If A is regular and $B/\mathfrak{m}_A B$ reduced, then B is reduced.*

Proof. Proceed by induction on $n = \dim A$. First, consider the case $n = 1$ and let t be a regular parameter for A. Let $x \in B$ be such that $x^r = 0$ for an integer $r \geqslant 0$. We get $x \in \mathfrak{m}_A B$, hence $x = tx_1$, $x_1 \in B$. The element t, as well as any power of it, is a nonzerodivisor in A, hence in B by the flatness hypothesis. Then we get $x_1^r = 0$.

Analogously, we get $x_1 = tx_2$, $x_2 \in B$. By the same reasoning one derives that $x \in \bigcap_{k \geqslant 0} t^k B$, hence $x = 0$ (by Krull's theorem).

Assume now $n \geqslant 2$ and the proposition proved for dimensions $< n$. Let $t_1, \ldots, t_n$ be a regular system of parameters for A. The morphism $A/t_1A \to B/t_1B$ is flat, A/t_1A is a regular ring of dimension $n-1$ and $(B/t_1B)/\mathfrak{m}_A (B/t_1B) \simeq B/\mathfrak{m}_A B$ is reduced. By the induction hypothesis, the ring B/t_1B is reduced. Since the multiplication by any power of t_1 is an injective map $B \to B$, it follows as above that B is reduced.

Let A be an analytic algebra, X a complex space and x a point of X such that $A = \mathcal{O}_{X,x}$. Denote by $A\{T_1, \ldots, T_r\}$, T_i being indeterminates, the analytic algebra $\mathcal{O}_{X \times \mathbb{C}^r,(x,0)}$. The projection $X \times \mathbb{C}^r \to X$ yields a canonical inclusion $A \hookrightarrow A\{T_1, \ldots, T_r\}$.

PROPOSITION 1.27. *Let $A \xrightarrow{f} B$ be a morphism of analytic algebras such that $\dim B = \dim A + \dim (B/\mathfrak{m}_A B)$. Then there exist an integer r and a finite morphism of analytic algebras of the same dimension $A\{T_1, \ldots, T_r\} \to B$ such that f is the composition of this with the inclusion $A \hookrightarrow A\{T_1, \ldots, T_r\}$.*

Proof. Let $r = \dim B/\mathfrak{m}_A B$. Choose elements $x_1, \ldots, x_r$ of B such that their classes form an ideal of definition in $B/\mathfrak{m}_A B$. The morphism $A\{T_1, \ldots, T_r\} \to B$ given by f and by the substitutions $T_i \mapsto x_i$ is quasi-finite, since a suitable power of $\mathfrak{m}_B$ is contained in the ideal generated in B by $\mathfrak{m}_A$ and the elements x_i. By Weierstrass preparation theorem, this morphism is finite. Moreover, $\dim A\{T_1, \ldots, T_r\} = \dim A + r = \dim A + \dim B/\mathfrak{m}_A B = \dim B$.

(f) Let U be an open subset of some space $\mathbb{R}^n$ and let x be one of its points. Denote by $\mathcal{S}_x$ the smallest family of germs of subsets of U in x, which is stable to finite unions, to finite intersections and to passing to complementary, and which contains in addition the germs of the form $\{x' \in U \mid f(x') < 0\}_x$ where f is a real analytic function defined in a neighbourhood of x. A part of U is called *semianalytic* if for any point x of U its germ in x is an element of $\mathcal{S}_x$.

We consider now an analytic subset X of an open set U of some space $\mathbb{C}^n \simeq \mathbb{R}^{2n}$. A part of X is called semianalytic if it is semianalytic in U in the previously defined meaning. By defining locally one easily obtains the notion of the *semianalytic subset* of a complex space. For instance, the analytic sets are semianalytic.

By the very definitions, every point of a complex space admits a fundamental system of neighbourhoods which are semianalytic and Stein compacts. Moreover, semianalytic compacts are stable to finite intersections and products, to images under isomorphisms and to inverse images under finite morphisms.

One of the remarkable properties of semianalytic sets is that they are locally connected [52]; in particular the compact semianalytic sets have a finite number of connected components.

§ 2. Algebraic and topological properties of the flat morphisms

We first give the transcription, in terms of flat morphisms of complex spaces, of some results of local algebra from § 1.

PROPOSITION 2.1. *Let $f: X \to Y$ be a flat surjective morphism of complex spaces. The correspondence $Y' \mapsto f^{-1}(Y') = X \times_Y Y'$, between the set of the subspaces of Y and the set of the subspaces of X, is injective.*

Proof. Let Y' be a subspace of Y. We have the equality of sets $Y' = f(f^{-1}(Y'))$. If V is an open set such that Y is a closed subspace of it, then $f^{-1}(V)$ is an open subset of X, $f^{-1}(Y')$ is a closed subspace in $f^{-1}(V)$ and the morphism $f^{-1}(V) \to V$ induced by f is flat and surjective.

In this way we can confine to closed subspaces. If Y' is determined by an ideal-sheaf $\mathfrak{J} \subset \mathcal{O}_Y$, then $f^{-1}(Y')$ is the subspace determined by $f^*(\mathfrak{J})\,\mathcal{O}_X$. The morphism f being flat, $f^*(\mathfrak{J})\,\mathcal{O}_X$ is identified with $f^*(\mathfrak{J})$.

If $\mathfrak{J}$ and $\mathfrak{J}'$ are two coherent ideal-sheaves of $\mathcal{O}_Y$ such that $f^*(\mathfrak{J}) = f^*(\mathfrak{J}')$, then we will show that $\mathfrak{J} = \mathfrak{J}'$ and the proposition will be proved. By the flatness hypothesis,

$$f^*(\mathfrak{J} + \mathfrak{J}'/\mathfrak{J}) \simeq f^*(\mathfrak{J} + \mathfrak{J}')/f^*(\mathfrak{J}) \simeq f^*(\mathfrak{J} + \mathfrak{J}')\,\mathcal{O}_X/f^*(\mathfrak{J})\mathcal{O}_X$$

$$\simeq (f^*(\mathfrak{J})\,\mathcal{O}_X + f^*(\mathfrak{J}')\,\mathcal{O}_X)/f^*(\mathfrak{J})\mathcal{O}_X = 0.$$

By passing to stalks and applying 1.18, one gets $\mathfrak{J} + \mathfrak{J}'/\mathfrak{J} = 0$; hence $\mathfrak{J}' \subset \mathfrak{J}$. Similarly, $\mathfrak{J} \subset \mathfrak{J}'$.

COROLLARY 2.2. *Let X and Y be two complex spaces over a complex space S and let $S' \to S$ be a flat surjective morphism of complex spaces. Then the map*

$$\mathrm{Hom}_S(X, Y) \to \mathrm{Hom}_{S'}(X \times_S S', Y \times_S S'),\ f \mapsto f \times_S S'$$

is injective.

Proof. Since $(X \times_S S') \times_{S'} (Y \times_S S') \simeq (X \times_S Y) \times_S S'$, the projection morphism $(X \times_S S') \times_{S'} (Y \times_S S') \xrightarrow{\pi} X \times_S Y$ is flat and surjective. The elements of $\mathrm{Hom}_S(X, Y)$ are in 1—1 correspondence with their graphs, which are subspaces of $X \times_S Y$. Analogously, one can characterize the elements of $\mathrm{Hom}_{S'}(X \times_S S', Y \times_S S')$. Moreover, if $f \in \mathrm{Hom}_S(X, Y)$, then $\Gamma_{f \times_S S'} = \pi^{-1}(\Gamma_f)$, where Γ means the graph. The corollary then follows from the proposition.

Here is some further addition to 2.1:

PROPOSITION 2.3. *Let* $X \xrightarrow{f} Y$ *be a flat surjective morphism of complex spaces and* $Y' \subset Y$ *a closed analytic subset. Then*

$$\operatorname{codim}(f^{-1}(Y'), X) = \operatorname{codim}(Y', Y).$$

Proof. If x is an arbitrary point of X, then we will prove the equality $\operatorname{codim}_x (f^{-1}(Y'), X) = \operatorname{codim}_{f(x)}(Y', Y)$ and the proposition follows. Let $\mathfrak{J}$ be the coherent ideal-sheaf associated to Y'. We have $\operatorname{codim}_x(f^{-1}(Y'), X) =$ the height of the $\mathcal{O}_x$-ideal $\mathfrak{J}_{f(x)}\mathcal{O}_x$ and $\operatorname{codim}_{f(x)}(Y', Y) =$ the height of the $\mathcal{O}_{f(x)}$-ideal $\mathfrak{J}_{f(x)}$. The conclusion follows from 1.22.

PROPOSITION 2.4. *Let* $f: X \to Y$ *be a morphism of complex spaces and* $\mathcal{F} \in \mathbf{Coh}(X)$, *which is flat with respect to* f. *For any point* $x \in \operatorname{Supp} \mathcal{F}$,

$$\dim \mathcal{F}_x = \dim \mathcal{O}_{f(x)} + \dim(\mathcal{F}_x/\mathfrak{m}_{f(x)}\mathcal{F}_x) \quad \textit{and}$$

$$\operatorname{prof}(\mathcal{F}_x) = \operatorname{prof} \mathcal{O}_{f(x)} + \operatorname{prof}(\mathcal{F}_x/\mathfrak{m}_{f(x)}\mathcal{F}_x).$$

In particular, if f *is flat, then for any* $x \in X$, *the following equalities hold*

$$\dim \mathcal{O}_x = \dim \mathcal{O}_{f(x)} + \dim(\mathcal{O}_x/\mathfrak{m}_{f(x)}\mathcal{O}_x) \quad \textit{and}$$

$$\operatorname{prof} \mathcal{O}_x = \operatorname{prof} \mathcal{O}_{f(x)} + \operatorname{prof}(\mathcal{O}_x/\mathfrak{m}_{f(x)}\mathcal{O}_x).$$

The assertions follow from 1.21.

COROLLARY 2.5. *Under the conditions of the proposition,*

$$\operatorname{coprof} \mathcal{F}_x = \operatorname{coprof} \mathcal{O}_{f(x)} + \operatorname{coprof}(\mathcal{F}_x/\mathfrak{m}_{f(x)}\mathcal{F}_x).$$

In particular, if f *is flat, then for any* $x \in X$,

$$\operatorname{coprof} \mathcal{O}_x = \operatorname{coprof} \mathcal{O}_{f(x)} + \operatorname{coprof}(\mathcal{O}_x/\mathfrak{m}_{f(x)}\mathcal{O}_x).$$

COROLLARY 2.6. *Let* $f: X \to Y$ *be a flat morphism of complex spaces and let* x *be a point of* X. *Then* $\mathcal{O}_x$ *is a Cohen-Macauley local ring if and only if the rings* $\mathcal{O}_{f(x)}$ *and* $\mathcal{O}_x/\mathfrak{m}_{f(x)}\mathcal{O}_x$ *enjoy the same property. In particular, if* f *is moreover surjective, then* X *is perfect if and only if* Y *and the fibers* X_y, $y \in Y$, *are perfect spaces.*

Further we use the next lemma.

LEMMA 2.7. *Suppose* $(X, \mathcal{O})$ *is a complex space,* $\mathcal{F} \in \mathbf{Coh}(X)$ *and* k *an integer. Then* $\{x \in X \mid \operatorname{dh} \mathcal{F}_x \geqslant k\}$ *is a closed analytic subset of* X.

Proof. First, consider the case $k = 1$ (for $k < 1$ the assertion is clear). The problem is local on X, hence one may assume the existence of an exact sequence of $\mathcal{O}$-modules of the form

$$0 \to \mathcal{G} \to \mathcal{O}^p \to \mathcal{F} \to 0.$$

If x is a point of X, $\mathcal{F}_x$ is $\mathcal{O}_x$-free if and only if this sequence splits in x. Then it follows that the set $\{x \in X \mid \mathcal{F}_x \text{ is not } \mathcal{O}_x\text{-free}\}$ coincides with the support of the cokernel of the map $Hom_{\mathcal{O}}(\mathcal{F}, \mathcal{O}^p) \to Hom_{\mathcal{O}}(\mathcal{F}, \mathcal{F})$, hence it is a closed analytic subset.

Now we consider the case $k \geqslant 2$. The problem being local in nature, one can assume that an exact sequence as above exists. For a point $x \in X$, the inequality $\operatorname{dh} \mathcal{F}_x \geqslant k$ is equivalent to $\operatorname{dh} \mathcal{G}_x \geqslant k - 1$, etc.

THEOREM 2.8. *Let $f: X \to Y$ be a morphism of complex spaces.*

For any coherent analytic sheaf $\mathcal{F}$ on X, flat with respect to f, and for any integer k, the sets $\{x \in X \mid \dim (\mathcal{F}_x/\mathfrak{m}_{f(x)} \mathcal{F}_x) \geqslant k\}$, $\{x \in X \mid \operatorname{prof} (\mathcal{F}_x/\mathfrak{m}_{f(x)} \mathcal{F}_x) \leqslant k\}$, $\{x \in X \mid \operatorname{coprof} (\mathcal{F}_x/\mathfrak{m}_{f(x)} \mathcal{F}_x) \geqslant k\}$ are closed analytic subsets of X.

Proof. Denote these sets respectively by $M_k(\mathcal{F};f)$, $S_k(\mathcal{F};f)$ and $D_k(\mathcal{F};f)$. Since coprof = dim − prof, it follows that

$$D_k(\mathcal{F};f) = \bigcup_l (M_l(\mathcal{F};f) \cap S_{l-k}(\mathcal{F};f)).$$

Moreover, this union is locally finite. Thus, it is sufficient to show the analyticity of the first two sets of the statement.

The problem is local on X and Y, hence we may assume that f can be decomposed into a closed immersion $X \xrightarrow{i} Y \times U$ and the projection $Y \times U \xrightarrow{\pi} Y$, U being an open of some numerical space $\mathbf{C}^n$. By considering the direct image $\widetilde{\mathcal{F}}$ of $\mathcal{F}$ under i, we have

$$i(M_k(\mathcal{F};f)) = M_k(\widetilde{\mathcal{F}};\pi), \qquad i(S_k(\mathcal{F};f)) = S_k(\widetilde{\mathcal{F}};\pi).$$

We may thus reduce the problem to the particular case $X = Y \times U$, $f: X \to Y$ the projection, U being an open subset of $\mathbf{C}^n$. Let $x = (y, z)$ be a point of X. One has

$$\mathcal{O}_{X,x}/\mathfrak{m}_y\mathcal{O}_{X,x} \simeq \mathcal{O}_{U,z}, \quad \operatorname{prof}(\mathcal{F}_x/\mathfrak{m}_y \mathcal{F}_x) = n - \operatorname{dh}_{\mathcal{O}_z}(\mathcal{F}_x/\mathfrak{m}_y \mathcal{F}_x).$$

By 1.25, $\operatorname{dh}_{\mathcal{O}_z}(\mathcal{F}_x/\mathfrak{m}_y F_x) = \operatorname{dh}_{\mathcal{O}_x} \mathcal{F}_x$. Therefore, the set

$$S_k(\mathcal{F};f) = \{x \in X \mid \operatorname{dh} \mathcal{F}_x \geqslant n - k\}$$

is a closed analytic subset of X (2.7).

Denote by $(Y_i)_i$ the family of the irreducible components of Y_{red}. Let $X_i = f^{-1}(Y_i)$, $f_i: X_i \to Y_i$ be the morphism induced by f and $\mathcal{F}_i$ the inverse image of $\mathcal{F}$ on X_i. The following equality

$$M_k(\mathcal{F};f) = \bigcup_i M_k(\mathcal{F}_i;f_i)$$

holds and the union is locally finite. In this way, in order to prove that $M_k(\mathcal{F};f)$ is a closed analytic subset, we may assume Y irreducible. Let $l = \dim Y$. By 2.4, for any $x \in X$,

$$\dim(\mathcal{F}_x/\mathfrak{m}_{f(x)} \mathcal{F}_x) = \dim \mathcal{F}_x - \dim \mathcal{O}_{f(x)} = \dim \mathcal{F}_x - l.$$

To conclude, it is enough to check that the sets $M_i(\mathfrak{F}) = \{x \in X \mid \dim \mathfrak{F}_x \geqslant i\}$ are closed analytic sets. But $M_i(\mathfrak{F})$ coincides with the union of the irreducible components of dimension $\geqslant i$ of Supp $\mathfrak{F}$.

Remark. The fact that the sets $\{x \in X \mid \dim \mathfrak{F}_x/\mathfrak{m}_x\mathfrak{F}_x \geqslant k\}$ are analytic can be proved without the hypothesis $\mathfrak{F}$ f-flat, as a consequence of the following theorem of Remmert [65]:

"For any morphism $f\colon X \to Y$ of complex spaces and for any integer k, the set $\{x \in X \mid \dim_x f^{-1} f(x) \geqslant k\}$ is closed analytic in X".

(One applies this result to the composition Supp $\mathfrak{F} \to X \to Y$ and realizes that Supp $\mathfrak{F} \cap f^{-1} f(x) = \text{Supp}\,(\mathfrak{F}/\hat{\mathfrak{m}}_{f(x)}\,\mathfrak{F})$, by Nakayama lemma).

The next corollaries show that the flatness hypothesis allows to convey some properties of a fiber to the neighbouring fibers.

COROLLARY 2.9. *Let $f\colon X \to Y$ be a proper morphism of complex spaces, $\mathfrak{F} \in$ **Coh** (X) flat over Y and k an integer. Then the sets* $\{y \in Y \mid \dim \mathfrak{F}_y \geqslant k\}$, $\{y \in Y \mid \text{prof}\, \mathfrak{F}_y \leqslant k\}$ and $\{y \in Y \mid \text{coprof}\, \mathfrak{F}_y \geqslant k\}$ *are closed analytic subsets of Y (if $f^{-1}(y) = \emptyset$, we agree to put* $\dim \mathfrak{F}_y = -\infty$, $\text{prof}\, \mathfrak{F}_y = +\infty$, $\text{coprof}\, \mathfrak{F}_y = 0$).

Proof. By the very definitions,

$$\dim \mathfrak{F}_y = \sup_{x \in f^{-1}(y)} \dim (\mathfrak{F}_x/\mathfrak{m}_y\mathfrak{F}_x), \quad \text{prof}\, \mathfrak{F}_y = \inf_{x \in f^{-1}(y)} \text{prof}\, (\mathfrak{F}_x/\mathfrak{m}_y\mathfrak{F}_x) \text{ and}$$

$$\text{coprof}\, \mathfrak{F}_y = \dim \mathfrak{F}_y - \text{prof}\, \mathfrak{F}_y,$$

whatever $y \in f(X)$, $\mathfrak{F}_y \neq 0$. The conclusion follows by the theorem and by Remmert projection theorem (III. 2.11).

COROLLARY 2.10. *Let $f\colon X \to Y$ be a proper morphism of complex spaces, $\mathfrak{F} \in$ **Coh** (X) flat over Y and y a point of Y. If $\mathfrak{F}_y$ is a Cohen-Macauley sheaf* (coprof $\mathfrak{F}_y = 0$), *then for any point y' of some neighbourhood of y, the sheaf $\mathfrak{F}_{y'}$ is Cohen-Macauley.*

COROLLARY 2.11. *Let $f\colon X \to Y$ be a flat proper morphism of complex spaces and y a point of Y such that the fiber X_y is a perfect space. Then, there exists a neighbourhood of y such that $X_{y'}$ is perfect for all points y' of that neighbourhood.*

THEOREM 2.12. *Let $f\colon X \to Y$ be a morphism of complex spaces and let $\mathfrak{F}$ be a coherent analytic sheaf on X, which is flat with respect to f. Then the restriction of f to* Supp $\mathfrak{F}$ *is an open map.*

In particular, any flat morphism is open.

Proof. If we replace X by (Supp $\mathfrak{F}$, $\mathcal{O}_X/\text{Ann}\, \mathfrak{F} \mid \text{Supp}\, \mathfrak{F}$), we can assume that Supp $\mathfrak{F} = X$. For any $y \in Y$, it follows that Supp $\mathfrak{F}_y = X_y$ by Nakayama lemma. If x is an arbitrary point of X, we derive the equalities

$$\dim \mathfrak{F}_x = \dim \mathcal{O}_x, \quad \dim (\mathfrak{F}_x/\mathfrak{m}_y\, \mathfrak{F}_x) = \dim (\mathcal{O}_x/\mathfrak{m}_{f(x)}\, \mathcal{O}_x).$$

By applying 2.4 we get

$$(*)\ \dim \mathcal{O}_x = \dim \mathcal{O}_{f(x)} + \dim (\mathcal{O}_x/\mathfrak{m}_{f(x)}\, \mathcal{O}_x), \qquad x \in X.$$

Let Y' be the normalization of Y_{red}, $g: Y' \to Y$ the normalization morphism composed with $Y_{\mathrm{red}} \hookrightarrow Y$, $X' = X \times_Y Y'$ and $X' \xrightarrow{f'} Y'$, $X' \xrightarrow{g} X$ the natural morphisms which appear.

If x' is a point of X' and x, y, y' are its images in X, Y, Y', then $\dim \mathcal{O}_{x'} = \dim \mathcal{O}_x$, $\dim \mathcal{O}_{y'} = \dim \mathcal{O}_y$, $\dim (\mathcal{O}_{x'}/\mathfrak{m}_{y'}\mathcal{O}_{x'}) = \dim (\mathcal{O}_x/\mathfrak{m}_y \mathcal{O}_x)$. Consequently, for the morphism $X' \xrightarrow{f'} Y'$ too,

$$\dim \mathcal{O}_{x'} = \dim \mathcal{O}_{f'(x')} + \dim (\mathcal{O}_{x'}/\mathfrak{m}_{f'(x')}\mathcal{O}_{x'}) \text{ for all } x' \in X'.$$

If U is an open subset of X, then $g^{-1}(f(U)) = f'(g'^{-1}(U))$. Since Y has the quotient topology of Y', if f' is an open map, then f itself enjoyes the same property. In this way, it is enough to prove the assertion: any map of complex spaces $f: X \to Y$ which satisfies the equalities (*) and such that Y is integer in every point is an open map.

Let x be an arbitrary point of X. We have to show that the image under f of a neighbourhood U' of x contains a neighbourhood of $y = f(x)$. Apply 1.27 in connection with the correspondence between the analytic algebras and the germs of analytic spaces. There exist a neighbourhood U of x, $U \subset U'$, a neighbourhood V of y, $f(U) \subset V$, an open subset W of some numerical space and a finite morphism $g: U \to W \times V$ such that the morphism $U \to V$ which is deduced from f by restriction coincides with the composition of g with the projection $W \times V \to V$ and such that $\dim U = \dim_x U = \dim (W \times V) = \dim_{g(x)} (W \times V)$.

The map $W \times V \to V$ being open, it is enough to prove that $g(U)$ contains a neighbourhood of $g(x)$. The morphism g being finite, $g(U)$ is a closed analytic subset of $W \times V$. One has $\dim_x U \leqslant \dim_{g(x)} g(U) \leqslant \dim_{g(x)} W \times V$, hence $\dim_{g(x)} g(U) = \dim_{g(x)} W \times V$. Since the analytic space $W \times V$ is integer in $g(x)$, it follows that the germs of $g(U)$ and $W \times V$ in $g(x)$ coincide and thus the proof is over.

A converse result would be the following:

THEOREM 2.13. *Let $f: X \to Y$ be an open morphism of complex spaces. Suppose in addition that Y is a manifold and the fibers X_y, $y \in Y$, are reduced. Then f is flat.*

Proof. We may assume that Y is an open subset of some numerical space. Proceed by induction on $\dim Y$. For an arbitrary point $x \in X$ we must prove that $\mathcal{O}_x$ is $\mathcal{O}_{f(x)}$-flat. One can also assume that Y contains the origin and that $f(x) = 0$.

The case $\dim Y = 1$. Let t be the coordinate function on Y which is null in the origin. In order to show that $\mathcal{O}_x$ is $\mathcal{O}_{Y.0} \simeq \mathbb{C}\{t\}$-flat it is sufficient to prove that it is torsionfree. We have to check that the multiplication by t, $\mathcal{O}_x \xrightarrow{t} \mathcal{O}_x$, is injective. Let $\varphi \in \mathcal{O}_x$ be such that $t\varphi = 0$. Let U be a neighbourhood of x where φ extends and where $t\varphi$ is the null section. Consider an arbitrary point x' of $U \cap X_0$. There exists an irreducible component T of U which passes through x' and which is not contained in X_0 (otherwise, X_0 would contain a neighbourhood of x', hence $f(X_0)$ would contain a neighbourhood of $f(x') = 0$, a contradiction).

For any $x'' \in T \setminus X_0$, $\varphi(x'') = 0$. So the function associated to φ vanishes on T, hence $\varphi(x') = 0$. In this way, the function associated to φ vanishes on $X_0 \cap U$. Since the analytic space $X_0 = (f^{-1}(0), \mathcal{O}_X/t\mathcal{O}_X \mid f^{-1}(0))$ is reduced, it follows that $\varphi = t\varphi_1$, $\varphi_1 \in \mathcal{O}_x$. Analogously, from the equality $t^2\varphi_1 = 0$ we can derive $\varphi_1 = t\varphi_2$, $\varphi_2 \in \mathcal{O}_x$, hence $\varphi = t^2\varphi_2$. By iterating the reasoning we get $\varphi \in \bigcap_{r \geqslant 0} t^r\mathcal{O}_x$, hence $\varphi = 0$ by Krull's theorem.

The case $n = \dim Y \geqslant 2$. Suppose the theorem proved for dimensions $< n$. Let $t_1, \ldots, t_n$ be the coordinate functions of Y (null in the origin), and let Y' be the submanifold of Y given by the equation $t_1=0$ and $X'=f^{-1}(Y')$. The morphism $X' \to Y'$ will be open and its fibers will be reduced (if $y' \in Y'$, then $X_{y'} = X'_{y'}$), hence by the induction hypothesis it will be flat. Moreover, X' is reduced (1.26).

We first prove that the multiplication by t_1, $\mathcal{O}_x \xrightarrow{t_1} \mathcal{O}_x$, is injective. Let $\varphi \in \mathcal{O}_x$ be such that $t_1\varphi = 0$. Consider a neighbourhood U of x on which φ extends and where $t_1\varphi = 0$. Let x' be a point of $U \cap X'$. There exists an irreducible component T of U passing through x' and which is not contained in X' (otherwise, X' would contain a neighbourhood of x', hence $f(X')$ would contain a neighbourhood of $f(x')$, a contradiction).

Since $t_1\varphi = 0$, it follows that $\varphi(x'') = 0$ for any $x'' \in T \setminus X'$. Therefore the function associated to φ is null on T, particularly $\varphi(x') = 0$. So we have proved that the function associated to φ vanishes on $U \cap X'$. Since X' is reduced and its structural sheaf is $\mathcal{O}_X/t_1\mathcal{O}_X$, it follows that $\varphi = t_1\varphi_1$, $\varphi_1 \in \mathcal{O}_x$. Similarly, one shows that $\varphi_1 = t_1\varphi_2$, $\varphi_2 \in \mathcal{O}_x$. The reasoning goes on and one finds that $\varphi \in \bigcap_{r \geqslant 0} t_1^r \mathcal{O}_x$, therefore $\varphi = 0$.

In this way we have checked that the morphism $\mathcal{O}_x \xrightarrow{t_1} \mathcal{O}_x$ is injective and since $\mathcal{O}_x/t_1 \mathcal{O}_x$ is a flat $\mathcal{O}_{Y,0}/t_1 \mathcal{O}_{Y,0}$-module (the morphism $X' \to Y'$ is flat!), $\mathcal{O}_x$ will result to be a flat $\mathcal{O}_{Y,0}$-module (1.4).

§ 3. A noetherianity theorem with respect to Stein compacts

The main result in the paragraph is the following theorem of Frisch [25], Grothendieck [36] and Siu [82].

THEOREM 3.1. *Let K be a Stein compact of the complex space $(X, \mathcal{O})$. Then $\Gamma(K, \mathcal{O})$ is a noetherian ring if and only if, for any analytic set Y defined in a neighbourhood of K, $Y \cap K$ has a finite number of connected components.*

First we will prove the following two lemmas.

LEMMA 3.2. *Let $f: X \to Y$ be a finite morphism of complex spaces which are normal and of the same dimension. Then f is an open map.*

Proof. Let x be a point of X and let D be one of its neighbourhoods. We must show that $f(D)$ is a neighbourhood of $f(x)$. We may assume that the closure $\bar{D}$ is compact and that $\bar{D} \cap f^{-1}f(x) = \{x\}$. Then $f(x) \notin f(\partial D)$, where ∂D is the boundary of D. Let G be a connected neighbourhood of $f(x)$ such that $G \cap f(\partial D) = \emptyset$. We will consider it as a complex space, namely, an open subspace of Y. Since Y is normal, G is irreducible. Let $H = D \cap f^{-1}(G)$ and $g: H \to G$ be the morphism induced by f. The morphism g is finite, hence $f(H)$ is an analytic subset of G, of dimension $n = \dim X = \dim Y$. Since G is irreducible and $\dim G = n$, it follows that $f(H) = G$, therefore $f(D) \supset G$.

LEMMA 3.3. *Let X be a complex space and let K be one of its compacts such that for any analytic set Y defined in a neighbourhood of K, $Y \cap K$ has finite number of connected components. If $f: Z \to X$ is a finite morphism of complex spaces, then $f^{-1}(K)$ has a finite number of connected components.*

Proof. We may assume X and Z reduced and finite-dimensional. We will use induction on dim Z. The case dim $Z = 0$ is trivial.

Let $n \geqslant 1$ and suppose the lemma proved whenever dim $Z < n$. We are going to verify the assertion of the lemma in the hypothesis dim $Z = n$. Consider the normalization morphism

$$\pi: \tilde{Z} \to Z.$$

It is enough to show that $(f\pi)^{-1}(K)$ has a finite number of connected components; so we can assume that Z is normal. Since $f^{-1}(K)$ does intersect a finite number of connected components of Z, we reduce the problem to the case when Z is moreover connected (that is irreducible, being normal). $Y = f(Z)$ is an analytic set of dimension n in X, which is irreducible. Let

$$\sigma: \tilde{Y} \to Y$$

be the normalization morphism. There exists a unique finite morphism $\tilde{f}: Z \to \tilde{Y}$ such that $f = \tilde{f}\sigma$. By the previous lemma, $\tilde{f}$ is an open map. From 1.2, it is enough to show that $\sigma^{-1}(K)$ has a finite number of connected components only. Let Y' be the set of the singular points of Y, $\tilde{Y}' = \sigma^{-1}(Y')$ and $\tau = \sigma \mid \tilde{Y}'$. Since dim $\tilde{Y}' < n$, by the induction hypothesis $\tau^{-1}(K)$ has a finite number of connected components only, $D_1, \ldots, D_k$. Let $\sigma^{-1}(K) = \bigcup_{i \in I} E_i$ be the decomposition into connected components. Since D_j is a connected subset of $\sigma^{-1}(K)$, hence $D_j \subset E_{i_j}$ for some index $i_j \in I$. Let $J = I \setminus \{i_1, \ldots, i_k\}$. For $i \in J$, $E_i \cap \tilde{Y}' = \emptyset$, hence E_i is a connected component of $\sigma^{-1}(K) \setminus \tilde{Y}'$. Since $\tilde{Y} \setminus \tilde{Y}'$ is isomorphic to $Y \setminus Y'$, $\sigma(E_i)$ are distinct connected components of $K \cap (Y \setminus Y')$, $i \in J$. By lemma 1.3, $\sigma(E_i)$ $(i \in J)$ are distinct connected components of $K \cap Y$. Accordingly, J and thereby I are finite sets.

The proof of the theorem. (*a*) Let K be a Stein compact such that $\Gamma(K, \mathcal{O})$ is a noetherian ring. Suppose on the contrary that there exists an analytic set Y defined in a neighbourhood U of K such that $Y \cap K$ has infinitely many connected components.

We will construct by means of induction on n subsets Y_n $(n \geqslant 1)$ in $Y \cap K$ such that:

(i) $Y_1 = Y \cap K$;

(ii) Y_{n+1} is a proper open and closed subset of Y_n;

(iii) Y_n has infinitely many connected components.

Let $Y_1 = Y \cap K$. Suppose $Y_1, \ldots, Y_n$ already constructed. Since Y_n is not connected, Y_n is the disjoint union of two proper subsets B and C simultaneously open and closed. Since Y_n has an infinity of connected components, at least one of these subsets, say B, inherits the same property. We will set $Y_{n+1} = B$ and the construction is thus completed.

Define $\mathfrak{a}_n = \{f \in \Gamma(K, \mathcal{O}) \,|\, f(x) = 0 \text{ for any } x \in Y_n\}$. $\mathfrak{a}_n$ is an ideal of $\Gamma(K, \mathcal{O})$ and $\mathfrak{a}_n \subset \mathfrak{a}_{n+1}$. We will show that for any n, $\mathfrak{a}_n \neq \mathfrak{a}_{n+1}$ and thus we will come to a contradiction. Fix an integer n. Y_{n+1} and $(Y \cap K) \setminus Y_{n+1}$ are disjoint compact subsets of U. Let G and H be two disjoint open subsets of U, which contain Y_{n+1} and $(Y \cap K) \setminus Y_{n+1}$ respectively. $(U \setminus Y) \cup G \cup H$ is a neighbourhood of K, hence it contains a Stein neighbourhood D of K. Clearly, $D \cap Y \subset G \cup H$. Let $\mathfrak{I}$ be the ideal-sheaf of Y in U. Let $s \in \Gamma(D \cap Y, \mathcal{O}/\mathfrak{I})$ be the section induced by the section $\bar{s} \in \Gamma(G \cup H, \mathcal{O})$ which is 0 on G and 1 on H. As D is Stein, s is induced by an element $\tilde{t} \in \Gamma(D, \mathcal{O})$. Let $t = \tilde{t}|K$. It follows immediately that $t \in \mathfrak{a}_{n+1} \setminus \mathfrak{a}_n$.

(b) Let K be a Stein compact satisfying the topological condition in the statement of the theorem. We will show that the ring $\Gamma(K, \mathcal{O})$ is noetherian. We may assume that X is of bounded dimension and let $n = \dim X$.

Consider an arbitrary ideal $\mathfrak{a}$ in $\Gamma(K, \mathcal{O})$. We will prove by descending induction on k, $0 \leq k \leq n+1$, the following assertion:

$(*)_k$: "There exist an open neighbourhood U_k of K and elements $f_1^{(k)}, \ldots, f_{q(k)}^{(k)} \in \Gamma(U_k, \mathcal{O})$ such that

(i) $f_i^{(k)}|K \in \mathfrak{a}$, $1 \leq i \leq q(k)$;

(ii) provided that $f \in \Gamma(U, \mathcal{O})$, U being a neighbourhood of K included in U_k, and $f|K \in \mathfrak{a}$, then in some neighbourhood U' of K in U, the analytic set

$$\left\{x \in U' \,|\, f_x \notin \sum_{i=1}^{q(k)} (f_i^{(k)})_x \mathcal{O}_x\right\}$$

is of dimension $< k$".

The assertion $(*)_{n+1}$ is utterly obvious. Suppose $(*)_k$ verified for $m < k \leq n+1$ and we are going to verify it for $k = m$. We consider the ideal-sheaf $\mathfrak{I} = \sum_{i=1}^{q(m+1)} f_i^{(m+1)} \mathcal{O}$ on U_{m+1}. The subsheaf $\mathcal{H}_m^0(\mathcal{O}/\mathfrak{I}) = (\mathcal{O}/\mathfrak{I})[m]$ of $\mathcal{O}/\mathfrak{I}$ defined by the sections whose supports are of dimension $\leq m$ is coherent and its support has the dimension $\leq m$ (II. 5.3 and II. 5.5). Let $\mathfrak{J}$ be its inverse image under the morphism $\mathcal{O} \to \mathcal{O}/\mathfrak{I}$. $\mathfrak{J}$ is a coherent ideal-sheaf containing $\mathfrak{I}$. If $x \in U_{m+1}$, then one can see that

$\mathfrak{J}_x = \{s \in \mathcal{O}_x |$ there exist a neighbourhood D of x, an analytic subset V of D of dimension $\leq m$ and $t \in \Gamma(D, \mathcal{O})$ such that $t_x = s$ and $t_y \in \mathfrak{I}_y$ for any $y \in U_{m+1} \setminus V\}$.

From the exact sequence

$$0 \to \mathfrak{I} \to \mathfrak{J} \to (\mathcal{O}/\mathfrak{I})[m] \to 0$$

there results that $Y = \{x \in U_{m+1} | \mathfrak{I}_x \neq \mathfrak{J}_x\}$ coincides with $\mathrm{Supp}\,((\mathcal{O}/\mathfrak{I})[m])$. Therefore Y is an analytic subset of U_{m+1} of dimension $\leq m$. Consider Y endowed with the structure of a reduced space and let

$$\pi: \tilde{Y} \to Y$$

be the normalization morhism. By 3.3, $\pi^{-1}(K \cap Y)$ has a finite number of connected components $C_1, \ldots, C_p$. Define $z_j \in C_j$ and $x_j = \pi(z_j)$, $1 \leqslant j \leqslant p$. Since the fibers $\mathcal{O}_x$ are noetherian rings, there exist a neighbourhood U_m of K in U_{m+1} and sections $g_1, \ldots, g_r \in \Gamma(U_m, \mathcal{O})$ such that $g_i|K \in \mathfrak{a}$, $1 \leqslant i \leqslant r$, and $\{(g_i)_{x_j} | 1 \leqslant i \leqslant r\}$ generate the same ideal in $\mathcal{O}_{x_j}$ as the set $\{f_{x_j} | f \in \mathfrak{a}\}$, $1 \leqslant j \leqslant p$.

Let $q(m) = q(m+1) + r$. Define $f_i^{(m)} = f_i^{(m+1)}|U_m$ for $1 \leqslant i \leqslant q(m+1)$ and $f_{j+q(m+1)}^{(m)} = g_j$ for $1 \leqslant j \leqslant r$. We will prove that these entities verifiy $(*)_m$.

Let $f \in \Gamma(U, \mathcal{O})$, U being a neighbourhood of K in U_m such that $f|K \in \mathfrak{a}$. By $(*)_{m+1}$, $f|U'' \in \Gamma(U'', \mathfrak{J})$ on a neighbourhood U'' of K in U. The analytic set

$$Z = \left\{ x \in U'' \,|\, f_x \notin \sum_{i=1}^{q(m)} (f_i^{(m)})_x \mathcal{O}_x \right\}$$

is contained in Y, hence it is of dimension $\leqslant m$. Let Z' be an irreducible component of Z of dimension m. We have to prove that $Z' \cap K = \emptyset$; if so, by considering $Z_1 =$ the union of all irreducible components of Z of dimension m, the open set $U' = U'' \setminus Z_1$ contains K, $\dim (Z \cap U') < m$ and $(*)_m$ then follows.

Suppose, on the contrary, that $Z' \cap K \neq \emptyset$. Since $\dim Z' = m$ and $\dim Y \leqslant m$, Z' coincides with an irreducible component of $Y \cap U''$, hence it is the image under π of a connected component T of $\pi^{-1}(Y \cap U'')$. Obviously, $T \cap \pi^{-1}(K \cap Y) \neq \emptyset$. Choose j, $1 \leqslant j \leqslant p$, such that $C_j \cap T \neq \emptyset$. Since $T \cap \pi^{-1}(K \cap Y)$ is a union of connected components of $\pi^{-1}(K \cap Y)$, then $C_j \subset T$. Hence $z_j \in T$ and consequently $x_j \in Z'$. This contradicts the very selection of sections $g_1, \ldots, g_r$. Thus the assertion $(*)_k$ is proved for any integer k, $0 \leqslant k \leqslant n+1$.

We now prove the elements $f_i^{(0)}|K$, $1 \leqslant i \leqslant q(0)$, generate the ideal $\mathfrak{a}$ and thereby the proof of the theorem will be concluded. Define $g \in \mathfrak{a}$. There exist a neighbourhood U of K in U_0 and a section $f \in \Gamma(U, \mathcal{O})$ such that $g = f|K$. By $(*)_0$ we can choose a neighbourhood U' of K in U so that $f_x \in \sum_{i=1}^{q(0)} (f_i^{(0)})_x \mathcal{O}_x$ for any $x \in U'$. Then one easily derives by choosing a Stein neighbourhood W of K in U' and by means of theorem B that $f|W \in \sum_{i=1}^{q(0)} (f_i^{(0)}|W)\Gamma(W, \mathcal{O})$. Therefore $g = f|K \in$

$$\in \sum_{i=1}^{q(0)} (f_i^{(0)}|K)\,\Gamma(K, \mathcal{O}).$$

COROLLARY 3.4. *Let K be a Stein compact of complex space $(X, \mathcal{O})$, which is contained in an analytic set of dimension* 1. *Then the ring $\Gamma(K, \mathcal{O})$ is noetherian if and only if the compact K has a finite number of connected components only.*

Proof. The inverse implication is an immediate consequence of the theorem.

We prove the direct implication. If Y is an analytic set defined in a neighbourhood U of K, we should prove that $Y \cap K$ has a finite number of connected components. Let $Z \subset X$ be an analytic subset of dimension 1, containing K. If $Z \cap U$ is irreducible, then $Y \cap K$ either coincides with K or is a finite set. In the general case the conclusion easily follows since only a finite number of irreducible components of $Z \cap U$ cut K.

The theorem can be applied to a Stein semianalytic compact K. Indeed, if Y is an analytic set defined in a neighbourhood of K, then $Y \cap K$ is a semianalytic subset, therefore it possesses only a finite number of connected components (§ 1, **f**).

In this way, each point of a complex space admits of a fundamental system of compact neighbourhoods K, which is stable to finite intersections and such that the rings $\mathcal{O}(K)$ are noetherian.

We are going to close this paragraph with another proof of the noetherianity of the ring $\mathcal{O}(K)$, where K is a Stein semianalytic compact subset of a complex space $(X, \mathcal{O})$.

We have to prove that any ascending sequence of ideals of finite type

$$\mathfrak{a}_1 \subset \mathfrak{a}_2 \subset \mathfrak{a}_3 \subset \dots$$

from $\Gamma(K, \mathcal{O})$ is stationary. Every ideal $\mathfrak{a}_n$ is of the form $\mathfrak{a}_n = \Gamma(K, \mathfrak{I}_n)$ where $\mathfrak{I}_n$ is a coherent ideal-sheaf, defined in a neighbourhood of K. By theorem A one has the sequence of inclusions

$$\mathfrak{I}_1|K \subset \mathfrak{I}_2|K \subset \mathfrak{I}_3|K \subset \dots$$

of coherent ideal-sheaves in $\mathcal{O}_K = \mathcal{O}|K$. The conclusion will follow from the compactness, provided that one verifies the assertion:

(*) "Let $\mathfrak{F}$ be a coherent $\mathcal{O}_K$-module and let $(\mathfrak{F}_n)$ be an ascending sequence of coherent $\mathcal{O}_K$-submodules of $\mathfrak{F}$. Then for any $x \in K$ there exists a neighbourhood U of x such that the sequence $(\mathfrak{F}_n|U \cap K)$ is stationary".

In this assertion we agree to call *coherent* $\mathcal{O}_K$*-module* the restriction to K of a coherent $\mathcal{O}_U$-module, U being a neighbourhood of K.

Let $x \in K$. The ring $\mathcal{O}_x$ is noetherian, hence there is an integer n_0 such that $(\mathfrak{F}_n)_x = (\mathfrak{F}_{n_0})_x$ for $n \geqslant n_0$. If we replace $\mathfrak{F}$ and the sequence $(\mathfrak{F}_n)_{n \geqslant n_0}$ by $\mathfrak{F}/\mathfrak{F}_{n_0}$ and $(\mathfrak{F}_n/\mathfrak{F}_{n_0})_{n \geqslant n_0}$ respectively, we may assume that $(\mathfrak{F}_n)_x = 0$ for any n.

Define $F = \mathfrak{F}_x$. F is an $\mathcal{O}_x$-module of finite type and consider a filtration

$$F = F^0 \supset F^1 \supset \dots \supset F^q = 0$$

such that quotients F^i/F^{i+1} are isomorphic to $\mathcal{O}_x$-modules of the form $\mathcal{O}_x/\mathfrak{p}^i$, $\mathfrak{p}^i \in \operatorname{Spec} \mathcal{O}_x$. We "localize in x" (that is, replace X by a suitable neighbourhood U of x and the compact K by $K \cap K'$, where K' is a Stein semianalytic compact neighbourhood of x in U). We can thus suppose that there exist a filtration of $\mathfrak{F}$ by coherent sheaves

$$\mathfrak{F} = \mathfrak{F}^0 \supset \mathfrak{F}^1 \supset \dots \supset \mathfrak{F}^q = 0$$

and coherent ideal-sheaves $\mathfrak{I}^i$ such that

$$\mathfrak{F}^i/\mathfrak{F}^{i+1} \simeq \mathcal{O}/\mathfrak{I}^i, \quad (\mathfrak{F}^i)_x = F^i, \quad (\mathfrak{I}^i)_x = \mathfrak{p}^i, \quad 0 \leqslant i \leqslant q.$$

To prove the assertion (*) with respect to the stationarity of the sequence $(\mathcal{F}_n)_n$ around x is to prove (*) in the point x for the sequences induced by this in the sheaves $\mathcal{F}^i/\mathcal{F}^{i+1} \simeq \mathcal{O}/\mathfrak{I}^i$. In other words, we may assume the sheaf $\mathcal{F}$ of the form $\mathcal{O}/\mathfrak{I}$ with $\mathfrak{I}_x$ a prime ideal.

If X is replaced by the subspace given by $\mathfrak{I}$ and K by its intersection with this subspace, we may assume $\mathcal{F} = \mathcal{O}$ and $\mathcal{O}_x$ an integral domain. Further, localizing in x, X can be supposed reduced. We will show that there exists a neighbourhood U of x such that $\mathcal{F}_n|U \cap K = 0$ for any n and the proof will be concluded.

Consider the normalization morphism

$$\pi: \tilde{X} \to X.$$

$\pi^{-1}(K)$ is a Stein semianalytic compact in $\tilde{X}$. For every integer n one obtains, by taking the inverse image, a coherent ideal-sheaf $\tilde{\mathcal{F}}_n$ of $\mathcal{O}_{\tilde{X}}$, defined in a neighbourhood of $\pi^{-1}(K)$. Clearly, $\mathcal{F}_n \subset \pi_*(\tilde{\mathcal{F}}_n)$. More precisely, $\pi_*(\tilde{\mathcal{F}}_n)$ coincides with the ideal-sheaf generated by $\mathcal{F}_n$ in the integral closure of $\mathcal{O}_X$ (which coincides with $\pi_*(\mathcal{O}_{\tilde{X}})$). If $\tilde{x} \in \pi^{-1}(x)$, then $(\tilde{\mathcal{F}}_n)_{\tilde{x}} = 0$. From these facts it follows that we can suppose X just normal in the assertion we need; in particular, X is integer in every point.

Let us fix an integer n. The sheaf $\mathcal{F}_n$ is the restriction to K of a coherent ideal-sheaf defined on a neighbourhood U_n, which is denoted also by $\mathcal{F}_n$.

Consider $A = \{x' \in U_n \mid (\mathcal{F}_n)_{x'} = 0\}$. A is an open subset. Define $x' \in U_n$ such that $(\mathcal{F}_n)_{x'} \neq 0$. Denote by Y the analytic set given by $\mathcal{F}_n$, $Y = \operatorname{Supp}(\mathcal{O}_X/\mathcal{F}_n)$. If $\mathfrak{p}$ is a prime ideal of $\mathcal{O}_{x'}$ which contains $(\mathcal{F}_n)_{x'}$, then its height is $\geqslant 1$ ($\mathcal{O}_{x'}$ is an integral domain!). It then results that $\operatorname{codim}_{x'}(Y, U_n) \geqslant 1$. The same property holds in a neighbourhood of x' ([15]). The stalks of $\mathcal{F}_n$ will be nonnull in the points of this neighbourhood. Accordingly, A is also a closed subset. Let T be the connected component of K containing x. Since $x \in A \cap K$, $T \subset A$. In this way $\mathcal{F}_n|T = 0$ for any n. Since K is locally connected (§ 1, **f**), T is of the form $U \cap K$, U open subset of X containing x.

§ 4. The flatness locus of a morphism

Let X be a complex space, $K \subset X$ a Stein semianalytic compact and $\mathcal{F} \in \mathbf{Coh}(X)$. The ring $\Gamma(K, \mathcal{O}_X)$ is noetherian (theorem 3.1) and $\Gamma(K, \mathcal{F})$ is a $\Gamma(K, \mathcal{O}_X)$-module of finite type. For a point $x \in K$, denote by $\mathfrak{m}(x)$ the maximal ideal of $\Gamma(K, \mathcal{O}_X)$ given by x (the kernel of the composite map $\Gamma(K, \mathcal{O}_X) \to \mathcal{O}_x \to \mathcal{O}_x/\mathfrak{m}_x \simeq \mathbb{C}$).

LEMMA 4.1. *The completion of $\Gamma(K, \mathcal{F})$ with respect to the $\mathfrak{m}(x)$-adic topology is canonically isomorphic to the completion of $\mathcal{F}_x$ with respect to the $\mathfrak{m}_x$-adic topology.*

Proof. Let $\mathfrak{I}$ be the coherent sheaf of ideals given by x. Since K is a Stein compact, we have natural isomorphisms

$$\Gamma(K, \mathcal{F})/\mathfrak{m}(x)^p\, \Gamma(K, \mathcal{F}) = \Gamma(K, \mathcal{F})/\Gamma(K, \mathfrak{I})^p\, \Gamma(K, \mathcal{F}) \simeq$$

$$\Gamma(K, \mathcal{F}/\mathfrak{I}^p\mathcal{F}) \simeq (\mathcal{F}/\mathfrak{I}^p\mathcal{F})_x \simeq \mathcal{F}_x/\mathfrak{I}^p_x\mathcal{F}_x = \mathcal{F}_x/\mathfrak{m}^p_x\mathcal{F}_x,$$

$p \geqslant 0$ being an arbitrary integer (generalities about Stein compacts can be found in Ch. VI, § 1, **b**). Passing to projective limits one obtains the required isomorphism.

Let now $f: X \to Y$ be a morphism of complex spaces and let $K \subset X$ and $L \subset Y$ be Stein semianalytic compacts such that $f(K) \subset L$ and $\mathcal{F} \in \mathbf{Coh}(X)$. $\Gamma(L, \mathcal{O}_Y)$ and $\Gamma(K, \mathcal{O}_X)$ are noetherian rings and denote by $\Gamma(f)$ the map induced by f between them. Consider a point x of K and let $y = f(x)$. Obviously, $\Gamma(f)^{-1}(\mathfrak{m}_y) = \mathfrak{m}_x$.

LEMMA 4.2. *$\mathcal{F}$ is f-flat in the point x if and only if $\Gamma(K, \mathcal{F})$ is $\Gamma(f)$-flat in $\mathfrak{m}(x)$ (that is, $\Gamma(K, \mathcal{F})_{\mathfrak{m}(x)}$ is $\Gamma(L, \mathcal{O})_{\mathfrak{m}(y)}$-flat).*

Proof. $\mathcal{F}_x$ is $\mathcal{O}_y$-flat if and only if $\hat{\mathcal{F}}_x$ (the completion with respect to the $\mathfrak{m}_x$-adic topology) is $\hat{\mathcal{O}}_y$ (the completion in the $\mathfrak{m}_y$-adic topology)-flat (1.6). Similarly, $\Gamma(K, \mathcal{F})_{\mathfrak{m}(x)}$ is $\Gamma(L, \mathcal{O}_Y)_{\mathfrak{m}(y)}$-flat if and only if the same property holds when passing to completions. The conclusion follows from the previous lemma.

THEOREM 4.3. *Let Y be a reduced complex space, D an open of some numerical space, $X = Y \times D$, $f: X \to Y$ the projection, $x = (y, z)$ a point of X and $\mathcal{F}$ a coherent analytic sheaf on X. Then there exist an open neighbourhood V of y and an analytic subset T of V for which* $\dim T < \dim V$, *such that $\mathcal{F}$ is f-flat in the point (y', z) for any $y' \in V \setminus T$.*

Proof. Let L be a Stein semianalytic compact of Y, neighbourhood of y. It is enough to find an analytic germ T around L, which is distinct of the germ defined by the whole space and such that $\mathcal{F}$ is f-flat in (y', z) for any $y' \in L \setminus T$.

Let $Y_1, \ldots, Y_r$ be the analytic sets defined in a neighbourhood of L which correspond to the minimal prime ideals of the noetherian ring $\Gamma(L, \mathcal{O}_Y)$. Since the germ defined around L by $\bigcup_{i \neq j} (Y_i \cap Y_j)$ has the dimension $< \dim V$, it is sufficient to find T in the case we do the base changes $Y_i \to Y$. We can thus suppose that the ring $A = \Gamma(L, \mathcal{O}_Y)$ is an integral domain. Let $\mathfrak{J} \subset \mathcal{O}_X$ be the ideal associated to $Y \times z$. For a point $x' = (y', z)$, $\mathfrak{J}_{x'} = \mathfrak{m}_z \mathcal{O}_{x'}$. Denote $K = L \times z$. K is a Stein semianalytic compact of X. Let $I = \Gamma(K, \mathfrak{J})$, $B = \Gamma(K, \mathcal{O}_X)$ and $M = \Gamma(K, \mathcal{F})$. Since the composite morphism

$$\Gamma(L, \mathcal{O}_Y) \to \Gamma(K, \mathcal{O}_X) \to \Gamma(K, \mathcal{O}_X)/\Gamma(K, \mathfrak{J})$$

is an isomorphism, we may apply lemma 1.17. Then there exists a nonnull element $\varphi \in A$ such that M_φ is A-flat in all prime ideals of $V(I_\varphi)$.

Consider the analytic germ T of analytic set around L defined by φ. Let $y' \in L \setminus T$. We will show that $\mathcal{F}$ is f-flat in the point $x' = (y', z)$ and the theorem will be proved. By 4.2 we have to show that the module $\Gamma(K, \mathcal{F})_{\mathfrak{m}(x')}$ is $\Gamma(L, \mathcal{O}_Y)_{\mathfrak{m}(y')}$-flat. But this fact easily results from the inclusion $\mathfrak{m}(y')_\varphi \supset \mathfrak{J}_\varphi$.

Let Y be a complex space and Y' the set of all regular points of Y_{red}. Y' is a dense open subset in Y and its complementary is analytic. A part A of Y is called *negligible* if $A \cap Y'$ is an at most countable union of the submanifolds of Y', which are locally closed with empty interior. From the definition it follows that an at most countable union of negligible sets is negligible. Since the complex spaces are Baire spaces, the interiors of negligible parts are empty and their complementaries are dense subsets.

LEMMA 4.4. *Let $f : X \to Y$ be a morphism of complex spaces. Suppose Y reduced and X with countable topology. Then $f(X)$ is a negligible set in Y if and only if f admits no sections $\sigma: V \to X$ ($f\sigma$ = inclusion) on any nonempty open subset V of Y.*

Proof. One can suppose Y a complex manifold and X a reduced space. X is a locally finite (hence countable) union of locally closed complex manifolds X_i such that the rank of $f \mid X_i$ is constant (one can reason by induction, making use of the singular locus of X!).

Accordingly, we may assume that X is a complex manifold and f is a holomorphic map of constant rank. Locally, f is then a composition of a submersion and an immersion. Thus we have to prove the lemma under the assumption that X and Y are complex manifolds, that $f(X)$ is a submanifold of Y and f a submersion of X on $f(X)$. In this case, the set $f(X)$ is negligible if and only if its interior is empty, that is if and only if there is no point $x \in X$ where f is a submersion. The required conclusion follows.

The main result in the paragraph is the following theorem of Frisch:

THEOREM 4.5. *Let $f: X \to Y$ be a morphism of complex spaces and $\mathfrak{F}$ a coherent analytic sheaf on X. Then the set of the points of X where $\mathfrak{F}$ is not f-flat is a closed analytic subset.*

If moreover Y is reduced and X has a countable topology, then its image in Y is negligible.

Proof. The first assertion is local in nature on X and Y. Therefore we may assume that f is the composition $X \xrightarrow{i} \mathbb{C}^n \times Y \xrightarrow{p} Y$, where i is a closed immersion and p the projection. By considering the sheaf $i_*(\mathfrak{F})$ we can easily reduce the problem to the particular case $X = \mathbb{C}^n \times Y$ and f the projection.

Let $x = (z, y)$ be an arbitrary point of X. Consider some Stein semianalytic compacts $L \subset Y$ and $Q \subset \mathbb{C}^n$ such that $y \in \mathring{L}$ and $z \in \mathring{Q}$. $K = Q \times L$ is a Stein semianalytic compact in X and x is one of its interior points. Denote $Z = X \times_Y X$ and let $Z \overset{q_1}{\underset{q_2}{\rightrightarrows}} X$ be the natural morphisms. $M = K \times_L K$ is a Stein semianalytic compact in Z. Denote by $\Delta: X \to Z$ the diagonal map. Δ is a closed immersion and let $\mathfrak{J}$ be the coherent ideal-sheaf of Z which defines it. We set $I = \Gamma(M, \mathfrak{J})$ and $\tilde{\mathfrak{F}} = q_2^*(\mathfrak{F})$. The factor-ring $\Gamma(M, \mathcal{O}_Z)/I$ is canonically identified (by the morphisms $\Gamma(q_1)$ and $\Gamma(q_2)$) with $\Gamma(K, \mathcal{O}_X)$. Apply theorem 1.16 to the morphism $\Gamma(q_1) : \Gamma(K, \mathcal{O}_X) \to \Gamma(M, \mathcal{O}_Z)$, to the ideal I and to the module $\Gamma(M, \tilde{\mathfrak{F}})$. Then there exists an ideal $\mathfrak{a}$ of $\Gamma(M, \mathcal{O}_Z)$ such that

$$V(\mathfrak{a}) \cap V(I) = \{\mathfrak{p} \in V(I) | \Gamma(M, \tilde{\mathfrak{F}}) \text{ is not } \Gamma(q_1)\text{-flat in } \mathfrak{p}\}.$$

By 4.2, the set

$$A = \{z \in \Delta(K) | \varphi(z) = 0 \text{ for any } \varphi \in \mathfrak{a}\}$$

coincides with the set of points of $\Delta(K)$ where $\tilde{\mathfrak{F}}$ is not q_1-flat. Since f is a flat morphism, $\mathfrak{F}$ is f-flat in a point $x \in X$ if and only if $\tilde{\mathfrak{F}}$ is q_1-flat in the point $\Delta(x)$

(§ 1, **d**). It then follows that $\Delta^{-1}(A)$ is the set of the points of K where $\mathcal{F}$ is not f-flat. The restriction of $\Delta^{-1}(A)$ to the interior of K is a closed analytic subset of $\mathring{K}$ and in this way the first assertion is proved.

Suppose now Y reduced and X with countable topology verifying the last assertion. The problem being local on X and Y, we may suppose that $X = Y \times D$ where D is an open subset of some numerical space $\mathbb{C}^m$ and f is the projection. We can also assume Y a complex manifold.

Let Z be the set of points of X where $\mathcal{F}$ is not f-flat. Apply lemma 4.4 to the composite morphism

$$Z \hookrightarrow X = Y \times D \xrightarrow{f} Y.$$

In order to achieve the proof we show that if V is a nonempty open subset of Y and $\varphi : V \to D$ a morphism, then the graph F of φ can not be contained in Z. Let D' be the image of $V \times D$ under the morphism

$$V \times D \to \mathbb{C}^n, \qquad (y, z) \mapsto z - \varphi(y).$$

The morphism

$$\alpha : V \times D \to V \times D', \qquad (y, z) \mapsto (y, z - \varphi(y))$$

is a V-isomorphism and transforms F into $V \times \{0\}$. The conclusion results from theorem 4.3 applied to the morphism $V \times D' \to V$ and to the sheaf $\alpha_*(\mathcal{F}|V \times D)$.

We now give some consequences of the theorem.

COROLLARY 4.6. *Let $f : X \to Y$ be a morphism of complex spaces, $\mathcal{F}$ and $\mathcal{F}'$ two coherent $\mathcal{O}_X$-modules and $u : \mathcal{F}' \to \mathcal{F}$ an $\mathcal{O}_X$-morphism. Suppose $\mathcal{F}$ is f-flat. Then the set of the points $x \in X$ where the induced morphism*

$$(\mathcal{F}'_{f(x)})_x = \mathcal{F}'_x/\mathfrak{m}_{f(x)}\mathcal{F}'_x \to (\mathcal{F}_{f(x)})_x = \mathcal{F}_x/\mathfrak{m}_{f(x)}\mathcal{F}_x$$

is injective is open in X and its complementary is analytic.

Proof. Let $\mathcal{N}$ and $\mathcal{K}$ be the kernel and the cokernel of u, respectively. By 1.5, the morphism

$$\mathcal{F}'_x/\mathfrak{m}_{f(x)}\mathcal{F}'_x \to \mathcal{F}_x/\mathfrak{m}_{f(x)}\mathcal{F}_x$$

is injective if and only if $\mathcal{N}_x = 0$ and $\mathcal{K}$ is f-flat in x. The conclusion then follows from the theorem and from the fact that the support of a coherent sheaf is a closed analytic set.

COROLLARY 4.7. *Let $f : X \to Y$ be a flat morphism of complex spaces and $\varphi \in \Gamma(X, \mathcal{O}_X)$. Then the set of the points $x \in X$ such that the image of φ_x in $\mathcal{O}_x/\mathfrak{m}_{f(x)}\mathcal{O}_x = \mathcal{O}_{X_{f(x)},x}$ is a nonzerodivisor is open and its complementary is analytic.*

The proof follows from the previous corollary applied to the endomorphism of $\mathcal{O}_X$ given by φ.

COROLLARY 4.8. *Let $f : X \to Y$ be a morphism of complex spaces, $\mathcal{F}$ a coherent $\mathcal{O}_X$-module, which is flat with respect to f, and $(\varphi_i)_{1 \leqslant i \leqslant r}$ a sequence of sections of $\Gamma(X, \mathcal{O}_X)$. Then the set of the points $x \in X$ such that the image of the sequence $(\varphi_{i,x})_{1 \leqslant i \leqslant r}$ in $\mathcal{O}_x/\mathfrak{m}_{f(x)}\mathcal{O}_x$ is a regular $\mathcal{F}_x/\mathfrak{m}_{f(x)}\mathcal{F}_x$-sequence is open in X and its complementary is analytic.*

Proof. Let U be the set from the statement. Denote $\mathcal{G} = \mathcal{F} \Big/ \sum_{i=1}^{r} \varphi_i \mathcal{F}$. Since $\mathcal{F}$ is f-flat, it follows from 1.9 that U is just the set of the points $x \in X$ enjoying the property that the sequence $(\varphi_{i,x})$ is a regular $\mathcal{F}_x$-sequence and $\mathcal{G}$ is f-flat in x. The conclusion follows from the theorem and from

LEMMA 4.9. *Let X be a complex space, $\mathcal{F} \in$ **Coh** (X) and $(\varphi_i)_{1 \leqslant i \leqslant r}$ a sequence of global sections of $\mathcal{O}_X$. Then the set of the points $x \in X$ such that the sequence $(\varphi_{i,x})_{1 \leqslant i \leqslant r}$ is $\mathcal{F}_x$-regular is open and its complementary is analytic.*

Proof. Consider the sheaves $\mathcal{G}_i = \mathcal{F} \Big/ \sum_{j=1}^{i-1} \varphi_j \mathcal{F}$ and the endomorphisms u_i : $\mathcal{G}_i \to \mathcal{G}_i$ given by φ_i, $1 \leqslant i \leqslant r$ (we take $\mathcal{F}$ as $\mathcal{G}_1$). If U stands for the set from the statement, then

$$X \setminus U = \bigcup_{i=1}^{r} \operatorname{Supp}(\operatorname{Ker} u_i).$$

We now reformulate the main result of the paragraph for proper morphisms.

If $X \to Y$ is a morphism of ringed spaces, y a point of Y and $\mathcal{F}$ an $\mathcal{O}_X$-module, then we say $\mathcal{F}$ *is f-flat in y* if $\mathcal{F}$ is f-flat in all points of $f^{-1}(y)$.

THEOREM 4.10. *Let $f : X \to Y$ be a proper morphism of complex spaces and $\mathcal{F}$ a coherent analytic sheaf on X. Then the set of the points of Y where $\mathcal{F}$ is not f-flat is a closed analytic subset.*

If moreover Y is reduced, then its dimension is greater than the dimension of the previous set.

Proof. Let $Z = \{x \in X \mid \mathcal{F}$ is not f-flat in $x\}$ and $T = \{y \in Y \mid \mathcal{F}$ is not f-flat in $y\}$. By theorem 4.5, Z is a closed analytic subset of X and hence $f(Z)$ is an analytic closed in Y in virtue of Remmert's projection theorem.

The first assertion follows from the equality $f(Z) = T$. In order to prove the second assertion, we may suppose Y with countable base. Since f is proper, X follows with countable base. By 4.5 the open $Y \setminus T$ is dense and the conclusion results.

COROLLARY 4.11. *Let $f : X \to Y$ be a proper morphism of complex spaces and $\mathcal{F} \in$ **Coh** (X). Suppose* $\dim Y < \infty$. *Then there exists a finite partition $(Y_i)_{1 \leqslant i \leqslant r}$ of Y with locally closed analytic subsets such that, for any i, the sheaf $\mathcal{F}_i = \mathcal{F} \times_Y Y_i$ is flat over Y_i where Y_i is considered with its reduced structure of complex subspace of Y.*

Proof. We may suppose Y reduced and proceed by induction on $n = \dim Y$. The case $n = 0$ is obvious.

Suppose the corollary proved in the case when the base is of dimension $< n$. By the theorem the set T of the points of Y where $\mathcal{F}$ is not f-flat is analytic and closed. Moreover, $\dim T < n$. We perform the base change $T \to Y$ and apply the induction hypothesis. The partition so obtained for T, to which we add the open subspace $Y \setminus T$, verifies the corollary.

Bibliographical indications

Theorem 2.12 is due to Douady and Kiehl ([19] and [45]); the proof given here is that from [45].

Theorem 2.13 and the proof presented here belong to Kerner [44]. Theorem 2.8 is due to the authors. The other results from § 2, as well as the corollaries from § 4, are from ([37], Ch. IV).

Theorem of noetherianity 3.1 (for Stein semianalytic compacts) belongs to Frisch [25] and Groethendieck [36]. The more general statement given here and the proof are from the paper of Siu [82]. The proof given at the end of the third paragraph in the semianalytic case is that from ([36]. SGA 2).

The results of § 4 were obtained by Frisch [25]. The proof of the main result (the first assertion of theorem 4.5) is given in Kiehl's paper [46]. Theorem 4.10 was already pointed out by Grauert in [29].

A presentation of the flatness in the case of the complex spaces can be found in Douady's paper [20].

Chapter VI

The formal completion of a complex space with respect to a subspace

Introduction

Let $(X, \mathcal{O})$ be a complex space, let X' be a closed analytic subset of X and let $\mathfrak{J} \subset \mathcal{O}$ be an ideal of definition of X'. Many problems suggest the consideration of the complex spaces $(X', \mathcal{O}/\mathfrak{J}^{k+1}|X')$ (the "infinitesimal neighbourhoods of X' in X"). The ringed space $\hat{X} = (X', (\varprojlim \mathcal{O}/\mathfrak{J}^{k+1})|X')$ is called the formal completion of X with respect to X'. For any $\mathcal{O}$-module $\mathcal{F}$, one may consider the $\mathcal{O}_{\hat{X}}$-module $\hat{\mathcal{F}} = \varprojlim (\mathcal{F}/\mathfrak{J}^{k+1}\mathcal{F})$, called the completion of $\mathcal{F}$ with respect to X'. If $f : X \to Y$ is a morphism of complex spaces and X', Y' are closed analytic subsets of X, Y so that $f(X') \subset Y'$, then one obtains a morphism $\hat{f} : \hat{X} \to \hat{Y}$.

Our aim in this chapter is to study the behaviour of the direct images $R^\bullet f_*$ with respect to completion.

Theorem 4.1 connects the invariants $R^\bullet f_*(\mathcal{F})^\wedge$, $R^\bullet \hat{f}_*(\hat{\mathcal{F}})$ and $\varprojlim R^\bullet f_*(\mathcal{F}/\mathfrak{J}^{k+1}\mathcal{F})$. The coherence of some graded sheaves is assumed; thereby one may derive conditions of Artin-Rees and Mittag-Leffler type and the assertions follow from the reasonings of projective systems. Suppose, for instance, f proper and $X' = f^{-1}(Y')$. In order to apply the theorem here, the difficulty in the case of complex spaces is to check the hypothesis of coherence of the $\bigoplus_{k \geqslant 0} \mathfrak{I}^k$-modules $\bigoplus_{k \geqslant 0} R^q f_*(\mathfrak{J}^k \mathcal{F})$($\mathfrak{I}$ is an ideal of definition of Y' and $\mathfrak{J} = f^*(\mathfrak{I})\mathcal{O}_X$). Theorem 4.4 ensures just this; its proof makes essential use of a finiteness theorem for ringed spaces of polynomials over complex spaces (theorem 3.1).

As an application of the above facts, one obtains the following result (4.6 and 4.7):

"Let $f : X \to Y$ be a proper morphism of complex spaces, y a point in Y and $\mathcal{F} \in \mathbf{Coh}(X)$. Then there exists an integer k_0 such that

$$H^q(X_y, \hat{\mathfrak{m}}_y^{k+r}\mathcal{F}) = \mathfrak{m}_y^r H^q(X_y, \hat{\mathfrak{m}}_y^k \mathcal{F}),$$

$$\operatorname{Ker}(H^q(X_y, \mathcal{F}) \to H^q(X_y, \mathcal{F}/\hat{\mathfrak{m}}_y^{k+r}\mathcal{F})) = \mathfrak{m}_y^r \operatorname{Ker}(H^q(X_y, \mathcal{F}) \to H^q(X_y, \mathcal{F}/\hat{\mathfrak{m}}_y^k \mathcal{F})),$$

$$\operatorname{Im}(H^q(X_y, \mathcal{F}/\hat{\mathfrak{m}}_y^{k+r}\mathcal{F}) \to H^q(X_y, \mathcal{F}/\hat{\mathfrak{m}}_y^r \mathcal{F})) = \operatorname{Im}(H^q(X_y, \mathcal{F}/\hat{\mathfrak{m}}_y^{k_0+r}\mathcal{F}) \to H^q(X_y, \mathcal{F}/\hat{\mathfrak{m}}_y^r \mathcal{F})),$$

for $q \geqslant 0$, $r \geqslant 0$ and $k \geqslant k_0$.

The first paragraph contains preliminaries and in § 2 one proves the elementary properties of the completion functor.

§ 1. Preliminaries

(a) Let A be a noetherian commutative ring, let $\mathfrak{a}$ be an ideal in A and M an A-module. Denote by $\hat{M}$ the completion of M with respect to the $\mathfrak{a}$-adic topology, $\hat{M} = \varprojlim_k (M/\mathfrak{a}^k M)$. $\hat{M}$ is a module over the completion $\hat{A} = \varprojlim_k A/\mathfrak{a}^k$ of A. Recall the following results ([10], Ch. III, § 3, n° = 4 and § 5, n° = 5):

— the ring $\hat{A}$ is noetherian and separated in the $\mathfrak{a}\hat{A}$-adic topology;

— if M is an A-module of finite type, then the canonical morphism $\hat{A} \otimes_A M \to \to \hat{M}$ is an isomorphism;

— if $M \to N \to P$ is an exact sequence of A-modules of finite type, then the sequence $\hat{M} \to \hat{N} \to \hat{P}$ is exact;

— the canonical morphism $A \to \hat{A}$ is flat;

— an $\hat{A}$-module, which is separated in the $\mathfrak{a}\hat{A}$-adic topology, is flat over A if and only if it is $\hat{A}$-flat.

LEMMA 1.1. *Let $A \to B$ be a flat morphism of noetherian rings, let $\mathfrak{a}$ be an ideal of A and let $\mathfrak{b}$ be an ideal of B such that $\mathfrak{a}B \subset \mathfrak{b}$. Denote by $\hat{A}$ and $\hat{B}$ the completions of A and B in the $\mathfrak{a}$-adic and $\mathfrak{b}$-adic topology respectively. Under these assumptions, the morphism $\hat{A} \to \hat{B}$ induced to completion is flat.*

Proof. $\hat{B}$ is separated in the $\mathfrak{b}\hat{B}$-adic topology, hence it its separated in the $\mathfrak{a}A$-adic topology. Since the morphisms $A \to B$ and $B \to \hat{B}$ are flat, the conclusion follows.

(b) Consider a complex space $(X, \mathcal{O})$. Let Q be a Stein compact, hence a compact which admits of a fundamental system of Stein neighbourhoods. If $\mathcal{F} \in \in \mathbf{Coh}(X)$, then from theorems A and B there results that the cohomology groups $H^q(Q, \mathcal{F})$ vanish if $q \geqslant 1$ and in addition, $\mathcal{F}$ admits an exact sequence of the form $\mathcal{O}^p \to \mathcal{O}^q \to \mathcal{F} \to 0$ on Q. In particular, if $\mathcal{F}' \to \mathcal{F} \to \mathcal{F}''$ is an exact sequence in $\mathbf{Coh}(X)$, then the sequence $\Gamma(Q, \mathcal{F}') \to \Gamma(Q, \mathcal{F}) \to \Gamma(Q, \mathcal{F}'')$ is exact.

If Q_1, Q_2 are Stein compacts, then $Q_1 \cap Q_2$ is also a Stein compact, as follows easily by means of diagonal morphism $X \to X \times X$. Consequently, the Stein compacts are stable to finite intersections.

LEMMA 1.2. *Let Q be a Stein compact in X. Then:*

(i) For any $\mathcal{F}, \mathcal{G} \in \mathbf{Coh}(X)$, the canonical morphism

$$\Gamma(Q, \mathcal{F}) \otimes_{\Gamma(Q, \mathcal{O})} \Gamma(Q, \mathcal{G}) \to \Gamma(Q, \mathcal{F} \otimes_{\mathcal{O}} \mathcal{G})$$

is an isomorphism.

(*ii*) *For any* $\mathcal{F} \in \mathbf{Coh}(X)$ *and whenever* $\mathcal{J}$ *is a coherent ideal-sheaf,*

$$\Gamma(Q, \mathcal{J}\mathcal{F}) = \Gamma(Q, \mathcal{J}).\Gamma(Q, \mathcal{F}).$$

Proof. (*i*) We may assume that there is an exact sequence of the form $\mathcal{O}^p \to \to \mathcal{O}^q \to \mathcal{F} \to 0$, by an eventual substitution of X for a neighbourhood of Q. The conclusion follows from the exact commutative diagram

$$\begin{array}{ccccc}
\Gamma(Q, \mathcal{O}^p) \otimes_{\Gamma(Q,\mathcal{O})} \Gamma(Q, \mathcal{G}) & \to & \Gamma(Q, \mathcal{O}^q) \otimes_{\Gamma(Q,\mathcal{O})} \Gamma(Q, \mathcal{G}) & \to & \Gamma(Q, \mathcal{F}) \otimes_{\Gamma(Q,\mathcal{O})} \Gamma(Q, \mathcal{G}) \to 0 \\
\downarrow & & \downarrow & & \downarrow \\
\Gamma(Q, \mathcal{O}^p \otimes_{\mathcal{O}} \mathcal{G}) & \to & \Gamma(Q, \mathcal{O}^q \otimes_{\mathcal{O}} \mathcal{G}) & \to & \Gamma(Q, \mathcal{F} \otimes_{\mathcal{O}} \mathcal{G}) \to 0
\end{array}$$

(*ii*) $\mathcal{J}\mathcal{F}$ is the image of the morphism $\mathcal{J} \otimes_{\mathcal{O}} \mathcal{F} \to \mathcal{F}$ and apply (*i*).

LEMMA 1.3. *Let* $Q' \subset Q$ *be Stein compacts. Then*

(*i*) *For any* $\mathcal{F} \in \mathbf{Coh}(X)$, *the canonical morphism*

$$\Gamma(Q, \mathcal{F}) \otimes_{\Gamma(Q,\mathcal{O})} \Gamma(Q', \mathcal{O}) \to \Gamma(Q', \mathcal{F})$$

is an isomorphism.

(*ii*) *The restriction map*

$$\Gamma(Q, \mathcal{O}) \to \Gamma(Q', \mathcal{O})$$

is flat.

Proof. (*i*) We may assume that there exists an exact sequence of the form $\mathcal{O}^p \to \to \mathcal{O}^q \to \mathcal{F} \to 0$. The conclusion will result from the exact commutative diagram

$$\begin{array}{ccccc}
\Gamma(Q, \mathcal{O}^p) \otimes_{\Gamma(Q,\mathcal{O})} \Gamma(Q', \mathcal{O}) & \to & \Gamma(Q, \mathcal{O}^q) \otimes_{\Gamma(Q,\mathcal{O})} \Gamma(Q', \mathcal{O}) & \to & \Gamma(Q, \mathcal{F}) \otimes_{\Gamma(Q,\mathcal{O})} \Gamma(Q', \mathcal{O}) \to 0 \\
\downarrow & & \downarrow & & \downarrow \\
\Gamma(Q', \mathcal{O}^p) & \to & \Gamma(Q', \mathcal{O}^q) & \to & \Gamma(Q', \mathcal{F}) \to 0.
\end{array}$$

(*ii*) If $\mathfrak{a} \subset \Gamma(Q, \mathcal{O})$ is an ideal of finite type, we have to show that the canonical morphism $\mathfrak{a} \otimes_{\Gamma(Q,\mathcal{O})} \Gamma(Q', \mathcal{O}) \to \Gamma(Q', \mathcal{O})$ is injective. There exists a coherent ideal-sheaf $\mathcal{J}$ on a neighbourhood of Q such that $\Gamma(Q, \mathcal{J}) = \mathfrak{a}$ and apply (*i*).

We shall sometimes use the notation

$$\mathcal{O}(Q) = \Gamma(Q, \mathcal{O}), \quad \mathcal{F}(Q) = \Gamma(Q, \mathcal{F}).$$

(c) Let X be a locally compact topological space and $(\mathcal{F}_i)_{i \in I}$ a family of sheaves of abelian groups on X. If Q is a compact subset of X, then the canonical morphism

$$\bigoplus_i \Gamma(Q, \mathcal{F}_i) \to \Gamma(Q, \bigoplus_i \mathcal{F}_i)$$

is bijective. By passing to cohomology (for instance, by means of resolutions with flabby sheaves) one gets isomorphisms

$$\bigoplus_i H^\bullet(Q, \mathcal{F}_i) \to H^\bullet(Q, \bigoplus_i \mathcal{F}_i).$$

From this fact, by lemma III. 1.3 one obtains

LEMMA 1.4. *Let $f : X \to Y$ be a proper map of locally compact topological spaces and $(\mathcal{F}_i)_{i \in I}$ a family of sheaves of abelian groups on X. Then the canonical morphisms*

$$\bigoplus_i R^q f_*(\mathcal{F}_i) \to R^q f_*(\bigoplus_i \mathcal{F}_i)$$

are isomorphisms.

Let $\mathcal{F} \in \mathbf{Ab}(X)$. Recall the construction of Godement's flabby resolution [26]. Denote by $\mathcal{C}^0 = \mathcal{C}^0(\mathcal{F})$ the sheaf $U \mapsto \prod_{x \in U} \mathcal{F}_x$, U open subset of X (the restriction maps are obvious). $\mathcal{C}^0$ is a flabby sheaf and there exists an injective morphism $\mathcal{F} \to \mathcal{C}^0$, $s \mapsto (s_x)_x$. Denote $\mathcal{C}^1 = \mathcal{C}^0(\mathrm{Coker}\,(\mathcal{F} \to \mathcal{C}^0))$, etc. One thus obtains a flabby resolution $\mathcal{C}^\bullet = \mathcal{C}^\bullet(\mathcal{F})$

$$0 \to \mathcal{F} \to \mathcal{C}^0 \to \mathcal{C}^1 \to \mathcal{C}^2 \to \dots$$

If $\mathcal{F} \to \mathcal{G}$ is a morphism in $\mathbf{Ab}(X)$, then one obtains a morphism of complexes $\mathcal{C}^\bullet(\mathcal{F}) \to$ $\to \mathcal{C}^\bullet(\mathcal{G})$. These correspondences give rise to an exact functor.

(d) ([37], Ch. 0_{III}, § 13.2.3). We say that a projective system $(A_n)_{n \geqslant 0}$ of abelian groups satisfies the ML condition (Mittag-Leffler) if for any $n \geqslant 0$ there exists an integer $m \geqslant n$ such that, for any $r \geqslant 0$

$$\mathrm{Im}(A_{m+r} \to A_n) = \mathrm{Im}\,(A_m \to A_n).$$

LEMMA 1.5. *Let $0 \to A_n \xrightarrow{u_n} B_n \xrightarrow{v_n} C_n \to 0$ be an exact sequence of projective systems of abelian groups. If the systems (A_n) and (C_n) satisfy the ML condition, then the same property holds for (B_n).*

Proof. Let $n \geqslant 0$. Choose $m \geqslant n$ such that $\mathrm{Im}\,(A_{m+r} \to A_n) = \mathrm{Im}\,(A_m \to A_n)$ for any $r \geqslant 0$. We then choose $p \geqslant m$ such that $\mathrm{Im}\,(C_{p+r} \to C_m) = \mathrm{Im}(C_p \to C_m)$ for all $r \geqslant 0$. A simple reasoning with exact sequences shows that

$$\mathrm{Im}\,(B_p \to B_n) = \mathrm{Im}\,(B_{p+r} \to B_n), \text{ for all } r \geqslant 0.$$

PROPOSITION 1.6. *Let*

$$0 \to A_n \to B_n \to C_n \to 0,\ n \geqslant 0,$$

be an exact sequence of projective systems of abelian groups. If $(A_n)_n$ satisfies the ML condition, then the sequence

$$0 \to \varprojlim_n A_n \to \varprojlim_n B_n \to \varprojlim_n C_n \to 0$$

is exact.

The proof can proceed exactly as for proposition III. 1.11.

PROPOSITION 1.7. *Let $(K_n^\cdot)_{n\geq 0}$ a projective system of complexes of abelian groups having the differentials of degree $+1$. For every m there exists a canonical morphism*

$$h_m: H^m(\varprojlim K_n^\cdot) \to \varprojlim H^m(K_n^\cdot).$$

If, for any degree m, the projective system $(K_n^m)_{n\geq 0}$ satisfies the ML condition, then all the morphisms h_m are surjective. If moreover, for a degree m, the projective system $(H^{m-1}(K_n^\cdot))$ satisfies the ML condition, then the morphism h_m is bijective.

Proof. Denote for an integer m, $K^m = \varprojlim_n K_n^m$. The morphisms h_m are to be deduced from the commutative diagrams

$$\begin{array}{ccccccccc} \dots \to & K^{m-1} & \to & K^m & \to & K^{m+1} & \to \dots \\ & \downarrow & & \downarrow & & \downarrow & \\ \dots \to & K_n^{m-1} & \to & K_n^m & \to & K_n^{m+1} & \to \dots, \end{array}$$

the differentials of $K^\cdot$ being the projective limits of the differentials of the complexes $K_n^\cdot$.

Consider the exact sequences

$$(*) \qquad 0 \to B^m(K_n^\cdot) \to Z^m(K_n^\cdot) \to H^m(K_n^\cdot) \to 0$$

$$(**) \qquad 0 \to Z^{m-1}(K_n^\cdot) \to K_n^{m-1} \to B^m(K_n^\cdot) \to 0$$

(as usual, B dessignates the group of coboundaries and Z the group of cocycles). By hypothesis one derives easily that the projective systems $(B^m(K_n^\cdot))_{n\geq 0}$ satisfy the ML condition for any m. By 1.6 we get the exact sequence

$$0 \to \varprojlim_{n\geq 0} B^m(K_n^\cdot) \to \varprojlim_{n\geq 0} Z^m(K_n^\cdot) \to \varprojlim_{n\geq 0} H^m(K_n^\cdot) \to 0.$$

From the left exactness of the projective limit one obtains the isomorphisms

$$Z^m(K^\cdot) \xrightarrow{\sim} \varprojlim Z^m(K_n^\cdot).$$

Then the fact that h_m are surjective results from the exact commutative diagram:

$$\begin{array}{ccccccccc} 0 \to & \varprojlim_{n\geqslant 0} B^m(K^{\cdot}_n) & \to & \varprojlim_{n\geqslant 0} Z^m(K^{\cdot}_n) & \to & \varprojlim_{n\geqslant 0} H^m(K^{\cdot}_n) & \to 0 \\ & \uparrow & & \wr\uparrow & & \uparrow h_m & \\ 0 \longrightarrow & B^m(K^{\bullet}) & \longrightarrow & Z^m(K^{\bullet}) & \longrightarrow & H^m(K^{\bullet}) & \to 0. \end{array}$$

Assume now that the projective system $(H^{m-1}(K^{\cdot}_n))$ enjoys the ML condition. From lemma 1.5 and the exact sequences (*) one derives that the system $(Z^{m-1}(K^{\cdot}_n))_{n\geqslant 0}$ also satisfies the ML condition. From (**) one then obtains the exact sequence

$$0 \to \varprojlim_{n\geqslant 0} Z^{m-1}(K^{\cdot}_n) \to \varprojlim_{n\geqslant 0} K^{m-1}_n \to \varprojlim_{n\geqslant 0} B^m(K^{\cdot}_n) \to 0.$$

Then it results that the morphism $B^m(K^{\bullet}) \to \varprojlim_{n\geqslant 0} B^m(K^{\cdot}_n)$ is bijective, etc.

Further we will give an application of proposition 1.7.

LEMMA 1.8. *Let X be a topological space and $(\mathcal{F}_n)_{n\geqslant 0}$ a projective system of flabby sheaves such that the maps $\mathcal{F}_{n+1} \to \mathcal{F}_n$ are epimorphisms and their kernels are flabby. Under these assumptions, the sheaf $\mathcal{F} = \varprojlim_n \mathcal{F}_n$ is flabby.*

Proof. $\Gamma(X, \mathcal{F}) \simeq \varprojlim \Gamma(X, \mathcal{F}_n)$. Let U be an open subset of X. $\Gamma(U, \mathcal{F}) \simeq \varprojlim \Gamma(U, \mathcal{F}_n)$. The hypothesis implies that the maps $\Gamma(X, \mathcal{F}_{n+1}) \to \Gamma(X, \mathcal{F}_n)$, $\Gamma(U, \mathcal{F}_{n+1}) \to \Gamma(U, \mathcal{F}_n)$, $n \geqslant 0$ are surjective. Let $s = (s_n)$ be an element of $\Gamma(U, \mathcal{F})$. Consider an element $t_0 \in \Gamma(X, \mathcal{F}_0)$ such that $t_0 | U = s_0$. Let ξ be an element of $\Gamma(X, \mathcal{F}_1)$ so that its image in $\Gamma(X, \mathcal{F}_0)$ is just t_0. The element $s_1 - \xi | U$ lies in the kernel of the map $\Gamma(U, \mathcal{F}_1) \to \Gamma(U, \mathcal{F}_0)$, hence there exists $\eta \in \operatorname{Ker}(\Gamma(X, \mathcal{F}_1) \to \Gamma(X, \mathcal{F}_0))$ such that $\eta | U = s_1 - \xi | U$. We set $t_1 = \xi + \eta \in \Gamma(X, \mathcal{F}_1)$. The image of t_1 in $\Gamma(X, \mathcal{F}_0)$ is t_0 and $t_1 | U = s_1$. The reasoning goes on and one can find an element t of $\Gamma(X, \mathcal{F})$ such that $t | U = s$.

PROPOSITION 1.9. *Let X be a topological space and let $(\mathcal{F}_n)_{n\geqslant 0}$ be a projective system of sheaves of abelian groups. Suppose the following conditions fulfilled:*

(i) There is a base $\mathcal{B}$ for the topology of X such that for any $U \in \mathcal{B}$, $H^q(U, \mathcal{F}_n) = 0$ for all $q \geqslant 1$, $n \geqslant 0$.

(ii) For any $U \in \mathcal{B}$, the morphisms $\Gamma(U, \mathcal{F}_{n+1}) \to \Gamma(U, \mathcal{F}_n)$, $n \geqslant 0$, are surjective.

Under these assumptions, for any $m \geqslant 0$, the canonical morphism

$$h_m : H^m(X, \varprojlim_{n\geqslant 0} \mathcal{F}_n) \to \varprojlim_{n\geqslant 0} H^m(X, \mathcal{F}_n)$$

is surjective; if moreover for some m the projective system $(H^{m-1}(X, \mathcal{F}_n))_{n\geqslant 0}$ satisfies the ML condition, then h_m is bijective.

Proof. For every sheaf $\mathcal{F}_n$ we consider Godement's flabby resolution $\mathcal{C}^\bullet(\mathcal{F}_n)$. Denote $\mathcal{F} = \varprojlim \mathcal{F}_n$ and $\mathcal{C}^\bullet = \varprojlim \mathcal{C}^\bullet(\mathcal{F}_n)$. Let $\mathcal{N}_{n+1} = \operatorname{Ker}(\mathcal{F}_{n+1} \to \mathcal{F}_n)$. From the exact sequence $0 \to \mathcal{N}_{n+1} \to \mathcal{F}_{n+1} \to \mathcal{F}_n \to 0$ one derives the exact sequence of complexes of sheaves

$$0 \to \mathcal{C}^\bullet(\mathcal{N}_{n+1}) \to \mathcal{C}^\bullet(\mathcal{F}_{n+1}) \to \mathcal{C}^\bullet(\mathcal{F}_n) \to 0.$$

Consequently, the kernels of the epimorphisms $\mathcal{C}^\bullet(\mathcal{F}_{n+1}) \to \mathcal{C}^\bullet(\mathcal{F}_n)$ are flabby sheaves. By lemma 1.8 it follows that the sheaves $\mathcal{C}^m = \varprojlim_n \mathcal{C}^m(\mathcal{F}_n)$ $(m \geqslant 0)$ are flabby.

For any open set $U \in \mathfrak{B}$, the condition (*i*) shows that the sequences

$$0 \to \Gamma(U, \mathcal{F}_n) \to \Gamma(U, \mathcal{C}^0(\mathcal{F}_n)) \to \Gamma(U, \mathcal{C}^1(\mathcal{F}_n)) \to \dots$$

are exact. By using (*ii*) and the surjectivity of the maps $\Gamma(U, \mathcal{C}^m(\mathcal{F}_{n+1})) \to \Gamma(U, \mathcal{C}^m(\mathcal{F}_n))$ $(m \geqslant 0,\ n \geqslant 0)$, we derive from 1.7 the exactness of the sequence

$$0 \to \Gamma(U, \mathcal{F}) \to \Gamma(U, \mathcal{C}^0) \to \Gamma(U, \mathcal{C}^1) \to \dots$$

So $\mathcal{C}^\bullet$ is a flabby resolution of $\mathcal{F}$ and therefore the cohomology groups $H^\bullet(X, \mathcal{F})$ are isomorphic to the cohomology groups of the complex $\Gamma(X, \mathcal{C}^\bullet) = \varprojlim_n (\Gamma(X, \mathcal{C}^\bullet(\mathcal{F}_n))$. The proof will be completed by applying again proposition 1.7

§ 2. Definition and elementary properties

Let $(X, \mathcal{O})$ be a complex space and let $X' \subset X$ be a closed analytic subset.

LEMMA 2.1. (*i*) *If* $\mathfrak{I} \subset \mathcal{O}$ *is a coherent ideal-sheaf such that* $\operatorname{Supp}(\mathcal{O}/\mathfrak{I}) = X'$, *then* $\operatorname{Supp}(\varprojlim_k \mathcal{O}/\mathfrak{I}^{k+1}) = X'$.

(*ii*) *If* $\mathfrak{J} \subset \mathcal{O}$ *is another coherent ideal-sheaf such that* $\operatorname{Supp}(\mathcal{O}/\mathfrak{J}) = X'$, *then there exists a natural isomorphism of* $\mathcal{O}$-*algebras*

$$\varprojlim_k (\mathcal{O}/\mathfrak{I}^{k+1}) \simeq \varprojlim_k (\mathcal{O}/\mathfrak{J}^{k+1}).$$

Proof. (*i*) Obviously, $\operatorname{Supp}(\varprojlim (\mathcal{O}/\mathfrak{I}^{k+1})) \subset X'$. Let $x \in X'$; we have to show that the canonical morphism $\mathcal{O} \to \varprojlim_k (\mathcal{O}/\mathfrak{I}^{k+1})$ is injective in x. Let f be a section defined around x so that the image of f_x in the projective limit is null. There exists a neighbourhood U of x such that the image of $f|U$ under the morphism

$$\Gamma(U, \mathcal{O}) \to \Gamma(U, \varprojlim_k (\mathcal{O}/\mathfrak{I}^{k+1})) = \varprojlim_k \Gamma(U, \mathcal{O}/\mathfrak{I}^{k+1})$$

s null. Therefore, $f_x \in \bigcap_k \mathfrak{I}_x^{k+1} \subset \bigcap_k \mathfrak{m}_x^{k+1} = 0$.

(*ii*) Replacing $\mathfrak{J}$ by $\mathfrak{I}+\mathfrak{J}$, we may assume moreover that $\mathfrak{I}\subset\mathfrak{J}$. Then there exists a natural morphism of $\mathcal{O}$-algebras

$$\varprojlim_k (\mathcal{O}/\mathfrak{I}^{k+1}) \to \varprojlim_k (\mathcal{O}/\mathfrak{J}^{k+1})$$

and we need to show that this is an isomorphism. By Nullstellensatz it follows that, locally, some power of $\mathfrak{J}$ is contained in $\mathfrak{I}$. The assertion we have to prove is local in nature, so its proof becomes clear.

Let us fix an ideal $\mathfrak{I}$ as in the lemma and let us call it an *ideal of definition of* X'. The ringed space

$$(X', \varprojlim_k (\mathcal{O}/\mathfrak{I}^{k+1}) \mid X')$$

is called *the formal completion of* X *with respect to* X' and it is denoted by $(\hat{X}, \mathcal{O}_{\hat{X}})$. In virtue of the above lemma this does not depend on the sheaf $\mathfrak{I}$ and Supp $\mathcal{O}_{\hat{X}} = \hat{X}$. The inclusion $X' \to X$ and the canonical morphism $\mathcal{O} \to \varprojlim_k (\mathcal{O}/\mathfrak{I}^{k+1})$ yield a morphism of ringed spaces $\hat{X} \to X$, which will be denoted by i.

Let $\mathfrak{F}$ be an $\mathcal{O}$-module. Denote $\hat{\mathfrak{F}} = \varprojlim_k \mathfrak{F}^{(k)}$, where $\mathfrak{F}^{(k)} = \mathfrak{F}/\mathfrak{I}^{k+1}\mathfrak{F}$. It is easy to check that $\hat{\mathfrak{F}}$ is independent of the ideal of the definition $\mathfrak{I}$ of X'. Clearly, Supp $\hat{\mathfrak{F}} \subset X'$ and by restriction one gets an $\mathcal{O}_{\hat{X}}$-module which is also denoted by $\hat{\mathfrak{F}}$. In particular, $\hat{\mathcal{O}} = \mathcal{O}_{\hat{X}}$. There exists a canonical morphism $\mathfrak{F} \to \hat{\mathfrak{F}}$. If U is an open subset of X, then $\Gamma(U, \hat{\mathfrak{F}}) = \varprojlim_k \Gamma(U, \mathfrak{F}^{(k)})$. Remark that if A is a subset of X, then $\varprojlim_k \Gamma(A, \mathfrak{F}^{(k)})$ does not coincide generally with $\Gamma(A, \hat{\mathfrak{F}})$ (even when A is reduced to a point!). If $A \subset B$ are two subsets of X, then the restriction map defines a natural morphism $\varprojlim_k \Gamma(B, \mathfrak{F}^{(k)}) \to \varprojlim_k \Gamma(A, \mathfrak{F}^{(k)})$. These morphisms are functorial with respect to inclusions.

LEMMA 2.2. (*i*) *For any Stein semianalytic compact* K, *the* $\Gamma(K, \mathcal{O})$-*module* $\varprojlim_k \Gamma(K, \mathfrak{F}^{(k)})$ *is canonically isomorphic to the completion* $\widehat{\Gamma(K, \mathfrak{F})}$ *of the* $\Gamma(K, \mathcal{O})$-*module* $\Gamma(K, \mathfrak{F})$ *with respect to the* $\Gamma(K, \mathfrak{I})$-*adic topology.*

(*ii*) *For any* $x \in K$ *there is a canonical isomorphism*

$$(\hat{\mathfrak{F}})_x \simeq \varinjlim_K \widehat{\Gamma(K, \mathfrak{F})},$$

K being a Stein semianalytic compact which contains x as an interior point.

Proof. (*i*) From theorems *A* and *B* we easily derive the isomorphisms

$$\varprojlim_k \Gamma(K, \mathcal{F}^{(k)}) = \varprojlim_k \Gamma(K, \mathcal{F}/\mathcal{I}^{k+1}\mathcal{F}) \simeq \varprojlim_k (\Gamma(K, \mathcal{F})/\Gamma(K, \mathcal{I}^{k+1}\mathcal{F})) \simeq$$

$$\simeq \varprojlim_k(\Gamma(K, \mathcal{F})/\Gamma(K, \mathcal{I}^{k+1})\,\Gamma(K, \mathcal{F})) \simeq \varprojlim_k (\Gamma(K, \mathcal{F})/\Gamma(K, \mathcal{I})^{k+1}\Gamma(K, \mathcal{F})).$$

(*ii*) $\hat{\mathcal{F}}_x = \varinjlim_U \Gamma(U, \varprojlim_k \mathcal{F}^{(k)}) = \varinjlim_U \varprojlim_k \Gamma(U, \mathcal{F}^{(k)}) = \varinjlim_K \varprojlim_k \Gamma(K, \mathcal{F}^{(k)}) \simeq$
$\simeq \varinjlim_K \widehat{\Gamma(K, \mathcal{F})}$, where U is a neighbourhood of x and K is as in the statement.

An $\mathcal{O}$-morphism $\mathcal{F} \to \mathcal{G}$ induces naturally an $\hat{\mathcal{O}}$-morphism $\hat{\mathcal{F}} \to \hat{\mathcal{G}}$; one can easily check that a functor is so obtained. For any $\mathcal{O}$-module $\mathcal{F}$ the canonical morphisms

$$\mathcal{F} \otimes_{\mathcal{O}} (\varprojlim_k \mathcal{O}/\mathcal{I}^{k+1}) \to \mathcal{F} \otimes_{\mathcal{O}} (\mathcal{O}/\mathcal{I}^{k+1}) \simeq \mathcal{F}^{(k)}$$

induce a morphism

$$i^*(\mathcal{F}) \to \hat{\mathcal{F}},$$

which is functorial in $\mathcal{F}$.

PROPOSITION 2.3. (*i*) *The functor* $\mathcal{F} \mapsto \hat{\mathcal{F}}$ *is exact on the coherent sheaves.*

(*ii*) *If* $\mathcal{F} \in \mathbf{Coh}(X)$, *then the morphism* $i^*(\mathcal{F}) \to \hat{\mathcal{F}}$ *is an isomorphism.*

Proof. (*i*) Let $0 \to \mathcal{F}' \to \mathcal{F} \to \mathcal{F}'' \to 0$ be an exact sequence in $\mathbf{Coh}(X)$. For any Stein compact K one obtains an exact sequence

$$0 \to \Gamma(K, \mathcal{F}') \to \Gamma(K, \mathcal{F}) \to \Gamma(K, \mathcal{F}'') \to 0$$

of $\mathcal{O}(K)$-modules of finite type. If moreover K is semianalytic, then $\mathcal{O}(K)$ is a noetherian ring (V. 3.1) and by completion with respect to $\Gamma(K, \mathcal{I})$-adic topology one gets an exact sequence

$$0 \to \widehat{\Gamma(K, \mathcal{F}')} \to \widehat{\Gamma(K, \mathcal{F})} \to \widehat{\Gamma(K, \mathcal{F}'')} \to 0.$$

The conclusion follows by the previous lemma.

(*ii*) The problem being local, we may assume that there is an exact sequence of the form $\mathcal{O}^p \to \mathcal{O}^q \to \mathcal{F} \to 0$. The morphism from the statement is obviously an isomorphism if $\mathcal{F} = \mathcal{O}$, hence if $\mathcal{F}$ is free of finite rank. The conclusion follows by means of the exact commutative diagram

$$\begin{array}{ccccccc} i^*(\mathcal{O}^p) & \to & i^*(\mathcal{O}^q) & \to & i^*(\mathcal{F}) & \to & 0 \\ \downarrow & & \downarrow & & \downarrow & & \\ (\mathcal{O}^p)^\wedge & \to & (\mathcal{O}^q)^\wedge & \to & \hat{\mathcal{F}} & \to & 0. \end{array}$$

COROLLARY 2.4. *If $\mathcal{F}$ and $\mathcal{G}$ are coherent $\mathcal{O}$-modules, then there are canonical isomorphisms which are functorial in $\mathcal{F}$ and $\mathcal{G}$,*

$$\hat{\mathcal{F}} \otimes_{\hat{\mathcal{O}}} \hat{\mathcal{G}} \rightrightarrows (\mathcal{F} \otimes_{\mathcal{O}} \mathcal{G})^{\wedge} \quad , \quad (Hom_{\mathcal{O}}(\mathcal{F}, \mathcal{G}))^{\wedge} \rightrightarrows Hom_{\hat{\mathcal{O}}}(\hat{\mathcal{F}}, \hat{\mathcal{G}}).$$

Proof. There are functorial canonical isomorphisms

$$i^*(\mathcal{F}) \otimes_{\hat{\mathcal{O}}} i^*(\mathcal{G}) \rightrightarrows i^*(\mathcal{F} \otimes_{\mathcal{O}} \mathcal{G}), \quad i^*(Hom_{\mathcal{O}}(\mathcal{F}, \mathcal{G})) \rightrightarrows Hom_{\hat{\mathcal{O}}}(i^*(\mathcal{F}), \ i^*(\mathcal{G})).$$

The first one holds for any morphism i of ringed spaces, without supplementary hypothesis on $\mathcal{F}$ and $\mathcal{G}$; as for the second, one applies the exactness of functor i^* on the coherent sheaves.

PROPOSITION 2.5. *The sheaf of rings $\hat{\mathcal{O}}$ is coherent.*

Proof. We have to show that the kernel of any $\hat{\mathcal{O}}$-morphism of the form $\hat{\mathcal{O}}^p \to \hat{\mathcal{O}}$, defined on an open subset U, is an $\hat{\mathcal{O}}$-module of finite type.

Let $x \in U$ and $K \subset U$ be a Stein semianalytic compact so that $x \in \mathring{K}$. As above, denote by " $\wedge$ " the completion of the $\Gamma(K, \mathcal{O})$-modules with respect to the $\Gamma(K, \mathcal{I})$-adic topology. Since $\widehat{\Gamma(K, \mathcal{O})}$ is a noetherian ring, there is a morphism of the form $\widehat{\Gamma(K, \mathcal{O})}^q \to \widehat{\Gamma(K, \mathcal{O})}^p$ such that the sequence

$$(*) \qquad \widehat{\Gamma(K, \mathcal{O})}^q \to \widehat{\Gamma(K, \mathcal{O})}^p \to \widehat{\Gamma(K, \mathcal{O})}$$

is exact. This morphism induces an $\hat{\mathcal{O}}$-morphism $\hat{\mathcal{O}}^q \to \hat{\mathcal{O}}^p$ on $\mathring{K}$. We are going to prove that the sequence $\hat{\mathcal{O}}^q \to \hat{\mathcal{O}}^p \to \hat{\mathcal{O}}$ is exact and so the proof will be over. It is sufficient to show for any Stein semianalytic compact $K' \subset \mathring{K}$, the exactness of the sequence

$$\widehat{\Gamma(K', \mathcal{O})}^q \to \widehat{\Gamma(K', \mathcal{O})}^p \to \widehat{\Gamma(K', \mathcal{O})}.$$

This can be obtained by tensoring $\otimes_{\widehat{\Gamma(K, \mathcal{O})}} \widehat{\Gamma(K', \mathcal{O})}$ the sequence (*) and by means of lemmas 1.1 and 1.3.

COROLLARY 2.6. *If $\mathcal{F} \in$ **Coh**(X), then $\hat{\mathcal{F}} \in$ **Coh**$(\hat{X})$.*

Proof. The problem is local, so we may assume the existence of an exact sequence of the form $\mathcal{O}^q \to \mathcal{O}^p \to \mathcal{F} \to 0$. One then applies 2.3 and 2.5.

PROPOSITION 2.7. *For any $\mathcal{F} \in$ **Coh**(X), the kernel of the canonical map*

$$\Gamma(X, \mathcal{F}) \to \Gamma(\hat{X}, \hat{\mathcal{F}})$$

is formed by all the sections which vanish on a neighbourhood of X'.

Proof. By the very definition of $\hat{\mathfrak{F}}$ the image of such a section is null. We now prove the converse. Let $s \in \Gamma(X, \mathfrak{F})$ be such that its image in $\Gamma(\hat{X}, \hat{\mathfrak{F}}) = \varprojlim_k \Gamma(X, \mathfrak{F}/\mathfrak{I}^{k+1}\mathfrak{F})$ is null. For every point $x \in X'$, $s_x \in \bigcap_k \mathfrak{I}_x^{k+1}\mathfrak{F}_x$. Since $\mathfrak{I}_x^{k+1} \subset \mathfrak{m}_x^{k+1}$, one derives by Krull's separation theorem that $s_x = 0$, etc.

COROLLARY 2.8. *For any $\mathfrak{F} \in \mathbf{Coh}(X)$, $\operatorname{Supp} \hat{\mathfrak{F}} = (\operatorname{Supp} \mathfrak{F}) \cap X'$, In particular, $\hat{\mathfrak{F}}$ is null if and only if there is a neighbourhood U of X' such that $\mathfrak{F} \mid U = 0$.*

COROLLARY 2.9. *Let $u : \mathfrak{F} \to \mathfrak{G}$ be a morphism of coherent $\mathcal{O}$-modules. The morphism $\hat{u} : \hat{\mathfrak{F}} \to \hat{\mathfrak{G}}$ is null if and only if u is null in a neighbourhood of X'.*

Proof. By 2.3, $(\operatorname{Im} u)^\wedge \simeq \operatorname{Im}(\hat{u})$ and apply the previous corollary.

COROLLARY 2.10. *Let $u : \mathfrak{F} \to \mathfrak{G}$ be a morphism in $\mathbf{Coh}(X)$. The morphism $\hat{u} : \hat{\mathfrak{F}} \to \hat{\mathfrak{G}}$ is a monomorphism (epimorphism, respectively) if and only if u is a monomorphism (epimorphism, respectively) in a neighbourhood of X'.*

Proof. Let $\mathfrak{N} = \operatorname{Ker} u$ and $\mathfrak{Q} = \operatorname{Coker} u$. By 2.3 it will follow that $\hat{\mathfrak{N}} = \operatorname{Ker}(\hat{u})$ and $\hat{\mathfrak{Q}} = \operatorname{Coker}(\hat{u})$. One then applies 2.8.

Let $f : X \to Y$ be a morphism of complex spaces and let X' and Y' be two analytic subsets of X and Y such that $f(X') \subset Y'$. If $\mathfrak{J} \subset \mathcal{O}_Y$ is a coherent sheaf of ideals such that $\operatorname{Supp}(\mathcal{O}_Y/\mathfrak{J}) = Y'$, and $\mathfrak{I}$ is an ideal-sheaf of $\mathcal{O}_X$ which is maximal with respect to the property $\operatorname{Supp}(\mathcal{O}_Y/\mathfrak{I}) = X'$, then $f^*(\mathfrak{J})\mathcal{O}_X \subset \mathfrak{I}$. For any integer k, one gets an inclusion $f^*(\mathfrak{J}^{k+1})\mathcal{O}_X \subset \mathfrak{I}^{k+1}$ and so, morphisms of ringed spaces $(X', \mathcal{O}_X/\mathfrak{I}^{k+1}) \to (Y', \mathcal{O}_Y/\mathfrak{J}^{k+1})$. By passing to the projective limit these yield a morphism

$$\hat{f} : \hat{X} \to \hat{Y}.$$

One can check easily the functorial dependence on this with respect to f and the commutativity of the diagram

$$\begin{array}{ccc} \hat{X} & \xrightarrow{\hat{f}} & \hat{Y} \\ \downarrow{\scriptstyle i_X} & & \downarrow{\scriptstyle i_Y} \\ X & \xrightarrow{f} & Y . \end{array}$$

PROPOSITION 2.11. *Let $f : X \to Y$ be a morphism of complex spaces and let X' and Y' be closed analytic subsets of X and Y respectively, such that $f(X') \subset Y'$. For any $\mathfrak{F} \in \mathbf{Coh}(Y)$ there exists an isomorphism (functorial in $\mathfrak{F}$) of $\mathcal{O}_{\hat{X}}$-modules*

$$f^*(\mathfrak{F})^\wedge \simeq \hat{f}^*(\hat{\mathfrak{F}}).$$

Proof. If $f^*(\mathfrak{F})^\wedge$ is identified with $i_X^*(f^*(\mathfrak{F}))$ and $\hat{\mathfrak{F}}$ with $i_Y^*(\mathfrak{F})$ (by 2.3), then the proposition follows from the equality $i_Y \hat{f} = f\, i_X$.

PROPOSITION 2.12. *Suppose X and Y are two complex spaces over a complex space S and let X' and Y' be two closed analytic subsets of X and Y, respectively and f, g two S-morphisms from X to Y such that $f(X') \subset Y'$ and $g(X') \subset Y'$. Then $\hat{f} = \hat{g}$ if and only if f and g coincide in a neighbourhood of X'.*

Proof. The direct implication follows straightforwardly from definitions. We now prove the converse. Suppose $\hat{f} = \hat{g}$. Let us fix a point x' of X'. We should prove that f and g coincide in a neighbourhood of x'. Obviously, $f(x') = g(x')$.

The problem is local, hence we may assume moreover that Y is a Stein space. We may also assume that there exists a finite number of sections $s_1, \ldots, s_r \in \Gamma(Y, \mathcal{O}_Y)$ so that the set of polynomials in these sections is dense in $\Gamma(Y, \mathcal{O}_Y)$. Apply theorem I. 4.11: the morphisms f and g correspond to two morphisms φ and ψ of topological $\mathbb{C}$-algebras from $\Gamma(Y, \mathcal{O}_Y)$ to $\Gamma(X, \mathcal{O}_X)$.

The morphisms $\Gamma(\hat{f}), \Gamma(\hat{g}) : \Gamma(\hat{Y}, \mathcal{O}_{\hat{Y}}) \to \Gamma(\hat{X}, \mathcal{O}_{\hat{X}})$ induced by $\hat{f}$ and $\hat{g}$ coincide. Then it results that the images of $\varphi(s_i)$ and $\psi(s_i)$, $1 \leqslant i \leqslant r$, in $\Gamma(\hat{X}, \mathcal{O}_{\hat{X}})$ coincide. By 2.7, there exists a neighbourhood U of X' such that $\varphi(s_i)|U = \psi(s_i)|U$ for any i. The composite maps $\Gamma(Y, \mathcal{O}_Y) \underset{\psi}{\overset{\varphi}{\rightrightarrows}} \Gamma(X, \mathcal{O}_X) \to \Gamma(U, \mathcal{O}_X)$ coincide on the elemens s_i, hence they are equal. By applying again I. 4.11, one derives that $\varphi = \psi$ on U. ∎

PROPOSITION 2.13. *Let X and Y be two complex spaces over a complex space S such that the structural morphisms $X \xrightarrow{g} S$ and $Y \xrightarrow{h} S$ are proper. Consider a closed analytic subset S' of S, $X' = g^{-1}(S')$ and $Y' = h^{-1}(S')$ the inverse images of S', $\hat{X}$, $\hat{Y}$ and $\hat{S}$ the corresponding formal completions, $f : X \to Y$ an S-morphism and let finally $\hat{f} : \hat{X} \to \hat{Y}$ be the morphism induced by f.*

In order to assure that $\hat{f}$ is an isomorphism (a closed immersion, respectively) it is necessary and sufficient to have a neighbourhood U of S' such that the morphism $g^{-1}(U) \to h^{-1}(U)$, induced by f, is an isomorphism (closed immersion, respectively).

Proof. The assertion concerning "the sufficient" follows immediately. For the converse assertion, it is enough to prove that every point $y \in Y'$ has a neighbourhood V_y such that the restriction $f^{-1}(V_y) \to V_y$ is an isomorphism (closed immersion, respectively). By hypothesis the fiber $f^{-1}(y)$ is reduced to a point. Since f is proper, then there is a neighbourhood V of y so that the restriction $f^{-1}(V) \to V$ of f is a finite morphism.

Thus we confine the question to proving the inverse assertion of the proposition under the supplementary hypothesis that f is finite. It results easily that $\hat{f}_*(\mathcal{O}_{\hat{X}})$ can be identified with $f_*(\mathcal{O}_X)^\wedge$($f_*$ is exact and commutes with the projective limits; much more, if $\mathfrak{J}$ is a coherent ideal-sheaf defining S', then $g^*(\mathfrak{J})\mathcal{O}_X$ and $h^*(\mathfrak{J})\mathcal{O}_Y$ define X' and Y', respectively) and the morphism $\mathcal{O}_{\hat{Y}} \to \hat{f}_*(\mathcal{O}_{\hat{X}})$ given by $\hat{f}$ becomes, after this identification, equal to the completion of the morphism $\mathcal{O}_Y \to f_*(\mathcal{O}_X)$ given by f. The conclusion required follows then from 2.10, by the fact that any neighbourhood of Y' contains a neighbourhood of the form $h^{-1}(U)$ for some neighbourhood U of S'.

§ 3. A finiteness theorem

Let $(X, \mathcal{O})$ be a complex space and let $N \geqslant 0$ be an integer. Consider a system $T = (T_1, \ldots, T_n)$ of indeterminates and denote by $\mathcal{C}[T]$ the associated sheaf of polynomials: if $Q \subset X$ is a compact, then

$$\Gamma(Q, \mathcal{C}[T]) = \Gamma(Q, \mathcal{O})[T] = \mathcal{C}(\mathcal{Q})[T].$$

$\mathcal{O}[T]$ has a natural structure of sheaf of graded rings. An $\mathcal{C}[T]$-module $\mathfrak{M}$ is called *graded* if it is isomorphic (over $\mathcal{O}$) to a direct sum $\bigoplus_{i \geqslant 0} \mathfrak{M}_i$ such that $\mathcal{O}[T]_p \cdot \mathfrak{M}_q \subset \subset \mathfrak{M}_{p+q}$, p and q integers $\geqslant 0$ ($\mathcal{O}[T]_p$ is the component of degree p in $\mathcal{C}[T]$).

Now we consider a morphism $X \xrightarrow{f} Y$ of complex spaces. We canonically get a morphism of ringed spaces $(X, \mathcal{O}_X[T]) \to (Y, \mathcal{O}_Y[T])$. If $\mathfrak{M}$ is an $\mathcal{O}_X[T]$-module, then the sheaves $R^q f_*(\mathfrak{M})$ display a structure of $\mathcal{O}_Y[T]$-modules.

THEOREM 3.1. *Let $f: X \to Y$ be a proper morphism of complex spaces, $T = (T_1, \ldots, T_N)$ a system of indeterminates and $\mathfrak{M}$ a graded coherent $\mathcal{O}_X[T]$-module. Then the sheaves $R^q f_*(\mathfrak{M})$ are coherent $\mathcal{O}_Y[T]$-modules for any $q \geqslant 0$.*

For the proof we need some preparations.

I. Use again the notation from the beginning of the paragraph.

LEMMA 3.2. *$\mathcal{O}[T]$ is a coherent sheaf of rings.*

Proof. Let $\mathcal{O}[T]^p \xrightarrow{\varphi} \mathcal{C}[T]$ be a morphism on an open set U and let x be a point of U. Consider a Stein semianalytic compact $Q \subset U$, neighbourhood of x. We have an exact sequence

$$0 \to \Gamma(Q, \operatorname{Ker} \varphi) \to \mathcal{O}(Q)[T]^p \to \mathcal{O}(Q)[T].$$

Since $\mathcal{O}(Q)[T]$ is a noetherian ring, there exists a surjection $\mathcal{O}(Q)[T]^q \to \Gamma(Q, \operatorname{Ker} \varphi) \to 0$. We get an exact sequence

$$\mathcal{O}(Q)[T]^q \to \mathcal{O}(Q)[T]^p \to \mathcal{O}(Q)[T]$$

and accordingly, a sequence of $\mathcal{O}[T]$-modules on Q

$$\mathcal{O}[T]^q \to \mathcal{O}[T]^p \to \mathcal{C}[T].$$

We will show that this is exact and the lemma will follow. Let $Q' \subset Q$ be a Stein compact. The required conclusion then results from 1.3 by the isomorphisms $\mathcal{O}(Q)[T]^r \otimes_{\mathcal{O}(Q)} \mathcal{O}(Q') \simeq \mathcal{O}(Q')[T]^r$, $r \geqslant 0$ being an arbitrary integer.

LEMMA 3.3. *Let $\mathfrak{M}$ be a coherent $\mathcal{C}[T]$-module. For any point $x \in X$ there exists a neighbourhood U enjoying the following property: for any Stein compact $Q \subset U$, $H^q(Q, \mathfrak{M}) = 0$ for $q \geqslant 1$.*

Proof. If $Q \subset X$ is a Stein compact, then $H^q(Q, \mathcal{C}[T]) = 0$ for $q \geqslant 1$. Indeed, $\mathcal{C}[T]$ is isomorphic to a direct sum of sheaves, each of them being isomorphic to $\mathcal{O}$, and apply theorem B to Stein compacts.

Let $x \in X$. In virtue of the coherence we get a resolution

$$(*) \qquad \mathcal{L}^d \to \ldots \to \mathcal{L}^0 \to \mathcal{M} \to 0,$$

on a neighbourhood U of x, where $\mathcal{L}^i$ $(0 \leqslant i \leqslant d)$ is a free $\mathcal{O}[T]$-module of finite rank and $d \geqslant 2\dim U$. For every Stein compact $Q \subset U$, $H^q(Q, \mathcal{M}) = 0$ for $q \geqslant 1$.

Remark. The lemma holds for any Stein compact Q on which $\mathcal{M}$ admits a resolution of the type $(*)$; we shall say that such a compact is *sufficiently small with respect to* $\mathcal{M}$.

COROLLARY 3.4. *Let $\mathcal{M}$ be a coherent $\mathcal{O}[T]$-module, $Q \subset X$ a compact and $\{Q_i\}_i$ a finite covering of Q by Stein compacts sufficiently small with respect to $\mathcal{M}$. Then the canonical map $H^\bullet(\{Q_i\}_i, \mathcal{M}) \to H^\bullet(Q, \mathcal{M})$ is an isomorphism (the first term represents the Čech cohomology).*

Proof. This follows from the lemma by using Leray's theorem (a finite intersection of Stein compacts is a Stein compact too; the condition of being sufficiently small with respect to $\mathcal{M}$ is obviously fulfilled).

LEMMA 3.5. *Let $X' \xrightarrow{i} X$ be a closed immersion of complex spaces and $\mathcal{M}$ an $\mathcal{O}_{X'}[T]$-module.*

(i) $\mathcal{M}$ is a coherent $\mathcal{O}_{X'}[T]$-module if and only if $i_*(\mathcal{M})$ is a coherent $\mathcal{O}_X[T]$-module.

(ii) If $\mathcal{M}$ admits a resolution on X'

$$\mathcal{L}^d \to \mathcal{L}^{d-1} \to \ldots \to \mathcal{L}^0 \to \mathcal{M} \to 0,$$

where $d = 2\dim X' + 2$ and if $\mathcal{L}^i (0 \leqslant i \leqslant d)$ are free $\mathcal{O}_{X'}[T]$-modules of finite rank, then:

(a) there exists an epimorphism $\mathcal{R}^0 \to i_*(\mathcal{M}) \to 0$ on any Stein semianalytic compact of X, where $\mathcal{R}^0$ is a free $\mathcal{O}_X[T]$-module of finite rank;

(b) for any $\mathcal{O}_X[T]$-morphism $\mathcal{R}^0 \to i_*(\mathcal{M})$, where $\mathcal{R}^0$ is a free $\mathcal{O}_X[T]$-module of finite rank, there exists an exact sequence of the form $\mathcal{R}^1 \to \mathcal{R}^0 \to i_*(\mathcal{M})$ on any Stein semianalytic compact, $\mathcal{R}^1$ being a free $\mathcal{O}_X[T]$-module of finite rank.

Proof. (*i*) It is sufficient to show that $i_*(\mathcal{O}_{X'}[T])$ is $\mathcal{O}_X[T]$-coherent. Since $i_*(\mathcal{O}_{X'}[T]) \simeq i_*(\mathcal{O}_{X'})[T]$, the conclusion results if we notice that $\mathcal{F} \otimes_{\mathcal{O}_X} \mathcal{O}_X[T]$ is $\mathcal{O}_X[T]$-coherent for any $\mathcal{F} \in \mathbf{Coh}(X)$.

(*ii*) (*a*). Let $Q \subset X$ be a Stein semianalytic compact, $Q' = i^{-1}(Q)$ is a Stein semianalytic compact. We get the exact sequence

$$(*) \qquad \Gamma(Q', \mathcal{L}^1) \to \Gamma(Q', \mathcal{L}^0) \to \Gamma(Q', \mathcal{M}) \to 0$$

(we use here the hypothesis $d = 2\dim X' + 2$). In particular, it follows that $\Gamma(Q, i_*(\mathcal{M})) = \Gamma(Q', \mathcal{M})$ is of finite type over $\mathcal{O}(Q')[T]$ hence of finite type over $\mathcal{O}(Q)[T]$. We thus obtain a surjection $\mathcal{O}(Q)[T]^{n_0} \to \Gamma(Q, i_*(\mathcal{M}))$. Denote $\mathcal{R}^0 = \mathcal{O}_X[T]^{n_0}$, hence we get a morphism $\mathcal{R}^0 \to i_*(\mathcal{M})$ on Q. We will show that this is an epimorphism.

Let $Q_1 \subset Q$ be a Stein compact. Obviously, $\Gamma(Q_1, \mathcal{R}^0) \simeq \Gamma(Q, \mathcal{R}^0) \otimes_{\mathcal{O}(Q)} \mathcal{O}(Q_1)$. There are also the isomorphisms $\Gamma(Q_1, i_*(\mathcal{M})) = \Gamma(Q_1', \mathcal{M}) \simeq \Gamma(Q', \mathcal{M}) \otimes_{\mathcal{O}(Q')} \mathcal{O}(Q_1') \simeq$ $\simeq \Gamma(Q, i_*(\mathcal{M})) \otimes_{\mathcal{O}(Q)} \mathcal{O}(Q_1)$, $Q_1' = i^{-1}(Q_1)$ (the former isomorphism derives from the sequence $(*)$ and from the similar sequence for Q_1'; for the latter we use the equality $\Gamma(Q_1, \mathfrak{I}) = \Gamma(Q, \mathfrak{I}) \cdot \mathcal{O}(Q_1)$, where $\mathfrak{I}$ is the ideal of the immersion i). We deduce that the map $\Gamma(Q_1, \mathcal{R}^0) \to \Gamma(Q_1, i_*(\mathcal{M}))$ is surjective, etc.

(*ii*) (*b*). Let $\mathcal{R}^0 \overset{\varphi}{\to} i_*(\mathcal{M})$ be such a morphism and Q a Stein semianalytic compact. $\Gamma(Q, \operatorname{Ker} \varphi)$ is a submodule of $\Gamma(Q, \mathcal{R}^0)$, hence it is of finite type over $\mathcal{O}(Q)[T]$. Then there exist a free $\mathcal{O}_X[T]$-module of finite rank $\mathcal{R}^1$ and a morphism $\mathcal{R}^1 \to \mathcal{R}^0$ on Q such that the sequence

$$\Gamma(Q, \mathcal{R}^1) \to \Gamma(Q, \mathcal{R}^0) \to \Gamma(Q, i_*(\mathcal{M}))$$

is exact. One can prove the exactness of the sequence $\mathcal{R}^1 \to \mathcal{R}^0 \to i_*(\mathcal{M})$, as in the proof of (*a*).

II. Let $\mathcal{M}$ and $\mathcal{N}$ be two graded $\mathcal{O}[T]$-modules. A morphism $\varphi: \mathcal{M} \to \mathcal{N}$ is called *homogeneous* if $\varphi(\mathcal{M}_p) \subset \mathcal{N}_p$ for any p. The graded $\mathcal{O}_X[T]$-modules, together with the homogeneous morphisms constitute an abelian category. Let d be an integer (we will consider only integers $d \leqslant 0$). As usual $\mathcal{O}[T]\,[d]$ stands for the graded $\mathcal{O}[T]$-module of components $\mathcal{O}[T]\,[d]_n = \mathcal{O}[T]_{n+d}$. A finite direct sum of modules of this type is called *free* graded $\mathcal{O}[T]$-module of finite rank. In lemma 3.5, if $\mathcal{M}$ is moreover graded, then the sheaves $\mathcal{R}^0$ and $\mathcal{R}^1$ may be supposed free graded of finite rank and the morphisms which appear may be supposed homogeneous.

If $X \overset{f}{\to} Y$ is a morphism of analytic spaces, $\mathcal{M}$ is a graded $\mathcal{O}_X[T]$-module and $\mathcal{N}$ a graded $\mathcal{O}_Y[T]$-module, then an $\mathcal{O}_Y[T]$-morphism $\mathcal{N} \to f_*(\mathcal{M})$ is called *homogeneous* if for any integer n one has the factorization

$$\begin{array}{ccc} \mathcal{N}_n & \to & f_*(\mathcal{M}_n) \\ \cap & & \cap \\ \mathcal{N} & \to & f_*(\mathcal{M}). \end{array}$$

III. Let $r > 0$ be a real number and $D(r) \subset \mathbb{C}^m$ the open polydisc of radius r; denote by $D(\leqslant r)$ the associated closed polydisc. Let Y be an open subset of some numerical space and $Q \subset Y$ a compact. Consider on $\Gamma(D(\leqslant r) \times Q, \mathcal{O}_{\mathbb{C}^m \times Y})$ the topology of the uniform convergence of germs: one obtains a DFS structure (I, § 1, *c*). Denote by $t_1, \ldots, t_m$ a system of coordinates in $\mathbb{C}^m$. Every $f \in \Gamma(D(\leqslant r) \times Q, \mathcal{O}_{\mathbb{C}^m \times Y})$ can be expanded in series

$$f = \sum a_\nu t^\nu,\ f_\nu \in \Gamma(Q, \mathcal{O}_Y),\ \nu = (\nu_1, \ldots, \nu_m),\ t^\nu = t_1^{\nu_1} \ldots t_m^{\nu_m}$$

(the expansion holds in a neighbourhood). For $r' \leqslant r$ and $Q' \subset Q$ compact, denote

$$\|f\|_{r', Q'} = \sum_\nu \|a_\nu\|_{Q'} \cdot r'^{\nu},$$

where $\|a_\nu\|_{Q'} = \sup\{|a_\nu(y)|,\ y \in Q'\}$ and $|\nu| = \nu_1 + \ldots + \nu_m$. One thus obtains a continuous seminorm on $\Gamma(D(\leqslant r) \times Q, \mathcal{O}_{\mathbb{C}^m \times Y})$.

LEMMA 3.6. *Let* $0 < r' < r'' < r$. *The family* $(t/r'')^\nu$, $\nu \in \mathbf{N}^m$, *enjoys the following properties on any compact* Q:

(*i*) *every* $f \in \Gamma(D(\leqslant r) \times Q, \mathcal{O}_{\mathbb{C}^m \times Y})$ *can be uniquely written* (*on a neighbourhood of* $D(\leqslant r) \times Q$) $f = \Sigma\, a_\nu(t/r'')^\nu$, *such that* $\|a_\nu\|_{Q'} \leqslant \|f\|_{r'', Q'}$ *for any compact* $Q' \subset Q$;

(*ii*)
$$\sum_\nu \|(t/r'')^\nu\|_{r', Q} < \infty$$

The proof is immediate.

Let us now consider a finite number of polydiscs $D_k(r) \subset \mathbb{C}^{m(k)}$. Denote

$$'K(\leqslant r, Q) = \prod_k \Gamma(D_k(\leqslant r) \times Q, \mathcal{O}_{\mathbb{C}^{m(k)} \times Y})\ (r > 0,\ Q \subset Y \text{ compact}).$$

Endowed with the product topology, $'K(\leqslant r, Q)$ is a DFS space. This topology coincides in fact with the topology inductive limit of Frechet spaces $'K(\rho, U)$ ($\rho > r$, $U \supset Q$ open)

$$'K(\rho, U) := \prod_k \Gamma(D_k(\rho) \times U, \mathcal{O}_{\mathbb{C}^{m(k)} \times Y}).$$

For $f = (f_k) \in {}'K(\leqslant r, Q)$ we set

$$\|f\|_{r', Q'} = \max_k \|f_k\|_{r', Q'} (r' \leqslant r,\ Q' \subset Q \text{ compact}).$$

From the previous lemma we derive

LEMMA 3.7. *Let* $0 < r' < r'' < r$ *and* $Q \subset Y$ *be a compact. Then there exists a countable family* $(e_i)_{i \in I}$ *of elements of* $'K(\leqslant r, Q)$ *such that*:

(*i*) *for every* $Q' \subset Q$, *any element* $f \in {}'K(\leqslant r, Q')$ *can be uniquely written* $f = \sum_i a_i e_i$ (*the convergence is in the above defined topology*), *where* $a_i \in \Gamma(Q', \mathcal{O}_Y)$ *and* $\|a_i\|_L \leqslant \|f\|_{r'', L}$ *for any compact* $L \subset Q'$;

(*ii*)
$$\sum_{i \in I} \|e_i\|_{r', Q} < \infty.$$

Remark. Suppose ρ chosen such that $r'' < \rho < r$. From the construction of the family (e_i), it follows that its image under the restriction map $'K(\leqslant r, Q) \to$ $\to {}'K(\leqslant \rho, Q)$ also satisfies the properties (*i*) and (*ii*).

It is useful to consider spaces of the following form:

$$K(\leqslant r, Q) = \prod_k \Gamma(D_k(\leqslant r) \times Q, \mathcal{O}_{\mathbb{C}^{m(k)} \times Y}[T][-d_k]),$$

where $T = (T_1, \ldots, T_N)$ is a system of indeterminates and $d_k \geqslant 0$ integers. These are graded $\mathcal{O}(Q)[T]$-modules (positive). By grouping conveniently the terms with respect to the integers d_k, it follows that there exists an integer $q \geqslant 1$, integers $0 \leqslant \leqslant d_1 < d_2 < \ldots < d_q$ and $\mathcal{O}(Q)$-submodules $'K_1(\leqslant r, Q), \ldots, 'K_q(\leqslant r, Q)$ of $K(\leqslant r, Q)$ of the above considered type, $'K_k(\leqslant r, Q)$ submodule of the component of degree d_k ($'K_1(\leqslant r, Q)$ coincides with this component) such that: any element $f \in K(\leqslant r, Q)$ can be written $f = \sum_\alpha T^\alpha f_\alpha$ (finite sum), $f_\alpha \in {}'K_1(\leqslant r, Q) \oplus \ldots \oplus {}'K_q(\leqslant r, Q)$. If f is homogeneous of degree d, then every f_α has the component in $'K_k(\leqslant r, Q)$ null whenever $d - |\alpha| \neq d_k$; in particular, $f_\alpha = 0$ for $|\alpha| > d$.

For $r' \leqslant r$ and $Q' \subset Q$ compact, we set

$$\|f\|_{r', Q'} = \max_\alpha \|f_\alpha\|_{r', Q'}.$$

For every α one has $\|T^\alpha f\|_{r', Q'} = \|f\|_{r' Q'}$. The homogeneous components of $K(\leqslant r, Q)$ are spaces of the type $'K(\leqslant r, Q)$ (hence they have a topological DFS structure) and the seminorms $\| \ \|_{r', Q'}$ induced from $K(\leqslant r, Q)$ are the seminorms defined previously. From lemma 3.7 one derives:

LEMMA 3.8. *Let $0 < r' < r'' < r$ and $Q \subset Y$ be a compact. Under the above conditions, there exist an integer q, integers $0 \leqslant d_1 < d_2 < \ldots < d_q$, and for every k, $1 \leqslant k \leqslant q$, there is a countable family $(e_i^k)_{i \in I}$ of homogeneous elements of degree d_k from $K(\leqslant r, Q)$ which enjoy the following properties:*

(i) for every compact $Q' \subset Q$, any element $f \in K(\leqslant r, Q')$ can be uniquely written

$$f = \sum_{\alpha, i, k} T^\alpha a_{i,\alpha}^k e_i^k, \; a_{i,\alpha}^k \in \mathcal{O}(Q')$$

(the sum with respect to α is finite) such that $\|a_{i,\alpha}^k\|_L \leqslant \|f\|_{r'',L}$ for any compact $L \subset Q'$. Morevoer, if f is homogeneous of degree d, then $a_{i,\alpha}^k = 0$ for (i, α, k) verifying $|\alpha| + d_k \neq d$;

(ii) $\sum_{i,k} \|e_i^k\|_{r', Q} < \infty$.

Remark. If ρ satisfies $r'' < \rho < r$, then the image of the family $(e_i^k)_{i,k}$ in $K(\leqslant \rho, Q)$ also verifies properties (i) and (ii).

The proof of the theorem. **(a)** By means of lemma 3.5 the problem is confined to the case when Y is an open subset of some numerical space. We may assume $\dim X < \infty$.

Let U_k, V_*, k_*, r_*, j_* be the entities from the proof of Grauert's theorem of finiteness (III, § 2). V_* can be supposed to be a polydisc. We may assume in addition that $\mathcal{M}$ admits a resolution of the type

$$\mathfrak{L}^d \to \ldots \to \mathfrak{L}^0 \to \mathcal{M} \to 0$$

on any open subset U_k, where $d = 2\dim X + 2$ and $\mathcal{L}^i (0 \leqslant i \leqslant d)$ are free sheaves of finite rank (over $\mathcal{O}_X[T]$).

For a compact $Q \subset V_*$ and for a number r, $r_* \leqslant r < 1$, denote $U_k(\leqslant r, Q) = j_k^{-1}(D_k(\leqslant r) \times Q))$, $1 \leqslant k \leqslant k_*$. If Q is Stein, then the compacts $U_k(\leqslant r, Q)$ are also Stein. We have

$$\bigcup_{k=1}^{k_*} U_k(\leqslant r, Q) = X(Q) = f^{-1}(Q)$$

and denote

$$\mathcal{U}(\leqslant r, Q) = (U_k(\leqslant r, Q))_{1 \leqslant k \leqslant k_*}.$$

$\mathcal{U}(\leqslant r, Q)$ is a finite covering of $f^{-1}(Q)$ by Stein compacts which is sufficiently small with respect to $\mathcal{M}$. By 3.4,

$$H^\bullet(C^\bullet(\mathcal{U}(\leqslant r, Q), \mathcal{M})) \simeq H^\bullet(X(Q), \mathcal{M}).$$

Let Δ_n, Δ, $U_\alpha(r, V)$, $D_\alpha(r_\alpha)$, j_α, $\pi_{\alpha\beta}$ be as in (III, § 2). Denote $U_\alpha(\leqslant r, Q) = \bigcap_{\nu=0}^{n} U_{k_\nu}(\leqslant r, Q)$ where $\alpha = (k_0, \ldots, k_n)$.

We shall use link systems of sheaves $\mathcal{N} = (\mathcal{N}_\alpha, \varphi_{\alpha\beta})$ of a particular type, namely *graded systems:*

(i) the sheaves $\mathcal{N}_\alpha$ are graded $\mathcal{O}_{D_\alpha(r) \times V}[T]$-modules;

(ii) the morphisms $\varphi_{\alpha\beta}$ are homogeneous.

One can easily see that $j_*(\mathcal{M})$ is such a system. If every $\mathcal{N}_\alpha$ is a free graded $\mathcal{O}_{D_\alpha(r) \times V}[T]$-module of finite rank, then $\mathcal{N}$ is called a *free graded system of finite rank*. By a *morphism* between two graded system we mean a family of homogeneous morphisms verifying the corresponding conditions of compatibility.

One thus obtains an abelian category.

For a compact $Q \subset V_*$ and for a real number r, $r_* \leqslant r < 1$, define

$$C^n(\leqslant r, Q; \mathcal{N}) = \prod_{\alpha \in \Delta_n} \Gamma(D_\alpha(\leqslant r) \times Q, \mathcal{N}_\alpha).$$

$C^n(\leqslant r, Q; \mathcal{N})$ is a graded $\mathcal{O}(Q)[T]$-module and the differential

$$\delta : C^n(\leqslant r, Q; \mathcal{N}) \to C^{n+1}(\leqslant r, Q; \mathcal{N})$$

is homogeneous. Similarly, for an open subset $V \subset V_*$, one can define

$$C^\bullet(\leqslant r, V; \mathcal{N}) = \prod_{\alpha \in \Delta_n} \Gamma(D_\alpha(\leqslant r) \times V, \mathcal{N}_\alpha)$$

(sections in the sense of the sheaf theory). There exist isomorphisms

$$C^\bullet(\mathfrak{U}(\leqslant r, Q), \mathfrak{M}) \simeq C^\bullet(\leqslant r, Q; j_*(\mathfrak{M})).$$

(b) LEMMA 3.9. *For any Stein open set $V' \subset \subset V_*$ and for every $r, r_* \leqslant \leqslant r' < 1$, there exists a resolution*

$$\ldots \to \mathfrak{R}^{k+1} \to \mathfrak{R}^k \to \ldots \to \mathfrak{R}^1 \to \mathfrak{R}^0 \to j_*(\mathfrak{M}) \to 0$$

on $(D_\alpha(r') \times V', \pi_{\alpha\beta})$, where $(\mathfrak{R}^k)_k$ are free graded systems of finite rank and the maps of the resolution are homogeneous.

Proof. We construct such a resolution for any Stein semianalytic compact Q, $V' \subset Q \subset V_*$: proceed by induction on k, as in (III, § 2). For $k = 0$ and $k = 1$, apply lemma 3.5 and for $k \geqslant 2$ use a reasoning as in the proof of 3.5 (ii).

So we may assume that there exists a polydisc, denoted also by V_*, a number $r_{**}, r_* < r_{**} < 1$ and a resolution like in the lemma on $(D_\alpha(r_{**}) \times V_*, \pi_{\alpha\beta})$. For a Stein compact $Q \subset V_*$ and $r, r_* \leqslant r \leqslant r_{**}$, consider the double complex $(C^l(\leqslant r, Q; \mathfrak{R}^k))_{l,k}$. All the components are graded $\mathcal{O}(Q)[T]$-modules and the differentials are homogeneous. Denote by

$$C^n(\leqslant r, Q) = \prod_{l-k=n} C^l(\leqslant r, Q; \mathfrak{R}^k)$$

the associated simple complex. Its components are also graded $\mathcal{O}(Q)[T]$-modules and the differentials

$$C^n(\leqslant r, Q) \to C^{n+1}(\leqslant r, Q)$$

are homogeneous. For every open $V \subset V_*$ one can define

$$C^n(\leqslant r, V) = \prod_{l-k=n} C^l(\leqslant r, V; \mathfrak{R}^k).$$

The correspondence $V \mapsto C^n(\leqslant r, V)$ gives rise to a sheaf on V_*, which is denoted by $C^n(\leqslant r)$. One can easily check the equality

$$\Gamma(Q, C^n(\leqslant r)) = C^n(\leqslant r, Q).$$

We have thus obtained a complex $C^\bullet(\leqslant r)$ whose components are graded $\mathcal{O}_{V_*}[T]$-modules and the differentials are homogeneous.

LEMMA 3.10. *Let $r, r_* \leqslant r \leqslant r_{**}$, and $Q \subset V_*$ be a Stein compact. The canonical morphism*

$$C^\bullet(\leqslant r, Q) \to C^\bullet(\leqslant r, Q; j_*(\mathfrak{M}))$$

is a quasi-isomorphism.

Proof. The lemma follows by the properties of the spectral sequences, by lemma 3.3 and the remark made after this lemma.

COROLLARY 3.11. *For every $r, r_* \leqslant r < r_{**}$, and for any Stein compact $Q \subset V_*$,*

$$H^\bullet(C^\bullet(\leqslant r, Q)) \simeq H^\bullet(X(Q), \mathfrak{M}).$$

COROLLARY 3.12. *For every pair (r, r'), $r_* \leqslant r' \leqslant r < r_{**}$, and for any Stein compact $Q \subset V_*$, the restriction*

$$C^\bullet(\leqslant r, Q) \to C^\bullet(\leqslant r', Q)$$

is a quasi-isomorphism.

(c) Consider a compact Q_* in V_*.

LEMMA *A(n). There exist a Stein open set V_n, $Q_* \subset V_n \subset V_*$, a real number r_n, $r_* < r_n \leqslant r_{**}$, a complex $\mathfrak{L}^\bullet$ on V_n of the form*

$$\ldots \to 0 \to \mathfrak{L}^n \xrightarrow{\alpha} \mathfrak{L}^{n+1} \to \ldots \xrightarrow{\alpha} \mathfrak{L}^{k_*} \to 0$$

of free graded $\mathcal{O}_Y[T]$-modules of finite rank and a homogeneous $\mathcal{O}_Y[T]$-morphism of complexes

$$\mathfrak{L}^\bullet \xrightarrow{\sigma} C^\bullet(\leqslant r_n)$$

such that for any Stein compact Q of V_n the morphism $\mathfrak{L}^\bullet(Q) \xrightarrow{\sigma(Q)} C^\bullet(\leqslant r_n, Q)$ is an n-quasi-isomorphism. In particular, σ is an n-quasi-isomorphism.

By corollary 3.12, for any r, $r_* \leqslant r \leqslant r_n$, the composite map $\mathfrak{L}^\bullet \xrightarrow{\sigma} C^\bullet(\leqslant r_n) \to$ $\to C^\bullet(\leqslant r)$, which will be denoted by σ too, satisfies also the assertion from lemma $A(n)$.

We use the hypothesis of lemma $A(n)$. Denote by $K^\bullet(\leqslant r)$ the cone of the composite map $\mathfrak{L}^\bullet \xrightarrow{\sigma} C^\bullet(\leqslant r)$ $(r_* \leqslant r \leqslant r_n)$ and for a compact Q of V_n, $K^\bullet(\leqslant r, Q) =$ $= \Gamma(Q, K^\bullet(\leqslant r))$. $K^\bullet(\leqslant r, Q)$ is a complex of graded $\mathcal{C}(Q)[T]$-modules. Its components $K^m(\leqslant r, Q)$ are of the same type as the modules studied in section III, hence their homogeneous components have a DFS structure. All morphisms defined till now are continuous with respect to these topologies.

LEMMA *B(n − 1). For every Stein compact $Q' \subset V_n$ and for any pair (r, r'), $r_* \leqslant r' < r \leqslant r_n$, there exists a homogeneous and continuous morphism on Q'*

$$\tau : K^{n-1}(\leqslant r) \to Z^{n-1}(K^\bullet(\leqslant r'))$$

such that the diagram

$$\begin{array}{ccc} K^{n-1}(\leqslant r) & \xrightarrow{\tau} & Z^{n-1}(K^\bullet(\leqslant r')) \\ & \nwarrow \quad \nearrow{\scriptstyle \text{restr.}} & \\ & Z^{n-1}(K^\bullet(\leqslant r)) & \end{array}$$

is commutative.

We also consider under the hypothesiss of lemma $A(n+1)$ the following

LEMMA $B^*(n)$. *For any Stein compact $Q' \subset V_{n+1}$ and for every pair (r, r'), $r_* \leqslant r' < r \leqslant r_{n+1}$, there is a homogeneous and continuous morphism on Q'*

$$\tau : K^n(\leqslant r) \to Z^n(K^\bullet(\leqslant r'))$$

such that the diagram

$$\begin{array}{ccc} K^n(\leqslant r) & \xrightarrow{\tau} & Z^n(K^\bullet(\leqslant r')) \\ & \nwarrow \quad \nearrow_{\text{restr.}} & \\ & Z^n(K^\bullet(\leqslant r)) & \end{array}$$

is commutative.

Moreover, there exist an integer $q > 0$, integers $0 \leqslant d_1 < \ldots < d_q$, countable families $(e_i^k)_{i \in I}$, $1 \leqslant i \leqslant q$, of homogeneous elements from $K^n(\leqslant r, Q')$ of degree d_k and a number $\tilde{r}$, $r' < \tilde{r} < r$, such that:

(i) for every compact $Q'' \subset Q'$, any element $f \in K^n(\leqslant r, Q'')$ can be uniquely written

$$f = \sum_{\alpha, i, k} T^\alpha a_{i,\alpha}^k e_i^k, \quad a_{i,\alpha}^k \in \mathcal{O}(Q'')$$

such that $\|a_{i,\alpha}^k\|_L \leqslant \|f\|_{\tilde{r}, L}$ for any compact $L \subset Q''$ (the sum with respect to α is finite). In addition, if f is homogeneous of degree d, then $a_{i,\alpha}^k = 0$ for systems (i, α, k) such that $|\alpha| + d_k \neq d$;

(ii) $\sum_{i,k} \|\tau e_i^k\|_{r', Q'} < \infty.$

PROPOSITION 3.13. *$A(n+1)$ and $B(n) \Rightarrow B^*(n)$.*

Proof. Let $Q' \subset V_{n+1}$ be a Stein compact and (r', r) such that $r_* \leqslant r' < r \leqslant r_{n+1}$. Let $\tilde{Q}$ be a Stein compact in V_{n+1} such that $Q' \subset\subset \tilde{Q}$ (that is $Q' \subset \mathring{\tilde{Q}}$) and $\rho', \rho, \tilde{r}$ so that $r' < \rho' < \rho < \tilde{r} < r$. Apply lemma $B(n)$ for $\tilde{Q}$ and (ρ', r') : let $\tilde{\tau} : K^n(\leqslant \rho') \to Z^n(K^\bullet(\leqslant r'))$ be the morphism so obtained on $\tilde{Q}$. The morphism

$$K^n(\leqslant r) \xrightarrow{\text{restr.}} K^n(\leqslant \rho') \xrightarrow{\tilde{\tau}} Z^n(K^\bullet(\leqslant r'))$$

has the required properties on Q'.

In order to prove this, apply lemma 3.8 for $0 < \rho < \tilde{r} < r$ and $K(\leqslant r, \tilde{Q})$. Hence there exist q, d_k $(1 \leqslant k \leqslant q)$, $(e_i^k)_{i \in I}$ such that the expansions from (i) take place and $\sum_{i,k} \|e_i^k\|_{\rho, \tilde{Q}} < \infty$. For every k, the map

$$\tilde{\tau} : K^n(\leqslant \rho', Q')_{d_k} \to Z^n(\leqslant r', Q')_{d_k}$$

is continuous.

There exist $M'_k > 0$ and a seminorm $\| \ \|$ on $K^n(\leqslant \rho', Q')_{d_k}$ such that

$$\| \tilde{\tau}(f) \|_{r',Q'} \leqslant M'_k \| f \|, \quad f \in K^n(\leqslant \rho', Q')_{d_k}.$$

There exists $M''_k > 0$ such that

$$\| \text{restr.} \ (g) \| \leqslant M''_k \cdot \| g \|_{\rho,\tilde{Q}}$$

for any $g \in K^n(\leqslant \rho, \tilde{Q})_{d_k}$ (restr. means the restriction, we will often omit to write it). Therefore for some $M > 0$,

$$\| \tilde{\tau} e_i^k \|_{r',Q'} \leqslant M \cdot \| e_i^k \|_{\rho,\tilde{Q}}$$

for all i and k. One then obtains (*ii*).

Remark. From the proof there results that under the assumptions of lemma $B^*(n)$ one can assume that there exist r'', $\tilde{r} < r'' < r$ and a morphism τ: $K^n(\leqslant r'') \to Z^n(K^\bullet (\leqslant r'))$ such that the morphism $K^n(\leqslant r) \to Z^n(K^\bullet (\leqslant r'))$ from $B^*(n)$ coincides with the composition of τ with restriction $K^n(\leqslant r) \to K^n(\leqslant r'')$.

The remark made on lemma 3.8 shows that for every r''', $r'' < r''' < r$, the morphism

$$K^n(\leqslant r''') \to K^n(\leqslant r'') \to Z^n(K^\bullet (\leqslant r')),$$

the integers $q, d_1, \ldots, d_q$, the number $\tilde{r}$ and the restriction of the family $(e_i^k)_{i,k}$ to $K^n(\leqslant r''', Q')$ also verify $B^*(n)$.

(d) Proposition 3.14. *$A(n)$ and $B(n) \Rightarrow B(n-1)$.*

Proof. One may suppose that $V_n \subset\subset V_{n+1}$ and that $r_n \leqslant r_{n+1}$. Let $Q' \subset V_n$ be a Stein compact and (r, r') such that $r_* \leqslant r' < r \leqslant r_n$. Choose Q and $\tilde{Q}$ Stein compacts such that $Q' \subset\subset Q \subset\subset \tilde{Q} \subset V_n$ and ρ', ρ real numbers, $r_* \leqslant r' < \rho' < \rho < r \leqslant r_n$. Apply the lemma $B^*(n)$ for $\tilde{Q}$ and for the pair (ρ, r); let $\tau, q, d_k, e_i^k, \tilde{r}(\rho < \tilde{r} < r)$ be the entities so obtained. We have $\sum_{i,k} \| \tau e_i^k \|_{\rho,\tilde{Q}} < \infty$. By lemma $A(n)$, the map

$$K^{n-1}(\leqslant \rho', Q) \xrightarrow{\delta} Z^n(K^n(\leqslant \rho', Q))$$

is surjective, hence for all k, $1 \leqslant k \leqslant q$, the map

$$K^{n-1}(\leqslant \rho', Q)_{d_k} \xrightarrow{\delta} Z^n(K^\bullet(\leqslant \rho', Q))_{d_k}$$

is also surjective. $Z^n(K^\bullet (\leqslant \rho', Q))_{d_k}$ is the kernel of the map $K^n(\leqslant \rho', Q)_{d_k} \xrightarrow{\delta} K^{n+1}(\leqslant \rho', Q)_{d_k}$, hence it is a closed subspace of $K^n(\leqslant \rho', Q)_{d_k}$, therefore a DFS space.

The continuous and surjective map $K^{n-1}(\leqslant \rho', Q)_{d_k} \overset{\delta}{\to} Z^n(K^\bullet(\leqslant \rho', Q))_{d_k}$ is open. Then there exists a seminorm $\| \; \|$ in $K^n(\leqslant \rho', Q)_{d_k}$, a constant $M'_k > 0$ and a family of elements $(\xi_i^k)_i$, $\xi_i^k \in K^{n-1}(\leqslant \rho', Q)_{d_k}$, such that $\delta\xi_i^k = \tau e_i^k$(we omit to write the restriction $K^n(\leqslant \rho, \tilde{Q}) \to K^n(\leqslant \rho', Q))$ and $\| \xi_i^k \|_{\rho', Q} \leqslant M'_k \cdot \|\tau e_i^k \|$, for all i. There exists also a constant $M''_k > 0$ so that $\|$ restr. $(f)\| \leqslant M''_k \cdot \| f \|_{\rho, \tilde{Q}}$ for any $f \in K^n(\leqslant \rho, \tilde{Q})_{d_k}$. For some $M > 0$, we get $\| \xi_i^k \|_{\rho', Q} \leqslant M \cdot \| \tau e_i^k \|_{\rho, Q}$ for any $i \in I$ and k, $1 \leqslant k \leqslant q$. Consequently, $\sum_{i,k} \| \xi_i^k \|_{\rho', Q} < \infty$. Define

$$h\colon K^n(\leqslant r) \to K^{n-1}(\leqslant r')$$

by the formula

$$f = \sum_{\alpha,i,k} T^\alpha a^k_{i,\alpha} e_i^k \mapsto h(f) = \sum_{\alpha,i,k} T^\alpha a^k_{i,\alpha} \xi_i^k .$$

The morphism h is continuous and makes the folowing diagram commutative

$$\begin{array}{ccc} K^n(\leqslant r) & \leftarrow & Z^n(K^\bullet(\leqslant r)) \\ \downarrow h & & \downarrow \text{restr.} \\ K^{n-1}(\leqslant r') & \overset{\delta}{\to} & Z^n(K^\bullet(\leqslant r')). \end{array}$$

The morphism $\tau\colon K^{n-1}(\leqslant r) \to Z^{n-1}(K^\bullet(\leqslant r'))$, $\tau = \text{restr.} - h\delta$, verifies $B(n-1)$.

PROPOSITION 3.15. *$A(n)$ and $B(n) \Rightarrow A(n-1)$.*

Proof. Let Q, Q' be Stein compacts such that $Q^* \subset\subset Q \subset\subset Q' \subset V_n$, and r_{n-1} such that $r_* < r_{n-1} < r_n$. Consider ρ', ρ enjoying the property $r_{n-1} < \rho' < \rho < r_n$. Apply lemma $B^*(n-1)$ for Q' and (ρ', ρ); define τ, q, d_k, $(e_i^k)_{i,k}$, $\tilde{\rho}$ $(\rho' < \tilde{\rho} < \rho)$ the entities so obtained. We have $\sum_{i,k} \| \tau e_i^k \|_{\rho' Q'} < \infty$. Define r', $r_{n-1} < r' < \rho'$. The map $C'(\leqslant \rho, Q) \to C^\bullet(\leqslant r', Q)$ is a quasi-isomorphism, hence the sum of the maps of the diagram

$$\begin{array}{ccc} & & Z^{n-1}(K^\bullet(\leqslant \rho, Q)) \\ & & \downarrow \\ C^{n-2}(\leqslant r', Q) & \to & Z^{n-1}(K^\bullet(\leqslant r', Q)) \end{array}$$

is a surjective map. The same fact will result also for the diagram

$$\begin{array}{ccc} & & K^{n-1}(\leqslant \rho, Q) \\ & & \downarrow \tau \\ C^{n-2}(\leqslant r', Q) & \overset{\delta}{\to} & Z^{n-1}(K^\bullet(\leqslant r', Q)) \end{array}$$

(we have denoted again by τ the composition of τ with the restriction $Z^{n-1}(K^\bullet(\leqslant \rho')) \to Z^{n-1}(K^\bullet(\leqslant r')))$. We now use the surjectivity for the homogeneous components: as in the proof of 3.14, it results that there are elements

$$\xi_i^k \in K^{n-1}(\leqslant \rho, Q)_{d_k}, \ \eta_i^k \in C^{n-2}(\leqslant r', Q)_{d_k}$$

and a constant $M > 0$ such that

$$\tau(\xi_i^k) + \delta(\eta_i^k) = \tau e_i^k \text{ and } \max(\| \xi_i^k \|_{\rho, Q}, \| \eta_i^k \|_{r', Q}) \leqslant M \cdot \| \tau e_i^k \|_{\rho', Q'}$$

for $i \in I$, $1 \leqslant k \leqslant q$. Then it will result that

$$(*) \qquad \sum_{i,k} \| \xi_i^k \|_{\rho, Q} < \infty, \ \sum_{i,k} \| \eta_i^k \|_{r', Q} =: M_1 < \infty.$$

Let $J \subset I$ be a finite part such that $\sum_{i \in I \setminus J, k} \| \xi_i^k \| \leqslant \frac{1}{2N_0}$, N_0 being an integer greater than the number of multi-indices α such that $|\alpha| \leqslant d_k$. Consider the free graded $\mathcal{O}(Q)[T]$-module of finite rank, whose basis is indexed over the set $\{(i, k) \mid i \in J, 1 \leqslant k \leqslant q\}$. Denote this sheaf by $\mathfrak{L}^{n-1}$ and by

$$\omega: \mathfrak{L}^{n-1} \to Z^{n-1}(K^\bullet(\leqslant r_{n-1}))$$

the homogeneous morphism $g_i^k \mapsto \tau\xi_i^k$, where g_i^k are the canonical generators. The morphism ω defines two morphisms on Q

$$\alpha^{n-1}: \mathfrak{L}^{n-1} \to \mathfrak{L}^n, \ \sigma^{n-1}: \mathfrak{L}^{n-1} \to C^{n-1}(\leqslant r_{n-1}).$$

We show that lemma $A(n-1)$ is fulfilled in the interior of Q. Just as in III, § 2, it is sufficient to prove that for any Stein compact $\tilde{Q} \subset\subset Q$, the sum of the maps

$$\begin{array}{c} \Gamma(\tilde{Q}, \mathfrak{L}^{n-1}) \\ \downarrow \omega \\ C^{n-1}(\leqslant r_{n-1}, \tilde{Q}) \xrightarrow{\delta} Z^{n-1}(K^\bullet(\leqslant r_{n-1}, \tilde{Q})) \end{array}$$

is surjective. By the remark made at the end of proposition 3.13, one may suppose the existence of some ρ'', $\tilde{\rho} < \rho'' < \rho$, such that the morphism τ: $K^{n-1}(\leqslant \rho) \to$ $\to Z^{n-1}(K^\bullet(\leqslant \rho'))$ is the composition of a morphism (denoted τ too) $K^{n-1}(\leqslant \rho'') \to$ $\to Z^{n-1}(K^\bullet(\leqslant \rho'))$ with the restriction $K^{n-1}(\leqslant \rho) \to K^{n-1}(\leqslant \rho'')$.

Just as in III, § 2, in order to complete the proof of the proposition, it is enough to prove the following:

LEMMA 3.16. *For every* $f \in K^{n-1}(\leqslant \rho'', \tilde{Q})$ *there exist* $g \in \Gamma(\tilde{Q}, \mathfrak{L}^{n-1})$ *and* $\eta \in C^{n-1}(\leqslant r_{n-1}, \tilde{Q})$ *such that*

$$\tau(f) = \omega(g) + \delta(\eta).$$

Proof. By using the expansions given by $(e_i^k)_{i,k}$, we may suppose f to be a homogeneous element of degree d, $d \leqslant d_q$. Let $\tilde{r}$, ρ^* be numbers such that $r_{n-1} < \tilde{r} < r'$, $\rho'' < \rho^* < \rho$ and let Q^* be a Stein compact, $\tilde{Q} \subset\subset Q^* \subset\subset Q$, such that $f \in K^{n-1}(\leqslant \rho^*, Q^*)_d$. We have the following expansion in convergent series

$$f = \sum_{\alpha, i, k} T^\alpha a_{\alpha, i}^k e_i^k,$$

where there is no term such that $|\alpha| > d$. Define

$$f_1 = \sum_{\alpha, k, i \in I \setminus J} T^\alpha a_{\alpha, i}^k \xi_i^k \in K^{n-1}(\leqslant \rho^*, Q^*)_d,$$

$$g_1 = \sum_{\alpha, k, i \in J} T^\alpha a_{\alpha, i}^k g_i^k \in \Gamma(Q^*, \mathfrak{L}^{n-1})_d$$

and

$$\eta_1 = \sum_{\alpha, k, i \in I} T^\alpha a_{\alpha, i}^k \eta_i^k \in C^{n-1}(\leqslant \tilde{r}, Q^*)_d$$

(the convergence is assured by $(*)$). Then it will result (in $K^{n-1}(\leqslant \tilde{r}, Q^*)_d$) that $\tau(f) = \omega(g_1) + \delta(\eta)_1 + \tau(f_1)$. We get the evaluations

$$\| f_1 \|_{\rho^*, Q^*} \leqslant N_0 \cdot \max_\alpha \| \sum_{k, i \in I \setminus J} a_{\alpha, i}^k \xi_i^k \|_{\rho^*, Q^*} \leqslant$$

$$\leqslant N_0 \cdot \max_\alpha \Big(\sum_{k, i \in I \setminus J} \| a_{\alpha, i}^k \|_{Q^*} \| \xi_i^k \|_{\rho^*, Q^*} \Big) \leqslant$$

$$\leqslant N_0 \cdot \Big(\sum_{k, i \in I \setminus J} \| f \|_{\tilde{\rho}, Q^*} \cdot \| \xi_i^k \|_{\rho, Q} \Big) \leqslant \frac{1}{2} \| f \|_{\tilde{\rho}, Q^*}$$

and similarly

$$\| g_1 \|_{Q^*} \leqslant \| f \|_{\tilde{\rho}, Q^*}, \quad \| \eta_1 \|_{\tilde{r}, Q^*} \leqslant N_0 \cdot M_1 \cdot \| f \|_{\tilde{\rho}, Q^*}.$$

In order to conclude, we make iterations; the above inequalities imply the convergence in $K^{n-1}(\leqslant \rho'', \tilde{Q})_d$, $\Gamma(\tilde{Q}, \mathfrak{L}^{n-1})_d$ and $C^{n-1}(\leqslant r_{n-1}, \tilde{Q})$.

Thereby lemmas $A(n)$ and $B(n)$ are completely proved (for n sufficiently large they are obviously satisfied).

The proof of the finiteness theorem can be concluded exactly as in III, § 2.

§ 4. The comparison theorem

Let $X \xrightarrow{f} Y$ be a morphism of complex spaces and let X', Y' be closed analytic subsets of X and Y, respectively, such that $f(X') \subset Y'$. Choose $\mathfrak{I} \subset \mathcal{O}_X$, $\mathfrak{J} \subset \mathcal{O}_Y$ as ideals of definition of X' and Y' such that $f^*(\mathfrak{J})\,\mathcal{O}_X \subset \mathfrak{I}$. Denote by $\hat{X}$, $\hat{Y}$ the respective formal completions and by $\hat{f}: \hat{X} \to \hat{Y}$ the extension of f. Recall that there are natural morphisms $i_X: \hat{X} \to X$ and $i_Y: \hat{Y} \to Y$ which agree with f and $\hat{f}$.

In the following formulae we will sometimes identify a sheaf defined on a topological subspace with its trivial extension.

Fix an $\mathcal{O}_X$-module $\mathfrak{F}$. For any integer $k \geqslant 0$ the morphism $\mathfrak{F} \to \mathfrak{F}^{(k)}(\mathfrak{F}^{(k)} = \mathfrak{F}/\mathfrak{I}^{k+1}\mathfrak{F})$ induces morphisms

$$R^{\bullet}f_*(\mathfrak{F}) \to R^{\bullet}f_*(\mathfrak{F}^{(k)}).$$

Since $\mathfrak{F}^{(k)}$ is an $\mathcal{O}_X/\mathfrak{I}^{k+1}$-module, $R^{\bullet}f_*(\mathfrak{F}^{(k)})$ is an $\mathcal{O}_Y/\mathfrak{J}^{k+1}$-module. One thus obtains the morphisms

$$R^{\bullet}f_*(\mathfrak{F}) \otimes_{\mathcal{O}_Y} (\mathcal{O}_Y/\mathfrak{J}^{k+1}) \to R^{\bullet}f_*(\mathfrak{F}^{(k)})$$

and by passing to limit, the morphisms

$$\varphi: R^{\bullet}f_*(\mathfrak{F})^{\wedge} \to \varprojlim_k R^{\bullet}f_*(\mathfrak{F}^{(k)}).$$

We now define morphisms

$$\psi: R^{\bullet}\hat{f}_*(\hat{\mathfrak{F}}) \to \varprojlim_k R^{\bullet}f_*(\mathfrak{F}^{(k)}).$$

There are isomorphisms $H^{\bullet}(\hat{X}, \hat{\mathfrak{F}}) \simeq H^{\bullet}(X, \hat{\mathfrak{F}}) = H^{\bullet}(X, \varprojlim_k \mathfrak{F}^{(k)})$ and these yield natural morphisms

$$H^{\bullet}(\hat{X}, \hat{\mathfrak{F}}) \to H^{\bullet}(X, \mathfrak{F}^{(k)}), \quad k \geqslant 0.$$

For every open set V of Y we thus obtain morphisms

$$\psi_V: H^{\bullet}(\hat{X} \cap f^{-1}(V), \hat{\mathfrak{F}}) \to \varprojlim_k H^{\bullet}(f^{-1}(V), \mathfrak{F}^{(k)}) \to \varprojlim_k \Gamma(V, R^{\bullet}f_*(\mathfrak{F}^{(k)})).$$

The family $(\psi_V)_V$ determines the required morphisms ψ.

Next we are going to construct a morphism

$$\rho\colon i_Y^*(R^\bullet f_*(\mathfrak{F})) \to R^\bullet \hat{f}_*(\hat{\mathfrak{F}}).$$

Let V be an open set of Y; by means of i_X one gets morphisms $H^\bullet(f^{-1}(V), \mathfrak{F}) \to$ $\to H^\bullet(i_X^{-1}(f^{-1}(V)), i_X^*(\mathfrak{F})) = H^\bullet(\hat{f}^{-1}(i_Y^{-1}(V)), i_X^*(\mathfrak{F}))$. By composition with $H^\bullet(\hat{f}^{-1}(i_Y^{-1}(V)), i_X^*(\mathfrak{F})) \to \Gamma(i_Y^{-1}(V), R^\bullet\hat{f}_*(i_X^*(\mathfrak{F})))$ we obtain morphisms $H^\bullet(f^{-1}(V), \mathfrak{F}) \to$ $\to \Gamma(i_Y^{-1}(V), R^\bullet\hat{f}_*(i_X^*(\mathfrak{F})))$ and since V is arbitrary we obtain morphisms $R^\bullet f_*(\mathfrak{F}) \to$ $\to (i_Y)_*(R^\bullet\hat{f}_*(i_X^*(\mathfrak{F})))$. By adjunction we derive morphisms $i_Y^*(R^\bullet f_*(\mathfrak{F})) \to R^\bullet\hat{f}_*(i_X^*(\mathfrak{F}))$. The composition of these with the morphisms $R^\bullet\hat{f}_*(i_X^*(\mathfrak{F})) \to R^\bullet\hat{f}_*(\hat{\mathfrak{F}})$ given by $i_X^*(\mathfrak{F}) \to$ $\to \hat{\mathfrak{F}}$ determines ρ.

We finally denote by $r\colon i_Y^*(R^\bullet f_*(\mathfrak{F})) \to R^\bullet f_*(\mathfrak{F})^\wedge$ the natural morphism defined in § 2 (here we have identified again the completion of an $\mathcal{O}_Y$-module with respect to Y' with its restriction to Y').

Thus we have obtained a diagram

$$\begin{array}{ccc} i_Y^*(R^q f_*(\mathfrak{F})) & \xrightarrow{\rho_q} & R^q\hat{f}_*(\hat{\mathfrak{F}}) \\ \big\downarrow{\scriptstyle r_q} & & \big\downarrow{\scriptstyle \psi_q} \\ R^q f_*(\mathfrak{F})^\wedge & \xrightarrow{\varphi_q} & \varprojlim\limits_{k} R^q f_*(\mathfrak{F}^{(k)}) \end{array}$$

for any integer $q \geqslant 0$. One can easily check the commutativity of this.

Consider the graded $\mathcal{O}_Y$-algebra

$$\mathfrak{S} = \mathcal{O}_Y \oplus \mathfrak{J} \oplus \mathfrak{J}^2 \oplus \dots$$

In virtue of lemma 4.3, $\mathfrak{S}$ is a coherent sheaf of rings. By $\mathfrak{S}[d]$ we mean the graded $\mathfrak{S}$-module of components $\mathfrak{S}[d]_n = \mathfrak{S}_{n+d}$. For an integer $q \geqslant 0$ consider the $\mathcal{O}_Y$-module

$${}^q\mathcal{H} = \bigoplus_{k \geqslant 0} R^q f_*(\mathfrak{I}^k \mathfrak{F}).$$

Let V be an open set of Y and let s be an element of $\Gamma(V, \mathfrak{J}^m)$. Denote by s' the image of s in $\Gamma(f^{-1}(V), f^*(\mathfrak{J}^m)\mathcal{O}_X) \subset \Gamma(f^{-1}(V), \mathfrak{I}^m)$.

For any k the homotety given by s' defines an $\mathcal{O}_X$-morphism on $f^{-1}(V)$

$$\mathfrak{I}^k\mathfrak{F} \to \mathfrak{I}^{k+m}\mathfrak{F},$$

hence a morphism $R^q f_*(\mathfrak{I}^k\mathfrak{F}) \to R^q f_*(\mathfrak{I}^{k+m}\mathfrak{F})$ on V. Thus on each ${}^q\mathcal{H}$ one obtains a structure of graded $\mathfrak{S}$-module.

Suppose in the following that $\mathfrak{F} \in \mathbf{Coh}(X)$.

THEOREM 4.1. *Let n be an integer. If the $\mathcal{S}$-module ${}^q\mathcal{H}$ is $\mathcal{S}$-coherent for $q = n-1$ and $q = n$, then:*

(i) The morphisms r_{n-1} and φ_{n-1} are isomorphisms, ρ_{n-1} is a monomorphism and ψ_{n-1} is an epimorphism.

(ii) The morphisms ρ_n and φ_n are monomorphisms and r_n is an isomorphism.

(iii) For any compact Q of Y, there exists an integer k_0 such that the following equalities hold on Q
$\mathrm{Ker}\,(R^qf_*(\mathcal{F}) \to R^qf_*(\mathcal{F}^{(k+r)})) = \mathcal{J}^r\ \mathrm{Ker}\,(R^qf_*(\mathcal{F}) \to R^qf_*(\mathcal{F}^{(k)}))$, *whenever* $r \geqslant 0$, $k \geqslant k_0$ *and* $q = n-1$, n *and* $\mathrm{Im}\,(R^qf_*(\mathcal{F}^{(k+r)}) \to R^qf_*(\mathcal{F}^{(r)})) = \mathrm{Im}\,(R^qf_*(\mathcal{F}^{(k_0+r)}) \to R^qf_*(\mathcal{F}^{(r)}))$, *whenever* $r \geqslant 0$, $k \geqslant k_0$ *and* $q = n-1$.

Proof. We first establish some facts required in this section.

(a) Suppose $q = n-1$ or $q = n$. Locally, ${}^q\mathcal{H}$ is generated as an $\mathcal{S}$-module by a finite number of sections, which can be considered homogeneous. Thus, locally on Y, there are homogeneous $\mathcal{S}$-epimorphisms of the form $\mathcal{L}^0 \to {}^q\mathcal{H} \to 0$, where $\mathcal{L}^0$ is isomorphic to a direct sum of sheaves of type $\mathcal{S}\,[d]$. The kernels of these morphisms are $\mathcal{S}$-modules of finite type; thus we can use the same reasoning. Consequently, locally on Y, there are exact sequences of the form $\mathcal{L}^1 \to \mathcal{L}^0 \to {}^q\mathcal{H} \to 0$, where $\mathcal{L}^1$ is of the same type as $\mathcal{L}^0$ and the morphisms are $\mathcal{S}$-homogeneous.

Passing to homogeneous components one derives that the *$\mathcal{O}_Y$-modules $R^qf_*(\mathcal{J}^k\mathcal{F})$ are coherent for all* $k \geqslant 0$. In particular, $R^{n-1}f_*(\mathcal{F})$ and $R^nf_*(\mathcal{F})$ are coherent $\mathcal{O}_Y$-modules, hence r_{n-1} and r_n are isomorphisms (2.3). We deduce that the sheaves $R^{n-1}f_*(\mathcal{F}^{(k)})$, $k \geqslant 0$, are coherent too.

(b) Fix a Stein semianalytic compact Q of Y. Denote $A = \Gamma\,(Q, \mathcal{O}_Y)$, $J = \Gamma\,(Q, \mathcal{J})$, $S = \bigoplus_{k\geqslant 0} J^k$ and ${}^qH = \bigoplus_{k\geqslant 0} \Gamma(Q, R^qf_*(\mathcal{J}^k\mathcal{F}))$. A is a noetherian ring, S is an A-algebra of finite type hence also a noetherian ring and ${}^qH$ is a graded S-module. The relation $\Gamma\,(Q, {}^q\mathcal{H}) = {}^qH$ holds. The components of ${}^q\mathcal{H}$ are $\mathcal{O}_Y$-coherent for $q = n-1$ and $q = n$, hence by theorem A there exists an epimorphism $\mathcal{L}^0 \to {}^q\mathcal{H} \to 0$ as above even on a neighbourhood of Q. The kernel of this epimorphism is a direct sum of coherent $\mathcal{O}_Y$-modules, therefore it is $\Gamma\,(Q, *)$-acyclic. The morphism $\Gamma\,(Q, \mathcal{L}^0) \to \Gamma\,(Q, {}^q\mathcal{H})$ is thus surjective, hence ${}^qH$ *is an S-module of finite type for $q = n-1$ and $q = n$.*

There exists an integer $k_1 = k_1\,(Q)$ such that $R^q_{k+r} = J^rR^q_k$ for $k \geqslant k_1$, $r \geqslant 0$, $q = n-1$, $q = n$, where for an arbitrary integer k, one has denoted

$$R^q_k = \mathrm{Ker}\,(\Gamma\,(Q, R^qf_*(\mathcal{F})) \to \Gamma\,(Q, R^qf_*(\mathcal{F}^{(k)}))).$$

In order to prove this, consider the exact sequence

$$R^qf_*(\mathcal{J}^{k+1}\mathcal{F}) \to R^qf_*(\mathcal{F}) \to R^qf_*(\mathcal{F}^{(k)}),\quad k \geqslant 0.$$

Since the first two $\mathcal{O}_Y$-modules are coherent ($q = n-1$, n) one derives easily the exact sequence

$$\Gamma\,(Q, R^qf_*\,(\mathcal{J}^{k+1}\mathcal{F})) \to \Gamma\,(Q, R^qf_*(\mathcal{F})) \to \Gamma\,(Q, R^qf_*(\mathcal{F}^{(k)})).$$

We can consider on the direct sum $\bigoplus_{k\geqslant 0} R_k^q$ the structure of graded S-module, since $J^m R_k^q \subset R_{k+m}^q$ (this follows easily from some suitable commutative diagrams given by the homoteties of the elements of J^m). By the above exact sequence, it follows that this S-module is isomorphic to a quotient of the S-submodule $\bigoplus_{k\geqslant 0} \Gamma(Q, R^q f_*(\mathfrak{J}^{k+1}\mathfrak{F}))$ of ${}^qH$, hence it is of finite type since S is noetherian. The assertion follows easily.

(c) Denote $M^q = \Gamma(Q, R^q f_*(\mathfrak{F}))$, $H_k^q = \Gamma(Q, R^q f_*(\mathfrak{F}^{(k)}))$. Consider the completion of M^q with respect to the J-adic topology, hence $(M^q)^\wedge = \varprojlim_k (M^q/J^k M^q)$.

The canonical morphism

$$(M^q)^\wedge \xrightarrow{s_q} \varprojlim_k (M^q/R_k^q)$$

is an isomorphism $(q = n-1, n)$.

This follows immediately in virute of **(b)**.

The canonical morphism

$$\varprojlim_k (M^q/R_k^q) \xrightarrow{t_q} \varprojlim_k H_k^q$$

is a monomorphism $(q = n-1,\, n)$, as follows easily from the left exactness of the projective limit.

(d) Let $N_k^q = \operatorname{Coker}(\Gamma(Q, R^q f_*(\mathfrak{F})) \to \Gamma(Q, R^q f_*(\mathfrak{F}^{(k)})))$; so we get a projective system of exact sequences

$$(*) \qquad 0 \to R_k^q \to M^q \to H_k^q \to N_k^q \to 0.$$

The homoteties given by the elements of J^m induce naturally commutative diagrams of the type

$$\begin{array}{ccc} \Gamma(Q, R^q f_*(\mathfrak{F})) & \to & \Gamma(Q, R^q f_*(\mathfrak{F}^{(k)})) \\ \downarrow & & \downarrow \\ \Gamma(Q, R^q f_*(\mathfrak{F})) & \to & \Gamma(Q, R^q f_*(\mathfrak{F}^{(k+m)})), \end{array}$$

hence morphisms $N_k^q \to N_{k+m}^q$. One thus obtains a structure of graded S-module on $N^q = \bigoplus_{k\geqslant 0} N_k^q$. It easily follows that $\alpha(J^{k+1}) \cdot N_k^q = 0$ ($k \geqslant 0$, q arbitrary), where $\alpha\colon J^{k+1} \to \Gamma(Q, \mathcal{O}_Y) \subset S$ is the inclusion; the notation α points out that in this formula the elements of J^{q+1} are regarded as elements of zero degree.

Suppose now $q = n-1$. The exact sequence of coherent $\mathcal{O}_Y$-modules

$$R^{n-1}f_*(\mathfrak{F}) \to R^{n-1}f_*(\mathfrak{F}^{(k)}) \to R^n f_*(\mathfrak{J}^{k+1}\mathfrak{F})$$

yields the exact sequence

$$\Gamma(Q, R^{n-1}f_*(\mathfrak{F})) \to \Gamma(Q, R^{n-1}f_*(\mathfrak{F}^{(k)})) \to \Gamma(Q, R^n f_*(\mathfrak{J}^{k+1}\mathfrak{F})).$$

Accordingly, N^{n-1} is an S-submodule of ${}^nH$, hence an S-module of finite type. *Then there exists an integer* $k_2 = k_2(Q)$ *such that* $N_{k+h}^{n-1} = J^h N_k^{n-1}$ *for* $k \geqslant k_2$ *and* $h \geqslant 0$. *There results* $\alpha(J^{k_2+1})\ N_k^{n-1} = 0$ *for* $k \geqslant k_2$. The integer $k_2(Q)$ depends only on the degrees of a finite system of homogeneous generators of N^{n-1}. If $Q' \subset Q$ is an arbitrary Stein semianalytic compact, then we can similarly construct integers $k_1(Q')$ and $k_2(Q')$. We notice that, by the construction of these integers, *one can assume that* $k_1(Q') = k_1(Q)$ *and* $k_2(Q') = k_2(Q)$ (we make use of the relations $\Gamma(Q', R^q f_*(\mathfrak{J}^{k+1}\mathfrak{F})) \simeq \Gamma(Q, R^q f_*(\mathfrak{J}^{k+1}\mathfrak{F})) \otimes_{\mathcal{O}(Q)} \mathcal{O}(Q')$ and $N_k^{n-1}(Q') \simeq N_k^{n-1}(Q) \otimes_{\mathcal{O}(Q)} \mathcal{O}(Q')$, which are consequences of lemma 1.3).

Let $r \geqslant 0$ and $k \geqslant 2k_2 + 1$. Obviously, $N_{r+k}^{n-1} = J^{k_2+1}\ N_{k'}^{n-1}$ where $k' = r + k - k_2 - 1 \geqslant k_2 + r$. Then one can easily see that *the image of the morphism* $N_{r+k}^{n-1} \to N_r^{n-1}$ coincides with the image of $\alpha(J^{k_2+1})\ N_{k'}^{n-1}$ under the morphism $N_{k'}^{n-1} \to N_r^{n-1}$, hence it *is null*. From the exact sequences $(*)$, one obtains the assertion:

for any $r \geqslant 0$ *and* $k \geqslant 2k_2 + 1$, $\operatorname{Im}(\Gamma(Q, R^{n-1}f_*(\mathfrak{F}^{(k+r)})) \to \Gamma(Q, R^{n-1}f_*(\mathfrak{F}^{(r)}))) =$

$$= \operatorname{Im}(\Gamma(Q, R^{n-1}f_*(\mathfrak{F}^{(2k_2+1+r)})) \to \Gamma(Q, R^{n-1}f_*(\mathfrak{F}^{(r)}))).$$

(e) By using again $(*)$, we get the exact sequences

$$0 \to M^{n-1}/R_k^{n-1} \to H_k^{n-1} \to N_k^{n-1} \to 0,$$

hence the exact sequence

$$0 \to \varprojlim_k (M^{n-1}/R_k^{n-1}) \xrightarrow{t_{n-1}} \varprojlim_k H_k^{n-1} \to \varprojlim_k N_k^{n-1}.$$

Since $\varprojlim_k N_k^{n-1} = 0$, it follows that *the morphism* t_{n-1} *is an isomorphism.*

Let us prove now the assertions of the theorem.

(*i*) The morphism r_{n-1} is an isomorphism as we have seen in **(a)**. By **(c)** and **(e)**, the canonical morphism

$$\Gamma(Q, R^{n-1}f_*(\mathfrak{F}))^\wedge \to \varprojlim_k \Gamma(Q, R^{n-1}f_*(\mathfrak{F}^{(k)})),$$

which coincides with the composition $t_{n-1}\ s_{n-1}$, is an isomorphism. Since Q is an arbitrary Stein semianalytic compact of Y, it results (inserting sufficiently small open sets between such compacts and using suitable canonical commutative diagrams) that φ_{n-1} is an isomorphism.

The assertions with respect to ρ_{n-1} and ψ_{n-1} follow from the equality $\varphi_{n-1}\ r_{n-1} = \psi_{n-1}\ \rho_{n-1}$.

(*ii*) We have already seen that r_n is an isomorphism. Since s_n and t_n are monomorphisms, the morphism φ_n is a monomorphism. Therefore ρ_n is also monomorphism.

(*iii*) It is enough to consider Stein semianalytic compacts. Let Q be such a compact and $k_0 = k_0(Q)$ an integer such that $k_0 \geqslant k_1(Q) + 2k_2(Q) + 1$. The first assertion follows from (**b**) and the second one from (**d**).

Thereby the theorem is proved.

COMPLEMENT TO THEOREM. *Under the conditions of the theorem, suppose in addition the sheaves $R^i f_*(\mathcal{F}^{(k)})$ coherent for $i \leqslant n - 1$ and $k \geqslant 0$. Then the morphism ψ_n is an isomorphism.*

Proof. We have to prove that the canonical morphism

$$R^n f_* (\varprojlim_k \mathcal{F}^{(k)}) \to \varprojlim_k R^n f_*(\mathcal{F}^{(k)})$$

is an isomorphism. The problem is local on Y, so we may assume the existence of an integer k_0 such that

$$\mathrm{Im}\,(R^{n-1}f_*(\mathcal{F}^{(k+r)}) \to R^{n-1}f_*(\mathcal{F}^{(r)})) = \mathrm{Im}\,(R^{n-1}f_*(\mathcal{F}^{(k_0+r)}) \to R^{n-1}f_*(\mathcal{F}^{(r)}))$$

for all $r \geqslant 0$ and $k \geqslant k_0$. For each Stein open subset V of Y there are isomorphisms

$$H^i(f^{-1}(V), \mathcal{F}^{(k)}) \rightleftarrows \Gamma(V, R^i f_*(\mathcal{F}^{(k)}))$$

for all $k \geqslant 0$ and $i \leqslant n$. Then we derive that

$$\mathrm{Im}\,(H^{n-1}(f^{-1}(V), \mathcal{F}^{(k+r)}) \to H^{n-1}(f^{-1}(V), \mathcal{F}^{(r)})) =$$

$$= \mathrm{Im}\,(H^{n-1}(f^{-1}(V), \mathcal{F}^{(k_0+r)}) \to H^{n-1}(f^{-1}(V), \mathcal{F}^{(r)}))$$

for all $k \geqslant k_0$, $r \geqslant 0$. In virtue of proposition 1.9, the canonical morphism

$$H^n(f^{-1}(V), \varprojlim_k \mathcal{F}^{(k)}) \to \varprojlim_k H^n(f^{-1}(V), \mathcal{F}^{(k)})$$

is an isomorphism. Since

$$\varprojlim_k H^n(f^{-1}(V), \mathcal{F}^{(k)}) \simeq \varprojlim_k \Gamma(V, R^n f_*(\mathcal{F}^{(k)})) = \Gamma(V, \varprojlim_k R^n f_*(\mathcal{F}^{(k)})),$$

the conclusion follows.

Now, we consider two cases when the theorem applies.

PROPOSITION 4.2. *Suppuse* $\mathcal{J}$ *is generated by a regular section* t' *of* $\mathcal{O}_Y$ *and* $\mathcal{I} = f^*(\mathcal{J})\, \mathcal{O}_X = t\mathcal{O}_X$, *where* t *is the image of* t' *in* $\mathcal{O}_X$. *Suppose the following conditions are fulfilled*:

(*i*) t *is* $\mathcal{F}$*-regular;*

(*ii*) $R^q f_*(\mathcal{F})$ *is coherent for* $q = n - 1$ *and* $q = n$.

Then the hypothesis of theorem 4.1 is fulfilled.

Proof. There exist some isomorphisms $\mathcal{I}^k \mathcal{F} \simeq \mathcal{F}$. One then deduces the isomorphisms ${}^q\mathcal{H} \simeq R^q f_*(\mathcal{F}) \otimes_{\mathcal{O}_Y} \mathcal{O}_Y [T]$, where T is an indeterminate. Since $\mathcal{S} \simeq \mathcal{O}_Y [T]$, the conclusion follows from lemma 3.2.

If $\mathcal{F}$ is f-flat, then condition (*i*) is satisfied. If f is an open immersion and $Y \setminus f(X)$ is an analytic set, then the finiteness theorem for local analytic cohomology from Chapter II gives conditions when the assertion (*ii*) holds.

The second case when the theorem can be applied is when f is proper.

Let $(X, \mathcal{O})$ be a complex space, let $\mathcal{I} \subset \mathcal{O}$ be a coherent ideal and $\mathcal{F} \in \mathbf{Coh}(X)$. Suppose there are sections $f_1, \ldots, f_N$ of $\Gamma(X, \mathcal{O})$ which generate $\mathcal{I}$. These yield naturally a structure of graded $\mathcal{O}[T]$-module on $\bigoplus_{k \geq 0} \mathcal{I}^k \mathcal{F} (T = (T_1, \ldots, T_N))$. We also remark that there exists a canonical epimorphism $\mathcal{O}[T] \to \bigoplus_{k \geq 0} \mathcal{I}^k \to 0$.

LEMMA 4.3. $\bigoplus_{k \geq 0} \mathcal{I}^k \mathcal{F}$ *is a coherent* $\mathcal{O}[T]$*-module; in particular,* $\bigoplus_{k \geq 0} \mathcal{I}^k$ *is a coherent sheaf of rings.*

Proof. Let $\mathcal{M} = \bigoplus_{k \geq 0} \mathcal{I}^k \mathcal{F}$. Consider a Stein semianalytic compact Q of X. Obviously,

$$\Gamma(Q, \mathcal{M}) = \bigoplus_{k \geq 0} \Gamma(Q, \mathcal{I}^k \mathcal{F}) = \bigoplus_{k \geq 0} \Gamma(Q, \mathcal{I})^k\, \Gamma(Q, \mathcal{F}).$$

$\Gamma(Q, \mathcal{M})$ is a module of finite type over the noetherian ring $\mathcal{O}(Q)[T]$, hence there exists an exact sequence of the form

$$\mathcal{O}(Q)[T]^p \to \mathcal{O}(Q)[T]^q \to \Gamma(Q, \mathcal{M}) \to 0.$$

We get a sequence of $\mathcal{O}[T]$-modules on Q and if we prove that this is exact, then the lemma will follow from 3.2. Consider a Stein compact $Q' \subset Q$; by lemma 1.3 we get exact sequence

$$\mathcal{O}(Q')[T]^p \to \mathcal{O}(Q')[T]^q \to \Gamma(Q', \mathcal{M}) \to 0, \text{ etc.}$$

THEOREM 4.4. *Let* $f \colon X \to Y$ *be a proper morphism of complex spaces,* $\mathcal{J} \subset \mathcal{O}_Y$ *a coherent ideal,* $\mathcal{I} = f^*(\mathcal{J})\, \mathcal{O}_X$ *and* $\mathcal{F}$ *a coherent sheaf on* X. *Under these assumptions,* $\bigoplus_{k \geq 0} R^q f_*(\mathcal{I}^k \mathcal{F})$ *is a coherent* $\bigoplus_{k \geq 0} \mathcal{J}^k$ *-module for any integer* q.

If Q *is a Stein compact of* Y, *then there exists an integer* $k_0 = k_0(Q)$ *such that*

$$H^q(f^{-1}(Q), \mathcal{I}^{k+r}\mathcal{F}) = \Gamma(Q, \mathcal{I})^r \cdot H^q(f^{-1}(Q), \mathcal{I}^k \mathcal{F})$$

for all integers $q \geq 0$, $r \geq 0$, $k \geq k_0$.

Proof. The problem is local on Y, hence we may assume that there exist sections $f_1, \ldots, f_N \in \Gamma(Y, \mathcal{J})$ such that $(f_1, \ldots, f_N)\, \mathcal{O}_Y = \mathcal{J}$. Then $(f_1, \ldots, f_N)\, \mathcal{O}_Y = \mathfrak{I}$ (we have denoted also by $f_1, \ldots, f_N$ the images under the morphism $\Gamma(Y, \mathcal{O}_Y) \to \Gamma(X, \mathcal{O}_X)$).

Let $\mathcal{M} = \bigoplus_{k\geqslant 0} \mathfrak{I}^k\mathcal{F}$. By lemma 4.3, $\mathcal{M}$ has a natural structure of coherent $\mathcal{O}_X\,[T]$-module. By the finiteness theorem 3.1, the sheaves $R^qf_*(\mathcal{M})$ are $\mathcal{O}_Y\,[T]$-coherent. The first assertion of the theorem follows from 4.3, since $R^\bullet f_*(\mathcal{M}) \simeq \bigoplus_{k\geqslant 0} R^\bullet f_*(\mathfrak{I}^k\mathcal{F})$.

Let Q be a Stein compact of Y. There exist canonical isomorphisms

$$H^\bullet(f^{-1}(Q),\, \mathfrak{I}^k\mathcal{F}) \simeq \Gamma(Q, R^\bullet f_*(\mathfrak{I}^k\mathcal{F})), \quad k \geqslant 0.$$

Then it easily follows that, in order to prove the second assertion of the theorem, we may suppose Q sufficiently small. Then by the coherence of the $\mathcal{O}_Y\,[T]$-modules $R^\bullet f_*(\mathcal{M})$, the $\mathcal{O}(Q)\,[T]$-modules

$$\Gamma(Q, R^\bullet f_*(\mathcal{M})) \simeq \bigoplus_{k\geqslant 0} \Gamma(Q,\, R^\bullet f_*(\mathfrak{I}^k\mathcal{M})) \simeq \bigoplus_{k\geqslant 0} H^\bullet(f^{-1}(Q),\, \mathfrak{I}^k\mathcal{F})$$

are of finite type and the required conclusion is obtained easily.

From theorem 4.1 and its complement we get the following

COROLLARY 4.5. *Let $f\colon X \to Y$ be a proper morphism of complex spaces, Y' a closed analytic subset of Y, $X' = f^{-1}(Y')$, $\mathcal{J} \subset \mathcal{O}_Y$ an ideal of definition of Y', $\mathfrak{I} = f^*(\mathcal{J})\mathcal{O}_X$ and $\mathcal{F} \in$ **Coh** (X). For an integer $k \geqslant 0$, denote $\mathcal{F}^{(k)} = \mathcal{F}/\mathfrak{I}^{k+1}\mathcal{F}$. Then:*

(*i*) *The morphisms from the diagram*

$$\begin{array}{ccc} i_Y^*(R^qf_*(\mathcal{F})) & \xrightarrow{\rho_q} & R^q\hat{f}_*(\hat{\mathcal{F}}) \\ \downarrow{\scriptstyle r_q} & & \downarrow{\scriptstyle \psi_q} \\ R^qf_*(\mathcal{F})^\wedge & \xrightarrow{\varphi_q} & \varprojlim_k R^q f_*(\mathcal{F}^{(k)}) \end{array}$$

are isomorphisms ($q \geqslant 0$).

(*ii*) *For each compact Q of Y there exists an integer $k_0 = k_0(Q)$ such that one has on Q*

$$\operatorname{Ker}(R^qf_*(\mathcal{F}) \to R^qf_*(\mathcal{F}^{(k+r)})) = \mathcal{J}^r \operatorname{Ker}(R^qf_*(\mathcal{F}) \to R^qf_*(\mathcal{F}^{(k)})),$$

$$\operatorname{Im}(R^qf_*(\mathcal{F}^{(k+r)}) \to R^qf_*(\mathcal{F}^{(r)})) = \operatorname{Im}(R^qf_*(\mathcal{F}^{(k_0+r)}) \to R^qf_*(\mathcal{F}^{(r)}))$$

for $r \geqslant 0$, $q \geqslant 0$ and $k \geqslant k_0$.

Remark. The isomorphism $R^\bullet f_*(\mathcal{F})^\wedge \xrightarrow{\rho\cdot r^{-1}} R^q\hat{f}_*(\hat{\mathcal{F}})$ shows that the direct image functors $R^\bullet f_*$ commute with the completion; for these reasons theorem 4.1 is called a comparison theorem (between the analytic theory and the formal theory).

In particular, from theorem 4.4 we get the following

COROLLARY 4.6. *Let $f: X \to Y$ be a proper morphism of complex spaces and let y be a point of Y and $\mathfrak{F} \in$ **Coh** (X). Then there exists an integer k_0 such that*

$$H^q(f^{-1}(y),\ \hat{\mathfrak{m}}_y^{k+r}\mathfrak{F}) = \mathfrak{m}_y^r H^q(f^{-1}(y),\ \hat{\mathfrak{m}}_y^k\mathfrak{F})$$

for all integers $q \geqslant 0$, $r \geqslant 0$ and $k \geqslant k_0$.

From 4.5 we find again, in a little stronger form (the explanation of the function F), theorem II 3.1.

COROLLARY 4.7. *Let $f: X \to Y$ be a proper morphism of complex spaces, y a point of Y and $\mathfrak{F} \in$ **Coh** (X). Denote $X_y = f^{-1}(y)$ and $\mathfrak{F}^{(k)} = \mathfrak{F}/\hat{\mathfrak{m}}_y^{k+1}\mathfrak{F}$. Then:*

(i) The canonical morphism

$$R^q f_*(\mathfrak{F})_y^{\wedge} \to \lim_{\overleftarrow{k}} H^q(X_y, \mathfrak{F}^{(k)})$$

is an isomorphism for any $q \geqslant 0$ (the completion is taken with respect to the $\mathfrak{m}_y$-adic topology).

(ii) There exists an integer k_0 such that

$$\operatorname{Ker}(H^q(X_y, \mathfrak{F}) \to H^q(X_y, \mathfrak{F}^{(k+r)})) = \mathfrak{m}_y^r \operatorname{Ker}(H^q(X_y, \mathfrak{F}) \to H^q(X_y, \mathfrak{F}^{(k)})),$$

$$\operatorname{Im}(H^q(X_y, \mathfrak{F}^{(k+r)}) \to H^q(X_y, \mathfrak{F}^{(r)})) = \operatorname{Im}(H^q(X_y, \mathfrak{F}^{(k_0+r)}) \to H^q(X_y, \mathfrak{F}^{(r)}))$$

for all $r \geqslant 0$, $q \geqslant 0$ and $k \geqslant k_0$.

Proof. We have to check only the isomorphisms

$$(R^q f_*(\mathfrak{F})^{\wedge})_y \simeq (R^q f_*(\mathfrak{F})_y)^{\wedge},$$

where the first completion is with respect to the ideal-sheaf defined by y and the second one with respect to the $\mathfrak{m}_y$-adic topology. This fact is achieved in following way: if V is a Stein neighbourhood of y and $\mathfrak{m}(y)$ is the maximal ideal of $\Gamma(V, \mathcal{O}_Y)$ which corresponds to y, then

$$\Gamma(V, R^{\bullet} f_*(\mathfrak{F}))/\mathfrak{m}(y)^{k+1}\,\Gamma(V, R^{\bullet} f_*(\mathfrak{F})) \simeq R^{\bullet} f_*(\mathfrak{F})_y/\mathfrak{m}_y^{k+1} R^{\bullet} f_*(\mathfrak{F})_y,$$

the sheaves $R^{\bullet} f_*(\mathfrak{F})$ being coherent.

Bibliographical indications

The results from § 2 are taken from [37] and the comparison theorem 4.1 from ([36], SGA 2, Exp. IX). The other results in this chapter are due to the first author [6].

Chapter VII

Duality on complex spaces

Introduction

Let X be a Riemann compact surface, Ω the sheaf of germs of holomorphic forms of type (1, 0) and $\mathfrak{L}$ an invertible sheaf on X. A classical result (Roch's half of the Riemann-Roch theorem) shows that the complex vectorial spaces $H^0(X, \mathfrak{L})$ and $H^1(X, \check{\mathfrak{L}} \otimes \Omega)$ are in duality. The generalization of this result for arbitrary dimension is the following duality theorem of Serre [75]:

"Let X be a compact complex manifold of dimension n, let Ω be the sheaf of germs of holomophic forms of maximal degree and $\mathfrak{F} \in \mathbf{Coh}(X)$ locally free. Then the complex (finite-dimensional) vectorial spaces $H^p(X, \mathfrak{F})$ and $H^{n-p}(X, \check{\mathfrak{F}} \otimes \Omega)$ are in algebraic duality".

If $\mathfrak{F}$ is an arbitrary coherent sheaf, then $H^{n-p}(X, \check{\mathfrak{F}} \otimes \Omega)$ should be replaced by $\mathrm{Ext}^{n-p}(X; \mathfrak{F}, \Omega)$. If the manifold X is not compact, then the invariants must be topologized and the algebraic duality becomes a topological one (moreover, H^{n-p} must be replaced by H_c^{n-p}).

The aim of this chapter is to extend the duality to the singular case. When compared with the case of manifolds, the difficulties are considerable. Grothendieck overcame them in algebraic geometry [39] and Ramis and Ruget for complex spaces.

Another way of studying the nonsingular case can be found in Andreotti and Kas' paper [4].

Let X be a complex space. In § 2 we construct a complex $K_X^\bullet$ of $\mathcal{O}_X$-modules which is called the dualizing complex and which plays the part of the sheaf Ω. The Ext's, whose arguments are complexes of sheaves defined in § 1, B, allow us to lend a meaning to the invariants $\mathrm{Ext}^\bullet(X; \mathfrak{F}, K_X^\bullet)$, $\mathfrak{F} \in \mathbf{Coh}(X)$. In the case of complex manifolds, the couple which achieves the duality can be obtained from the exterior product of the forms and from the trace map $H_c^n(X, \Omega) \to \mathbb{C}$, which is deduced from the integration of the forms of maximal degree. In the singular case one defines similarly a trace map $H_c^0(X, K_X^\bullet) \to \mathbb{C}$. The last mentioned map and the Yoneda bilinear maps define bilinear maps

$$\mathrm{H}^q(X, \mathfrak{F}) \times \mathrm{Ext}_c^{-p}(X; \mathfrak{F}, K_X^\bullet) \to \mathbb{C},$$

$$H_c^p(X, \mathfrak{F}) \times \mathrm{Ext}^{-p}(X; \mathfrak{F}, K_X^\bullet) \to \mathbb{C}.$$

In § 3 one defines naturally some topologies on these invariants (generally nonseparated) and proves that these pairs achieve topological dualities between the associated separated spaces (theorems 3.7 and 3.10). The separation theorems concerning the above invariants, which are proved in [63], stand for a complement to these results and they are necessary for applications.

In § 4 we consider the case of manifolds. The proofs given in § 3 are resumed in this case (now the simplifications are essential since the dualizing complex and the formalism of the trace map are not in the least necessary). Moreover, we provide a minute presentation of Serre and Malgrange's proof by means of the Dolbeault resolutions (which, however, makes use of the results from [54]).

The chapter is concluded by the presentation of the dualizing sheaves introduced by Andreotti and Kas [4].

In § 1 we give the construction of the Cousin complex of a ring and the definition of the derived functors of functors whose arguments are complexes of objects and we also present some topological facts required in the study of duality.

§ 1. Preliminaries

(A) In this section we prove some facts concerning the local cohomology of the rings and make the construction of the Cousin complex ([36] SGA 2, [39]).

(1) Fix a topological space X. Let $Z \subset X$ be an arbitrary subset. For any $\mathcal{F} \in \mathbf{Ab}(X)$, denote

$$\Gamma_Z(X, \mathcal{F}) = \{s \in \Gamma(X, \mathcal{F}) \mid \text{Supp } s \subset Z\} \text{ and}$$

$\underline{\Gamma}_Z(\mathcal{F})$ = the sheaf associated to the presheaf $U \mapsto \Gamma_{U \cap Z}(U, \mathcal{F})$. Thus, we get two functors and the associated derived functors are denoted by $H_Z^{\boldsymbol{\cdot}}(X, \mathcal{F})$ and $\mathcal{H}_Z^{\boldsymbol{\cdot}}(\mathcal{F})$ respectively. H_Z^0 coincides with Γ_Z and $\mathcal{H}_Z^0$ with $\underline{\Gamma}_Z$. In Chapter II, § 1, we have considered the particular case when Z is a locally closed subset (moreover, Γ_Z and $\underline{\Gamma}_Z$ were defined in a little different way; if Z is closed set both definitions coincide).

Let now $Z' \subset Z$ be two subsets of Z. For any sheaf of abelian groups $\mathcal{F}$ on X, denote

$$\Gamma_{Z/Z'}(X, \mathcal{F}) = \Gamma_Z(X, \mathcal{F})/\Gamma_{Z'}(X, \mathcal{F}).$$

A morphism $\mathcal{F} \to \mathcal{G}$ induces naturally a homorphism of groups

$$\Gamma_{Z/Z'}(X, \mathcal{F}) \to \Gamma_{Z/Z'}(X, \mathcal{G}).$$

It is obvious that these correspondences yield a functor

$$\Gamma_{Z/Z'}(X, *) : \mathbf{Ab}(X) \to \mathbf{Ab}$$

and its (right) derived functors are denoted by $H^{\bullet}_{Z/Z'}(X, *)$. Generally, the functor $\Gamma_{Z/Z'}$ is not left exact and hence $H^0_{Z/Z'}$ can be different from $\Gamma_{Z/Z'}$.

Now, define invariants of local nature which are associated to the inclusion $Z' \subset Z$. If $\mathcal{F} \in \mathbf{Ab}(X)$, by $\underline{\Gamma}_{Z/Z'}(\mathcal{F})$ we mean the sheaf associated to the presheaf $U \mapsto \Gamma_{U\cap Z/U\cap Z'}(U, \mathcal{F})$. One obtains a functor

$$\underline{\Gamma}_{Z/Z'}: \mathbf{Ab}(X) \to \mathbf{Ab}(X)$$

and its derived functors are denoted by $\mathcal{H}^{\bullet}_{Z/Z'}$. For any $\mathcal{F} \in \mathbf{Ab}(X)$, one gets exact sequences

$$0 \to H^0_{Z'}(X, \mathcal{F}) \to H^0_Z(X, \mathcal{F}) \to H^0_{Z/Z'}(X, \mathcal{F}) \to H^1_{Z'}(X, \mathcal{F}) \to \dots,$$

$$0 \to \mathcal{H}^0_{Z'}(\mathcal{F}) \to \mathcal{H}^0_Z(\mathcal{F}) \to \mathcal{H}^0_{Z/Z'}(\mathcal{F}) \to \mathcal{H}^1_{Z'}(\mathcal{F}) \to \dots$$

Finally, we will define invariants of punctual nature, which will allow us to compute the invariants $\mathcal{H}^{\bullet}_{Z/Z'}$. Let x be a point of X. If $\mathcal{F} \in \mathbf{Ab}(X)$, we may consider the subgroup $\Gamma_x(\mathcal{F})$ of $\mathcal{F}_x$ formed by the element 0 and by those elements s_x which admit a representative s in some neighbourhood U of x such that $\operatorname{Supp} s = \overline{\{x\}} \cap U$. The derived functors of the functor thus obtained are denoted by $\mathcal{H}^{\bullet}_x(\mathcal{F})$. One can easily see that

$$H^{\bullet}_x(\mathcal{F}) \simeq (\mathcal{H}^{\bullet}_Z(\mathcal{F}))_x, \quad \text{if } Z = \overline{\{x\}}.$$

Recall some topological facts ([10], Ch. II, § 4 and [37], 0_I), which are straightforward consequences of the definitions. In fact, we have already used some of them (IV.3.1, V.1.13).

A topological space is called *noetherian* if any descending sequence of closed subsets is stationary. Any open subset of such a space is also noetherian. A topological space is called *locally noetherian* if any point owns a neighbourhood which is a noetherian space.

The topological space X is called *irreducible* if it cannot be written as a union of two of its proper closed subsets. A part of X is called *irreducible* if it is an irreducible space with respect to the induced topology. A part of X is irreducible if and only if its closure enjoies this property. If X is noetherian, then any closed subset can be written as a finite union of irreducible closed subsets and whenever this representation is reduced, it is unique (modulo order).

If $Y \subset X$ is a closed set, then a point $y \in Y$ is called *generic* if $Y = \overline{\{y\}}$. A topological space is called *sober* if any irreducible closed subset has a generic point and only one. Any open set of such a space is also sober. The space X is said to be *locally sober* if each of its points owns a neighbourhood which is sober with respect to the induced topology.

A *specialization* of a point x is a point x' such that $x' \in \overline{\{x\}}$. We express this symbolically by $x \to x'$. A subset $Z \subset X$ is called *stable to specialization* if, whenever $x \in Z$ and $x \to x'$, $x' \in Z$.

PROPOSITION 1.1. *Let X be a locally noetherian topological space, which is also locally sober. Consider two subsets $Z' \subset Z$ stable to specialization and so that any point of $Z \setminus Z'$ is maximal in Z ($x \in Z \setminus Z'$, $x \to x' \Rightarrow$ either $x' = x$ or $x' \in Z'$). Under these assumptions, there exist functorial isomorphisms*

$$\mathcal{H}^{\bullet}_{Z/Z'}(\mathfrak{F}) \simeq \bigoplus_{x \in Z \setminus Z'} i_x(H^{\bullet}_x(\mathfrak{F})), \quad \mathfrak{F} \in \mathbf{Ab}(X),$$

where for a group G, $i_x(G)$ means the sheaf on X which is equal to G on $\overline{\{x\}}$, and is null on $X \setminus \overline{\{x\}}$.

Proof. It will be enough to prove that for any $\mathfrak{F} \in \mathbf{Ab}(X)$ there exists a functorial isomorphism

$$\underline{\Gamma}_{Z/Z'}(\mathfrak{F}) \simeq \bigoplus_{x \in Z \setminus Z'} i_x(\Gamma_x(\mathfrak{F})).$$

For any open set U which is sufficiently small (hence noetherian and sober), define the map

$$\Gamma_{Z \cap U}(U, \mathfrak{F}) \to \bigoplus_{x \in (Z \setminus Z') \cap U} \Gamma_x(\mathfrak{F}),$$

which associates to a section the system of its germs in the points of $(Z \setminus Z') \cap U$. The definition is meaningfull: indeed, if $s \in \Gamma_{Z \cap U}(U, \mathfrak{F})$, then by the hipothesis $\operatorname{Supp} s = \overline{\{z_1\}} \cup \ldots \cup \overline{\{z_n\}}$ and one can easily see that no element $z' \in (Z \setminus Z') \cap U$, which is distinct of $z_1, \ldots, z_n$, belongs to $\operatorname{Supp} s$.

A section s has null image if and only if it belongs to $\Gamma_{Z' \cap U}(U, \mathfrak{F})$. Thus, we obtain an injective map

$$\Gamma_{Z \cap U}(U, \mathfrak{F})/\Gamma_{Z' \cap U}(U, \mathfrak{F}) \to \bigoplus_{x \in (Z \setminus Z') \cap U} \Gamma_x(\mathfrak{F}).$$

One then derives a monomorphism of sheaves

$$\underline{\Gamma}_{Z/Z'}(\mathfrak{F}) \to \bigoplus_{x \in Z \setminus Z'} i_x(\Gamma_x(\mathfrak{F})),$$

which is functorial in $\mathfrak{F}$. It remains only to prove that this is an epimorphism. It is sufficient to check that any germ s_x of $\Gamma_x(\mathfrak{F})$, $x \in Z \setminus Z'$, has a preimage; but by the definition of $\Gamma_x(\mathfrak{F})$ there exists a representative s on a neighbourhood U such that $\operatorname{Supp} s = \overline{\{x\}} \cap U$, etc.... The proposition is proved.

Consider a filtration

$$X = Z^0 \supset Z^1 \supset Z^2 \supset \ldots$$

of X by arbitrary subsets. Associate to any $\mathfrak{F} \in \mathbf{Ab}(X)$ a complex of sheaves on X, which is called the Cousin complex with respect to this filtration. Let $\mathfrak{J}^{\cdot}$ be an injective resolution of $\mathfrak{F}$. Deduce a filtration of complexes of sheaves

$$\mathfrak{J}^{\bullet} = \underline{\Gamma}_{Z^0}(\mathfrak{J}^{\bullet}) \supset \underline{\Gamma}_{Z^1}(\mathfrak{J}^{\bullet}) \supset \underline{\Gamma}_{Z^2}(\mathfrak{J}^{\bullet}) \supset \ldots$$

For any $p \geqslant 0$ we have an exact sequence

$$0 \to \Gamma_{Z^{p+1}}(\mathcal{J}^\bullet) \to \Gamma_{Z^p}(\mathcal{J}^\bullet) \to \Gamma_{Z^p/Z^{p+1}}(\mathcal{J}^\bullet) \to 0,$$

hence an exact sequence

$$\ldots \to \mathcal{H}^p_{Z^p}(\mathcal{F}) \to \mathcal{H}^p_{Z^p/Z^{p+1}}(\mathcal{F}) \to \mathcal{H}^{p+1}_{Z^{p+1}}(\mathcal{F}) \to \ldots$$

For the index $p+1$ we obtain similarly the exact sequence

$$\ldots \to \mathcal{H}^{p+1}_{Z^{p+1}}(\mathcal{F}) \to \mathcal{H}^{p+1}_{Z^{p+1}/Z^{p+2}}(\mathcal{F}) \to \mathcal{H}^{p+2}_{Z^{p+2}}(\mathcal{F}) \to \ldots$$

From these two exact sequences one obtains by composition a morphism

$$\mathcal{H}^p_{Z^p/Z^{p+1}}(\mathcal{F}) \to \mathcal{H}^{p+1}_{Z^{p+1}/Z^{p+2}}(\mathcal{F}).$$

From the above exact sequences written for $p = 0$, one obtains a morphism

$$\mathcal{F} \to \mathcal{H}^0_{Z^0/Z^1}(\mathcal{F}).$$

One can easilly prove that these morphisms do not depend on the resolution $\mathcal{J}^\bullet$ and they have a functorial character in $\mathcal{F}$.

PROPOSITION 1.2. *Let X be a topological space, $X = Z^0 \supset Z^1 \supset Z^2 \supset \ldots$ a filtration and $\mathcal{F} \in$ **Ab**(X). Under these assumptions, the above constructions yield a complex of sheaves on X*

$$0 \to \mathcal{H}^0_{Z^0/Z^1}(\mathcal{F}) \to \mathcal{H}^1_{Z^1/Z^2}(\mathcal{F}) \to \mathcal{H}^2_{Z^2/Z^3}(\mathcal{F}) \to \ldots,$$

together with an augmentation morphism $\mathcal{F} \to \mathcal{H}^0_{Z^0/Z^1}(\mathcal{F})$; moreover, these associations are functorial in $\mathcal{F}$.

If the filtration is finite and if $\mathcal{H}^q_{Z^p/Z^{p+1}}(\mathcal{F}) = 0$ for $q \neq p$, then the above complex is a resolution of $\mathcal{F}$.

Proof. Only the last assertion presents some difficulty. We first show that $\mathcal{H}^q_{Z^p}(\mathcal{F}) = 0$ for $q < p$. Proceed by descending induction on p. For p sufficiently large, $Z^p = \emptyset$ and the assertion is obvious. The induction step follows in virtue of the exact sequence

$$\ldots \to \mathcal{H}^q_{Z^{p+1}}(\mathcal{F}) \to \mathcal{H}^q_{Z^p}(\mathcal{F}) \to \mathcal{H}^q_{Z^p/Z^{p+1}}(\mathcal{F}) \to \ldots$$

Now we prove that $\mathcal{H}^q_{Z^p}(\mathcal{F}) = 0$ for $q > p$. Proceed by ascending induction on p. The case $p = 0$ is clear since $Z^0 = X$. The induction step will result from the exact sequence

$$\ldots \to \mathcal{H}^{q-1}_{Z^{p-1}/Z^p}(\mathcal{F}) \to \mathcal{H}^q_{Z^p}(\mathcal{F}) \to \mathcal{H}^q_{Z^{p-1}}(\mathcal{F}) \to \ldots$$

Consequently, $\mathcal{H}^q_{Z^p}(\mathcal{F}) = 0$ for $q \neq p$ and the proof of the proposition can be completed by use of the definition merely of the morphisms $\mathcal{F} \to \mathcal{H}^0_{Z^0/Z^1}(\mathcal{F})$, $\mathcal{H}^p_{Z^p/Z^{p+1}}(\mathcal{F}) \to \mathcal{H}^{p+1}_{Z^{p+1}/Z^{p+2}}(\mathcal{F})$.

(2) Concerning the properties of the Zariski topology associated to a ring one can consult ([10], Ch. II, § 4 or [37], Ch. I, § 1.1). Recall the following facts. Let A be a commutative and unitary ring. Spec A is a sober space. The irreducible closed subsets are of the form $V(\mathfrak{p})$, $\mathfrak{p}$ being a prime ideal of A. Here $\mathfrak{p}$ stands for the unique generic point of $V(\mathfrak{p})$. For an element $f \in A$ one may consider *the main open set* $D(f) = \{\mathfrak{p} \in \operatorname{Spec} A \mid f \notin \mathfrak{p}\}$, which is identified with Spec A_f. The family $D(f)$, $f \in A$, constitutes a basis of open sets in Spec A. If f and g are elements of A, then $D(f) \cap D(g) = D(fg)$. Any open covering of Spec A can be refined to a finite covering with main open sets. Moreover, if $f_1, \ldots, f_r$ are elements of A such that $\operatorname{Spec} A = \bigcup_{i=1}^{r} D(f_i)$, then the ideal generated by these elements equals A.

If A is noetherian, then Spec A is a noetherian space and consequently, any open set of Spec A is a finite union of main open sets.

We denote for convenience $X = \operatorname{Spec} A$ and $X_f = D(f)$, $f \in A$. Let M be an A-module. For an element f of A, M_f means the module of quotients with respect to the multiplicative system of the powers f^n, $n \geqslant 0$. M_f is an A_f-module. If $D(f) \subset D(g)$, then a suitable power of f lies in the ideal generated by g. One then derives a canonical morphism $M_g \to M_f$. In particular, if $D(f) = D(g)$, then $M_g \simeq M_f$.

Let us denote by $\tilde{M}$ the sheaf associated to the presheaf on X

$$X_f \mapsto M_f, \ X_f \subset X_g \mapsto M_g \to M_f.$$

If $M = A$ one obtains a sheaf of rings $\tilde{A}$. $\tilde{M}$ has a natural structure of $\tilde{A}$-module. One can easily check that the stalk of $\tilde{M}$ in the point which corresponds to the prime ideal $\mathfrak{p}$ is canonically isomorphic to the module of quotients $M_{\mathfrak{p}}$.

THEOREM 1.3. ([37], Ch. I, 1.3.7). *For any A-module M and for any $f \in A$, the morphism*

$$\theta_f : M_f \to \Gamma(X_f, \tilde{M})$$

is bijective. In particular, M can be identified with $\Gamma(X, \tilde{M})$ by θ_1.

Proof. We first show that θ_f is injective. Let $s \in M_f$ be so that $\theta_f(s) = 0$. For each ideal $\mathfrak{p} \in X_f = D(f)$ the image of s under the morphism $M_f \to M_{\mathfrak{p}}$ is null. Consequently, there exists $h \notin \mathfrak{p}$ such that $hs = 0$. Thus, the annihilator of s (consider M_f as an A_f-module) is not contained in any ideal of Spec A_f; therefore, it equals A_f. Then $s = 0$.

Now we prove the surjectivity. Notice the following: if g is an element of A such that $D(g) \subset D(f)$, then M_g is canonically isomorphic to the module of quotients $(M_f)_{\bar{g}}$, where $\bar{g}$ is the image of g in A_f ([10], Ch. II, § 2, n° 3, prop. 7).

If we identify X_f by Spec A_f, then by this remark there results that the sheaf $\tilde{M}|X_f$ can be canonically identified with $(M_f)^\sim$. As a consequence, we reduce the problem to proving the surjectivity of the map $\theta = \theta_1$. Let $s \in \Gamma(X, \tilde{M})$. There exist a finite covering $(D(f_i))_{i \in I}$ of X and elements $s_i = t_i'/f_i^{n_i'}$ of M_{f_i} such that $\theta_{f_i}(s_i) = s\,|X_{f_i}$. We may assume that all integers n_i' are equal to an integer n'. The injectivity of the maps θ_f assures the existence of an integer m_{ij} such that $(f_if_j)^{m_{ij}}(f_j^{n'}t_i' - f_i^{n'}t_j') = 0$ for each pair (i,j). Since I is finite, we can assume all m_{ij} equal to an integer m.

Let $n = n' + m$ and $t_i = f_i^m t_i'$. Obviously, $s_i = t_i/f_i^n$, $f_j^n t_i = f_i^n t_j$. We also have $D(f_i^n) = D(f)$, $i \in I$. Since $D(f_i)$ cover X, the ideal generated by the elements f_i^n coincides with A. Therefore, there exist elements $g_i \in A$ so that $\Sigma g_i f_i^n = 1$. The element $t = \Sigma g_i t_i$ of M has the property $\theta(t) = s$ (both sections coincide on any X_{f_i}) and the theorem is proved.

(3) Let A be a ring and $\underline{f} = (f_i)_{1 \leq i \leq r}$ a system of r elements of A. We will denote by $K.(\underline{f})$ the Koszul complex associated to $\underline{f}$ ([37], Ch. III, § 1 and [79], Ch. IV, A). Recall its definition. For every element f_i, consider the complex $K_\bullet(f_i)$ of components

$$K_0(f_i) = K_1(f_i) = A, \quad K_n(f_i) = 0 \text{ for } n \neq 0,1,$$

whose differentials are defined by the multiplication by f_i. Then $K_\bullet(\underline{f})$ is the tensor product of complexes $K_\bullet(f_1) \otimes K_\bullet(f_2) \otimes \ldots \otimes K_\bullet(f_r)$, and it is endowed with the total degree ([26], I.2.7).

Denote by e_i the unit element of A, regarded in $K_1(f_i)$.

Then $K_p(\underline{f})$ is a free A-module of base $e_{i_1} \otimes \ldots \otimes e_{i_p}$, $i_1 < i_2 < \ldots < i_p$; in particular, it is isomorphic to the exterior product $\wedge^p(A^r)$. $K_1(\underline{f})$ is isomorphic to A^r and by this isomorphism the elements $e_1, \ldots, e_r$ are identified with the canonical basis of A^r. Via this identification we shall write sometimes $e_{i_1} \wedge \ldots \wedge e_{i_p}$ instead of $e_{i_1} \otimes \ldots \otimes e_{i_p}$. The differential d of $K_\bullet(\underline{f})$ is determined by the formula

$$d(e_{i_1} \otimes \ldots \otimes e_{i_p}) = \sum_k (-1)^{k+1} f_{i_k} e_{i_1} \otimes \ldots \otimes \hat{e}_{i_k} \otimes \ldots \otimes e_{i_p}.$$

In fact, one could construct the complex $K_\bullet(f)$ taking into account the formula $K_p(\underline{f}) = \wedge^p(A^r)$ and the above differentials.

For any A-module M define *the Koszul complex of chains*

$$K_\bullet(\underline{f}, M) = K_\bullet(\underline{f}) \otimes_A M$$

and *the Koszul complex of cochains*

$$K^\bullet(\underline{f}, M) = \operatorname{Hom}_A(K_\bullet(\underline{f}), M).$$

The module $K_p(\underline{f}, M)$ is the direct sum of the modules $e_{i_1}\otimes \ldots \otimes e_{i_p}\otimes M$, $i_1 < < i_2 < \ldots < i_p$ and the differential $d_p : K_p(\underline{f}, M) \to K_{p-1}(\underline{f}, M)$ is given by the formula

$$d(e_{i_1}\otimes \ldots \otimes e_{i_p}\otimes m) = \sum_k (-1)^{k+1} e_{i_k}\otimes \ldots \otimes \hat{e}_{i_k}\otimes \ldots \otimes e_{i_p}\otimes (f_{i_k}m).$$

The elements of $K^p(\underline{f}, M)$ are identified with the alternating maps from $[1, r]^p$ to M by the correspondence

$$g \in K^p(\underline{f}, M) \mapsto g(i_1, \ldots, i_p) = g(e_{i_1}\wedge \ldots \wedge e_{i_p}).$$

By this identification, the differential $d^p : K^p(\underline{f}, M) \to K^{p+1}(\underline{f}, M)$ becomes

$$(d^p g)(i_1, i_2, \ldots, i_{p+1}) = \sum_{k=1}^{p+1} (-1)^{k-1} f_{i_k} g(i_1, \ldots, \hat{i}_k, \ldots, i_{p+1}).$$

Define

$$H_{\bullet}(\underline{f}, M) = H_{\bullet}(K_{\bullet}(\underline{f}, M)),$$

$$H^{\bullet}(\underline{f}, M) = H^{\bullet}(K^{\bullet}(\underline{f}, M)).$$

There exist A-isomorphisms $K_{\bullet}(\underline{f}, M) \simeq K^{\bullet}(\underline{f}, M)$ which associate to any chain $z = \Sigma(e_{i_1}\wedge \ldots \wedge e_{i_p})\otimes z_{i_1\ldots i_p}$ the cochain g_z given by $g_z(j_1, \ldots, j_{r-p}) = \varepsilon z_{i_1\ldots i_p}$, where $(j_k)_{1\leqslant k\leqslant r-p}$ is the strictly ascending sequence which is complementary to the sequence $(i_k)_{1\leqslant k\leqslant p}$ in $[1, r]$ and ε is the signature of the permutation $[1, r] \to \to (i_1, \ldots, i_p, j_1, \ldots, j_{r-p})$. One can verify the compatibility with the differentials, and thus obtain isomorphisms

$$H^p(\underline{f}, M) \xrightarrow{\sim} H_{r-p}(\underline{f}, M).$$

One can easily deduce the equalities

$$H^0(\underline{f}, M) = (0 : (f_1, \ldots, f_r))_M \text{ and } H^r(\underline{f}, M) = M/(f_1, \ldots, f_r)M.$$

We shall use the following result ([37], Ch. III, Prop. 1.1.4 or [79], Ch. IV, A, Prop. 2): *If $(f_1, \ldots, f_r)$ is M-regular, then $H^p(\underline{f}, M) = 0$ for $p \neq r$.*

PROPOSITION 1.4. *If $(f_1, \ldots, f_r)$ generate the whole A, then the complex $K_{\bullet}(\underline{f})$ is homotopically trivial. Moreover, for any A-module M, the complexes $K_{\bullet}(\underline{f}, M)$ and $K^{\bullet}(\underline{f}, M)$ are also homotopically trivial.*

Proof. The assertions with respect to the module M follow from the first assertion by additivity. Let $g_1, \ldots, g_r$ be so that $1 = f_1g_1 + \ldots + f_rg_r$. The morphisms $K_p(\underline{f}) \to K_{p+1}(\underline{f})$, given by the formulae

$$z = \Sigma(e_{i_1}\wedge \ldots \wedge e_{i_p})\, z_{i_1\ldots i_p} \mapsto \Sigma(e_{i_1}\wedge \ldots \wedge e_{i_{p+1}})\, t_{i_1\ldots i_{p+1}},$$

where $t_{i_1\ldots i_{p+1}} = \sum_{k=1}^{p+1} (-1)^k g_{i_k} z_{i_1\ldots\hat{i}_k\ldots i_p}$, define a homotopy between the identity map and the null map of $K_\bullet(\underline{f})$.

For each integer $n \geqslant 0$ we denote by $\underline{f}^n$ the system $(f_1^n, \ldots, f_r^n)$. If $m \geqslant n \geqslant 0$, then the "multiplication" by $\underline{f}^{m-n}$ defines morphisms of complexes

$$\theta_{nm} : K_\bullet(\underline{f}^m) \to K_\bullet(\underline{f}^n).$$

If $K_\bullet(\underline{f}^m)$ and $K_\bullet(\underline{f}^n)$ are identified with the tensor products $K_\bullet(f_1^m) \otimes \ldots \otimes K_\bullet(f_r^m)$ and $K_\bullet(f_1^n) \otimes \ldots \otimes K_\bullet(f_r^n)$ respectively, then θ_{nm} is the tensor product of the morphisms $\theta_{nm}^i : K_\bullet(f_i^m) \to K_\bullet(f_i^n)$ which are reduced to the identity in degree 0 and to the multiplication by f_i^{m-n} in degree 1. As a matter of fact, if we denote by e_i^n the unit element of $K_1(f_1^n) = A$, then we may directly set

$$\theta_{nm}(e_{i_1}^m \wedge \ldots \wedge e_{i_p}^m) = (e_{i_1}^n \wedge \ldots \wedge e_{i_p}^n) f_{i_1}^{m-n} \ldots f_{i_p}^{m-n}.$$

The morphisms θ_{nm} yield morphisms of complexes

$$\theta^{mn} : K^\bullet(\underline{f}^n, M) \to K^\bullet(\underline{f}^m, M).$$

S o we obtain an inductive system and denote

$$C^\bullet((\underline{f}), M) = \varinjlim_n K^\bullet(\underline{f}^n, M).$$

$$H^\bullet((\underline{f}), M) = H^\bullet(C^\bullet((\underline{f}), M)) \simeq \varinjlim_n H^\bullet(\underline{f}^n, M).$$

By the previous proposition, we deduce

COROLLARY 1.5. *If $(f_1, \ldots, f_r)$ generate the whole A, then $H^\bullet(\underline{f}, M) = 0$ and $H^\bullet((\underline{f}), M) = 0$ for any A-module M.*

Indeed, for any $n \geqslant 0$, the ideal $(f_1^n, \ldots, f_r^n)$ coincides with A.

(4) We will use the above results to prove the triviality of the cohomology of the affine schemas. Use the previous notation. Let in addition $U_i = D(f_i)$, $U = \bigcup_{i=1}^r U_i$ and $\mathfrak{U}$ be the covering $(U_i)_{1 \leqslant i \leqslant r}$ of U. For any $(i_0, \ldots, i_p)$,

$$U_{i_0\ldots i_p} = \bigcap_{k=0}^p U_{i_k} = X_{f_{i_0}\ldots f_{i_p}}, \text{ hence } \Gamma(U_{i_0\ldots i_p}, \tilde{M}) = M_{f_{i_0}\ldots f_{i_p}}.$$

Fix a system $(i_0, \ldots, i_p)$. For any $n \geqslant 0$, denote $M_{i_0\ldots i_p}^{(n)} = M$. For $m \geqslant n$ consider the morphism

$$\varphi^{mn} : M_{i_0\ldots i_p}^{(n)} \to M_{i_0\ldots i_p}^{(m)},$$

which is given by the multiplication by $(f_{i_0} \ldots f_{i_p})^{m-n}$. One thus obtains an inductive system and it can be easily seen that $M_{f_{i_0} \ldots f_{i_p}}$ is identified with $\varinjlim_n M^{(n)}_{i_0 \ldots i_p}$ (one can consult [37], 0_{I}, 1.6).

If $C^p_{(n)}(M)$ means the set of the alternating maps from $[1, r]^{p+1}$ to M, then one can consider maps

$$\varphi^{mn} : C^p_{(n)}(M) \to C^p_{(m)}(M), \ m \geqslant n \geqslant 0,$$

as above. If $C^p(\mathfrak{U}, \tilde{M})$ stands for the group of the alternate p-cochains of $\mathfrak{U}$ with respect to the sheaf $\tilde{M}$, then by the previous ones we get isomorphisms

$$C^p(\mathfrak{U}, M) \xrightarrow{\sim} \varinjlim_n C^p_{(n)}(M).$$

But each $C^p_{(n)}(M)$ is identified with $K^{p+1}(\underline{f}^n, M)$ and by these identifications the maps θ^{mn} and φ^{mn} correspond to each other.

One thus obtains for any $p \geqslant 0$, an isomorphism

$$C^p(\mathfrak{U}, M) \simeq C^{p+1}((\underline{f}), M),$$

which is functorial in M. Moreover these isomorphisms are compatible with the coboundary operators.

PROPOSITION 1.6. *By the above notations, there exist isomorphisms functorial in M*

$$H^p(\mathfrak{U}, \tilde{M}) \simeq H^{p+1}((\underline{f}), M), \quad p \geqslant 1$$

and an exact sequence, which is functorial in M,

$$0 \to H^0((\underline{f}), M) \to M \to H^0(\mathfrak{U}, \tilde{M}) \to H^1((\underline{f}), M) \to 0.$$

Proof. The isomorphisms from the statement are deduced from the isomorphisms already established. On the other hand, $C^0(\mathfrak{U}, M) \simeq C^1((\underline{f}), M)$, and by this isomorphism $H^0(\mathfrak{U}, \tilde{M})$ is identified with the subgroup of cocycles of dimension 1 of $C^\bullet((\underline{f}), M)$. Also $M = C^0((\underline{f}), M)$ and the sequence from the proposition follows by the definitions.

THEOREM 1.7. ([37], Ch. III, 1.3.1). *Let A be a commutative and unitary ring, let M be an A-module and X =* Spec *A. Then* $H^p(X, \tilde{M}) = 0$ *for any* $p > 0$.

Proof. Let $\mathfrak{U}$ be a finite covering of X by main open sets $X_{f_i} = D(f_i)$, $1 \leqslant i \leqslant r$. The ideal generated by the elements $f_1, \ldots, f_r$ coincides with A. From 1.5 and 1.6 we then deduce that $H^p(\mathfrak{U}, \tilde{M}) = 0$ for $p > 0$. Since any open covering can be refined to a covering as above, it follows that the Čech cohomology groups $\check{H}^p(X, \tilde{M})$ vanish for $p \geqslant 1$. In particular, $\check{H}^p(X_{f_{i_0}} \cap \ldots \cap X_{f_{i_k}}, \tilde{M}) = 0$, for $p \geqslant 1$ and for any $(i_0, \ldots, i_k)$. The conclusion follows from ([26], Ch. II, 4.9.2).

COROLLARY 1.8. *For any open subset U of X, the cohomology groups $H^\bullet(U,\tilde{M})$ can be calculated by means of a covering with main open sets.*

(5) We now connect the Koszul complexes to the local cohomology.

THEOREM 1.9. *Let A be a commutative unitary ring, $\underline{f} = (f_1,\ldots,f_r)$ a finite family of elements of A, $X = \operatorname{Spec} A$, $Y =$ the closed subset of X defined by $\underline{f}$, and let M be an A-module. Under these assumptions there are functorial isomorphisms*

$$H_Y^\bullet(X, \tilde{M}) \simeq H^\bullet((\underline{f}), M) \left(\simeq \varinjlim_n H^\bullet(\underline{f}^n, M)\right).$$

Proof. By theorem 1.7 we get the exact sequence

$$0 \to H_Y^0(X, \tilde{M}) \to H^0(X, \tilde{M}) \to H^0(X \setminus Y, \tilde{M}) \to H_Y^1(X, \tilde{M}) \to 0$$

and isomorphisms

$$H_Y^p(X, \tilde{M}) \simeq H^{p-1}(X \setminus Y, \tilde{M}), \quad p \geqslant 2.$$

If 1.6 and 1.8 are applied to the open set $U = X \setminus Y$ and to the covering $\mathfrak{U} = (D(f_i))_{1 \leqslant i \leqslant r}$ of U we get the exact sequence

$$0 \to H^0((\underline{f}), M) \to M \to H^0(X \setminus Y, \tilde{M}) \to H^1((\underline{f}), M) \to 0$$

and isomorphisms

$$H^{p-1}(X \setminus Y, \tilde{M}) \simeq H^p((\underline{f}), M), \quad p \geqslant 2.$$

By making use of these facts (and of the isomorphism $M \simeq H^0(X, \tilde{M})$), the proof can be easily concluded.

COROLLARY 1.10. *The invariants $H^\bullet((\underline{f}), M)$ depend only on the closed set determined by the ideal generated by $f_1,\ldots,f_r$.*

Consider in what follows a special case. Suppose A is a local noetherian ring and let $\mathfrak{m}$ be its maximal ideal. We will denote $H_{\mathfrak{m}}^\bullet(M) = H_{\{\mathfrak{m}\}}^\bullet(\operatorname{Spec} A, \tilde{M})$. Every $H_{\mathfrak{m}}^p(M)$ has a natural structure of A-module.

COROLLARY 1.11. *If A is a Cohen-Macauley local ring of dimension n and $\mathfrak{m}$ is its maximal ideal, then $H_{\mathfrak{m}}^p(A) = 0$ for $p \neq n$.*

Proof. Let $f_1,\ldots,f_n \in \mathfrak{m}$ be a regular sequence. The ideal generated by these elements is an ideal of definition, hence the associated closed set $V((f_1,\ldots,f_n))$ s reduced to $\mathfrak{m}$. In virtue of theorem 1.9, $H_{\mathfrak{m}}^p(A) \simeq H^p((\underline{f}), A) \simeq \varinjlim_k H^p(\underline{f}^k, A)$.

For an integer $k \geqslant 0, \underline{f}^k = (f_1^k,\ldots,f_n^k)$ is a regular A-sequence; therefore, $H^p(\underline{f}^k, A) = 0$ whenever $p \neq n$.

Let A be a commutative and unitary ring and M an A-module. Recall that an injective A-module I together with a monomorphism $0 \to M \to I$ is called an *injective envelope of* M ([12], [59]) if the following condition is fulfilled: for any submodule P of I, $P \cap M = 0 \Rightarrow P = 0$. One proves that any module admits an injective envelope. A resolution $0 \to M \to I_0 \to I_1 \to I_2 \to \dots$ of M, where I_0 is the injective envelope of M and any I_{k+1} is the injective envelope of $\operatorname{Im}(I_k \to \to I_{k+1})$, $k \geqslant 0$, is called *a minimal injective resolution*. Its length coincides with the injective dimension of M. The next theorem requires the following assertion:

"Let A be a noetherian local ring, M an A-module and f an element of the maximal ideal of A, A-regular and M-regular. Then $\operatorname{di}_{A/fA}(M/fM) \leqslant \operatorname{di}_A(M) - 1$ (di = the injective dimension)".

The proof will proceed as follows ([60], vol. II, p. 233): Let $0 \to M \to I_0 \to \to I_1 \to I_2 \to \dots$ be a minimal injective resolution of M. Denote by N the image of the morphism $I_0 \to I_1$. Consider the exact commutative diagram

$$\begin{array}{ccccccccc} 0 & \to & M & \to & I_0 & \to & N & \to & 0 \\ & & \downarrow{\scriptstyle u} & & \downarrow{\scriptstyle v} & & \downarrow{\scriptstyle w} & & \\ 0 & \to & M & \to & I_0 & \to & N & \to & 0, \end{array}$$

where the vertical arrows are just the multiplications by f. By hypothesis, the first vertical arrow is injective. Since $\operatorname{Ker} v \cap M = 0$, it follows that $\operatorname{Ker} v = 0$, hence the map v is also injective. We show now its surjectivity. Let $y \in I_0$. Since $fA \simeq A$, there exists a morphism $\varphi : xA \to I_0$ such that $\varphi(x) = y$. From the injectivity of I_0, φ extends to a morphism $\psi : A \to I_0$. If $y' = \psi(1)$ then $fy' = y$.

Thereby w is also surjective. From the serpent lemma we deduce the isomorphism $\operatorname{Coker} u \simeq \operatorname{Ker} w$, hence $M/fM \simeq \operatorname{Hom}_A(A/fA, N)$. If I is an injective A-module, then $\operatorname{Hom}_A(A/fA, I)$ is an injective A/fA-module; for there exists a canonical identification $\operatorname{Hom}_{A/fA}(P, \operatorname{Hom}_A(A/fA, I)) \simeq \operatorname{Hom}_A(P, I)$ for any A/fA-module P.

Thereby the proof of the assertion will be completed as soon as we prove the exactness of the sequence

$$0 \to \operatorname{Hom}_A(A/fA, N) \to \operatorname{Hom}_A(A/fA, I_1) \to \operatorname{Hom}_A(A/fA, I_2) \to \dots$$

From the exact sequence $0 \to A \xrightarrow{f} A \to A/fA \to 0$, it follows that $\operatorname{Ext}_A^k(A/fA, \cdot) = 0$ for $k \geqslant 2$. By splitting the exact sequence $0 \to M \to I_0 \to I_1 \to I_2 \to \dots$ from left to right into short exact sequences, the proof is concluded by applying the functor $\operatorname{Hom}_A(A/fA, \cdot)$, etc.

THEOREM 1.12. *If A is a regular local ring of dimension n and $\mathfrak{m}$ is its maximal ideal, then $H^n_{\mathfrak{m}}(A)$ is an injective A-module.*

Proof. Let $\underline{f} = (f_1, \dots, f_n)$ be a regular system of parameters. $H^n_{\mathfrak{m}}(A) \simeq \simeq \varinjlim_k H^n(\underline{f}^k, A) \simeq \varinjlim_k A/(f_1^k, \dots, f_n^k)A$. The maps of the inductive system are

$$A/(\underline{f}^k)A \to A/(\underline{f}^{k+1})A, \text{ the class of } a \mapsto \text{the class of } f_1 \dots f_n a,\ a \in A.$$

Since $(f_1, \dots, f_n)$ is a regular A-sequence, it easily follows by a recursive reasoning that these maps are injective. Since A is regular, hence its injective dimension is n, by applying the above assertion it follows that the rings $A/(\underline{f}^k)A$ are injective (in fact $A/(\underline{f}^k)A$ is a Gorenstein ring of dimension zero, hence it is injective!).

We now prove the assertion of the theorem. Let $\psi : M \to N$ be a monomorphism of A-modules, supposed of finite type, and let $\varphi : M \to \varinjlim_k A(\underline{f}^k)A$ be an arbitrary A-morphism. Since M is of finite type, there exists an integer k such that φ factorizes by a morphism $\varphi_1 : M \to A/(\underline{f}^k)A$. From the Artin-Ress theorem there is an integer s which can be supposed larger than k so that

$$\psi((\underline{f}^k)M) \supset (\underline{f}^s)\, N \cap \psi(M).$$

Hence the map $\psi_1 : M/(\underline{f}^k)\, M \to N/(\underline{f}^s)\, N + \psi((\underline{f}^k)M)$ induced by ψ is injective. The map φ_1 factorizes by a map $M/(\underline{f}^k)\, M \to A/(\underline{f}^k)\; A$, which, composed with the map $A/(\underline{f}^k)A \to A/(\underline{f}^s)A$, gives rise to a map $\theta : M/(\underline{f}^k)M \to A/(\underline{f}^s)A$. The morphisms ψ_1 and θ are A-linear and the corresponding modules are annihilated by $(\underline{f}^s)A$. Since $A/(\underline{f}^s)A$ is an injective $A/(\underline{f}^s)A$-module, there exists an A-linear map $\theta_1 : N/(\underline{f}^s)N + \psi(\underline{f}^k M) \to A/(\underline{f}^s)A$ such that $\theta_1\psi_1 = \theta$. It will then result that the composed morphism

$$N \to N/(\underline{f}^s)\, N + \psi(\underline{f}^k)\, M \xrightarrow{\theta_1} A/(\underline{f}^s)\, A \to \varinjlim_t A/(\underline{f}^t)\, A$$

extends φ.

(6) Let A be a commutative unitary ring and $X = \operatorname{Spec} A$. For an integer p write

$$Z^p = \{\alpha \in \operatorname{Spec} A \,|\, ht\, \alpha = \dim A_\alpha \geqslant p\}.$$

Obviously, $X = Z^0$, $Z^p \supset Z^{p+1}$. The sets Z^p are stable to specialization and any point of $Z^p \setminus Z^{p+1}$ is maximal with respect to specialization.

By (1), there exists a complex of $\tilde{A}$-modules which is called *the Cousin complex of A associated to the filtration given by codimension*

$$0 \to \mathcal{H}^0_{Z^0/Z^1}(\tilde{A}) \to \mathcal{H}^1_{Z^1/Z^2}(\tilde{A}) \to \ldots,$$

together with an augmentation map $\tilde{A} \to \mathcal{H}^0_{Z^0/Z^1}(\tilde{A})$. By applying the functor $\Gamma(X, \cdot)$, one obtains a complex of A-modules which is called *the Cousin complex associated to A and to the filtration given by codimension.*

LEMMA 1.13. *If A is noetherian, then the sheaves of the Cousin complex are flabby.*

Proof. The conditions of 1.1 are fulfilled, hence for any $q \geqslant 0$ we have isomorphisms

$$(*) \qquad \mathcal{H}^q_{Z^p/Z^{p+1}}(\tilde{A}) \simeq \bigoplus_{\alpha \in Z^p \setminus Z^{p+1}} i_\alpha(H^q_\alpha(\tilde{A})).$$

Each sheaf $i_\alpha(H^q_\alpha(\tilde{A}))$ is flabby, since it is constant on the irreducible set $\overline{\{\alpha\}}$ and null on its complementary. The conclusion follows from the fact that a direct sum of flabby sheaves on a noetherian space is flabby.

We recall that a noetherian ring A (not necessarily local) is called Cohen-Macauley (respectively regular) if the localizations $A_\mathfrak{p}$ are Cohen-Macauley (regular, respectively) rings, $\mathfrak{p} \in \operatorname{Spec} A$.

THEOREM 1.14. *If A is a Cohen-Macauley ring, then the Cousin complex is a resolution of $\tilde{A}$; moveover, if A is regular, then the stalks of sheaves of this resolution are injective.*

Proof. The filtration is finite. We will show that $\mathcal{H}^q_{Z^p/Z^{p+1}}(\tilde{A}) = 0$ for any $q \neq p$ and the first assertion will result from proposition 1.2. In virtue of the isomorphisms (*) and by the proof of the previous lemma, it is enough to show that $H^q_\alpha(\tilde{A}) = 0$ for $\alpha \in Z^p \setminus Z^{p+1}$. It is easy to prove the isomorphisms $H^q_\alpha(\tilde{A}) \simeq H^q_{\alpha A_\alpha}(A_\alpha)$ and the conclusion follows from 1.11.

The second assertion will follow from theorem 1.12, again by the isomorphisms (*).

COROLLARY 1.15. *Let A be a Cohen-Macauley ring. Then the Cousin complex of A is a resolution of A. Moreover, if A is a regular ring, then the components of this resolution are A-injective.*

The proof follows from 1.7, 1.13 and 1.14 if we notice that

$$\Gamma(X, \mathcal{H}^q_{Z^p/Z^{p+1}}(\tilde{A})) \simeq \bigoplus_{\alpha \in Z^p \setminus Z^{p+1}} H^q_\alpha(\tilde{A}) \simeq \bigoplus_{\alpha \in Z^p \setminus Z^{p+1}} H^q_{\alpha A_\alpha}(A_\alpha).$$

Remark. Let A be a regular local ring of dimension 1. $Z^0 = \operatorname{Spec} A = \{0, \mathfrak{m}\}$, $Z^1 = \{\mathfrak{m}\}$. We have $\mathcal{H}^0_{Z^0/Z^1}(\tilde{A}) \simeq \bigoplus_{\alpha \in Z^0 \setminus Z^1} i_\alpha(H^0_\alpha(\tilde{A}))$, hence it is isomorphic to the constant sheaf given by the field M of the quotients of A. The term $\mathcal{H}^1_{Z^1}(\tilde{A})$ is isomorphic to $\mathcal{H}^0_{Z^0/Z^1}(\tilde{A})/\tilde{A}$, hence the Cousin complex is in this case

$$0 \to \tilde{M} \to \tilde{M}/\tilde{A} \to 0,$$

which explains the terminology!

(B) In this section we will define derived functors of the functors whose arguments are complexes of objects [39].

(1) Let A be an abelian category. For convenience one may suppose that A is the category of modules over a ring or the category of sheaves of modules over a sheaf of rings, the only cases considered below.

Denote by $C(A)$ the *category of complexes over* A, where the objects are complexes of objects of A and the morphisms are classes, modulo the homotopy relation, of morphisms of complexes (of zero degree). We have a canonical functor $A \to C(A)$, which assigns to any object X of A the complex $X^\bullet$ whose components are all null but $X^0 = X$.

Denote by $T : C(A) \to C(A)$ the functor (automorphism) which translates the degrees one step to the left and changes the sign of the differentials. By T^n (n arbitrary integer) one denotes the corresponding iteration of T. Hence for an object $X^\bullet$ of $C(A)$, $T^n(X^\bullet)$ is the complex of components $T^n(X^\bullet)^p = X^{n+p}$ and of differentials $d_{T^n(X^\bullet)} = (-1)^n d_{X^\bullet}$. We sometimes write $X^\bullet[n]$ instead of $T^n(X^\bullet)$.

By H^n we mean the functor $C(A) \to A$ which associates the cohomology object in the dimension n.

Recall some definitions (although repeating some facts stated in III.1.e!). An object of $C(A)$ is called *acyclic* if its objects of cohomology are null. Let $u : X^\bullet \to Y^\bullet$ be a morphism in $C(A)$; u is said to be a *quasi-isomorphism* if it induces isomorphisms between the cohomology objects; hence the morphisms $H^n(u) : H^n(X^\bullet) \to H^n(Y^\bullet)$ are isomorphisms for each integer n. In this case we may say that $\xrightarrow{u} Y^\bullet$ is a *resolution of* $X^\bullet$. Let X be an object of A. A complex $0 \to Y^0 \to Y^1 \to Y^2 \to \dots$ of objects of A, together with an augmentation morphism $X \xrightarrow{\varepsilon} Y^0$, is a resolution of X in the usual sense (i.e. the sequence $0 \to \to X \xrightarrow{\varepsilon} Y^0 \to Y^1 \to \dots$ is exact) if and only if the canonical morphism $X^\bullet \to Y^\bullet$ given by ε is a quasi-isomorphism (that is if and only if $\xrightarrow{\varepsilon} Y^\bullet$ is a resolution of $X^\bullet$ in the above sense). The notion of quasi-isomorphism can be reduced to the simpler notion of acyclicity. Recall first a definition.

Let $u : X^\bullet \to Y^\bullet$ be a morphism in $C(A)$. *The cone of* u is the complex $C^\bullet(u)$ of objects $T(X^\bullet) \oplus Y^\bullet$ whose differential is given by the matrix

$$\begin{pmatrix} T(d_X) & 0 \\ T(u) & d_Y \end{pmatrix}$$

(one can easily see that a complex is so obtained).

There exists an exact sequence of complexes

$$0 \to Y^\bullet \to C^\bullet(u) \to T(X^\bullet) \to 0.$$

From the exact cohomological sequence associated to this, one obtains (III. 1.8):

LEMMA 1.16. *A morphism u is a quasi-isomorphism if and only if its cone $C^\bullet(u)$ is acyclic.*

We have assumed, as we will do in the following, that the differentials have the degree $+1$; similar considerations can be made in the case when the differentials have the degree -1.

(2) Suppose now A is an abelian category with enough injective objects. Denote by $C^+(A)$ the full subcategory of $C(A)$ whose objects are the complexes bounded below. So a complex $X^\bullet$ is in $C^+(A)$ if $X^n = 0$ for any n sufficiently small. A complex $I^\bullet$ is called *injective* if all of its components I^n are injective objects. A morphism $X^\bullet \xrightarrow{f} I^\bullet$ is called *an injective resolution of* $X^\bullet$ if $\xrightarrow{f} I^\bullet$ is a resolution (hence f is a quasi-isomorphism) and $I^\bullet$ is injective.

LEMMA 1.17. *Let* $X^\bullet \xrightarrow{f} I^\bullet$ *be a morphism of complexes of* $C^+(A)$, *where* $X^\bullet$ *is acyclic and* $I^\bullet$ *injective. Then* f *is homotopic to zero (hence a null morphism in* $C^+(A)$).

Proof. We may assume that $X^n = I^n = 0$ for $n < 0$. By hypothesis, we have a commutative diagram with the first line exact and the second one formed by injective objects

$$\begin{array}{ccccccccc} 0 & \to & X^0 & \to & X^1 & \to & X^2 & \to & \dots \\ & & \downarrow{\scriptstyle f^0} & & \downarrow{\scriptstyle f^1} & & \downarrow{\scriptstyle f^2} & & \\ 0 & \to & I^0 & \to & I^1 & \to & I^2 & \to & \dots \end{array}$$

Since I^0 is injective, there is a morphism $g^1 : X^1 \to I^0$ such that $g_1 d^0_{X^\bullet} = f^0$. The morphism $f^1 - d^2_{I^\bullet} g^1$ vanishes on $\mathrm{Im}\,(d^0_{X^\bullet})$, hence it factorizes through $X^1/\mathrm{Im}\, d^0_{X^\bullet} = X^1/\mathrm{Ker}\, d^1_{X^\bullet} \simeq \mathrm{Im}\,(d^1_{X^\bullet})$. Since I^1 is an injective object, this factorization can be extended to a morphism $g^2 : X^2 \to I^1$, etc. ... One thus obtains a family of morphisms $(g^n)_{n\geq 1}$, $g^n : X^n \to I^{n-1}$ which define the stated homotopy.

PROPOSITION 1.18. *Each complex* $X^\bullet$ *of* $C^+(A)$ *admits an injective resolution and two such resolutions are homotopically equivalent.*

Proof. Suppose, for convenience, that $X^n = 0$ whenever $n < 0$. Choose a monomorphism $0 \to X^0 \xrightarrow{u^0} I^0$, where I^0 is an injective object. We set $I^n = 0$ for $n < 0$. Consider the fibred sum of X^1 and I^0 over X_0, $Y = X^1 \sum_{X^0} I^0 \simeq X^1 \oplus I^0/\mathrm{Im}\,(i_1 d^0_{X^\bullet} - i_2 u^0)$, where i_1 and i_2 are the canonical morphisms $X^1 \to X^1 \oplus I^0$, $I^0 \to X^1 \oplus I^0$. Let $0 \to Y \to I^1$ be a monomorphism such that I^1 is injective. Denote by $d^0_{I^\bullet}$ (respectively u^1) the composition of the maps $I^0 \to Y \to I^1$ ($X^1 \to Y \to I^1$, respectively).

The reasoning can be repeated and one thus obtains an injective complex $I^\bullet$ and a morphism of complexes $u : X^\bullet \to I^\bullet$. One can easily check that u induces an isomorphism between the cohomology objects. The uniqueness from the statement follows from:

PROPOSITION 1.19. *Let* $X^\bullet \xrightarrow{u} Y^\bullet$ *be a morphism in* $C^+(A)$ *and let* $X^\bullet \xrightarrow{i} I^\bullet$, $Y^\bullet \xrightarrow{j} J^\bullet$ *be injective resolutions. Then there exists a morphism* $v : I^\bullet \to J^\bullet$ *such that* $vi = gu$ *in* $C^+(A)$. *Moreover, two such morphisms are homotopic.*

Proof. By 1.16 the cone $C^\bullet(i)$ is acyclic. By lemma 1.17 the composition

$$C^\bullet(i) \xrightarrow{\text{projection}} T(X^\bullet) \xrightarrow{T(u)} T(Y^\bullet) \xrightarrow{T(j)} T(J^\bullet)$$

is homotopic to zero. Consider the maps $C^n(i) = X^{n+1} \oplus I^n \to T(J^\bullet)^{n-1} = J^n$ which define this homotopy. By composition with the canonical injections we obtain morphisms $\tau^n : X^n \to J^{n-1}$, $v^n : I^n \to J^n$. One can easily see that the family $(v^n)_n$ defines a morphism of complexes and that the maps vi and ju are homotopic (by means of the maps τ_n).

The last assertion of the proposition is a consequence of the following:

LEMMA 1.20. *Suppose* $X^\bullet \xrightarrow{i} Y^\bullet$, $Y^\bullet \xrightarrow{j} I^\bullet$ *are two morphisms in* $C^+(A)$ *such that i is a quasi-isomorphism, ji is homotopic to zero and* $I^\bullet$ *injective. Then j is homotopic to zero.*

Proof. Let $(s^n)_n$ be a family of morphisms which achieves the homotopy between 0 and ji, $s^n : X^n \to I^{n-1}$. One has $j^n i^n = d_{I^\bullet}^{n-1} s^n + s^{n+1} d_{X^\bullet}^n$. Let $Z^\bullet = C^\bullet(i)$. Consider, for any integer n, the map $w : Z^n = X^{n+1} \oplus Y^n \to I^n$, the sum of the maps s^{n+1} and j^n. One immediately verifies that the maps w^n commute with the differentials, hence they define a morphism $w : Z^\bullet \to I^\bullet$. Since $Z^\bullet$ is acyclic and $I^\bullet$ is injective, w is homotopic to zero by lemma 1.17. One has the equality $j = wk$, where k is the inclusion morphism $Y^\bullet \to Z^\bullet$ and the conclusion follows.

(3) Let A and B be two abelian categories, A with enough injective objects and $F : A \to B$ an additive functor which is left exact. F extends naturally to a functor $C^+(A) \to C^+(B)$ which is also denoted by F.

By 1.18, for any object $X^\bullet$ of $C^+(A)$ there exists an injective resolution $X^\bullet \to \to I^\bullet$. The cohomology objects of the complex $F(I^\bullet)$ do not depend on this resolution: indeed, in accordance with proposition 1.18, two such resolutions $I^\bullet$ and $J^\bullet$ are homotopic and hence the complexes $F(I^\bullet)$ and $F(J^\bullet)$ are homotopic (F is supposed additive). We shall denote by $R^\bullet F(X^\bullet)$ these cohomology objects:

$$R^i F(X^\bullet) = H^i(F(I^\bullet)) \in B.$$

If $X^\bullet \to Y^\bullet$ is a morphism in $C^+(A)$, then by means of 1.19 we get morphisms

$$R^i F(X^\bullet) \to R^i F(Y^\bullet).$$

In this way one obtains a family $(R^i F)_i$ of functors defined on the category $C^+(A)$ with values in B, and which are called *the right derived functors of F*. The compositions of functors $A \to C^+(A) \xrightarrow{R^i F} B$ coincide with the derived functors of F in the classical acceptance, as follows easily by the definitions. If $X^\bullet$ is an object of $C^+(A)$ and n is an integer, then $R^i F(X^\bullet[n]) \simeq R^{n+i} F(X^\bullet)$. If $X^\bullet \xrightarrow{u} Y^\bullet$ is a quasi-isomorphism in $C^+(A)$, then the morphisms $R^i F(u) : R^i F(X^\bullet) \to R^i F(Y^\bullet)$ are isomorphisms.

An object Z of A is called *F-acyclic* if $R^i F(Z) = 0$ for any $i > 0$. Any injective object is F-acyclic.

LEMMA 1.21. *If* $X^\bullet$ *is an acyclic complex of* $C^+(A)$ *formed by F-acyclic objects, then* $F(X^\bullet)$ *is acyclic; if u is a quasi-isomorphism between two complexes of* $C^+(A)$ *formed by F-acyclic objects, then* $F(u)$ *is a quasi-isomorphism.*

Proof. Let $X^\bullet$ be as in the statement. We may assume $X^n = 0$ for $n < 0$. Since $\operatorname{Coker} d_{X^\bullet}^0 = X^1/\operatorname{Im} d_{X^\bullet}^0 = X^1/\operatorname{Ker} d_{X^\bullet}^1 \simeq \operatorname{Im} d_{X^\bullet}^1$ and since $H^0(X^\bullet) = 0$, we get the exact sequence $0 \to X^0 \to X^1 \to \operatorname{Im}(d_{X^\bullet}^1) \to 0$. Then it will result that $\operatorname{Im} d_{X^\bullet}^1$ is F-acyclic and that the sequence

$$0 \to F(X^0) \to F(X^1) \to F(\operatorname{Im} d_{X^\bullet}^1) \to 0$$

is exact. Analogously, by using the exact sequence $0 \to \mathrm{Im} d^1_{X^\bullet} \to X^2 \to \mathrm{Im} d^2_{X^\bullet} \to 0$ one derives that $\mathrm{Im}\, d^2_{X^\bullet}$ is F-acyclic and that the sequence

$$0 \to F(\mathrm{Im}\, d^1_{X^\bullet}) \to F(X^2) \to F(\mathrm{Im}\, d^2_{X^\bullet}) \to 0$$

is exact. By iterating the reasoning, we get a sequence

$$0 \to F(X^0) \to F(X^1) \to F(X^2) \to \dots$$

which is exact; hence the complex $F(X^\bullet)$ is acyclic.

Let now $u : X^\bullet \to Y^\bullet$ be as in the statement of the lemma. If $Z^\bullet$ is the cone of u, then by the additivity of F it follows that $F(Z^\bullet)$ is the cone of $F(u)$. The required conclusion is obtained by the first part and by 1.16.

From this lemma we deduce that the derived functors of F can be calculated by means of resolutions with F-acyclic objects (the analogous of de Rham's abstract theorem).

LEMMA 1.22. *Let $X^\bullet$ be an object of $C^+(A)$ and $X^\bullet \to Z^\bullet$ a resolution such that the components of $Z^\bullet$ are F-acyclic. Then for any integer i, $H^i(F(Z^\bullet)) \simeq$ $\simeq R^iF(X^\bullet)$.*

Proof. Let $X^\bullet \to I^\bullet$ be an injective resolution. By 1.19 there exists a morphism $Z^\bullet \to I^\bullet$ which makes (*modulo* homotopy) the diagram commutative

$$\begin{array}{ccc} X^\bullet & \longrightarrow & Z^\bullet \\ & \searrow \quad \swarrow & \\ & I^\bullet & \end{array}$$

It follows easily that this morphism is a quasi-isomorphism and the assertion is a consequence of the previous lemma.

(4) We explain the previous facts in the case of functors Hom. Let A be an abelian category with enough injective objects and let X be an object. Consider the functor $\mathrm{Hom}\,(X, \cdot) : A \to \mathbf{Ab}$. Its derived functors $R^i\mathrm{Hom}(X, \cdot) : C^+(A) \to \mathbf{Ab}$ are denoted by $\mathrm{Ext}^i(X, \cdot)$. If $Y^\bullet \in C^+(A)$ and $Y^\bullet \to I^\bullet$ is an injective resolution, then $\mathrm{Ext}^\bullet(X, Y^\bullet)$ are the cohomology groups of the complex

$$\dots \to \mathrm{Hom}\,(X, Y^{i-1}) \to \mathrm{Hom}(X, Y^i) \to \mathrm{Hom}(X, Y^{i+1}) \to \dots .$$

For any integer n, $\mathrm{Ext}^i(X, Y^\bullet[n]) \simeq \mathrm{Ext}^{n+i}(X, Y^\bullet)$. If $Y^\bullet$ is the natural complex associated to an object Y of A, then the usual Ext's are obtained.

Recall Yoneda's method of calculating the functors Ext. Let X, Y be two objects of A and $I^\bullet$ ($J^\bullet$, respectively) an injective resolution of X (respectively Y). Define a complex of abelian groups $\mathrm{Hom}^\bullet(I^\bullet, J^\bullet)$ by formulae

$$\mathrm{Hom}^q(I^\bullet, J^\bullet) = \prod_p \mathrm{Hom}(I^p, J^{p+q}),$$

$$df = (d_{J^\bullet}^{p+q} f^p + (-1)^{q+1} f^{p+1} d^p_{I^\bullet})_{p \geqslant 0}, \quad f = (f^p)_p \in \mathrm{Hom}^q(I^\bullet, J^\bullet).$$

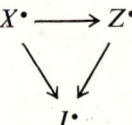

One can easily see that in this complex the q-cocycles are just the morphisms of degree q from $I^\bullet$ to $J^\bullet$ (equivalently, the morphisms of complexes $I^\bullet \to J^\bullet[q]$) and the q-boundaries are exactly the morphisms of degree q which are homotopic to zero. The augmentation morphism $X \xrightarrow{i} I^0$ determines a morphism of complexes $\mathrm{Hom}^\bullet(I^\bullet, J^\bullet) \to \mathrm{Hom}(X, J^\bullet)$ (in the degree q, $f = (f^p)_p \mapsto f^0 i$), hence by passing to cohomology, morphisms

$$H^q(\mathrm{Hom}^\bullet(I^\bullet, J^\bullet)) \to H^q(\mathrm{Hom}(X, J^\bullet)) = \mathrm{Ext}^q(X, Y).$$

PROPOSITION 1.23. *Let X, Y be objects of A and $I^\bullet$ ($J^\bullet$, respectively) an injective resolution of X (Y, respectively). Under these assumptions, the morphisms*

$$H^q(\mathrm{Hom}^\bullet(I^\bullet, J^\bullet)) \to \mathrm{Ext}^q(X, Y)$$

are isomorphisms.

Proof. We first prove the injectivity. Let $f = (f^p)_p \in \mathrm{Hom}^q(I^\bullet, J^\bullet)$ be a q-cocycle such that the image of the associated cohomology class is null. This image is the cohomology class of $f^0 i \in \mathrm{Hom}(X, J^q)$, hence there exists a morphism $v^0 : X \to J^{q-1}$ such that the diagram

$$\begin{array}{ccc} X & \xrightarrow{f^0 i} & J^q \\ & {\scriptstyle v^0} \searrow \quad \swarrow {\scriptstyle d^q_{J^\bullet}} & \\ & J^{q-1} & \end{array}$$

is commutative.

Consider the acyclic complex

$$0 \to X \to I^0 \to I^1 \to I^2 \to \ldots,$$

where the object X has the degree (-1).

The family of morphisms $(f^p)_{p \geq 0}$, to which we add the morphism $(-1)^q v_0$ in the degree (-1), yields a morphism from the above acyclic complex to the injective complex $J^\bullet[q]$. According to lemma 1.20, this morphism is homotopic to zero. Then f is homotopic to zero, hence a q-boundary.

We now prove the surjectivity of the maps from the statement. Let $g : X \to J^q$ be such that $d^q_{J^\bullet} \cdot g = 0$. Choose a morphism f^0 such that $g = f^0 i$. We set $f^i = 0$ for any $i \geq 1$ and it is easy to verify that the family $f = (f^p)_p \in \mathrm{Hom}^q(I^\bullet, J^\bullet)$ is a cocycle and the image of its cohomology class under the morphism $H^q(\mathrm{Hom}^\bullet(I^\bullet, J^\bullet)) \to$ $\to H^q(\mathrm{Hom}(X, J^q))$ equals the cohomology class of g.

Now we are going to define the operations of composition for the functors Ext.

Let X, Y, Z be three objects of A and let $I^\bullet, J^\bullet$ and $K^\bullet$ be some injective resolutions of theirs. The composition of maps defines a morphism

$$\mathrm{Hom}^\bullet(I^\bullet, J^\bullet) \otimes \mathrm{Hom}^\bullet(J^\bullet, K^\bullet) \to \mathrm{Hom}^\bullet(I^\bullet, K^\bullet)$$

which agrees with the differentials.

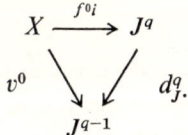

Passing to cohomology one gets the maps

$$\mathrm{Ext}^p(X, Y) \times \mathrm{Ext}^q(Y, Z) \to \mathrm{Ext}^{p+q}(X, Z)$$

which are called *the Yoneda bilinear maps*. One can show that these maps are well defined, functorial and agree with the short exact sequences.

Remark. One can see that an analogous result with respect to proposition 1.23 also holds if $Y \in C^+(A)$. Accordingly, the above bilinear maps are also defined when Z is in $C^+(A)$.

(5) Let X be a topological space and Φ a family of supports. The functor $\mathcal{F} \mapsto \Gamma_\Phi(X, \mathcal{F})$, from the category $\mathbf{Ab}(X)$ to the category $\mathbf{Ab}$, extends to a functor $C^+(\mathbf{Ab}(X)) \to C^+(\mathbf{Ab})$. The associated derived functors will be denoted by $H^{\bullet}_\Phi(X, \cdot)$.

Let $\mathcal{F}^\bullet$ be a complex bounded below of sheaves of abelian groups and let $\mathcal{F}^\bullet \to \mathcal{J}^\bullet$ be an injective resolution. By definition $H^{\bullet}_\Phi(X, \mathcal{F}^\bullet)$ are the cohomology groups of the complex

$$\ldots \to \Gamma_\Phi(X, \mathcal{J}^{q-1}) \to \Gamma_\Phi(X, \mathcal{J}^q) \to \Gamma_\Phi(X, \mathcal{J}^{q+1}) \to \ldots .$$

If $\mathcal{F}^\bullet$ is the complex associated to a sheaf $\mathcal{F}$, one obtains the cohomology groups of $\mathcal{F}$ with supports in Φ.

As in I. 4.16 we get the Mayer-Vietoris exact sequence:

LEMMA 1.24. *Let X be a paracompact space, which is a union of two open subsets U and V. For any $\mathcal{F}^\bullet \in C^+(\mathbf{Ab}(X))$, the following exact sequence of natural maps*

$$\ldots \to H^q_c(U \cap V, \mathcal{F}^\bullet) \to H^q_c(U, \mathcal{F}^\bullet) \oplus H^q_c(V, \mathcal{F}^\bullet) \to H^q_c(X, \mathcal{F}^\bullet) \to H^{q+1}_c(U \cap V, \mathcal{F}^\bullet) \to \ldots$$

holds.

Suppose now $(X, \mathcal{O})$ is a ringed space and $\mathcal{F}$ an $\mathcal{O}$-module. The functor $\mathcal{G} \mapsto \mathrm{Hom}_\mathcal{O}(\mathcal{F}, \mathcal{G})$, from the category $\mathbf{Mod}(\mathcal{O})$ to the category $\mathbf{Ab}$, extends to a functor $C^+(\mathbf{Mod}(\mathcal{O})) \to C^+(\mathbf{Ab})$. The associated derived functors will be denoted by $\mathrm{Ext}^{\bullet}_\mathcal{O}(X; \mathcal{F}, \cdot)$. If $\mathcal{G}^\bullet \in C^+(\mathbf{Mod}(\mathcal{O}))$ and $\mathcal{G}^\bullet \to \mathcal{J}^\bullet$ is an injective resolution, then $\mathrm{Ext}^{\bullet}_\mathcal{O}(X; \mathcal{F}, \mathcal{G}^\bullet)$ stands for the cohomology of the complex

$$\ldots \to \mathrm{Hom}_\mathcal{O}(\mathcal{F}, \mathcal{J}^{q-1}) \to \mathrm{Hom}_\mathcal{O}(\mathcal{F}, \mathcal{J}^q) \to \mathrm{Hom}_\mathcal{O}(\mathcal{F}, \mathcal{J}^{q+1}) \to \ldots .$$

Let Φ be a family of supports. Starting with the functor $\mathcal{G} \mapsto \mathrm{Hom}_{\Phi,\mathcal{O}}(\mathcal{F}, \mathcal{G}) = \Gamma_\Phi(X, \mathcal{H}om_\mathcal{O}(\mathcal{F}, \mathcal{G}))$, one can define the invariants $\mathrm{Ext}^{\bullet}_{\Phi,\mathcal{O}}(X; \mathcal{F}, \mathcal{G}^\bullet)$, $\mathcal{G}^\bullet \in C^+(\mathbf{Mod}(\mathcal{O}))$. We will consider only the particular cases: Φ = the family of all closed sets and Φ = the family of all compact sets (in the latter case we denote, as usual, by $\mathrm{Ext}_{c,\mathcal{O}}$ the obtained invariants).

If $\mathcal{F} = \mathcal{O}$, then there exist remarkable isomorphisms

$$\mathrm{Ext}^{\bullet}_{\Phi,\mathcal{O}}(X; \mathcal{O}, \mathcal{G}^\bullet) \simeq H^{\bullet}_\Phi(X, \mathcal{G}^\bullet),\ \mathcal{G}^\bullet \in C^+(\mathbf{Mod}(\mathcal{O})).$$

We define now composite maps for $\mathrm{Ext}_{\Phi,\mathcal{O}}$. Fix two $\mathcal{O}$-modules $\mathcal{F}$ and $\mathcal{G}$ ($\mathcal{G}$ can be in fact an object of $C^+(\mathbf{Mod}(\mathcal{O}))$) and let $\mathfrak{J}^\bullet$, $\mathfrak{j}^\bullet$ be injective resolutions of theirs. Consider the complex of abelian groups $\mathrm{Hom}^\bullet_{\Phi,\mathcal{O}}(\mathfrak{J}^\bullet, \mathfrak{j}^\bullet)$ given by the formulae

$$\mathrm{Hom}^q_{\Phi,\mathcal{O}}(\mathfrak{J}^\bullet, \mathfrak{j}^\bullet) = \prod_p \mathrm{Hom}_{\Phi,\mathcal{O}}(\mathfrak{J}^p, \mathfrak{j}^{p+q}),$$

$$df = (d^{p+q}f^q + (-1)^{q+1}f^{p+1}d^p)_p, \quad f = (f^p) \in \mathrm{Hom}^q_{\Phi,\mathcal{O}}(\mathfrak{J}^\bullet, \mathfrak{j}^\bullet).$$

The augmentation $\mathcal{F} \to \mathfrak{J}^0$ defines a morphism of complexes

$$\mathrm{Hom}^\bullet_{\Phi,\mathcal{O}}(\mathfrak{J}^\bullet, \mathfrak{j}^\bullet) \to \mathrm{Hom}^\bullet_{\Phi,\mathcal{O}}(\mathcal{F}, \mathfrak{j}^\bullet),$$

hence morphisms

$$H^q(\mathrm{Hom}^\bullet_{\Phi,\mathcal{O}}(\mathfrak{J}^\bullet, \mathfrak{j}^\bullet)) \to H^q(\mathrm{Hom}_{\Phi,\mathcal{O}}(\mathcal{F}, \mathfrak{j}^\bullet)) = \mathrm{Ext}^q_{\Phi,\mathcal{O}}(X; \mathcal{F}, \mathcal{G}).$$

LEMMA 1.25. *For any q, the map*

$$H^q(\mathrm{Hom}^\bullet_{\Phi,\mathcal{O}}(\mathfrak{J}^\bullet, \mathfrak{j}^\bullet)) \to \mathrm{Ext}^\bullet_{\Phi,\mathcal{O}}(X; \mathcal{F}, \mathcal{G})$$

is bijective.

The proof is analogous to the proof of 1.23, by making use of the following

LEMMA 1.26. *Let $0 \to \mathcal{M} \xrightarrow{i} \mathcal{N}$ be a monomorphism in* $\mathbf{Mod}(\mathcal{O})$, *let $\mathfrak{J}$ be an injective $\mathcal{O}$-module and $f : \mathcal{M} \to \mathfrak{J}$ an $\mathcal{O}$-morphism. Under these assumptions, there exists an $\mathcal{O}$-morphism $g : \mathcal{N} \to \mathfrak{J}$ such that $f = gi$ and $\mathrm{Supp}\, g = \mathrm{Supp}\, f$.*

Proof. Let $U = X \setminus Y$, where $Y = \mathrm{Supp} f$. Consider the sheaf $\mathcal{M}_U$($\mathcal{N}_U$, respectively), equal to $\mathcal{M}$($\mathcal{N}$, respectively) on U and with stalks null on Y. $\mathcal{M}_U$, $\mathcal{N}_U$ are subsheaves of $\mathcal{M}$, $\mathcal{N}$ and $f(\mathcal{M}_U) = 0$ (one can reason on stalks). As $\mathcal{N}_U \cap i(\mathcal{M}) = i(\mathcal{M}_U)$, f can be extended to a morphism $f' : \mathcal{N}_U \cup i(\mathcal{M}) \to \mathfrak{J}$, by putting $f' = 0$ on $\mathcal{N}_U$. Any extension $g : \mathcal{N} \to \mathfrak{J}$ of f' satisfies the assertions of the lemma.

The following result will be useful.

COROLLARY 1.27. *If Y is a closed subset of X and $\mathfrak{J}$ is an injective $\mathcal{O}$-module, then $\mathcal{H}^0_Y\mathfrak{J}$ is an injective $\mathcal{O}$-module.*

Proof. For any $\mathcal{O}$-module $\mathcal{M}$, $\mathrm{Hom}_{\mathcal{O}}(\mathcal{M}, \mathcal{H}^0_Y\mathfrak{J}) = \mathrm{Hom}_{Y,\mathcal{O}}(\mathcal{M}, \mathfrak{J})$ and apply the previous lemma.

We now consider a third $\mathcal{O}$-module $\mathcal{H}$, and a new family of supports Ψ. Let $\mathcal{K}^\bullet$ be an injective resolution of $\mathcal{H}$. The composition of maps defines a morphism

$$\mathrm{Hom}^\bullet_{\Phi,\mathcal{O}}(\mathfrak{J}^\bullet, \mathfrak{j}^\bullet) \otimes \mathrm{Hom}^\bullet_{\Psi,\mathcal{O}}(\mathfrak{j}^\bullet, \mathcal{K}^\bullet) \to \mathrm{Hom}^\bullet_{\Phi\cap\Psi,\mathcal{O}}(\mathfrak{J}^\bullet, \mathcal{K}^\bullet),$$

which agrees with the differentials.

Passing to cohomology we get the Yoneda bilinear maps (in fact $\Gamma(X, \mathcal{O})$-bilinear)

$$\mathrm{Ext}^p_{\Phi.\mathcal{O}}(X; \mathcal{F}, \mathcal{G}) \times \mathrm{Ext}^q_{\Psi.\mathcal{O}}(X; \mathcal{G}, \mathcal{H}) \to \mathrm{Ext}^{p+q}_{\Phi\cap\Psi.\mathcal{O}}(X; \mathcal{F}, \mathcal{H}).$$

In the case when X is paracompact, $\Phi =$ the family of closed sets of X, $\Psi =$ the family of compact sets of X (or conversely) one obtains the bilinear maps

$$\mathrm{Ext}^p_{\mathcal{O}}(X; \mathcal{F}, \mathcal{G}) \times \mathrm{Ext}^q_{c.\mathcal{O}}(X; \mathcal{G}, \mathcal{H}^{\bullet}) \to \mathrm{Ext}^{p+q}_{c.\mathcal{O}}(X; \mathcal{F}, \mathcal{H}^{\bullet}),$$

$$\mathrm{Ext}^p_{c.\mathcal{O}}(X; \mathcal{F}, \mathcal{G}) \times \mathrm{Ext}^q_{\mathcal{O}}(X; \mathcal{G}, \mathcal{H}^{\bullet}) \to \mathrm{Ext}^{p+q}_{c.\mathcal{O}}(X; \mathcal{F}, \mathcal{H}^{\bullet}).$$

Remark. In the above considerations we may replace the sheaf $\mathcal{H}$ by a complex $\mathcal{H}^{\bullet}$ of sheaves, bounded below. If $\mathcal{F} = \mathcal{O}$ then one obtains the bilinear maps

$$H^p(X, \mathcal{G}) \times \mathrm{Ext}^q_{c.\mathcal{O}}(X; \mathcal{G}, \mathcal{H}) \to H^{p+q}_c(X, \mathcal{H}),$$

$$H^p_c(X, \mathcal{G}) \times \mathrm{Ext}^q_{\mathcal{O}}(X; \mathcal{G}, \mathcal{H}) \to H^{p+q}_c(X, \mathcal{H}).$$

We conclude with some words on the local *Ext*'s. Fix again an $\mathcal{O}$-module $\mathcal{F}$. Starting with the functor $\mathcal{G} \mapsto \mathcal{H}om_{\mathcal{O}}(\mathcal{F}, \mathcal{G})$, we can define for any $\mathcal{G}^{\bullet} \in C^+(\mathbf{Mod}\ \mathcal{O})$), the $\mathcal{O}$-modules $\mathcal{E}xt^{\bullet}_{\mathcal{O}}(\mathcal{F}, \mathcal{G}^{\bullet})$. In this case one can also check easily that each $\mathcal{E}xt^q_{\mathcal{O}}(\mathcal{F}, \mathcal{G}^{\bullet})$ is the sheaf associated to the presheaf $U \mapsto \mathrm{Ext}^q_{\mathcal{O}|U}(U; \mathcal{F}\,|\,U, \mathcal{G}^{\bullet}\,|\,U)$. If $\mathcal{F} = \mathcal{O}$, one obtains isomorphisms

$$\mathcal{E}xt^q_{\mathcal{O}}(\mathcal{O}, \mathcal{G}^{\bullet}) \simeq H^q(\mathcal{G}^{\bullet}).$$

(**C**) In this section we recall some facts concerning topological vector spaces ([61], [75]). We have denoted by FS (DFS, respectively) the spaces of Fréchet-Schwartz type (strong duals of Fréchet-Schwartz spaces, respectively). By QFS and QDFS we mean quotients of such spaces (according to I, § 1, **C**).

(1) Let M, N be two topological vector spaces defined over the complex field and $u : M \to N$ a linear and continuous map. Recall that u is called *strict* (or *topological homomorphism*) if the quotient topology on $u(M)$ coincides with the induced topology from N. If M and N are FS or DFS spaces, then u is strict if and only if $u(M)$ is a closed subspace of N. This will be used in the following form: if $M \xrightarrow{u} N \xrightarrow{v} P$ is a complex of FS (or DFS) spaces, then the cohomology space $\mathrm{Ker}\, v/\mathrm{Im}\, u$ is separated (in the natural topology induced on it) if and only if u is strict.

LEMMA 1.28. *Let $u : M \to N$ be a linear and continuous map between FS spaces. Then u is strict if and only if the transposed map $u' : N' \to M'$ is strict.*

Proof. We easily reduce the problem to the case when u is injective or surjective and then the assertion follows by properties of reflexivity.

LEMMA 1.29. *Let $u : M \to N$ be a linear and continuous map between FS spaces. If $u(M)$ is of finite codimension in N, then u is strict.*

Proof. Let P be an algebraic complement of $u(M)$ in N. Consider on it the induced topology from N; by hypothesis there results that P is also FS. The map

$$v : M \times P \to N, \quad (m, p) \mapsto u(m) + p$$

is continuous and surjective. By the Banach theorem it is strict. If W is an open subset of M, then $u(W) = v(W \times P) \cap u(M)$; therefore u is strict.

(2) Two topological vector spaces M and N are by definition *in topological duality* if there exists a bilinear and separately continuous map $M \times N \to \mathbf{C}$, such that the induced maps $M \to N'$, $N \to M'$ are topological isomorphisms (the accent means the topological strong dual).

LEMMA 1.30. *Let $M \xrightarrow{u} N \xrightarrow{v} P$ be a sequence of FS spaces and linear continuous maps such that $vu = 0$; let $M' \xleftarrow{u'} N' \xleftarrow{v'} P$ be the transposed sequence. Then the duality between N and N' induces a topological duality between* $\mathrm{Ker}v/\overline{\mathrm{Im}u}$ *and* $\mathrm{Ker}u'/\overline{\mathrm{Im}v'}$, *where these spaces are endowed with their natural topologies.*

Proof. For convenience we denote by $\langle\ \rangle$ the canonical bilinear maps $M \times M' \to \mathbb{C}$, $N \times N' \to \mathbb{C}$, $P \times P' \to \mathbb{C}$. By restriction one obtains a bilinear map $\mathrm{Ker}\, v \times \mathrm{Ker}\, u' \to \mathbb{C}$. One can see easily that this vanishes on $\mathrm{Im}\, u \times \mathrm{Ker}\, u'$ and $\mathrm{Ker}\, v \times \mathrm{Im}\, u'$; by continuity we get a bilinear map

$$\mathrm{Ker}v/\overline{\mathrm{Im}u} \times \mathrm{Ker}u'/\overline{\mathrm{Im}\, v'} \to \mathbb{C}$$

and we will prove that the duality stated in the lemma is fulfilled by this map.

Let $\mathrm{Ker}u'/\overline{\mathrm{Im}\, v'} \to (\mathrm{Ker}v/\overline{\mathrm{Im}\, u})'$ be the induced map. We check its injectivity Let $n' \in \mathrm{Ker}u'$ be such that the image of the associated class is null, hence $\langle n, n'\rangle = 0$ for all $n \in \mathrm{Ker}v$. In order to show that $n' \in \overline{\mathrm{Im}\, v'}$ it is sufficient by the Hahn-Banach theorem to prove that whenever L is a linear continuous functional which is null on $\mathrm{Im}\, v'$, one has $L(n') = 0$. The functional L has the form $\langle n, \cdot\rangle$ for some $n \in N$. We have to prove that $\langle n, n'\rangle = 0$ and this follows as soon as $n \in \mathrm{Ker}v$. But for any $p' \in P'$, $\langle v(n), p'\rangle = \langle n, v'(p')\rangle = L(v'(p')) = 0$.

The surjectivity of the map $\mathrm{Ker}u'/\overline{\mathrm{Im}\, v'} \to (\mathrm{Ker}\, v/\overline{\mathrm{Im}\, u})'$ is immediate in virtue of the Hahn-Banach theorem. Its continuity is merely a consequence of the fact that any bounded subset of $\mathrm{Ker}v/\overline{\mathrm{Im}\, u}$ is the image of a bounded subset of $\mathrm{Ker}v$ ($\mathrm{Ker}v$ is an FS space); then the map will result a topological isomorphism, by open map theorem for DFS spaces.

Since the space $\mathrm{Ker}v/\overline{\mathrm{Im}\, u}$ is FS, hence reflexive, one derives also the topological isomorphism $\mathrm{Ker}v/\overline{\mathrm{Im}\, u} \xrightarrow{\sim} (\mathrm{Ker}u'/\overline{\mathrm{Im}\, v'})'$. The lemma is proved.

We now prove a result which ensures the uniqueness of the topologies of the spaces in duality.

LEMMA 1.31. *Let M, N be two QFS spaces and P, Q two QDFS spaces. Let $\lambda : M \times P \to \mathbb{C}$ and $\nu : N \times Q \to \mathbb{C}$ be bilinear maps which put in topological duality the separated spaces associated to M and P, to N and Q, respectively.*

Under these assumptions, any linear maps $u : M \to N$ and $v : Q \to P$ which agree with λ and ν are continuous.

Proof. By hypothesis the map $M \to P'$ given by λ is surjective and its kernel is $\overline{\{0_M\}}$. A similar assertion holds for the map $N \to Q'$ given by ν. Then $u(\overline{\{0_M\}}) \subset \subset \overline{\{0_N\}}$; analogously, one deduces that $v(\overline{\{0_Q\}}) \subset \overline{\{0_P\}}$. Hence the maps u and v induce maps $\hat{u}$ and $\hat{v}$ between the associated separated spaces. We will show that these are continuous, whence one derives the continuity of u and v. Let Γ be the graph of $\hat{u}$; we show that it is a closed subspace of $\hat{M} \times \hat{N}$. Then let $(\hat{m}_0, \hat{n}_0)$ be an element of its closure; we will show that $\hat{u}(\hat{m}_0) = \hat{n}_0$. If L is a linear continuous functional on $\hat{N}$, then $L\hat{u}$ is continuous ($\hat{u}$ and $\hat{v}$ are transposed, *modulo* the isomorphisms given by λ and ν). We will prove that $L(\hat{u}(\hat{m}_0) - \hat{n}_0) = 0$ and the assertion concerning Γ will follow. The functional $(\hat{m}, \hat{n}) \mapsto L(\hat{n}) - Lu(\hat{m})$, defined on $\hat{M} \times \hat{N}$, is continuous; it vanishes on Γ, hence at $(\hat{m}_0, \hat{n}_0)$, etc. By the closed graph theorem we derive that $\hat{u}$ is continuous; similarly one may check the continuity of $\hat{v}$.

(3) We consider *complexes of topological vector spaces* (over the complex field), namely, complexes whose components are TVS and the differentials are continuous $\mathbb{C}$-linear maps.

In this case the cohomology groups can be endowed with structures of topological vector spaces which are generally nonseparated. A *continuous morphism between two* TVS *complexes* is a morphism of complexes (of zero degree) whose components are continuous and $\mathbb{C}$-linear. If $u : M^\bullet \to N^\bullet$ is such a morphism, then the linear maps $H^n(u) : H^n(M^\bullet) \to H^n(N^\bullet)$ are continuous. The cone of a continuous morphism is naturally a TVS complex.

LEMMA 1.32. *Let $u : M^\bullet \to N^\bullet$ be a continuous morphism between two FS complexes (the components are FS spaces). If u is an algebraic quasi-isomorphism, then it is a topological quasi-isomorphism (induces topological isomorphism to cohomology).*

Proof. Let $Z^q(M^\bullet)$ ($Z^q(N^\bullet)$, respectively) be the space of cocycles of dimension q of $M^\bullet$ ($N^\bullet$, respectively) and $H^q(M^\bullet)$ ($H^q(N^\bullet)$, respectively) the corresponding cohomology groups. Define the continuous map

$$v^q : Z^q(M^\bullet) \oplus N^{q-1} \to Z^q(N^\bullet) \quad , \quad (m, n) \mapsto u(m) + d^{q-1}(n).$$

Since $H^q(u)$ is an algebraic isomorphism, v^q is surjective. The source and the target are FS; hence v^q is strict. One can conclude by means of the commutative diagram

$$\begin{array}{ccc} Z^q(M^\bullet) \oplus N^{q-1} & \xrightarrow{\pi} & H^q(M^\bullet) \\ {\scriptstyle v^q}\downarrow & & \downarrow{\scriptstyle H^q(u)} \\ Z^q(N^\bullet) & \xrightarrow{\pi'} & H^q(N^\bullet) \end{array}$$

where π is the composition of the first projection with the map $Z^q(M^\bullet) \to H^q(M^\bullet)$ and π' represents the passing to quotients.

LEMMA 1.33. *In the preceding statement let the complexes $M^\bullet$ and $N^\bullet$ be DFS. Then the conclusion of the above lemma remains true.*

Proof. One preceeds similarly, by using the properties of permanence of the DFS spaces and the open map theorem for such spaces.

We now restate lemma I.1.4.

LEMMA 1.34. *Let $M^\bullet$ be an acyclic complex of FS or DFS spaces. Then the transposed complex $M'^\bullet$ is acyclic.*

Proof. Let $L \in (M^q)'$ be so that $Ld^{q-1} = 0$. Since L vanishes on $\operatorname{Ker} d^q = \operatorname{Im} d^{q-1}$, it factors through the space $\operatorname{Im} d^q$, which is endowed with the quotient topology from M^q. Since $\operatorname{Im} d^q = \operatorname{Ker} d^{q+1}$, by Banach theorem it results that the quotient topology coincides with the one induced from M^{q+1}. By applying the Hahn-Banach theorem we can find a functional $L' \in (M^{q+1})'$ such that $L'd^q = L$ (in fact, the lemma is a mere consequence of 1.28 and 1.30).

LEMMA 1.35. *Let u: $M^\bullet \to N^\bullet$ be a topological quasi-isomorphism between FS or DFS complexes. Under these assumptions, the transposed morphism u' is a topological quasi-isomorphism.*

Proof. The cone of u is a complex as in the previous lemma. Its transposed complex will be acyclic. The conclusion follows from 1.32 by the permutability of the operations of transposition and of taking the cone.

§ 2. The construction of the dualizing complex

In the first part of the paragraph we construct a complex of $\mathcal{O}_{V,x}$-modules for any germ (V, x) of manifold, which will be the candidate for the stalk at the point x of the dualizing complex $K_V^\bullet$. For an immersion $(V, x) \to (W, y)$ one will deduce a natural formula of transformation between the associated complexes.

In the second part one builds the dualizing complex of an open subset of some numerical space. We will glue the complexes punctually constructed by means of the Frisch theorem of finiteness (V.3.1). For an immersion one will obtain a transformation formula between the corresponding complexes.

These facts will allow the construction of the dualizing complex for an arbitrary complex space (the third part of the paragraph).

(a) Let (V, x) be a germ of complex manifold. Denote by A the ring of germs of holomorphic functions at the point x and let $n = \dim A$. By $\mathfrak{m}$ we denote the maximal ideal of A. For an integer p, let

$$Z^p = \{\alpha \in \operatorname{Spec} A \mid \operatorname{ht} \alpha \geqslant p\}.$$

In § 1 we have constructed a complex of flabby sheaves on Spec A

$$0 \quad \mathcal{H}^0_{Z^0/Z^1}(\tilde{A}) \to \mathcal{H}^1_{Z^1/Z^2}(\tilde{A}) \to \dots \to \mathcal{H}^n_{Z^n/Z^{n+1}}(\tilde{A}) = \mathcal{H}^n_{\mathfrak{m}}(\tilde{A}) \to 0$$

(the Cousin complex of $\tilde{A}$ with respect to the filtration given by codimension) and an augmentation morphism

$$\tilde{A} \to \mathcal{H}^0_{Z^0/Z^1}(\tilde{A}).$$

Since A is a regular ring, the Cousin complex of $\tilde{A}$ is a resolution of $\tilde{A}$ by sheaves whose stalks are injective modules (1.14). The Cousin complex of A-modules associated to A, obtained by taking the global sections (in this case we actually localize in the maximal ideal $\mathfrak{m}$), is a resolution of A by injective modules (1.15), which will be denoted by $L^\bullet_{V,x}$. For instance, if V is a Riemann surface, the complex thus obtained is $0 \to \mathcal{M}_x \to \mathcal{M}_x/\mathcal{O}_x \to 0$, where $\mathcal{M}_x$ is the ring of germs of meromorphic functions in x (§ 1, **A**, 6).

Define

$$K^\bullet_{V,x} = L^\bullet_{V,x} \otimes_A \Omega^\bullet_{V,x}[n]$$

where Ω_V is the sheaf of germs of holomorphic forms of maximal degree on V.

LEMMA 2.1. *For any immersion of germs of complex manifolds* $(V, x) \xrightarrow{f} (W, y)$ *there exists a natural isomorphism*

$$\bar{f}: K^\bullet_{V,x} \simeq \operatorname{Hom}_{\mathcal{O}_{W,y}}(\mathcal{O}_{V,x}, K^\bullet_{W,y}).$$

Moreover, the correspondence $f \mapsto \bar{f}$ *agrees with the composition of immersions.*

Proof. If f is a local isomorphism, then the construction of $\bar{f}$ is clear.

First of all we consider the case when the codimension of f is 1. Denote $A = \mathcal{O}_{V,x}$, $B = \mathcal{O}_{W,y}$ and let $\rho: B \to A$ be the surjection which corresponds to f. Let t be an element of B which defines the subgerm $\operatorname{Im} f$. If $\mu = \mu_t$ is the multiplication by t, then the sequence $0 \to B \xrightarrow{\mu} B \xrightarrow{\rho} A \to 0$ is exact. The morphism ρ nduces a closed immersion $\tilde{\rho}$ between the prime spectra. In virtue of definitions, the sequence of sheaves on Spec B

$$(*) \qquad 0 \to \tilde{B} \to \tilde{B} \to \tilde{\rho}_*(\tilde{A}) \to 0$$

is exact. Denote by Z^p (T^p, respectively), the set of the prime ideals from A (B, respectively), of height $\geqslant p$. For any integer p, $\tilde{\rho}^{-1}(T^p) = Z^{p-1}$. In order to construct $\bar{f}$ we have to interpret the complex $\operatorname{Hom}_{\mathcal{O}_{W,y}}(\mathcal{O}_{V,y}, L^\bullet_{W,y})$. By 1.1, the p-th component is

$$\operatorname{Hom}_B(A, \Gamma\mathcal{H}^p_{T^p/T^{p+1}}(\tilde{B})) \simeq \operatorname{Hom}_B(A, \bigoplus_{\beta \in T^p \setminus T^{p+1}} H^p_\beta(\tilde{B})) \simeq \bigoplus_{\beta \in T^p \setminus T^{p+1}} \operatorname{Hom}_B(A, H^p_\beta(\tilde{B})).$$

From the left exactness of the functor Hom it follows that $\operatorname{Hom}_B(A, H^p_\beta(\tilde{B}))$ is identified with the kernel of the map $\mu: H^p_\beta(\tilde{B}) \to H^p_\beta(\tilde{B})$; this map can be inserted in the exact sequence

$$H^{p-1}_\beta(\tilde{B}) \to H^{p-1}_\beta(\tilde{\rho}_*(\tilde{A})) \xrightarrow{\partial} H^p_\beta(\tilde{B}) \xrightarrow{\mu} H^p_\beta(\tilde{B})$$

given by sequence (*). Since B is regular, $H_{\beta}^{p-1}(\tilde{B}) = 0$ (1.11). Accordingly,

$$\mathrm{Hom}_B(A, H_{\beta}^{p}(\tilde{B})) \simeq H_{\beta}^{p-1}(\tilde{\rho}_*(\tilde{A})), \ \beta \in T^p \setminus T^{p+1}.$$

If $\beta \notin \tilde{\rho}$ (Spec A), then Spec $B \setminus \tilde{\rho}$ (Spec A) is a neighbourhood of β where $\tilde{\rho}_*(A)$ is null; hence $\mathrm{Hom}_B(A, H_{\beta}^{p}(\tilde{B})) = 0$.

Let $\beta \in \tilde{\rho}$ (Spec A); thereby H^{p-1} (Spec B, $\tilde{\rho}_*(\tilde{A})$) can be identified with H_{α}^{p-1} (Spec A, $\tilde{A}$), where α is the element of $Z^{p-1} \setminus Z^p$ so that $\tilde{\rho}(\alpha) = \beta$ (the invariants with respect to a subspace and to a sheaf coincide with the invariants with respect to whole space and to the trivial extension of that sheaf!). By the antecedence it follows that $\bigoplus_{\beta \in T^p \setminus T^{p+1}} \mathrm{Hom}_B(A, H_{\beta}^{p}(\tilde{B}))$ is identified with $\bigoplus_{\alpha \in Z^{p-1} \setminus Z^p} H_{\alpha}^{p-1}(\tilde{A})$, which is isomorphic to $\Gamma \mathcal{H}_{Z^{p-1} \setminus Z^p}^{p-1}(\tilde{A})$ (1.1).

One thus obtains an isomorphism, which depends on the local equation t,

$$\Delta_t^p : L_{V,x}^{p-1} \simeq \mathrm{Hom}_B(A, L_{W,y}^{p}).$$

By means of these isomorphisms, for any integer q we define an isomorphism

$$\bar{f}^q : K_{V,x}^q \simeq \mathrm{Hom}_B(A, K_{W,y}^q).$$

We have $K_{V,x}^q = L_{V,x}^{q+\dim V} \otimes_A \Omega_{V,x}$ and $\mathrm{Hom}_B(A, K_{W,y}^q) = \mathrm{Hom}_B(A, L_{W,y}^{q+\dim W} \otimes_B \Omega_{W,y}) = \mathrm{Hom}_B(A, L_{W,y}^{q+\dim V+1} \otimes_B \Omega_{W,y})$. Next, define $\bar{f}^q$ by means of Δ_t and of the ρ-morphism $\Omega_{W,y} \to \Omega_{V,x}$ associating the form $2\pi i \Psi \mid V$ to the form $\Psi \wedge dt$.

If we replace t by another local parameter t', then $\Delta_{t'} = ((t/t') \mid V).\ \Delta_t$: this results from the fact that $\Delta_{t'}$ and Δ_t, are obtained by means of the exact sequences of local invariants associated to the exact sequence of the type (*) given by the multiplications μ_t and $\mu_{t'}$. On the other hand, as one can easily remark, the morphism $\Omega_{W,y} \to \Omega_{V,x}$ associated to t' differs from that associated to t by the multiplication with $(t'/t) \mid V$. Thereby the morphisms $\bar{f}^q$ are independent of t. Obviously, $\bar{f}^q$ are isomorphisms and $\bar{f} = (\bar{f}^q)_q$ is an isomorphism of complexes. In this way $\bar{f}$ is constructed for immersions of codimension 1.

If f is an arbitrary immersion, then the construction of $\bar{f}$ can be made by the decomposition of f into a finite number of immersions of codimension 1. The independence of the decomposition can be proved without difficulty as soon as one has, in the above notations, an explicit formula of identification for the isomorphisms $H_{\alpha}^{p-1}(\tilde{A}) \simeq \mathrm{Hom}_B(A, H_{\beta}^{p}(\tilde{B}))$, $\alpha \in Z^{p-1} \setminus Z^p$ and $\beta \in \rho^{-1}(\alpha)$ (deduced from the coboundary operator). For instance, one may proceed as follows. Let $a_1, \ldots, a_{p-1}$ be elements of B such that their images in A_{α} form a regular system of parameters. We have the identifications

$$H_{\alpha}^{p-1}(\tilde{A}) \simeq H_{\alpha A_{\alpha}}^{p-1}(A_{\alpha}) \simeq H^{p-1}((a_1, \ldots, a_{p-1}), A_{\alpha}) \simeq \varinjlim_k H^{p-1}(a_1^k, \ldots, a_{p-1}^k; A_{\alpha})$$

and

$$H^p_\beta(\tilde{B}) \simeq H^p_{\beta B_\beta}(B_\beta) \simeq H^p((t, a_1, \ldots, a_{p-1}), B_\beta) \simeq \varinjlim_k H^p(t^k, a_1^k, \ldots, a_{p-1}^k; B_\beta).$$

Consider an element ξ of $H^{p-1}_\alpha(\tilde{A})$. It corresponds to an element of some $H^{p-1}(a_1^k, \ldots, a_{p-1}^k; A_\alpha)$, hence to the cohomology class associated to an element of the form $a.a_1^k \wedge \ldots \wedge a_{p-1}^k \in K^{p-1}(a_1^k, \ldots, a_{p-1}^k; A_\alpha)$ $(\simeq \overset{p-1}{\wedge}(A_\alpha^{p-1}))$, where $a \in A_\alpha$. Let $b \in B_\beta$ be such that $\rho(\beta) = \alpha$. One has $d(b.a_1^k \wedge \ldots \wedge a_{p-1}^k) = t^k b(t^k \wedge \wedge a_1^k \wedge \ldots \wedge a_{p-1}^k) = t(t^{k-1}b.t^k \wedge a_1^k \wedge \ldots \wedge a_{p-1}^k)$.

Let η be the element of $H^p_\beta(\tilde{B})$ assigned to the cohomology class of $t^{k-1}b.t^k \wedge \wedge a_1^k \wedge \ldots \wedge a_{p-1}^k \in K^p(t^k, a_1^k, \ldots, a_{p-1}^k; B_\beta)$ in $H^p(t^k, a_1^k, \ldots, a_{p-1}^k; B_\beta)$. The correspondence $\xi \mapsto (1 \mapsto \eta)$ stands for the required explicitation (a consequence of the construction of the coboundary operator!). In the above formulae we abused notation as we denoted by the same letters the images under ρ of the elements a_i.

The functoriality of the correspondence $f \mapsto \bar{f}$ follows by using its independence of the decomposition into immersions of codimension 1.

(b) Consider an open set U of some $\mathbb{C}^n$.

Let $K \subset U$ be a Stein semianalytic compact. By theorem V.3.1 the ring $\mathcal{O}(K)$ of germs of holomorphic functions on K is noetherian. Each maximal ideal $\mathfrak{m}$ of $\mathcal{O}(K)$ is finitely generated, hence it is associated to a point of K; moreover, the localization $\mathcal{O}(K)_\mathfrak{m}$ is a regular ring (lemma V.4.1). Consequently the Cousin complex associated to the ring $\mathcal{O}(K)$, which will be denoted $L_K^\bullet$, is a resolution of $\mathcal{O}(K)$ (1.15).

Recall some facts about the analytic sets. A *germ of analytic set in the neighbourhood* of K is a pair (D, M), where D is an open subset such that $D \supset K$ and M is a closed analytic subset of D. We identify two such pairs (D_1, M_1), (D_2, M_2) whenever there exists an open set $D \subset D_1 \cap D_2$, $D \supset K$, such that $M_1 \cap D = M_2 \cap D$. A germ of analytic set in the neighbourhood of K is called *irreducible* if it cannot be written as the union of two proper subgerms (the notion of subgerm, union and intersection of germs, ... are defined naturally).

If we associate to each germ the ideal in $\mathcal{O}(K)$ of the functions which vanish on it, then one obtains a 1—1 correspondence between the germs of analytic sets in the neighbourhood of K and the radical ideals of $\mathcal{O}(K)$ (equivalently, the closed sets of Spec $\mathcal{O}$(K)). In this correspondence the irreducible germs correspond to the prime ideals. If $x \in K$, then we denote by $\rho_{Kx}\colon \mathcal{O}(K) \to \mathcal{O}_x$ the natural morphism and by $\tilde{\rho}_{Kx}\colon \operatorname{Spec} \mathcal{O}_x$ Spec $\mathcal{O}(K)$ the morphism induced between the prime spectra. If α is an ideal of $\mathcal{O}(K)$, then for convenience, we denote by $\rho_{Kx}(\alpha)$ the ideal generated by the image of α under ρ_{Kx}. If α corresponds to a germ of analytic set (D, M) in the neighbourhood of K, then $\rho_{Kx}(\alpha)$ determines the germ of analytic set given by M in the neighbourhood of x. Indeed, let $f_1, \ldots, f_s$ be a system of generators for α; $f_{i,x} = \rho_{Kx}(f_i)$, $1 \leqslant i \leqslant s$, is a system of generators for $\rho_{Kx}(\alpha)$ and the assertion follows from the fact that the passing from an ideal to the associated germ of analytic set is made by taking the common zeros of a system of generators.

LEMMA 2.2. *Let* $\alpha \in \operatorname{Spec} \mathcal{O}(K)$ *be so that* $\rho_{Kx}(\alpha) \neq \mathcal{O}_x$. *There exist a finite number of prime ideals* β, $\beta \in \operatorname{Spec} \mathcal{O}_x$, *which are minimal such that* $\beta \supset \rho_{Kx}(\alpha)$. *Moreover, for such an ideal* β, $\operatorname{ht} \beta = \operatorname{ht} \alpha$.

Proof. The first assertion is a general property of the noetherian rings, consequence of Lasker-Noether decomposition. If α corresponds to the germ (D, M), then the ideals β correspond in fact to the decomposition of the germ M_x into irreducible components. One can easily see that all these irreducible components are of the same dimension, equal to $\dim (D, M) = \dim (\mathcal{O}(K)/\alpha)$. It follows that $\operatorname{ht} \alpha = \dim \mathcal{O}(K) - \dim (\mathcal{O}(K)/\alpha) = \dim \mathcal{O}_x - \dim (\mathcal{O}_x/\beta) = \operatorname{ht} \beta$ for any β.

For a Stein semianalytic compact $K' \subset K$ one can do the same for the restriction morphism $\rho_{KK'}\colon \mathcal{O}(K) \to \mathcal{O}(K')$.

LEMMA 2.3. *Let* $K' \subset K$ *be Stein semianalytic compacts,* x *a point of* K' *and* $L_K^\bullet$, $L_{K'}^\bullet$, $L_x^\bullet$ *the Cousin complexes associated to the noetherian rings* $\mathcal{O}(K)$, $\mathcal{O}(K')$ *and* $\mathcal{O}_x$. *Then, by the construction which follows, there exist natural morphisms of complexes* $L_K^\bullet \to L_x^\bullet$, $L_{K'}^\bullet \to L_x^\bullet$, $L_K^\bullet \to L_{K'}^\bullet$, *which agree with* ρ_{Kx}, $\rho_{K'x}$, $\rho_{KK'}$ *and such that the diagram*

$$\begin{array}{ccc} L_K^\bullet & \longrightarrow & L_{K'}^\bullet \\ & \searrow \quad \swarrow & \\ & L_x^\bullet & \end{array}$$

is commutative.

Proof. We shall indicate the construction of the first morphism. For the other one may proceed analogously. The commutativity of the diagram can be verified canonically.

For any $p \geqslant 0$ we need to construct a ρ_{Kx}-morphism $L_K^p \to L_x^p$. We shall use the following notations: $A = \mathcal{O}_x$, $B = \mathcal{O}(K)$, $\rho = \rho_{Kx}$, $Z^p = \{\beta \in \operatorname{Spec} A \mid \operatorname{ht}\beta \geqslant p\}$, $T^p = \{\alpha \in \operatorname{Spec} B \mid \operatorname{ht} \alpha \geqslant p\}$. There exist isomorphisms

$$L_K^p \simeq \bigoplus_{\alpha \in T^p \setminus T^{p+1}} H_\alpha^p(\tilde{B}),\ L_x^p \simeq \bigoplus_{\beta \in Z^p \setminus Z^{p+1}} H_\beta^p(\tilde{A}).$$

If α is an ideal of Spec $\mathcal{O}(K)$ of height p, then we shall construct a ρ_{Kx}-morphism

$$\lambda\colon H_\alpha^p(\tilde{B}) \to \bigoplus_{\beta \in Z^p \setminus Z^{p+1}} H_\beta^p(\tilde{A}).$$

These maps will determine the required morphism $L_K^p \to L_x^p$.

We first consider the case $p \geqslant 2$; this hypothesis will allow us to express the local cohomological invariants in terms of usual invariants. By § 1, $H_\alpha^p(\tilde{B})$ is the stalk in $\alpha \in \operatorname{Spec} B$ of the sheaf $\mathcal{H}_{\overline{\{\alpha\}}}^p(\tilde{B})$; hence $\varinjlim_U H_{\overline{\{\alpha\}}}^p(U, \tilde{B})$, U being a neighbourhood of α in Spec B. In this inductive limit we will consider only affine open sets. We have the exact sequence

$$\ldots \to H^{p-1}(U, \tilde{B}) \to H^{p-1}(U \setminus \bar{\alpha}, \tilde{B}) \to H_{\bar{\alpha}}^p(U, \tilde{B}) \to H^p(U, \tilde{B}) \to \ldots$$

Since U is affine, $H^q(U, \tilde{B}) =$ for $q \geqslant 1$ and hence ($p \geqslant 2$), $H^p_{\bar{\alpha}}(U, B) \simeq H^{p-1}(U \setminus \bar{\alpha}, \tilde{B})$. Therefore $H^p_{\bar{\alpha}}(\tilde{B}) \simeq \varinjlim_U H^{p-1}(U \setminus \bar{\alpha}, \tilde{B})$. We will define the maps

$$\lambda_U : H^{p-1}(U \setminus \bar{\alpha}, \tilde{B}) \rightarrow \bigoplus_{\beta \in Z^p \setminus Z^{p+1}} H^p_{\beta}(\tilde{A}),$$

which are compatible with inclusions $U \subset V$ and thus one obtains λ.

Let U be an open set which appears in the inductive limit. The set $U \setminus \bar{\alpha}$ is open in Spec B, hence there exist a finite number of elements $f_j \in B$ such that $U \setminus \bar{\alpha} = \cup D_{f_j}$. The principal open sets D_{f_j} are affine and $\Gamma(D_{f_j}, \tilde{B}) = B_{f_j}$. For a finite part of them $f_{i_1}, \ldots, f_{i_k}$, $D_{f_{i_1}} \cap \ldots \cap D_{f_{i_k}} = D_{f_{i_1} \ldots f_{i_k}}$ and $\Gamma(D_{f_{i_1} \ldots f_{i_k}}, \tilde{B}) = B_{f_{i_1} \ldots f_{i_k}}$. By the Leray theorem, $H^{p-1}(U \setminus \bar{\alpha}, \tilde{B}) \simeq$ $\simeq H^{p-1}(\{D_{f_j}\}_j, \tilde{B})$. Let $c \in H^{p-1}(U \setminus \bar{\alpha}, \tilde{B})$ and $\{g_{j_1 \ldots j_p}\}$ be a cocycle which represents c and which corresponds to the covering $(D_{f_j})_j$. In the case when the ideal $\rho(\alpha)$ equals A we set $\lambda_U = 0$ (and hence $\lambda \mid H^p_{\bar{\alpha}}(\tilde{B}) = 0$). Otherwise, let $\beta_1, \ldots, \beta_s$ be the prime ideals of Spec $\mathcal{O}_x$ which contain $\rho(\alpha)$ and are minimal with this property. For each β_i, ht $\beta_i =$ htα. The open sets $D_{\rho(f_j)}$ give an affine covering for $\tilde{\rho}^{-1}(U) \setminus (\bar{\beta}_1 \cup \ldots \cup \bar{\beta}_s)$. The elements $\rho(g_{j_1 \ldots j_p}) \in A_{\rho(f_{j_1}) \ldots \rho(f_{j_p})}$ (denote by the same letter ρ the morphisms induced to the rings of quotients) yield a cocycle with respect to this covering, hence an element γ in $H^{p-1}(\tilde{\rho}^{-1}(U) \setminus \bigcup_{i=1}^{s} \bar{\beta}_i, \tilde{A})$. For any integer q one can find an affine neighbourhood V of β_q contained in $\tilde{\rho}^{-1}(U) \setminus \bigcup_{i \neq q} \bar{\beta}_i$. The restriction of γ to $H^{p-1}(V \setminus \bar{\beta}_q, \tilde{A}$ induces an element in $H^p_{\bar{\beta}_q}(V, \tilde{A})$, hence an element γ_q in $H^p_{\beta_q}(\tilde{A})$.

Define $\lambda_U(c) = \sum_{q=1}^{s} \gamma_q$. One can easily check that the definition of $\lambda_U(c)$ is independent of all choice and also that the maps λ_U agree with the inclusions $U \subset V$. For $p \geqslant 2$, we have thus obtained the required morphism λ.

For $p = 0$ and $p = 1$, the invariants $H^0_{\bar{\alpha}}(\tilde{B})$ and $H^1_{\bar{\alpha}}(\tilde{B})$ can be computed by means of suitable inductive limits and exact sequences of the form

$$0 \rightarrow H^0_{\bar{\alpha}}(U, \tilde{B}) = \Gamma_{\bar{\alpha}}(U, \tilde{B}) \rightarrow \Gamma(U, \tilde{B}) \rightarrow \Gamma(U \setminus \bar{\alpha}, \tilde{B}) \rightarrow H^1_{\bar{\alpha}}(U, \tilde{B}) \rightarrow 0.$$

The morphisms λ_0 and λ_1 can be defined as in the case $p \geqslant 2$ by using these sequences which allow us to express the invariants $H^0_{\bar{\alpha}}$ and $H^1_{\bar{\alpha}}$ in terms of invariants $H^0 = \Gamma$.

As soon as the morphisms λ are constructed, they extend as we have showed to morphisms $L^p_K \rightarrow L^p_x$ and it is easy to verify the compatibility with the differentials.

For an arbitrary point $x \in U$ we have the following

LEMMA 2.4. *The natural map of complexes*

$$\varinjlim_{K, x \in \mathring{K}} L_K^\bullet \to L_x^\bullet,$$

deduced from the previous lemma, is bijective.

Proof. We first prove the surjectivity. We will do this for degrees $p \geqslant 2$, the case $p \leqslant 1$ being similar.

Fix $\beta \in Z^p \setminus Z^{p-1}$ and $\gamma \in H_\beta^p(\tilde{A})$. There exists a closed polydisc K which is a neighbourhood of x, such that $\rho_{Kx}\rho_{Kx}^{-1}(\beta) = \beta$. For any $K' \subset K$, K' containing x, we also have $\rho_{K'x}\rho_{K'x}^{-1}(\beta) = \beta$. Then by lemma 2.2 there results that the prime ideals $\alpha_{K'} = \rho_{K'x}^{-1}(\beta)$ have the same codimension as α.

$$H_\beta^p(\tilde{A}) \simeq \varinjlim_{V, \beta \in V} H_{\overline{\beta}}^p(V, \tilde{A}) \simeq \varinjlim_{V, \beta \in V} H^{p-1}(V \setminus \overline{\beta}, \tilde{A}).$$

There exists an element $\psi \in A \setminus \beta$ such that the cohomology class γ is represented by an element of $H^{p-1}(D_\psi \setminus \overline{\beta}, \tilde{A})$.

If $\varphi_1, \ldots, \varphi_m$ is a system of generators of the ideal α_K, then one can easily see that $\alpha_K \notin \bigcup_{i=1}^{n} D_{\varphi_i}$ and $D_\psi \setminus \overline{\beta} = \bigcup_{i=1}^{m} D_{\psi\rho_{Kx}(\varphi_i)}$; similar relations hold for any $K' \subset K$ ($x \in K'$), if we replace the elements φ_i by $\varphi_i' = \rho_{KK_i}'(\varphi_i)$. Then it will follow that the element γ can be represented by a cocycle $\{\gamma_{i_1 \ldots i_p}\}$, $\gamma_{i_1 \ldots i_p} \in A_{\psi\rho_{Kx}(\varphi_{i_1}) \ldots \rho_{Kx}(\varphi_{i_p})}$. Choose a closed polydisc L, $L \subset \mathring{K}$ and $x \in \mathring{L}$, such that the elements ψ and $\gamma_{i_1 \ldots i_p}$ can be "lifted" to elements $\overline{\psi} \in \mathcal{O}(L)$, $\overline{\gamma}_{i_1 \ldots i_p} \in \mathcal{O}(L)_{\overline{\psi}\rho_{KL}(\varphi_{i_1}) \ldots \rho_{KL}(\varphi_{i_p})}$, respectively.

The family $\{\overline{\gamma}_{i_1 \ldots i_p}\}$ defines a cocycle with respect to the covering formed by the open sets $D_{\overline{\psi}\rho_{KL}(\varphi_i)}$ (it is enough to remark that the map $\rho_{Lx}: \mathcal{O}(L) \to \mathcal{O}_x$ is injective). One can easily show that this cocycle induces an element $\overline{\gamma}$ in $H_{\alpha_L}^p(\widetilde{\mathcal{O}(L)})$ and that $\lambda(\overline{\gamma}) = \gamma$. Thereby the surjectivity is proved.

The injectivity can be proved similarly.

We will define a complex $L_U^\bullet$ as follows: a section over an open set V is a collection $(s_x)_{x \in V}$, $s_x \in L_x^\bullet$, which is locally induced by an element of $L_K^\bullet$. By the previous lemma, the stalks of the complex of sheaves $L_U^\bullet$ are just the complexes $L_x^\bullet$.

Denote

$$K_U^\bullet = L_U^\bullet \otimes_{\mathcal{O}_U} \Omega_U^\bullet[n]$$

and call it the dualizing complex of U.

LEMMA 2.5. *If U and V are open subsets of some numerical spaces and $f: U \to V$ is a closed (or open) immersion, then there exists a natural isomorphism*

$$\overline{f}: K_U^\bullet \simeq Hom_{\mathcal{O}_V}(f_*(\mathcal{O}_U), K_V^\bullet) \,|\, U.$$

Moreover, the correspondence $f \mapsto \overline{f}$ is compatible with the composition of immersions.

Proof. One shows that the isomorphisms $\bar{f}_x(x \in U)$ given by lemma 2.1 vary "continuously" with respect to x; hence they extend to a morphism of sheaves.

Remark. We have put $Hom_{\mathcal{O}_V}(f_*(\mathcal{O}_U), K_V^\bullet)|U$ instead of $f^{-1}(Hom_{\mathcal{O}_V}(f_*(\mathcal{O}_U), K_V^\bullet))$ (there is a constant abuse of notation in the followings). Since the ideal which defines the immersion f annihilates this sheaf, we may replace in fact f^{-1} by f^*.

(**c**) In this section we prove the following *theorem of existence of the dualizing complex* of Ramis and Ruget:

THEOREM 2.6. *For each complex space X, there exists a complex $K_X^\bullet$ of $\mathcal{O}_X$-modules such that the following assertions are valid:*

(*i*) *If X is a manifold of dimension n, then the complex $K_X^\bullet$ is a resolution of $\Omega_X^\bullet[n]$ and for any point $x \in X$ and for any integer p, the stalks $K_{X,x}^p$ are injective $\mathcal{O}_{X,x}$-modules.*

(*ii*) *If $f: X \to Y$ is a closed* (*or open*) *immersion of complex spaces, then there exists an isomorphism*

$$\bar{f}: K_X^\bullet \simeq Hom_{\mathcal{O}_Y}(f_*(\mathcal{O}_X), K_Y^\bullet)\,|\,X.$$

In addition, the correspondence $f \mapsto \bar{f}$ agrees with the composition of immersions.

(*iii*) *If X is finite-dimensional, then the complex $K_X^\bullet$ is bounded; more precisely, $K_X^p = 0$ for $p < -\dim X$ and $p > 0$.*

(*iv*) *The sheaves of cohomology $\mathcal{H}^p(K_X^\bullet)$ are coherent. Moreover, $\mathcal{H}^p(K_X^\bullet) = 0$ for $p <$-prof $(\mathcal{O}_X)$.*

Proof. Consider an arbitrary complex space X. For an open set U of X such that there exists an immersion $\varphi: U \to V$, V open set in some numerical space, denote $K_U^\bullet = Hom_{\mathcal{O}_V}(\varphi_*(\mathcal{O}_U), K_V^\bullet)\,|\,U$, where $K_V^\bullet$ is the dualizing complex of V defined in the preceding section. $K_U^\bullet$ is a complex of $\mathcal{O}_U$-modules.

If U' is another open set enjoying the same property, then both immersions of $U \cap U'$ derived from φ and φ' can be refined by means of a third; from lemma 2.5 we deduce an isomorphism

$$\tau_{U,U'}: K_{U'}^\bullet|\,U \cap U' \simeq K_U^\bullet|\,U \cap U'.$$

Consider a covering of X by open sets as above. The isomorphisms τ verify the usual relations of compatibility (2.5) and hence the complexes $K_U^\bullet$ glue together to a complex $K_X^\bullet$. This complex is independent (up to isomorphism) of the considered covering.

We now verify the assertions of the theorem. Let U be as in section (**b**). The morphisms $\mathcal{O}_x \to L_x^0$ define morphisms $\Omega_{U,x}^\bullet[n] \to K_x^\bullet$ and these define a morphism $\Omega_U^\bullet[n] \to K_U^\bullet$. The isomorphisms $\bar{f}$ from lemma 2.5 (applied to open immersions) are compatible with these morphisms. The first assertion follows from section (**a**) and from the very way of constructing $K_X^\bullet$. The second results from the construction of $K_X^\bullet$, by 2.5. The assertion (*iii*) follows from the next lemma by passing to stalks.

LEMMA 2.7. *Let A be a regular noetherian local ring, $L_A^\bullet$ the associated Cousin complex and M an A-module of finite type. Then* $\mathrm{Hom}_A(M, L_A^p) = 0$ *for* $p <$ $\dim A - \dim M$.

Proof. We have $L_A^p = \bigoplus_{\alpha \in Z^p \setminus Z^{p+1}} H_\alpha^p(\tilde{A})$, Accordingly, it is enough to show that $\mathrm{Hom}_A(M, H_\alpha^p(\tilde{A})) = 0$ for $p = \mathrm{ht}\, \alpha < \dim A - \dim M$. By means of a composition series, we may assume M of the form A/β. We have $\beta \not\subset \alpha$ and let $a \in \beta - \alpha$. $\mathrm{Hom}_A(A/\beta, H_\alpha^p(\tilde{A})) = \{\xi \in H_\alpha^p(\tilde{A}) \mid \beta\xi = 0\}$. The element a is invertible in A_α and the homotety defined by it in the A_α-module $H_\alpha^p(\tilde{A})$ is injective. The conclusion follows.

The last assertion will be a consequence of the corollary 3.5 from the next paragraph and of corollary I.1.15.

Thereby theorem 2.6 is concluded.

The complex $K_X^\bullet$ is called *the dualizing complex of* X.

§ 3. Theorems of absolute duality

(**a**) Let V be a paracompact complex manifold of dimension n. $\mathcal{O}, \Omega$ and $\mathcal{E}^{p,q}$ ($\mathcal{K}^{p,q}$, respectively) stand as usual for the structural sheaf, the sheaf of germs of holomorphic forms in the maximal degree and the sheaf of germs of differential forms by the type (p, q) with coefficients C^∞ functions (distributions respectively). The Dolbeault resolution

$$0 \to \Omega \to \mathcal{K}^{n,0} \xrightarrow{d''} \cdots \xrightarrow{d''} \mathcal{K}^{n,n} \to 0$$

allows, by applying the functor $\Gamma_c(V, *)$, the computation of the invariants $H_c^\bullet(V, \Omega)$. The integration on V of the forms of the type (n, n) with compact supports will thus induce a linear form

$$H_c^n(V, \Omega) \to \mathbb{C}.$$

This form multiplied by $(-1)^n$ will be denoted by T_V and it will be called *the trace map on V*.

Let now $\mathcal{F}$ be an $\mathcal{O}$-module. For each integer p consider the Yoneda maps

$$H^p(V, \mathcal{F}) \times \mathrm{Ext}_c^{n-p}(V; \mathcal{F}, \Omega) \to H_c^n(V, \Omega)$$

$$H_c^p(V, \mathcal{F}) \times \mathrm{Ext}^{n-p}(V; \mathcal{F}, \Omega) \to H_c^n(V, \Omega).$$

By composition with T_V: $H_c^n(V, \Omega) \to \mathbb{C}$ we get pairings:

(*) $$H^p(V, \mathcal{F}) \times \mathrm{Ext}_c^{n-p}(V; \mathcal{F}, \Omega) \to \mathbb{C}$$

(**) $$H_c^p(V, \mathcal{F}) \times \mathrm{Ext}^{n-p}(V; \mathcal{F}, \Omega) \to \mathbb{C}.$$

The invariants $H_c^\bullet(V, \Omega)$ are isomorphic to the cohomology groups of the complex

$$0 \to \Gamma_c(V, \mathcal{K}^{n,0}) \xrightarrow{d''} \Gamma_c(V, \mathcal{K}^{n,1}) \to \ldots \xrightarrow{d''} \Gamma_c(V, \mathcal{K}^{n,n}) \to 0.$$

The spaces $\Gamma_c(V, \mathcal{K}^{n,p})$ are isomorphic to the topological duals of the spaces $\Gamma(V, \mathcal{E}^{0,n-p})$, hence they have a DFS structure. The differentials d'' are continuous in this topology and we thus get a natural structure QDFS on $H_c^\bullet(V, \Omega)$. One can easily check the continuity of the trace map.

LEMMA 3.1. *If V is a Stein manifold, then $H_c^p(V, \Omega)$ is null for $p \neq n$ and the space $H_c^n(V, \Omega)$ is separated and the map* (∗) *defines a topological duality between it and the space $\Gamma(V, \mathcal{O})$ (the latter is endowed with its natural FS structure).*

Proof. The invariants $H^\bullet(V, \mathcal{O})$ are isomorphic to the cohomology groups of the topological complex of FS spaces

$$0 \to \Gamma(V, \mathcal{E}^{0,0}) \xrightarrow{d''} \dots \xrightarrow{d''} \Gamma(V, \mathcal{E}^{0,n}) \to 0.$$

This complex is in topological duality with the complex $\Gamma_c(V, \mathcal{K}^{n,*})$ and the conclusion follows straightforwardly from 1.30 and theorem B.

The invariants $H_c^\bullet(V, \mathcal{O})$ can be computed by means of the complex $\Gamma_c(V, \mathcal{K}^{0,*})$; hence they have a natural QDFS structure. Just as in the above case, one can prove

LEMMA 3.2. *If V is a Stein manifold, then $H_c^p(V, \mathcal{O})$ is null for $p \neq n$ and the space $H_c^n(V, \mathcal{O})$ is separated and the map* (∗) *yields a topological duality between it and $\Gamma(V, \Omega)$ (the latter having its natural FS structure).*

Remark. If in lemmas 3.1 and 3.2 one drops the hypothesis that V is Stein, then (in virtue of the duality between C^∞ functions and distributions, extended to forms and by applying lemmas 1.28 and 1.30) one obtains topological dualities between the separated spaces associated to the spaces $(H_c^p(V, \Omega), H^{n-p}(V, \mathcal{O}))$ and $(H_c^p(V, \mathcal{O}), H^{n-p}(V, \Omega))$, respectively. Moreover, the separation of $H_c^p(V, \Omega)$ is equivalent to the separation of $H^{n-p+1}(V, \mathcal{O})$ (and a similar assertion for the second pair of invariants). This fact is a special case of the first duality theorem and it will be used in the proof of the second duality theorem; we will point it out in order to make this theorem independent on the first one.

LEMMA 3.3. *If V is a Stein manifold and $\mathcal{F}$ is an $\mathcal{O}$-module which admits of a finite resolution with free $\mathcal{O}$-modules of finite rank, the* $\mathrm{Ext}_c^p(V; \mathcal{F}, \Omega)$ *is null for $p \neq n$. Moreover, the spaces $\Gamma(V, \mathcal{F})$ and* $\mathrm{Ext}_c^n(V; \mathcal{F}, \Omega)$ *have natural FS and DFS structures, respectively, such that the bilinear map* (∗) *yields a topological duality between them.*

Proof. One proceeds by induction on the length of the resolution of $\mathcal{F}$.

(b) In this section we define the trace map on a complex space. First of all we prove the following:

PROPOSITION 3.4. *Let X be a complex space, V an n-dimensional manifold, $f: X \to V$ a closed immersion and $\mathcal{F} \in$* **Coh**(X). *Then for any integer p there exist natural isomorphisms*

$$\mathrm{Ext}_c^p(X; \mathcal{F}, K_X^\bullet) \simeq \mathrm{Ext}_c^{n+p}(V; f_*(\mathcal{F}), \Omega_V)$$

$$\mathrm{Ext}^p(X; \mathcal{F}, K_X^\bullet) \simeq \mathrm{Ext}^{n+p}(V; f_*(\mathcal{F}), \Omega_V)$$

which are functorial in $\mathcal{F}$ and compatible with the short exact sequences. Moreover, if $V' \subset V$ is an open set, then these isomorphisms are also compatible with the iso-

morphisms given by the immersion $X' = f^{-1}(V') \to V'$ and with the morphisms given by the inclusions $V' \subset V$, $X' \subset X$.

Proof. We will show the way of establishing the isomorphisms from the statement; the verification of the other assertions will be then tamely done.

Let $\mathfrak{I}^\bullet$ be an injective resolution of the $\mathcal{O}_V$-complex $K_V^\bullet$. This is an injective resolution for the complex $\Omega_V^\bullet[n]$, hence the invariants on the right hand of the statement of the proposition are the cohomology groups of the complexes

$$\mathrm{Hom}_c(V; f_*(\mathfrak{F}), \mathfrak{I}^\bullet), \text{ respectively } \mathrm{Hom}\,(V; f_*(\mathfrak{F}), \mathfrak{I}^\bullet).$$

From the natural isomorphism

$$Hom_{\mathcal{O}_V}(f_*(\mathfrak{F}), \mathfrak{I}^\bullet) \simeq Hom_{\mathcal{O}_V}(f_*(\mathfrak{F}), Hom_{\mathcal{O}_V}(f_*(\mathcal{O}_X), \mathfrak{I}^\bullet))$$

one derives that these complexes are isomorphic to the complexes

$$\mathrm{Hom}_c(X; \mathfrak{F}, Hom_{\mathcal{O}_V}(f_*(\mathcal{O}_X), \mathfrak{I}^\bullet) \,|\, X)$$

and

$$\mathrm{Hom}(X; \mathfrak{F}, Hom_{\mathcal{O}_V}(f_*(\mathcal{O}_X), \mathfrak{I}^\bullet) \,|\, X).$$

respectively.

In order to finish the proof it is enough to show that $Hom_{\mathcal{O}_V}(f_*(\mathcal{O}_X), \mathfrak{I}^\bullet) \,|\, X$ is a resolution of $K_X^\bullet$ with objects $\mathrm{Hom}_c(X; \mathfrak{F}, *)$ and $\mathrm{Hom}(X; \mathfrak{F}, *)$-acyclic. If $\mathcal{H}$ is an arbitrary $\mathcal{O}_X$-module, then there are canonical isomorphisms

$$\mathrm{Hom}_{\mathcal{O}_X}(\mathcal{H}, Hom_{\mathcal{O}_V}(f_*(\mathcal{O}_X), \mathfrak{I}^r) \,|\, X) \simeq \mathrm{Hom}_{\mathcal{O}_V}(f_*(\mathcal{H}), \mathfrak{I}^r),$$

which are functorial in $\mathcal{H}$. As a consequence, the sheaves $Hom_{\mathcal{O}_V}(f_*(\mathcal{O}_X), \mathfrak{I}^r) \,|\, X$ are $\mathcal{O}_X$-injective. The morphism $K_V^\bullet \to \mathfrak{I}^\bullet$ induces a morphism

$$Hom_{\mathcal{O}_V}(f_*(\mathcal{O}_X), K_V^\bullet) \to Hom_{\mathcal{O}_V}(f_*(\mathcal{O}_X), \mathfrak{I}^\bullet)$$

and the proof of the proposition will follow by showing that the latter is a quasi-isomorphism. By 1.21 it is enough to show that the sheaves K_V^r and $\mathfrak{I}^r$ are $Hom_{\mathcal{O}_V}(f_*(\mathcal{O}_X), *)$-acyclic. Since $f_*(\mathcal{O}_X)$ is a coherent $\mathcal{O}_V$-module, we conclude the proof by the fact that the stalks $K_{V,x}^r$ and $\mathfrak{I}_x^r$ are injective $\mathcal{O}_{V,x}$-modules for all $x \in V$.

COROLLARY 3.5. *Under the assumptions of the proposition, there exist isomorphisms*

$$Ext^p(\mathfrak{F}, K_X^\bullet) \simeq Ext^{n+p}(f_*(\mathfrak{F}), \Omega_V)|X,$$

which are functorial in $\mathfrak{F}$ and compatible with the short exact sequences. In particular,

$$\mathcal{H}^p(K_X^\bullet) \simeq Ext^{n+p}(f_*(\mathcal{O}_X), \Omega_V)|\, X.$$

COROLLARY 3.6. *Under the assumptions of the proposition, suppose moreover that V is Stein and that $f_*(\mathcal{O}_X)$ admits a finite resolution by free $\mathcal{O}_V$-modules of finite rank. Then for any $p > 0$,*

$$H_c^p(X, K_X^\bullet) = 0.$$

Proof. Obviously, $H_c^p(X, K_X^\bullet) \simeq \mathrm{Ext}_c^p(X; \mathcal{O}_X, K_X^\bullet) \simeq \mathrm{Ext}_c^{n+p}(V; f_*(\mathcal{O}_X), \Omega_V)$ and apply lemma 3.3.

Remark. If L is a compact of V and $K = f^{-1}(L)$, then, as in the proof of the proposition, one can show that

$$\mathrm{Ext}_K^p(X; \mathcal{F}, K_X^\bullet) \simeq \mathrm{Ext}_L^{n+p}(V; f_*(\mathcal{F}), \Omega_V).$$

Let V be a paracompact complex manifold of dimension n. Since $K_V^\bullet$ is quasi-isomorphic to $\Omega[n]$, $H_c^n(V, \Omega) \simeq H_c^0(V, K_V^\bullet)$. Consequently, the trace map can be considered as a linear functional on $H_c^0(V, K_V^\bullet)$, which is also denoted by T_V.

We now consider a closed immersion $V \xrightarrow{f} W$, V and W being paracompact complex manifolds. Let $n = \dim W$. By applying 3.4 we get the isomorphism

$$H_c^0(V, K_V^\bullet) (\simeq \mathrm{Ext}_c^0(V; \mathcal{O}_V, K_V^\bullet)) \simeq \mathrm{Ext}_c^n(W; f_*(\mathcal{O}_V), \Omega_W) \simeq \mathrm{Ext}_c^0(W; f_*(\mathcal{O}_V), K_W^\bullet).$$

By comsposition with the morphism

$$\mathrm{Ext}_c^0(W; f_*(\mathcal{O}_V), K_W^\bullet) \to \mathrm{Ext}_c^0(W; \mathcal{O}_W, K_W^\bullet) \overset{\sim}{\to} H_c^0(W, K_W^\bullet)$$

given by the map $\mathcal{O}_W \to f_*(\mathcal{O}_V)$, we obtain a linear map

$$T_{V/W}\colon H_c^0(V, K_V^\bullet) \to H_c^0(W, K_W^\bullet).$$

A calculation shows that this is compatible with the trace maps T_V and T_W.

Let now X be a paracompact complex space. If U is a Stein open subset of X and $f\colon U \to V$ is a closed immersion where V is an open set of some numerical space, then $H_c^0(U, K_V^\bullet) \simeq \mathrm{Ext}_c^0(V; f_*(\mathcal{O}_U), K_V^\bullet)$. The morphism $\mathcal{O}_V \to$ $\to f_*(\mathcal{O}_U)$ induces a morphism

$$\mathrm{Ext}_c^0(V; f_*(\mathcal{O}_U), K_V^\bullet) \to \mathrm{Ext}_c^0(V; \mathcal{O}_V, K_V^\bullet) \simeq H_c^0(V, K_V^\bullet).$$

We thus obtain a morphism $H_c^0(U, K_X^\bullet) \to H_c^0(V, K_V^\bullet)$ and hence, by composition with T_V, a morphism

$$T_U\colon H_c^0(U, K_X^\bullet) \to \mathbb{C}.$$

This morphism is independent of the immersion f.

Consider a locally finite covering $\mathcal{U}$ of X by Stein open subsets which are sufficiently small. If U and U' are in $\mathcal{U}$, then the composite morphisms

$$H_c^0(U \cap U', K_X^\bullet) \to H_c^0(U, K_X^\bullet) \xrightarrow{T_U} \mathbb{C}$$

and

$$H_c^0(U \cap U', K_X^\bullet) \to H_c^0(U', K_X^\bullet) \xrightarrow{T_{U'}} \mathbb{C}$$

are equal. By 3.6, $H_c^1(U \cap U', K_X^\bullet) = 0$. From these facts and from the Mayer-Vietoris exact sequence

$$H_c^0(U \cap U', K_X^\bullet) \to H_c^0(U, K_X^\bullet) \oplus H_c^0(U', K_X^\bullet) \to H_c^0(U \cup U', K_X^\bullet) \to H_c^1(U \cap U', K_X^\bullet),$$

it will result that T_U and $T_{U'}$ determine naturally a linear map on $H_c^0(U \cup U', K_X^\bullet)$.

The above reasoning can be iterated for finite unions and this thing is sufficient to enable us to extend the maps T_U to a linear map on $H_c^0(X, K_X^\bullet)$, denoted T_X. This is independent of the covering $\mathcal{U}$ and we call it *the trace map of* X.

Let $\mathcal{F}$ be an analytic sheaf on X. For any integer p there exist bilinear maps of type Yoneda

$$H^p(X, \mathcal{F}) \times \mathrm{Ext}_c^{-p}(X; \mathcal{F}, K_X^\bullet) \to H_c^0(X, K_X^\bullet)$$

$$H_c^p(X, \mathcal{F}) \times \mathrm{Ext}^{-p}(X; \mathcal{F}, K_X^\bullet) \to H_c^0(X, K_X^\bullet).$$

By composition with the trace map $T_X: H_c^0(X, K_X^\bullet) \to \mathbb{C}$ we get bilinear maps

$$(*) \qquad H^p(X, \mathcal{F}) \times \mathrm{Ext}_c^{-p}(X; \mathcal{F}, K_X^\bullet) \to \mathbb{C}$$

$$(**) \qquad H_c^p(X, \mathcal{F}) \times \mathrm{Ext}^{-p}(X; \mathcal{F}, K_X^\bullet) \to \mathbb{C}.$$

These are functorial in $\mathcal{F}$ and compatible with the short exact sequences and with the "localization" (i.e. with the spectral sequences associated to some coverings, which connect the global Ext's and the Ext's associated to the open sets of the covering) and also with the inclusions of open subsets of X.

(c) In this section we prove *the first theorem of absolute duality* of Ramis and Ruget.

THEOREM 3.7. *Let X be a finite-dimensional complex space with countable basis. For any coherent analytic sheaf $\mathcal{F}$ on X and for any integer p there exist a unique QFS structure on $H^p(X,\mathcal{F})$ and a unique QDFS structure on $\mathrm{Ext}_c^{-p}(X;\mathcal{F},K_X^\bullet)$ such that the trace map T_X induces a topological duality between the separated spaces associated to these spaces. Moreover, the separation of $H^p(X,\mathcal{F})$ is equivalent to that of* $\mathrm{Ext}_c^{1-p}(X; \mathcal{F}, K_X^\bullet)$.

We first consider topologies on the invariants $H^{\bullet}(X, \mathcal{F})$. Let $\mathfrak{U}$ be a locally finite covering by Stein open subsets of X. The complex of cochains $C^{\bullet}(\mathfrak{U}, \mathcal{F})$ is naturally an FS complex. $Z^p(\mathfrak{U}, \mathcal{F})$ is a closed subspace of $C^p(\mathfrak{U}, \mathcal{F})$. So $H^p(X, \mathcal{F}) \simeq Z^p(\mathfrak{U}, \mathcal{F})/B^p(\mathfrak{U}, \mathcal{F})$ has a QFS structure. We show that this is independent of $\mathfrak{U}$. Let $\mathfrak{V}$ be another locally finite covering by Stein open sets. There exists the third covering $\mathfrak{W}$, $\mathfrak{W} < \mathfrak{U}$, $\mathfrak{W} < \mathfrak{V}$. The morphisms of complexes

$$C^{\bullet}(\mathfrak{U}, \mathcal{F}) \to C^{\bullet}(\mathfrak{W}, \mathcal{F}),\ C^{\bullet}(\mathfrak{V}, \mathcal{F}) \to C^{\bullet}(\mathfrak{W}, \mathcal{F}),$$

obtained by means of some refinement functions, are continuous. Then the assertion follows by lemma 1.32. The topology thus defined is called *the natural QFS topology on $H^{\bullet}(X, \mathcal{F})$* (according to [93] in a more general situation).

In order to prove the theorem we make some preparations.

LEMMA 3.8. *Let X be a complex space and $\mathcal{F} \in$* **Coh**(X). *Suppose X is embedded in a Stein manifold V by means of an immersion f such that $f_*(\mathcal{F})$ admits a finite resolution with free $\mathcal{O}_V$-modules of finite type. Then* $\mathrm{Ext}_c^p(X; \mathcal{F}, K_X^{\bullet})$ *is null for $p \neq 0$ and* $\mathrm{Ext}_c^0(X; \mathcal{F}, K_X^{\bullet})$ *has a natural structure of DFS space such that the bilinear map* $(*)$ *yields a topological duality between it and* $\Gamma(X, \mathcal{F})$ (*the latter being endowed with the natural FS structure*).

The proof follows by 3.3 and 3.4.

We now recall some facts about precosheaves. A precosheaf of sets (abelian groups...) over a topological space X is a covariant functor defined on the category of the open subsets of X with values in the category of sets (the category of abelian groups ...), hence a system $(\mathfrak{D}(U), \rho_V^U)_{U, U \subset V}$ where $\mathfrak{D}(U)$ are sets (abelian groups...) and $\rho_V^U\colon \mathfrak{D}(U) \to \mathfrak{D}(V)$ are maps (group homomorphisms ...) such that $\rho_U^U =$ the identity and $\rho_W^U = \rho_W^V \rho_V^U (U \subset V \subset W)$.

If $\mathfrak{M}$ is a sheaf of abelian groups on X, then the assignments $\{U \mapsto \Gamma_c(U, \mathfrak{M}),$ $U \subset V \mapsto \rho_V^U =$ the trivial extension of sections$\}$ define a precosheaf of abelian groups denoted by $\mathfrak{M}_c$.

Let $\mathfrak{D}$ be a precosheaf of abelian groups and $\mathfrak{U}$ an open covering of X. Consider the complex $C_{\bullet}(\mathfrak{U}, \mathfrak{D})$ of finite chains given by the formula

$$C_p(\mathfrak{U}, \mathfrak{D}) = \bigoplus_{(i_0, \dots, i_p)} \mathfrak{D}(U_{i_0} \cap \dots \cap U_{i_p}),$$

whose differentials are defined naturally by alternate sums. Thus, one obtains the very complex of finite chains associated to the nerve of the covering $\mathfrak{U}$, which is defined by the system given by $\mathfrak{D}$. The homology groups of this complex are denoted by $H_{\bullet}(\mathfrak{U}, \mathfrak{D})$. The morphisms $\rho_X^{U_i}\colon \mathfrak{D}(U_i) \to \mathfrak{D}(X)$ define an augmentation morphism $C_0(\mathfrak{U}, \mathfrak{D}) = \bigoplus_i \mathfrak{D}(U_i) \to \mathfrak{D}(X)$, hence a morphism $H_0(\mathfrak{U}, \mathfrak{D}) \to \mathfrak{D}(X)$.

LEMMA 3.9. *Let X be a paracompact space, $\mathfrak{U}$ a locally finite open covering, and $\mathfrak{M}$ a flabby sheaf on X. Then $H_p(\mathfrak{U}, \mathfrak{M}_c) = 0$ for $p \geq 1$ and $H_0(\mathfrak{U}, \mathfrak{M}_c) = \mathfrak{M}_c(X) = \Gamma_c(X, \mathfrak{M})$.*

Proof. The complex $C_{\bullet}(\mathfrak{U}, \mathfrak{M}_c)$ is the inductive limit of the complexes $C_{\bullet}(\mathfrak{U}_f, \mathfrak{M}_c)$ where $\mathfrak{U}_f$ are the finite parts of $\mathfrak{U}$. Therefore $H_{\bullet}(\mathfrak{U}, \mathfrak{M}_c) \simeq \varinjlim H_{\bullet}(\mathfrak{U}_f, \mathfrak{M}_c)$.

On the other hand, $\Gamma_c(X, \mathcal{M}) = \varinjlim \Gamma_c(\bigcup_{U \in \mathcal{U}_f} U, \mathcal{M})$. We may thus assume the covering from the statement finite, namely, $\mathcal{U} = (U_i)_{0 \leq i \leq n}$. We may also consider only alternate chains. For each simplex $s = (i_0, \ldots, i_p)$, denote $U_s = U_{i_0} \cap \ldots \cap U_{i_p}$ and $\mathcal{M}_s = \mathcal{M}_{U_s}$. Let $\mathcal{C}_p(\mathcal{U}, \mathcal{M}) = \bigoplus_{\dim s = p} \mathcal{M}_s$ and $d: \mathcal{C}_p(\mathcal{U}, \mathcal{M}) \to \mathcal{C}_{p-1}(\mathcal{U}, \mathcal{M})$ be the morphism given by $d|\mathcal{M}_s = \sum_{k=0}^{p} (-1)^k j_k$, where $j_k: \mathcal{M}_{i_0 \ldots i_p} \to \mathcal{M}_{i_0 \ldots \hat{i}_k \ldots i_p}$ is the inclusion map.

One thus obtains a complex of sheaves. The inclusions $\mathcal{M}_i \subset \mathcal{M}$ yield an augmentation morphism $\mathcal{C}_0(\mathcal{U}, \mathcal{M}) \to \mathcal{M}$. In this way we get a resolution of $\mathcal{M}$: indeed, if x is an arbitrary point of X, then the complex $\mathcal{C}_\bullet(\mathcal{U}, \mathcal{M})_x$ coincides with the complex of chains associated to the nerve of $\mathcal{U}^x$ whose coefficients are in the group $\mathcal{M}_x$, where $\mathcal{U}^x$ consists of the open sets $U \in \mathcal{U}$ such that $x \in U$. We therefore have an exact sequence

$$0 \to \mathcal{C}_n(\mathcal{U}, \mathcal{M}) \xrightarrow{d} \ldots \xrightarrow{d} \mathcal{C}_0(\mathcal{U}, \mathcal{M}) \to \mathcal{M} \to 0.$$

The objects of this sequence are $\Gamma_c(X, *)$-acyclic. It is sufficient to remark that if $\mathcal{M}$ is flabby, then for any open set U the sheaf $\mathcal{M}_U$ is $\Gamma_c(X, *)$-acyclic (consequence of the long cohomology with compact supports sequence associated to the exact sequence $0 \to \mathcal{M}_U \to \mathcal{M} \to \mathcal{M}_{X \setminus U} \to 0$ and of the fact that $\mathcal{M}_{X \setminus U}$ is a soft sheaf). One can easily check the equality $\Gamma_c(X, \mathcal{C}_\bullet(\mathcal{U}, \mathcal{M})) = C_\bullet(\mathcal{U}, \mathcal{M}_c)$ and the conclusion of the lemma follows.

The proof of the theorem. Let $\mathcal{F} \in \mathbf{Coh}(X)$. Consider a locally finite covering $\mathcal{U}$ of X by Stein open sets. Suppose the covering sufficiently fine such that lemma 3.8 can be applied for any $U \in \mathcal{U}$. This lemma applies also for finite intersections of open sets of $\mathcal{U}$. The group $H^\bullet(X, \mathcal{F})$ are the cohomology groups of the complex $C^\bullet(\mathcal{U}, \mathcal{F})$ of FS spaces. By 3.8 the trace map yields a topological duality between this complex and the DFS complex of finite chains

$$C_\bullet(\mathcal{U}, Ext_c^0(\mathcal{F}, K_X^\bullet)),$$

where $Ext_c^0(\mathcal{F}, K_X^\bullet)$ is the precosheaf $U \mapsto \mathrm{Ext}_c^0(U; \mathcal{F}, K_X^\bullet)$ (the corestriction morphisms are defined in a canonical way). We show that the homology in the dimension p of $C_\bullet(\mathcal{U}, Ext_c^0(\mathcal{F}, K_X^\bullet))$ coincides with $\mathrm{Ext}_c^{-p}(X; \mathcal{F}, K_X^\bullet)$; then the theorem will follow by lemma 1.28, 1.30 and 1.31. In order to prove the above assertion consider an injective resolution $\mathcal{I}^\bullet$ of $K_X^\bullet$ and the bicomplex

$$C^{pq} = C_{-p}(\mathcal{U}, Hom_{\mathcal{O}_X}(\mathcal{F}, \mathcal{I}^q)_c).$$

We show that both associated spectral sequences degenerate. One has $'E_1^{p,q} = C_{-p}(\mathcal{U}, Ext_c^q(\mathcal{F}, K_X^\bullet))$, hence by 3.8,

$$'E_1^{p,q} = \begin{cases} 0 & \text{if } q \neq 0 \\ C_{-p}(\mathcal{U}, Ext_c^0(\mathcal{F}, K_X^\bullet)) & \text{if } q = 0. \end{cases}$$

Accordingly,

$$'E_2^{p,q} = \begin{cases} 0 & \text{if } q \neq 0 \\ H_{-p}(\mathfrak{U}, \mathcal{E}xt_c^0(\mathcal{F}, K_X^\bullet)) & \text{if } q = 0. \end{cases}$$

The second spectral sequence has the former term

$$''E_1^{p,q} = H_{-p}(\mathfrak{U}, \mathcal{H}om_{\mathcal{O}_X}(\mathcal{F}, \mathcal{J}^q)_c).$$

Since the sheaves $\mathcal{H}om_{\mathcal{O}_X}(\mathcal{F}, \mathcal{J}^q)$ are flabby, by lemma 3.9 we get

$$''E_1^{p,q} = \begin{cases} 0 & \text{if } p \neq 0 \\ \mathrm{Hom}_c(X; \mathcal{F}, \mathcal{J}^q) & \text{if } p = 0, \end{cases}$$

therefore,

$$''E_2^{p,q} = \begin{cases} 0 & \text{if } p \neq 0 \\ \mathrm{Ext}_c^q(X; \mathcal{F}, K_X^\bullet) & \text{if } p = 0. \end{cases}$$

The conclusion will follow from the properties of the spectral sequence.

(d) In this section we prove *the second theorem of absolute duality* of Ramis and Ruget.

THEOREM 3.10. *Let X be a finite-dimensional complex space with countable basis. For any coherent analytic sheaf $\mathcal{F}$ on X and for any integer p, there exists a unique QDFS structure on $H_c^p(X, \mathcal{F})$ and a unique QFS structure on* $\mathrm{Ext}^{-p}(X; \mathcal{F}, K_X^\bullet)$ *such that the trace map T_X induces a topological duality between their separated spaces. Moreover, the separation on $H_c^p(X, \mathcal{F})$ is equivalent to that of* $\mathrm{Ext}^{1-p}(X; \mathcal{F}, K_X^\bullet)$.

For the proof of this theorem we need some preparations. Some of them are consequences of the results in Chapter I; here we give a straightforward proof to make the duality stated in 3.10 independent of that chapter.

ASSERTION 3.11. *Let V be an open set of some numerical space $\mathbb{C}^n$, $n \geqslant 1$. Then $H^n(V, \mathcal{O}) = 0$.*

This fact, which has already been applied in the proof of theorem I.2.9, follows from the properties of the Laplace operator (see Ch. I).

LEMMA 3.12. *Let V be a complex manifold and let K be a compact of V. Then the canonical map*

$$\mathcal{O}(K) \to H_c^1(V \setminus K, \mathcal{O})$$

is continuous ($\mathcal{O}(K)$ is endowed with its natural LF structure and $H_c^1(V \setminus K, \mathcal{O})$ with the topology given by the Dolbeault resolution).

Proof. We must show that for any neighbourhood U of K the composite map $\mathcal{O}(U) \to \mathcal{O}(K) \to H_c^1(V \setminus K, \mathcal{O})$ is continuous. From the commutative diagram

$$\begin{array}{ccc} \mathcal{O}(U) & \to & H_c^1(U \setminus K, \mathcal{O}) \\ \downarrow & & \downarrow \\ \mathcal{O}(K) & \to & H_c^1(V \setminus K, \mathcal{O}) \end{array}$$

there results that it is enough to prove the continuity of the map

$$\mathcal{O}(U) \to H_c^1(U \setminus K, \mathcal{O}),$$

hence we may suppose $U = V$. Let $(f_n)_n \to f$ in $\mathcal{O}(V)$. Consider a function $\varphi \in C_c^\infty(V)$, equal to 1 on a neighbourhood of K. One can easily see that $d''(\varphi f_n)$, $d''(\varphi f)$ are cycles of $\Gamma_c(V \setminus K, \mathcal{E}^{0,1})$ and their classes in $H_c^1(V \setminus K, \mathcal{O})$ are just the images of the elements f_n, f under the morphism $\mathcal{O}(V) \to H_c^1(V \setminus K, \mathcal{O})$. But $d''(\varphi f_n) \to d''(\varphi f)$ and the lemma is concluded.

LEMMA 3.13. *Let V be a Stein open set of some space $\mathbb{C}^n$ and K a Stein compact of V. Then $H_K^p(V, \Omega) = 0$ for $p \neq n$ and the space $H_K^n(V, \Omega)$ can be endowed with a natural FS structure such that its strong topological dual is isomorphic to the space $\mathcal{O}(K)$ (the latter being endowed with the natural LF structure).*

Proof. We first consider the case $n \geqslant 2$. We have the exact sequence

$$(*) \qquad \ldots \leftarrow H^p(K, \mathcal{O}) \leftarrow H_c^p(V, \mathcal{O}) \leftarrow H_c^p(V \setminus K, \mathcal{O}) \leftarrow H^{p-1}(K, \mathcal{O}) \leftarrow \ldots$$

Since $H^q(K, \mathcal{O}) = 0$ for $q \geqslant 1$, it follows that the continuous map

$$H_c^n(V, \mathcal{O}) \leftarrow H_c^n(V \setminus K, \mathcal{O})$$

is bijective (the topologies are induced by the Dolbeault resolution).

From lemma 3.2 and the remark which follows it, there results that $H_c^n(V \setminus K, \mathcal{O})$ is separated and that the map $\Gamma(V, \Omega) \to \Gamma(V \setminus K, \Omega)$ is bijective.

From the exact sequence

$$(**) \qquad \ldots \to H_K^p(V, \Omega) \to H^p(V, \Omega) \to H^p(V \setminus K, \Omega) \to H_K^{p+1}(V, \Omega) \to \ldots$$

we deduce that $H_K^p(V, \Omega) = 0$ if $p = 0$ and $p = 1$.

By $(*)$, the continuous maps $H_c^p(V, \mathcal{O}) \leftarrow H_c^p(V \setminus K, \mathcal{O})$ are bijective for any integer p, $2 \leqslant p \leqslant n-1$. We then deduce from 3.2 that $H_c^p(V \setminus K, \mathcal{O}) = 0$ for these integers. By applying again the remark from lemma 3.2 it follows that $H^p(V \setminus K, \Omega) = 0$ for $1 \leqslant p \leqslant n-2$ (the separation of the space $H_c^n(V \setminus K, \mathcal{O})$ has already been proved !). Consequently, by means of $(**)$, $H_K^p(V, \Omega) = 0$ for $2 \leqslant p \leqslant n-1$. Moreover, by 3.11, one has $H_K^p(V, \Omega) = 0$ for $p > n$.

It remains only to prove the last assertion of the lemma. By assertion 3.11 one will deduce that $H_c^1(V \setminus K, \mathcal{O})$ is separated, hence DFS. The map $\mathcal{O}(K) \to$ $\to H_c^1(V \setminus K, \mathcal{O})$ is continuous (3.12) and bijective in accordance with the sequence (*) ($n \geqslant 2$). By the closed graph theorem it will be a topological isomorphism. The exact sequence (**) shows that the map $H^{n-1}(V \setminus K, \Omega) \to H_K^n(V, \Omega)$ is bijective. The space $H^{n-1}(V \setminus K, \Omega)$ is separated (since $H_c^2(V \setminus K, \mathcal{O})$ enjoys this property) and it is in duality with $H_c^1(V \setminus K, \mathcal{O})$.

The proof of the lemma in the case $n \geqslant 2$ will be concluded obviously. The case $n = 1$ can be treated similarly.

Remark. In particular, it follows in virtue of this lemma that the natural LF structure on $\mathcal{O}(K)$ is DFS. This fact can be readily extended to spaces of the form $\mathcal{F}(K)$, $\mathcal{F} \in \mathbf{Coh}(V)$; then there result the same property for the Stein compacts of an arbitrary complex space.

LEMMA 3.14. *Let V be a Stein open set of some space $\mathbb{C}^n$ and K a Stein compact of V. If $\mathcal{F}$ is an $\mathcal{O}$-module which admits a finite resolution with free $\mathcal{O}$-modules of finite rank, then $Ext_K^p(V; \mathcal{F}, \Omega) = 0$ for $p \neq n$ and the space $Ext_K^n(V; \mathcal{F}, \Omega)$ can be endowed with a natural FS structure such that its strong topological dual is isomorphic to $\mathcal{F}(K)$.*

This follows by the previous lemma, through induction on the length of the resolution.

LEMMA 3.15. *Let X be a complex space, K a Stein compact of X and $\mathcal{F}$ an $\mathcal{O}_X$-module. Suppose that X is embedded in a domain of holomorphy V by an immersion f such that $f_*(\mathcal{F})$ has a finite resolution with free $\mathcal{O}_V$-modules of finite rank. Then* $\mathrm{Ext}_K^p(X; \mathcal{F}, K_X^{\bullet}) = 0$ *for $p \neq 0$ and the space* $\mathrm{Ext}_K^0(X; \mathcal{F}, K_X^{\bullet})$ *can be endowed with an FS structure such that its strong topological dual is isomorphic to $\mathcal{F}(K)$.*

This follows from lemma 3.14 and the remark made to proposition 3.4.

We close the preparations for the proof of the theorem by a general fact of the sheaf theory and by indicating how the invariants $H_c^{\bullet}(X, \mathcal{F})$ topologize.

Let X be a paracompact space, let $\mathfrak{K} = (K_i)_i$ be a locally finite covering of X by compacts and let $\mathcal{M}$ be a sheaf of abelian groups on X.

For an integer p, denote

$$C_p(\mathfrak{K}, \mathcal{M}_{\mathfrak{K}}) = \prod_{\dim s = p} \Gamma_{K_s}(X, \mathcal{M}).$$

The inclusion maps $\Gamma_K(X, \mathcal{M}) \subset \Gamma_{K'}(X, \mathcal{M})$ for $K \subset K'$ define naturally maps

$$C_p(\mathfrak{K}, \mathcal{M}_{\mathfrak{K}}) \to C_{p-1}(\mathfrak{K}, \mathcal{M}_{\mathfrak{K}}),$$

by means of alternate sums.

One can easily see that in this way we get a complex $C_{\bullet}(\mathfrak{K}, \mathcal{M}_{\mathfrak{K}})$, the complex of chains associated to the nerve of the covering $\mathfrak{K}$, and to the system of coefficients $K \mapsto \Gamma_K(X, \mathcal{M})$.

The inclusion maps $\Gamma_K(X, \mathcal{M}) \subset \Gamma(X, \mathcal{M})$ define an augmentation morphism

$$C_0(\mathfrak{K}, \mathcal{M}_{\mathfrak{K}}) \to \Gamma(X, \mathcal{M}).$$

LEMMA 3.16. *Let X be a paracompact space, $\mathfrak{K} = (K_i)_i$ a locally finite family of compacts whose interiors cover X and let $\mathcal{M}$ be a flabby sheaf on X. Then the homology groups of the complex $C_\bullet(\mathfrak{K}, \mathcal{M}_{\mathfrak{K}})$ vanish in all dimensions except the dimension zero which is canonically isomorphic to $\Gamma(X, \mathcal{M})$.*

Proof. Let $\mathcal{M} \to \mathcal{J}^\bullet$ be an injective resolution. By using the spectral sequences associated by the double complex

$$E^{p,q} = \prod_{\dim s = p} \Gamma_{K_s}(X, \mathcal{J}^q),$$

one can easily derive that, in order to prove the lemma, we may assume $\mathcal{M}$ injective.

Let i be an index of the covering. Since $\mathcal{H}^0_{K_i}(\mathcal{M})$ is injective (1.27), there exists a morphism $\varphi_i : \mathcal{M} \to \mathcal{H}^0_{K_i}(\mathcal{M})$ such that the diagram

$$\begin{array}{ccc} 0 \to \mathcal{H}^0_{K_i}(\mathcal{M}) & \longrightarrow & \mathcal{M} \\ \downarrow \text{id} & \swarrow \varphi_i & \\ \mathcal{H}^0_{K_i}(\mathcal{M}) & & \end{array}$$

is commutative. The morphism $\varphi : \mathcal{M} \to \prod_i \mathcal{H}^0_{K_i}(\mathcal{M})$, defined by the morphisms φ_i, is injective (the interiors of the sets K_i cover X!). There exists $\psi : \prod_i \mathcal{H}^0_{K_i}(\mathcal{M}) \to \mathcal{M}$ such that $\psi\varphi = \text{id}$. Denote by ψ_i the restriction of ψ to $\mathcal{H}^0_{K_i}(\mathcal{M})$. The following equality holds $1 = \sum_i \psi_i\varphi_i$ ($\prod_i \mathcal{H}_{K_i}(\mathcal{M}) = \oplus_i \mathcal{H}_{K_i}(\mathcal{M})$, the family of supports being locally finite). Define

$$\theta_p : \prod_{\dim s = p} \Gamma_{K_s}(X, \mathcal{M}) \to \prod_{\dim s = p+1} \Gamma_{K_s}(X, \mathcal{M})$$

and

$$\theta_{-1} : \Gamma(X, \mathcal{M}) \to \prod_i \Gamma_{K_i}(X, \mathcal{M})$$

by the formulae

$$\theta_p(f)_{i_0 \ldots i_{p+1}} = \sum_{l=0}^{p+1} \psi_{i_l}\varphi_{i_l}(f_{i_0 \ldots \hat{i}_l \ldots i_{p+1}})$$

$$(\text{Supp}\, \psi_{i_l}\varphi_{i_l}(f_{i_0 \ldots \hat{i}_l \ldots i_{p+1}}) \subset K_{i_0 \ldots i_{p+1}}),$$

and

$$\theta_{-1}(f) = (\psi_i\varphi_i(f))_i, \text{ respectively.}$$

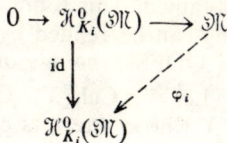

Thus, one obtains a homotopy for the complex

$$\dots \to C_1(\mathcal{R}, \mathcal{M}_\mathcal{R}) \to C_0(\mathcal{R}, \mathcal{M}_\mathcal{R}) \to \Gamma(X, \mathcal{M}) \to 0$$

and the lemma is proved.

We now consider topologies on the invariants $H_c^{\cdot}(X, \mathcal{F})$ (X being a complex space with countable basis and $\mathcal{F} \in \mathbf{Coh}(X)$). For any Stein compact K the space $\mathcal{F}(K) = \Gamma(K, \mathcal{F})$ has a natural LF structure given by the equality $\mathcal{F}(K) = \varinjlim_{U \supset K} \mathcal{F}(U)$. By a result due to Grothendieck (I.2.10 or the remark to lemma 3.13), this topological structure is DFS.

Let $\mathcal{R}$ be a locally finite covering of X by Stein compacts. Since $H^q(K, \mathcal{F}) = 0$ for $q \geq 1$ and for any Stein compact K, $H_c^{\cdot}(X, \mathcal{F})$ are the cohomology groups of the complex of finite cochains $C_c^{\cdot}(\mathcal{R}, \mathcal{F})$. For an integer p, $C_c^p(\mathcal{R}, \mathcal{F}) = \bigoplus_{\dim s = p} \mathcal{F}(K_s)$. Accordingly, each $C_c^p(\mathcal{R}, \mathcal{F})$ has a natural DFS structure and one can easily show that the differentials $C_c^p(\mathcal{R}, F) \to C_c^{p+1}(\mathcal{R}, \mathcal{F})$ are continuous. Thus on the invariants $H_c^{\cdot}(X, \mathcal{F})$ a QDFS topology is obtained. This topology is independent of the covering $\mathcal{R}$ (indeed, two such covering can be refined by a third one and apply lemma 1.33) and it is called *the natural QDFS topology on* $H_c^{\cdot}(X, \mathcal{F})$.

The proof of the theorem. Let $\mathcal{F} \in \mathbf{Coh}(X)$. Consider a locally finite family $\mathcal{R} = (K_i)_i$ of Stein compacts of X whose interiors cover X. Suppose $\mathcal{R}$ sufficiently fine such that for any $K \in \mathcal{R}$ there exists a neighbourhood U enjoying the property that one can apply lemma 3.15 for $(U, K, \mathcal{F}|U)$. There exist natural isomorphisms

$$\mathrm{Ext}_K^{\cdot}(X; \mathcal{F}, K_X^{\cdot}) \simeq \mathrm{Ext}_K^{\cdot}(U; \mathcal{F}, K_X^{\cdot}).$$

Thereby $\mathrm{Ext}_K^p(X; \mathcal{F}, K_X^{\cdot}) = 0$ for $p \neq 0$. Moreover, the space $\mathrm{Ext}_K^0(X; \mathcal{F}, K_X^{\cdot})$ has a natural FS structure and it is topologically isomorphic to the strong dual of $\mathcal{F}(K)$.

These facts remain true if K is replaced by an intersection of compacts from $\mathcal{R}$ and one has comatibility with inclusions $K_s \subset K_t$, $t \subset s$.

The groups $H_c^{\cdot}(X, \mathcal{F})$ are the cohomology groups of the complex of finite cochains $C_c^{\cdot}(\mathcal{R}, \mathcal{F})$ of DFS spaces. In accordance with the above said the topological dual of $C_c^{\cdot}(\mathcal{R}, \mathcal{F})$ is isomorphic to the FS complex of infinite chains $C_{\cdot}^{\infty}(\mathcal{R}, \mathrm{Ext}_\mathcal{R}^0(X; \mathcal{F}, K_X^{\cdot}))$, where $\mathrm{Ext}_\mathcal{R}^0(X; \mathcal{F}, K_X^{\cdot})$ is the system of coefficients $K \mapsto \mathrm{Ext}_K^0(X; \mathcal{F}, K_X^{\cdot})$ (the connecting morphisms are canonically defined).

As in the case of theorem 3.7, the proof ends by showing that the homology of this FS complex is $\mathrm{Ext}^{\cdot}(X; \mathcal{F}, K_X^{\cdot})$ (modulo the sign of the dimension). This thing follows by using the spectral sequences associated to the double complex $C^{pq} = C_{-p}(\mathcal{R}, \mathrm{Hom}_{\mathcal{O}_X}(\mathcal{F}, \mathcal{I}^q)_\mathcal{R})$, where $\mathcal{I}^{\cdot}$ is an injective resolution of $K_X^{\cdot}$.

The fact that the duality isomorphisms of theorems 3.7 and 3.10 derive from the Yoneda maps can be verified first on the particular dualities from lemmas 3.1, 3.2, 3.13, and for the general case one can use the properties of functoriality and compatibility with spectral sequences.

§ 4. Duality on complex manifolds

(a) For the case of manifolds we restate the results obtained in the previous paragraph and obtain *the duality theorems on complex manifolds* due to Serre and Malgrange.

THEOREM 4.1. *Let X be a complex manifold with countable basis of dimension n and Ω the sheaf of germs of holomorphic forms of maximal degree.*

Then for any coherent analytic sheaf $\mathcal{F}$ on X and for any integer p there exist a unique QFS structure on $H^p(X, \mathcal{F})$ and a unique QDFS structure on $\operatorname{Ext}_c^{n-p}(X; \mathcal{F}, \Omega)$ such that the trace map induces a topological duality between the associated separated spaces. Moreover, $H^p(X, \mathcal{F})$ is separated if and only if $\operatorname{Ext}_c^{n-p+1}(X; \mathcal{F}, \Omega)$ is separated.

THEOREM 4.2. *Let X be a complex manifold with countable basis of dimension n and Ω the sheaf of the germs of holomorphic forms of maximal degree.*

Then for any coherent analytic sheaf $\mathcal{F}$ on X and for any integer p there exist a unique QDFS structure on $H_c^p(X, \mathcal{F})$ and a unique QFS structure on $\operatorname{Ext}^{n-p}(X; \mathcal{F}, \Omega)$ such that the trace map induces a topological duality between the separated spaces associated to these spaces. Moreover, $H_c^p(X, \mathcal{F})$ is separated if and only if $\operatorname{Ext}^{n-p+1}(X; \mathcal{F}, \Omega)$ is separated.

Proof. The dualizing complex $K_X^\bullet$ is a resolution of $\Omega[n]$, hence

$$\operatorname{Ext}_{\mathcal{O}}^q(X; \mathcal{F}, K_X^\bullet) \simeq \operatorname{Ext}_{\mathcal{O}}^{n+q}(X; \mathcal{F}, \Omega)$$

and

$$\operatorname{Ext}_{\mathcal{O},c}^q(X; \mathcal{F}, K_X^\bullet) \simeq \operatorname{Ext}_{\mathcal{O},c}^{n+q}(X; \mathcal{F}, \Omega).$$

One applies then 3.7 and 3.10.

Remarks. (1) The topologies on the invariants $H^\bullet(X, \mathcal{F})$ ($H_c^\bullet(X, \mathcal{F})$, respectively) are the natural ones, defined in the previous paragraph by means of the Čech complex associated to a locally finite covering by Stein open (respectively compact) sets. In the next section we will show that these topologies coincide with those induced from the Dolbeault-Grothendieck resolutions.

A similar assertion holds for the invariants $\operatorname{Ext}^\bullet$, $\operatorname{Ext}_c^\bullet$.

(2) If $\mathcal{F}$ is a locally free sheaf of finite rank, then $\operatorname{Ext}^\bullet(X; \mathcal{F}, \Omega) \simeq H^\bullet(X, \check{\mathcal{F}} \otimes \Omega)$ and $\operatorname{Ext}_c^\bullet(X; \mathcal{F}, \Omega) \simeq H_c^\bullet(X, \check{\mathcal{F}} \otimes \Omega)$, where $\check{\mathcal{F}} = Hom_{\mathcal{O}}(\mathcal{F}, \mathcal{O})$ is the dual of $\mathcal{F}$; the duality results can be expressed in cohomological terms only.

(3) Theorems 4.1 and 4.2 can be proved straightforwardly, without the construction of the dualizing complex (which uses the Frisch theorem of noetherianity) and the formalism of the trace map on the singular case. In order to prove this we need only lemmas 3.3, 3.9, 3.14, 3.16 and some simple considerations of spectral sequence! One proceeds exactly as in the proof of theorems 3.7 and 3.10. We give a sketch for 4.1 (according to [87]).

Let $\mathfrak{U}$ be a locally finite covering of X by Stein open sets. The invariants $H^\bullet(X, \mathcal{F})$ are the cohomology groups of the Fréchet-Schwartz complex $C^\bullet(\mathfrak{U}, \mathcal{F})$. This is isomorphic to the strong dual of the DFS complex of finite chains

$$C_\bullet(\mathfrak{U}, \mathcal{E}xt_c^n(\mathcal{F}, \Omega)),$$

where $\mathcal{E}xt_c^n(\mathcal{F}, \Omega)$ is the precosheaf $U \mapsto \mathrm{Ext}^n_{\mathcal{O},c}(U; \mathcal{F}, \Omega)$ (the corestriction morphisms are canonically defined). Let $\mathcal{J}^\bullet$ be an injective resolution of Ω. Consider the complex

$$C^{pq} = C_{-p}(\mathfrak{U}, \mathcal{H}om_{\mathcal{O}}(\mathcal{F}, \mathcal{J}^q)_c).$$

The associated spectral sequences degenerate. Thereby the homology of the complex $C_\bullet(\mathfrak{U}, \mathcal{E}xt_c^n(\mathcal{F}, \Omega))$ coincides with $\mathrm{Ext}_c^\bullet(X; \mathcal{F}, \Omega)$, etc.

We now particularize 4.1 and 4.2 for the case of compact and Stein manifolds.

THEOREM 4.3. *Let X be a compact complex manifold of dimension n, let Ω be the sheaf of germs of holomorphic forms of type $(n, 0)$, and $\mathcal{F}$ a coherent analytic sheaf on X.*

Then for any integer p, the complex vectorial spaces $H^p(X, \mathcal{F})$ and $\mathrm{Ext}^{n-p}_{\mathcal{O}}(X; \mathcal{F}, \Omega)$ are finite-dimensional and the trace map yields a duality between them; in particular these spaces are of the same dimension.

Proof. By Cartan-Serre's theorem (III. 2.10) the vectorial spaces $H^\bullet(X, \mathcal{F})$ are finite-dimensional. Let $\mathfrak{U}$ be a locally finite covering of X by Stein open sets. By applying 1.29 to the morphism $C^{p-1}(\mathfrak{U}, \mathcal{F}) \to Z^p(\mathfrak{U}, \mathcal{F})$ we deduce that $H^p(X, \mathcal{F})$ is separated. The assertion from the statement follows now by 4.1.

THEOREM 4.4. *Let X be a Stein manifold of dimension n, Ω the sheaf of germs of holomorphic forms of type $(n, 0)$, and $\mathcal{F}$ a coherent analytic sheaf on X.*

(*i*) $\mathrm{Ext}_c^p(X; \mathcal{F}, \Omega) = 0$ *for any $p \neq n$ and* $\mathrm{Ext}_c^n(X; \mathcal{F}, \Omega)$ *is isomorphic (via the trace map) to the topological dual of the Fréchet-Schwartz space* $\Gamma(X, \mathcal{F})$.

(*ii*) *For any integer p, the space $H_c^p(X, \mathcal{F})$ is separated (hence DFS) and its topological dual is isomorphic (via the trace map) to* $\mathrm{Ext}^{n-p}(X; \mathcal{F}, \Omega)$.

Moreover, the FS topology induced on $\mathrm{Ext}^{n-p}(X; \mathcal{F}, \Omega)$ *from this duality coincides with that induced from the coherent sheaf $\mathcal{E}xt^{n-p}(\mathcal{F}, \Omega)$ by means of the isomorphism* $\mathrm{Ext}^{n-p}(X; \mathcal{F}, \Omega) \simeq \Gamma(X, \mathcal{E}xt^{n-p}(\mathcal{F}, \Omega))$.

Proof. The assertion (*i*) follows from 4.1 by theorem *B*. We now prove (*ii*).

Let $(U_r)_r$ be an exhaustion of X by relatively compact Stein open sets such that the restriction maps $\Gamma(U_{r+1}, \mathcal{O}) \to \Gamma(U_r, \mathcal{O})$ are dense. Suppose moreover that the topological dual of $\Gamma(U_r, \mathcal{E}xt^{n-p}(\mathcal{F}, \Omega))$ is algebraically isomorphic (via the trace map!) to $H_c^p(U_r, \mathcal{F})$ for any r.

If $\mathcal{H} \in \mathbf{Coh}(X)$, then the maps $\Gamma(X, \mathcal{H}) \to \Gamma(U_r, \mathcal{H})$, $\Gamma(U_{r+1}, \mathcal{H}) \to \Gamma(U_r, \mathcal{H})$ are also dense. From the topological isomorphism

$$\Gamma(X, \mathcal{H}) \xrightarrow{\sim} \varprojlim_r \Gamma(U_r, \mathcal{H})$$

one will derive the algebraic isomorphism

$$\varinjlim_{r} (\Gamma(U_r, \mathcal{H}))' \rightrightarrows (\Gamma(\dot{X}, \mathcal{H}))'.$$

Apply this to $\mathcal{H} = Ext^{n-p}(\mathcal{F}, \Omega)$. It results that the topological dual of $\Gamma(X, Ext^{n-p}(\mathcal{F}, \Omega))$ is algebraically isomorphic to $\varinjlim H_c^p(U_r, \mathcal{H})$. Consider on $\mathrm{Ext}^{n-p}(X; \mathcal{F}, \Omega) \simeq \Gamma(X, Ext^{n-p}(\mathcal{F}, \Omega))$ the FS topology from $Ext^{n-p}(\mathcal{F}, \Omega)$, and on the space $H_c^p(X, \mathcal{F}) = \varinjlim_{r} H_c^p(U_r, \mathcal{F})$, the DFS topology obtained by means of the algebraic isomorphisms

$$\Gamma(X, Ext^{n-p}(\mathcal{F}, \Omega))' \simeq \varinjlim_{r} H_c^p(U_r, \mathcal{F}).$$

The pairing

$$H_c^p(X, \mathcal{F}) \times \mathrm{Ext}^{n-p}(X; \mathcal{F}, \Omega) \to \mathbb{C}$$

defined by this isomorphism coincides with that given by the trace map, since it is obtained from the pairings $H_c^p(U_r, \mathcal{F}) \times \mathrm{Ext}^{n-p}(U_r, \mathcal{F}, \Omega) \to \mathbb{C}$. The conclusion of (*ii*) follows by 4.2.

It remains only to prove that for any r, $H_c^p(U_r, \mathcal{F})$ is algebraically isomorphic, by means of the trace map, to the topological dual of $\Gamma(U_r, Ext^{n-p}(\mathcal{F}, \Omega))$. If $\mathcal{F}$ is locally free then it follows easily by (*i*) (there is a single nonnull *Ext*, namely, $Hom(\mathcal{F}, \Omega) \simeq \check{\mathcal{F}} \otimes \Omega \ldots$). For the general case one considers a finite resolution of $\mathcal{F}$ by locally free sheaves of finite rank on each U_r and one proceeds by induction on the length of this resolution. The theorem is thereby proved.

Remark. Assertion (*ii*) makes theorem I.2.1 more precise: the topology on the invariants $H_c^\bullet(X, \mathcal{F})$ obtained by duality from the invariants $\mathrm{Ext}^\bullet(X; \mathcal{F}, \Omega) = \Gamma(X, Ext^\bullet(\mathcal{F}, \Omega))$ coincides with the natural topology obtained by means of a locally finite covering by Stein compact sets.

(b) In this section we will present the duality on complex manifolds following Serre [75] and Malgrange [53].

(I) Let X be a complex manifold with countable basis of dimension n. We still use the notation $\mathcal{O}, \mathcal{E}^{p,q}, \mathcal{K}^{p,q}, \Omega^p$. By $\mathcal{E}$ we mean the sheaf of germs of C^∞ functions. Recall the following result [54]:

For any point $x \in X$, $\mathcal{E}_x$ is a flat $\mathcal{O}_x$-module.

Let $\mathcal{F}$ be an $\mathcal{O}$-module. By Dolbeault-Grothendieck resolution

$$0 \to \Omega^p \to \mathcal{E}^{p,0} \xrightarrow{d''} \ldots \xrightarrow{d''} \mathcal{E}^{p,n} \to 0,$$

we get by tensoring $\otimes_{\mathcal{O}} \mathcal{F}$ an exact sequence

$$(*) \qquad 0 \to \Omega^p \otimes_{\mathcal{O}} \mathcal{F} \to \mathcal{E}^{p,0} \otimes_{\mathcal{O}} \mathcal{F} \xrightarrow{d''} \ldots \xrightarrow{d''} \mathcal{E}^{p,n} \otimes_{\mathcal{O}} \mathcal{F} \to 0.$$

The sheaves $\mathcal{E}^{p,*} \otimes_{\mathcal{O}} \mathcal{F}$ have structure of $\mathcal{E}$-modules, hence they are $\Gamma(X, *)$ and $\Gamma_c(X, *)$-acyclic.

For convenience we denote:

$$\mathcal{E}^{p,q}(\mathcal{F}) = \mathcal{E}^{p,q} \otimes_{\mathcal{O}} \mathcal{F}, \qquad E^{p,q}(\mathcal{F}) = \Gamma(X, \mathcal{E}^{p,q} \otimes_{\mathcal{O}} \mathcal{F})$$

and

$$E_c^{p,q}(\mathcal{F}) = \Gamma_c(X, \mathcal{E}^{p,q} \otimes_{\mathcal{O}} \mathcal{F}).$$

Accordingly, $H^\bullet(X, \Omega^p \otimes_{\mathcal{O}} \mathcal{F})$ are the cohomology groups of the complex

$$0 \to E_c^{p,0}(\mathcal{F}) \xrightarrow{d''} E^{p,1}(\mathcal{F}) \to \ldots \xrightarrow{d''} E^{p,n}(\mathcal{F}) \to 0$$

and $H_c^\bullet(X, \Omega^p \otimes_{\mathcal{O}} \mathcal{F})$ are the cohomology groups of the complex

$$0 \to E_c^{p,0}(\mathcal{F}) \xrightarrow{d''} E_c^{p,1}(\mathcal{F}) \to \ldots \xrightarrow{d''} E_c^{p,n}(\mathcal{F}) \to 0.$$

In particular, $H^\bullet(X, \mathcal{F})$ and $H_c^\bullet(X, \mathcal{F})$ stand for the cohomology of the complexes

$$0 \to E^{0,0}(\mathcal{F}) \to E_c^{0,1}(\mathcal{F}) \to \ldots \to E^{0,n}(\mathcal{F}) \to 0,$$

$$0 \to E_c^{0,0}(\mathcal{F}) \to E_c^{0,1}(\mathcal{F}) \to \ldots \to E_c^{0,n}(\mathcal{F}) \to 0.$$

Suppose now $\mathcal{F}$ is coherent and consider TVS structures. Use of the following result [54]:

If U is an open subset of some numerical space and

$$\Gamma(U, \mathcal{E})^r \xrightarrow{\varphi} \Gamma(U, \mathcal{E})^s$$

is a morphism given by a holomorphic matrix, then $\operatorname{Im} \varphi$ *is a closed subspace* (consider on $\mathcal{E}$ the FS structure given by the uniform convergence of functions and all their derivarions on any compact).

Let U be an open subset of X, contained in a cart and such that one has a finite presentation

$$\mathcal{O}^r|U \to \mathcal{O}^s|U \to \mathcal{F}|U \to 0.$$

By tensoring $\mathcal{E}^{p,q} \otimes_{\mathcal{O}}$ one derives the exact sequence of $\mathcal{E}|U$-modules

$$(\mathcal{E}^{p,q}|U)^r \to (\mathcal{E}^{p,q}|U)^s \to (\mathcal{E}^{p,q}(\mathcal{F}))|U \to 0.$$

Since the cohomology of the $\mathcal{E}$-modules is trivial, one gets the exact sequence

$$\Gamma(U, \mathcal{E}^{p,q})^r \to \Gamma(U, \mathcal{E}^{p,q})^s \to \Gamma(U, \mathcal{E}^{p,q}(\mathcal{F})) \to 0,$$

where $\Gamma(U, \mathcal{E}^{p,q})^r$ and $\Gamma(U, \mathcal{E}^{p,q})^s$ are free $\Gamma(U, \mathcal{E})$-modules of finite rank; hence they own natural FS structures. Moreover, the former morphism is continuous and its image is closed. One thus obtains an FS structure on $\Gamma(U, \mathcal{E}^{p,q}(\mathcal{F}))$ and by the Banach open map theorem it is obvious that this topology is independent of the sequence $\mathcal{O}^r|U \to \mathcal{O}^s|U \to \mathcal{F}|U \to 0$.

For an open subset V of X we define on $\Gamma(V, \mathcal{E}^{p,q} \otimes_{\mathcal{O}} \mathcal{F})$ the smallest topology making the restriction maps $\Gamma(V, \mathcal{E}^{p,q} \otimes_{\mathcal{O}} \mathcal{F}) \to \Gamma(U, \mathcal{E}^{p,q} \otimes_{\mathcal{O}} \mathcal{F})$ continuous, where $U \subset V$ is an open set as above. Thus, $\mathcal{E}^{p,q}(\mathcal{F})$ becomes an FS sheaf. A family of seminorms giving the topology of $E^{p,q}(\mathcal{F}) = \Gamma(X, \mathcal{E}^{p,q} \otimes_{\mathcal{O}} \mathcal{F})$ is defined in the following manner. Consider an open set U as above, an epimorphism $\mathcal{O}^s|U \to \mathcal{F}|U$, a compact $K \subset U$ and an operator of derivation D. For $f \in E^{p,q}(\mathcal{F})$ denote

$$p_{U,K,\varphi,D}(f) = \inf_{\omega} p_{K,D}(\omega),$$

where ω is an element of $\Gamma(U, \mathcal{E}^{p,q}(\mathcal{O}^s))$ whose image in $\Gamma(U, \mathcal{E}^{p,q}(\mathcal{F}))$ under φ coincides with the restriction of f and $p_{K,D}$ is the seminorm "sup" given by K and D (one identifies $\Gamma(U, \mathcal{E}^{p,q}(\mathcal{O}^s))$ with a free $\Gamma(U, \mathcal{E})$-module of finite rank ...). The space $E^{p,q}_c(\mathcal{F})$ is dense in $E^{p,q}(\mathcal{F})$: indeed, any $f \in E^{p,q}(\mathcal{F})$ arises as the limit of a sequence $(f_n)_n$, $f_n = \sum_{i=0}^{n} \rho_i f$, where $(\rho_i)_{i \geq 0}$ is a suitable partition of unity.

If $\mathcal{F} \to \mathcal{G}$ is a morphism in $\mathbf{Coh}(X)$, then we obtain continuous morphisms $\mathcal{E}^{p,q}(\mathcal{F}) \to \mathcal{E}^{p,q}(\mathcal{G})$ (by means of suitable carts and finite presentations for $\mathcal{F}$ and $\mathcal{G}$). Moreover, the maps $E^{p,q}(\mathcal{F}) \to E^{p,q}(\mathcal{G})$ are strict, as follows easily from the exact sequences

$$E^{p,q}(\mathcal{F}) \to E^{p,q}(\mathcal{G}) \to E^{p,q}(\mathcal{H}),$$

where $\mathcal{H} = \operatorname{Coker}(\mathcal{F} \to \mathcal{G})$.

By the local use of epimorphisms $\mathcal{O}^s \to \mathcal{F} \to 0$ one readily show that the maps $d'' : E^{p,q}(\mathcal{F}) \to E^{p,q+1}(\mathcal{F})$ are continuous.

In this way $E^{p,\bullet}(\mathcal{F})$ is an FS complex and we can consider QFS structures on the invariants $H^\bullet(X, \Omega^p \otimes_{\mathcal{O}} \mathcal{F})$, peculiarly on $H^\bullet(X, \mathcal{F})$. For each compact K of X, $\Gamma_K(X, \mathcal{E}^{p,q}(\mathcal{F}))$ is a closed subspace of $E^{p,q}(\mathcal{F})$. On $E^{p,q}_c(\mathcal{F}) = \Gamma_c(X, \mathcal{E}^{p,q}(\mathcal{F})) = \varinjlim_K \Gamma_K(X, \mathcal{E}^{p,q}(\mathcal{F}))$ consider the topology of inductive limit. $E^{p,q}_c(\mathcal{F})$ is a strict inductive limit of FS spaces and the inclusion $E^{p,q}_c(\mathcal{F}) \subset E^{p,q}(\mathcal{F})$ is continuous. A sequence $(f_n)_n$ of elements of $E^{p,q}_c(\mathcal{F})$ converges to an element $f \in E^{q,q}_c(\mathcal{F})$ if and only if there exists a compact K of X such that $\operatorname{Supp} f_n$, $\operatorname{Supp} f \subset K$ and $(f_n) \to f$ in $E^{p,q}(\mathcal{F})$. One easily checks the continuity of the differentials $E^{p,q}_c(\mathcal{F}) \xrightarrow{d''} E^{p,q+1}_c(\mathcal{F})$. Thus, $E^{p,\bullet}_c(\mathcal{F})$ is a complex of locally-convex TVS.

We can consider on the invariants $H^\bullet_c(X, \Omega^p \otimes_{\mathcal{O}} \mathcal{F})$, in particular on $H^\bullet_c(X, \mathcal{F})$, associated structures of locally-convex TVS (generally nonseparated).

Proposition 4.5 (cf. [51]). *Let X be a complex manifold with countable basis and $\mathcal{F}$ a coherent analytic sheaf on X. Then the natural topologies on $H^\bullet(X, \mathcal{F})$ and $H^\bullet_c(X, \mathcal{F})$, obtained by means of the Čech cohomology, coincide with the topologies obtained by means of the Dolbeault-Grothendieck resolutions.*

To prove this we need some preparations.

Let

$$\begin{array}{ccccccc}
A^{00} & \xrightarrow{d} & A^{01} & \xrightarrow{d} & A^{02} & \xrightarrow{d} & \dots \\
\downarrow \delta & & \downarrow \delta & & \downarrow \delta & & \\
A^{10} & \xrightarrow{d} & A^{11} & \xrightarrow{d} & A^{12} & \xrightarrow{d} & \dots \\
\downarrow \delta & & \downarrow \delta & & \downarrow \delta & & \\
A^{20} & \xrightarrow{d} & A^{21} & \xrightarrow{d} & A^{22} & \xrightarrow{d} & \dots \\
\downarrow \delta & & \downarrow \delta & & \downarrow \delta & & \\
\vdots & & \vdots & & \vdots & &
\end{array}$$

be a double complex of abelian groups where the rows and the columns are exact sequence. Denote by $B^{\bullet}$ the complex $\operatorname{Ker}(A^{0,\bullet} \to A^{1,\bullet})$, whose differential is induced by d and let $C^{\bullet}$ be the complex $\operatorname{Ker}(A^{\bullet,0} \to A^{\bullet,1})$ whose differential is induced by δ.

Let $p \geqslant 0$ be an integer. Consider an element $\dot{\xi} \in H^p(B^{\bullet})$, where $\xi \in Z^p(B^{\bullet})$. Denote $\xi^{0,p} = \xi$. Since $d\xi^{0,p} = 0$, there exists $\eta^{0,p-1} \in A^{0,p-1}$ so that $d\eta^{0,p-1} = \xi^{0,p}$. Let $\xi^{1,p-1} = \delta\eta^{0,p-1}$. Since $d\xi^{1,p-1} = 0$, there exists $\eta^{1,p-2} \in A^{1,p-2}$ such that $d\eta^{1,p-2} = \xi^{1,p-1}$. Denote $\xi^{2,p-2} = \delta\eta^{1,p-2}$; we can go on and finally find an element $\xi^{p,0}$. Let $\eta = \xi^{p,0}$. The class of η in $H^p(C^{\bullet})$ is independent of all choice and the assignment $\dot{\xi} \mapsto \dot{\eta}$ defines an isomorphism $H^p(B^{\bullet}) \rightrightarrows H^p(C^{\bullet})$.

Suppose now that $A^{p,q}$ are TVS and the differentials d, δ are continuous. The complexes $B^{\bullet}$ and $C^{\bullet}$ become naturally complexes of TVS. Consider the topologies induced on $H^{\bullet}(B^{\bullet})$ and $H^{\bullet}(C^{\bullet})$. By the above construction there follows

LEMMA 4.6. *If the differentials $A^{p,q} \xrightarrow{d} A^{p,q+1}$, $A^{p,q} \xrightarrow{\delta} A^{p+1,q}$ are strict, then there exist natural topological isomorphisms $H^{\bullet}(B^{\bullet}) \simeq H^{\bullet}(C^{\bullet})$.*

We shall also use the following

LEMMA 4.7. *Let $(A_n)_n$, $(B_n)_n$ be two sequences of locally-convex TVS and $f_n : A_n \to B_n$ a sequence of strict morphisms. Then the morphism $f : \oplus A_n \to \oplus B_n$ induced by $(f_n)_n$ is strict.*

Proof. Let (C_n) be a sequence of locally-convex TVS. If V_n is a neighbourhood of the origin in C_n, which is convex and disked, then $\oplus V_n$ is a neighbourhood of the origin in $C = \oplus C_n$. We will show that the neighbourhood of this kind constitute a fundamental system and the conclusion of the lemma will easily follow. Let V be a neighbourhood of the origin in C. There exist neighbourhoods V_n of the origin in C_n which are convex and disked such that V contains the disked convex envelope $\Gamma(\cup V_n)$ of the set $\cup V_n$ in C (we have identified the elements of C_n with the corresponding elements defined by them in C). The conclusion follows by the inclusion $\oplus 1/2^n V_n \subset \Gamma(\cup V_n)$.

The proof of the proposition. We first study the simpler case of the invariants $H^{\bullet}(X, \mathcal{F})$. Let $\mathfrak{U}$ be a locally finite covering of X by Stein open sets. Consider the double complex $C^p(\mathfrak{U}; \mathcal{E}^{0,q} \otimes_{\mathcal{O}} \mathcal{F})$ of the Fréchet-Schwartz spaces and the associated simple complex $C^{\bullet}$.

The connection morphisms

$$C^{\bullet}(\mathfrak{U}, \mathcal{F}) \to C^{\bullet} \text{ and } E^{0,\bullet}(\mathcal{F}) \to C^{\bullet}$$

are continuous quasi-isomorphisms. The required conclusion follows by lemma 1.32.

Let now $\mathfrak{U} = (U_i)_{i\in I}$ be a locally finite covering of X by relatively compact Stein open sets such that each $\overline{U_i}$ is a Stein compact. The cohomology of the complex of finite chains $C_c^{\bullet}(U, \mathcal{F})$ coincides with $H_c^{\bullet}(X, \mathcal{F})$. We consider on each $C_c^p(\mathfrak{U}, \mathcal{F}) = \oplus\, \mathcal{F}(U_{i_0\ldots i_p})$ the topology of direct sum (in the category of locally-convex TVS). Thus $C_c^{\bullet}(\mathfrak{U}, \mathcal{F})$ becomes a TVS complex. Consider on $H_c^{\bullet}(X, \mathcal{F})$ the induced topology and show that it is the same as the natural topology given by the coverings of Stein compacts.

There exist two locally finite coverings $\mathcal{K}' = (K_i')_{i\in I}$, $\mathcal{K} = (K_i)_{i\in I}$ by Stein compact sets such that the inclusions $K_i \subset U_i \subset K_i'$ take place. The restriction maps $C_c^{\bullet}(\mathcal{K}', \mathcal{F}) \to C_c^{\bullet}(\mathfrak{U}, \mathcal{F}) \to C_c^{\bullet}(\mathcal{K}, \mathcal{F})$ are continuous; hence the morphisms induced to cohomology are also continuous. Since $H^{\bullet}(C_c^{\bullet}(\mathcal{K}', \mathcal{F})) \to H^{\bullet}(C_c^{\bullet}(\mathcal{K}, \mathcal{F}))$ is a topological isomorphism, the assertion is proved.

Actually, we compare the topology given by $C_c^{\bullet}(\mathfrak{U}, \mathcal{F})$ on $H_c^{\bullet}(X, \mathcal{F})$ with the topology obtained from the complex $\Gamma_c(X, \mathcal{E}^{0,\bullet} \otimes \mathcal{F})$. And for this we apply lemma 4.6. Consider the double complex of components $C_c^p(\mathfrak{U}, \mathcal{E}^{0,q} \otimes \mathcal{F})$ on which we consider the topologies of direct sums (the differentials are obviously continuous in these topologies). For any Stein open set U, the FS topology on $\Gamma(U, \mathcal{F})$ coincided with that induced from $\Gamma(U, \mathcal{E}^{0,0} \otimes_{\mathcal{O}} \mathcal{F})$, as follows by the exact sequence

$$0 \to \Gamma(U, \mathcal{F}) \to \Gamma(U, \mathcal{E}^{0,0} \otimes_{\mathcal{O}} \mathcal{F}) \to \Gamma(U, \mathcal{E}^{0,1} \otimes_{\mathcal{O}} \mathcal{F})$$

of FS spaces and continuous maps.

The kernel of the morphism of complexes $C_c^{\bullet}(\mathfrak{U}, \mathcal{E}^{0,0} \otimes_{\mathcal{O}} \mathcal{F}) \to C_c^{\bullet}(\mathfrak{U}, \mathcal{E}^{0,1} \otimes_{\mathcal{O}} \mathcal{F})$ is algebraically isomorphic to $C_c^{\bullet}(\mathfrak{U}, \mathcal{F})$, and from lemma 4.7 we can deduce that the topology of $C_c^{\bullet}(\mathfrak{U}, \mathcal{F})$ coincides with the topology induced from $C_c^{\bullet}(\mathfrak{U}, \mathcal{E}^{0,0} \otimes_{\mathcal{O}} \mathcal{F})$.

The kernel of the morphism of complexes $C_c^0(\mathfrak{U}, \mathcal{E}^{0,\bullet} \otimes_{\mathcal{O}} \mathcal{F}) \to C_c^1(\mathfrak{U}, \mathcal{E}^{0,\bullet} \otimes_{\mathcal{O}} \mathcal{F})$ is algebraically isomorphic to $\Gamma_c(X, \mathcal{E}^{0,\bullet} \otimes_{\mathcal{O}} \mathcal{F})$. The maps $\Gamma_c(X, \mathcal{E}^{0,p} \otimes_{\mathcal{O}} \mathcal{F}) \to \to \bigoplus_i \Gamma(U_i, \mathcal{E}^{0,p} \otimes_{\mathcal{O}} \mathcal{F})$ are continuous, since the composite maps (obtained by restriction) $\Gamma_K(X, \mathcal{E}^{0,p} \otimes_{\mathcal{O}} \mathcal{F}) \to \bigoplus_i \Gamma(U_i, \mathcal{E}^{0,p} \otimes_{\mathcal{O}} \mathcal{F})$ are continuous ($K \subset X$ being an arbitrary compact). Prove that the topology of $\Gamma_c(X, \mathcal{E}^{0,p} \otimes_{\mathcal{O}} \mathcal{F})$ coincides with the topology induced from $C_c^0(\mathfrak{U}, \mathcal{E}^{0,p}(\mathcal{F})) = \bigoplus_i \Gamma(U_i, \mathcal{E}^{0,p} \otimes_{\mathcal{O}} \mathcal{F})$. It is enough to construct a continuous map $C_c^0(\mathfrak{U}, \mathcal{E}^{0,p}(\mathcal{F})) \to \Gamma_c(X, \mathcal{E}^{0,p} \otimes_{\mathcal{O}} \mathcal{F})$ whose composition to the left with the map $\Gamma_c(X, \mathcal{E}^{0,p} \otimes_{\mathcal{O}} \mathcal{F}) \to C_c^0(\mathfrak{U}, \mathcal{E}^{0,p} \otimes_{\mathcal{O}} \mathcal{F})$ gives rise to the identical map. Let (ρ_i) be a C^∞ partition of unity which is subordinate to the covering $\mathfrak{U}$. The assignment $(s_i)_i \mapsto \sum \rho_i s_i$ is the very required map. Thus the topology of $\Gamma_c(X, \mathcal{E}^{0,\bullet} \otimes_{\mathcal{O}} \mathcal{F})$ coincides with that induced from the complex $C_c^0(\mathfrak{U}, \mathcal{E}^{0,\bullet} \otimes_{\mathcal{O}} \mathcal{F})$.

In order to conclude the proof of the proposition, it is sufficient to show that both differentials

$$C_c^p(\mathfrak{U}, \mathcal{E}^{0,q} \otimes \mathcal{F}) \to C_c^p(\mathfrak{U}, \mathcal{E}^{0,q+1} \otimes \mathcal{F}), C_c^p(\mathfrak{U}, \mathcal{E}^{0,q} \otimes \mathcal{F}) \to C_c^{p+1}(\mathfrak{U}, \mathcal{E}^{0,q} \otimes \mathcal{F})$$

are strict. For the former, the conclusion follows by 4.7 from the exact sequences of FS spaces

$$\Gamma(U, \mathcal{E}^{0,q} \otimes \mathcal{F}) \to \Gamma(U, \mathcal{E}^{0,q+1} \otimes \mathcal{F}) \to \Gamma(U, \mathcal{E}^{0,q+2} \otimes \mathcal{F}).$$

As for the latter, we will construct a continuous homotopy $C_c^{p+1}(\mathfrak{U}, \mathcal{E}^{0,q} \otimes \mathcal{F}) \to$ $\to C_c^p(\mathfrak{U}, \mathcal{E}^{0,q} \otimes \mathcal{F})$ and the conclusion follows. Let $(\rho_i)_i$ be the above partition. The map

$$s = (s_{i_0 \ldots i_{p+1}}) \mapsto t = (t_{i_0 \ldots i_p}), \text{ where } t_{i_0 \ldots i_p} = \sum_i \rho_i s_{i i_0 \ldots i_p}$$

is the one required.

Remarks. 1) By the proposition it follows in particular that the topological structure on $H_c^{\bullet}(X, \mathcal{F})$ obtained from $\Gamma_c(X, \mathcal{E}^{0,\bullet} \otimes_{\mathcal{O}} \mathcal{F})$ is QDFS!

2) If $\mathcal{F}$ is furthermore locally free, then the invariants $H^{\bullet}(X, \mathcal{F})$ can be also computed by means of the complex $\Gamma(X, \mathcal{F} \otimes_{\mathcal{O}} \mathcal{K}^{0,\bullet}) \simeq K^{0,\bullet}(\mathcal{F})$. The topology thus defined on $H^{\bullet}(X, \mathcal{F})$ coincides with the existent one. A similar assertion holds for $H_c^{\bullet}(X, \mathcal{F})$. The proof is the same as in the proposition ([51]).

(II). Let $\mathcal{F}$ be an $\mathcal{O}$-module. For convenience, denote

$$K^{p,q}(\check{\mathcal{F}}) = \Gamma(X, \mathcal{H}om_{\mathcal{O}}(\mathcal{F}, \mathcal{K}^{p,q})) = \mathrm{Hom}_{\mathcal{O}}(\mathcal{F}, \mathcal{K}^{p,q}),$$

$$K_c^{p,q}(\check{\mathcal{F}}) = \Gamma_c(X, \mathcal{H}om_{\mathcal{O}}(\mathcal{F}, \mathcal{K}^{p,q})) = \mathrm{Hom}_{\mathcal{O},c}(\mathcal{F}, \mathcal{K}^{p,q}),$$

$$K^{p,q} = K^{p,q}(\check{\mathcal{O}}) = \Gamma(X, \mathcal{K}^{p,q}) \text{ and } K_c^{p,q} = K_c^{p,q}(\check{\mathcal{O}}) = \Gamma_c(X, \mathcal{K}^{p,q}).$$

These assignements are functorial.

If $\mathcal{F}$ and $\mathcal{G}$ are in $\mathbf{Mod}(O)$, then there exists a canonical $\mathcal{O}$-morphism

$$\mathcal{F} \otimes_{\mathcal{O}} \mathcal{H}om_{\mathcal{O}}(\mathcal{F}, \mathcal{G}) \to \mathcal{G}, \quad (f, \varphi) \mapsto \varphi(f).$$

The exterior product also defines morphisms

$$\mathcal{E}^{p,q} \otimes_{\mathcal{O}} \mathcal{K}^{p',q'} \to \mathcal{K}^{p+p',q+q'}.$$

Thus, one deduces for any $\mathcal{F} \in \mathbf{Mod}(\mathcal{O})$ morphisms

$$(\mathcal{E}^{p,q} \otimes_{\mathcal{O}} \mathcal{F}) \otimes_{\mathcal{O}} \mathcal{H}om_{\mathcal{O}}(\mathcal{F}, \mathcal{K}^{p',q'}) \to \mathcal{K}^{p+p',q+q'},$$

thereby, the maps

$$\varepsilon : E^{p,q}(\mathcal{F}) \otimes_{\Gamma(X,\mathcal{O})} K^{p',q'}(\check{\mathcal{F}}) \to K^{p+p',q+q'}.$$

If $\omega \in E^{p,q}(\mathcal{F})$ and $T \in K^{p',q'}(\check{\mathcal{F}})$, then we set $\omega \wedge T \in K^{p+p',q+q'}$ instead of $\varepsilon(\omega \otimes T)$. The operation $\wedge$ will be called *exterior product*. For any open set U of X, $(\omega \wedge T)|U = (\omega|U) \wedge (T|U)$. If ω or T has a compact support, then $\omega \wedge T$ has also a compact support. Thus, ε induces maps denoted by ε too,

$$E_c^{p,q}(\mathcal{F}) \otimes K^{p',q'}(\check{\mathcal{F}}) \to K_c^{p+p',q+q'} \qquad E^{p,q}(\mathcal{F}) \otimes K_c^{p',q'}(\check{\mathcal{F}}) \to K^{p+p',q+q'}.$$

If $\mathcal{F} = \mathcal{O}^r$ and if we identify ω with r forms $(\omega_i)_{1 \leq i \leq r}$ and T with r forms $(T_i)_{1 \leq i \leq r}$, then $\omega \wedge T = \sum_{i=1}^{r} \omega_i \wedge T_i$.

If $\varphi : \mathcal{F} \to \mathcal{G}$ is a morphism of $\mathcal{O}$-modules, then $\omega \wedge \varphi(T) = \varphi(\omega) \wedge T$ for any $\omega \in E^{p,q}(\mathcal{F})$ and $T \in K^{p',q'}(\mathcal{G})$. We suppose in the following that $\mathcal{F} \in \mathbf{Coh}(\mathcal{O})$. For $p' = n - p$ and $q' = n - q$ one obtains particularly the maps

$$E_c^{p,q}(\mathcal{F}) \otimes K^{n-p,n-q}(\check{\mathcal{F}}) \to K_c^{n,n},\ E^{p,q}(\mathcal{F}) \otimes K_c^{n-p,n-q}(\check{\mathcal{F}}) \to K_c^{n,n}.$$

If $\omega \in E_c^{p,q}(\mathcal{F})$ and $T \in K^{n-p,n-q}(\check{\mathcal{F}})$ (or $\omega \in E^{p,q}(\mathcal{F})$ and $T \in K_c^{n-p,n-q}(\check{\mathcal{F}})$), then we put

$$\langle \omega, T \rangle = \int_X \omega \wedge T.$$

For T fixed in $K_c^{n-p,n-q}(\check{\mathcal{F}})$ (in $K^{n-p,n-q}(\check{\mathcal{F}})$, respectively), the assignment $\omega \mapsto \langle \omega, T \rangle$ yields a linear functional

$$L_T : E^{p,q}(\mathcal{F}) \to \mathbb{C} \text{ (respectively } L_T : E_c^{p,q}(\mathcal{F}) \to \mathbb{C}).$$

If $T \in K_c^{n-p,n-q}(\check{\mathcal{F}})$, then by regarding T in $K^{n-p,n-q}(\check{\mathcal{F}})$ the corresponding functional $E_c^{p,q}(\mathcal{F}) \to \mathbb{C}$ coincides with the restriction of the functional $E^{p,q}(\mathcal{F}) \to \mathbb{C}$ given by T.

LEMMA 4.8. *The functionals $L_T : E^{p,q}(\mathcal{F}) \to \mathbb{C}$ (respectively $L_T : E_c^{p,q}(\mathcal{F}) \to \mathbb{C}$) are continuous for any $T \in K_c^{n-p,n-q}(\check{\mathcal{F}})$ (respectively $T \in K^{n-p,n-q}(\check{\mathcal{F}})$).*

Proof. Let $T \in K_c^{n-p,n-q}(\check{\mathcal{F}})$. Consider a locally finite covering $\mathcal{U} = (U_i)_i$ of X by relatively compact Stein open sets and $(\rho_i)_i$ a C^∞ partition of unity which is subordinate to this covering. Suppose that $\mathcal{F}$ admits a finite presentation on each U_i. By the properties of the integral it follows that

$$\int_X \omega \wedge T = \sum_i \int_{U_i} \omega \wedge (\rho_i T)$$

(the sum is finite since T has a compact support). Since the topology on $E^{p,q}(\mathfrak{F}) = \Gamma(X, \mathcal{E}^{p,q} \otimes \mathfrak{F})$ is defined by means of the topologies on $\Gamma(U_i, \mathcal{E}^{p,q} \otimes_{\mathcal{O}} \mathfrak{F})$, we reduce the problem to proving the continuity of L_T when X coincides with one of U_i 's.

Thus we may assume that we have an exact sequence of the form $\mathcal{O}^r \to \mathfrak{F} \to 0$. We obtain a monomorphism $Hom_{\mathcal{O}}(\mathfrak{F}, \mathcal{K}^{n-p,n-q}) \to Hom_{\mathcal{O}}(\mathcal{O}^r, \mathcal{K}^{n-p,n-q})$, hence a monomorphism $K_c^{n-p,n-q}(\check{\mathfrak{F}}) \to K_c^{n-p,n-q}(\check{\mathcal{O}}^r)$. Let T' be the image of T under this. One has the commutative diagram

$$\begin{array}{ccc} E^{p,q}(\mathcal{O}^r) & \longrightarrow & E^{p,q}(\mathfrak{F}) \\ & {\scriptstyle L_{T'}} \searrow \quad \swarrow {\scriptstyle L_T} & \\ & \mathbb{C} & \end{array}$$

where the horizontal arrow is surjective. We thus reduce the question to the case $\mathfrak{F} = \mathcal{O}^r$ and, by a previous remark, to the case $\mathfrak{F} = \mathcal{O}$. In this case the conclusion follows from the continuity of the integral.

Take now $T \in K^{n-p,n-q}(\check{\mathfrak{F}})$. For each compact K of X we must prove the continuity of the composite map

$$\Gamma_K(X, \mathcal{E}^{p,q} \otimes \mathfrak{F}) \hookrightarrow E_c^{p,q}(\mathfrak{F}) \xrightarrow{L_T} \mathbb{C}.$$

Let ρ be a C^∞ function on X with compact support which is equal to 1 on a neighbourhood of K. The proof is completed by the commutative diagram

$$\begin{array}{ccc} \Gamma_K(X, \mathcal{E}^{p,q} \otimes \mathfrak{F}) & \longrightarrow & E_c^{p,q}(\mathfrak{F}) \\ \downarrow & & \downarrow {\scriptstyle L_T} \\ \Gamma(X, \mathcal{E}^{p,q} \otimes \mathfrak{F}) & \xrightarrow{L_{\rho T}} & \mathbb{C}. \end{array}$$

Thus, we have obtained $\mathbb{C}$-linear maps $T \mapsto L_T$ which are functorial in $\mathfrak{F}$

$$K_c^{n-p,n-q}(\check{\mathfrak{F}}) \to (E^{p,q}(\mathfrak{F}))' \qquad K^{n-p,n-q}(\check{\mathfrak{F}}) \to (E_c^{p,q}(\mathfrak{F}))'.$$

Moreover, we have the commutative diagram

$$\begin{array}{ccc} K_c^{n-p,n-q}(\check{\mathfrak{F}}) & \to & (E^{p,q}(\mathfrak{F}))' \\ \downarrow & & \downarrow \\ K^{n-p,n-q}(\check{\mathfrak{F}}) & \to & E_c^{p,q}(\mathfrak{F}))' \end{array}$$

(the injectivity of the second vertical arrow follows by the density of $E_c^{p,q}(\mathcal{F})$ in $E^{p,q}(\mathcal{F})$).

LEMMA 4.9. *The above maps*

$$K_c^{n-p,\,n-q}(\check{\mathcal{F}}) \to (E^{p,q}(\mathcal{F}))',\quad K^{n-p,\,n-q}(\check{\mathcal{F}}) \to (E_c^{p,q}(\mathcal{F}))'$$

are algebraic isomorphisms.

Proof. We first consider the case when X is an open set in $\mathbf{C}^n$ and we have an exact sequence of the form $\mathcal{O}^r \to \mathcal{O}^s \to \mathcal{F} \to 0$. We get an exact and commutative diagram

$$\begin{array}{ccccccc} 0 \to & K_c^{n-p,\,n-q}(\check{\mathcal{F}}) & \to & K_c^{n-p,\,n-q}(\check{\mathcal{O}}^s) & \to & K_c^{n-p,\,n-q}(\check{\mathcal{O}}^r) \\ & \downarrow & & \downarrow & & \downarrow \\ 0 \to & (E^{p,q}(\mathcal{F}))' & \to & (E^{p,q}(\mathcal{O}^s))' & \to & (E^{p,q}(\mathcal{O}^r))' \end{array}$$

(for the line below we have used the exact sequence of continuous, hence strict, maps of FS spaces, $E^{p,q}(\mathcal{O}^r) \to E^{p,q}(\mathcal{O}^s) \to E^{p,q}(\mathcal{F}) \to 0$). The second and third vertical are isomorphisms, hence the first is an isomorphism.

We show now that the map $K^{n-p,\,n-q}(\check{\mathcal{F}}) \to (E_c^{p,q}(\mathcal{F}))'$ is an isomorphism. Consider a partition of unity $\rho = (\rho_i)_i$ which is subordinate to a locally finite covering $\mathcal{U} = (U_i)_{i \in I}$ with relatively compact open sets. Let $T \in K^{n-p,\,n-q}(\check{\mathcal{F}})$ be such that $L_T = 0$. For any $\omega \in E^{p,q}(\mathcal{F})$, $L_{\rho_i T}(\omega) = L_T(\rho_i\omega) = 0$. Accordingly, $\rho_i T = 0$ for any i, hence $T = 0$. Let now $L : E_c^{p,q}(\mathcal{F}) \to \mathbf{C}$ be a continuous and linear functional. For any i the linear functionals $\omega \in \Gamma(U_i,\ \mathcal{E}^{p,q} \otimes \mathcal{F}) \mapsto L(\rho_i\omega)$ are continuous. Then there exist $T_i \in \Gamma_c(U_i,\ Hom_{\mathcal{O}}(\mathcal{F}, \mathcal{K}^{n-p,\,n-q})) \subset K^{n-p,\,n-q}(\check{\mathcal{F}})$ such that $L_{T_i}(\omega) = L(\rho_i\omega)$. The element $T = \sum T_i \in K^{n-p,\,n-q}(\check{\mathcal{F}})$ is well defined and $L = L_T$. Indeed, let $\omega \in E_c^{p,q}(\mathcal{F})$ and U be a relatively compact neighbourhood of Supp ω. The set $J = \{i \in I \mid U_i \cap U \neq 0\}$ is finite and the conclusion follows by means of the relations

$$L(\omega) = L(\sum_{i \in J} \rho_i\omega) = \sum_{i \in J} L(\rho_i\omega) = \sum_{i \in J} L_{T_i}(\omega) = \sum_{i \in J} \int \omega \wedge T_i =$$

$$\int \omega \wedge (\sum_{i \in J} T_i) = \int_U (\omega \mid U) \wedge (\sum_{i \in J} T_i) \mid U = \int_U \omega \mid U \wedge T \mid U = \int \omega \wedge T = L_T(\omega).$$

Consider now the case when X is an arbitrary manifold. Let $\mathcal{U} = (U_i)_i$ be a locally finite covering of X by affine open sets where $\mathcal{F}$ admits finite presentations and $(\rho_i)_i$ be a partition of unity subordinate to this covering. We prove the bijectivity of the first map of the lemma. Let $T \in K_c^{n-p,\,n-q}(\check{\mathcal{F}})$ be such that $L_T = 0$; we will show that $T = 0$. It is enough to show that each $T \mid U_i \in K^{n-p,\,n-q}(\check{\mathcal{F}} \mid U_i)$

is null. But $L_{T|U_i}: E_c^{p,q}(\mathcal{F}|U_i) \to \mathbf{C}$ coincides with the composition $E_c^{p,q}(\mathcal{F}|U_i) = \Gamma_c(U_i, \mathcal{E}^{p,q}(\mathcal{F})) \hookrightarrow E^{p,q}(\mathcal{F}) \xrightarrow{L_T} \mathbf{C}$ and the conclusion follows.

Let now $L \in (E^{p,q}(\mathcal{F}))'$. For any i there exists $T_i \in \Gamma(U_i, Hom(\mathcal{F}, \mathcal{K}^{n-p,n-q}))$ such that L_{T_i} coincides with the composition $\Gamma_c(U_i, \mathcal{E}^{p,q} \otimes \mathcal{F}) \hookrightarrow E^{p,q}(\mathcal{F}) \xrightarrow{L} \mathbf{C}$. For each pair (i,j), $L_{T_i|U_i \cap U_j} = L_{T_j|U_i \cap U_j}$, hence $T_i | U_i \cap U_j = T_j | U_i \cap U_j$. Let T bc the element of $K^{n-p,n-q}(\check{\mathcal{F}})$ which is obtained by gluing together $(T_i)_i$. We show that T has a compact support, that is $T_\iota = 0$ but a finite number of indices. Since L is continuous, there exist $C > 0$ and a finite number of seminorms $p_{U^\alpha, K^\alpha, \varphi^\alpha, D^\alpha}$ such that

$$|L(\omega)| \leqslant C \cdot \sup_\alpha p_\alpha(\omega), \qquad \omega \in E^{p,q}(\mathcal{F}).$$

The compacts K^α cut a finite number of open sets U_i. If U_i is not such an open set, then one can easily see that $T_i = 0$. It remains only to prove that $L_T = L$. If $\omega \in E^{p,q}(\mathcal{F})$, then $\omega = \sum_i \rho_i\omega$ and the convergence is in the FS topology of $E^{p,q}(\mathcal{F})$. The following equalities hold

$$L(\omega) = \sum_i L(\rho_i\omega) = \sum_i L_{T_i}(\rho_i\omega) = \sum_i L_T(\rho_i\omega) = L_T(\sum_i \rho_i\omega) = L_T(\omega).$$

The bijectivity of the maps $K^{n-p,n-q}(\check{\mathcal{F}}) \to E_c^{p,q}(\mathcal{F})$ can be proved in the same manner as the particular case considered in the beginning.

LEMMA 4.10. *The transposed maps of the morphisms*

$$d'' : E^{p,q}(\mathcal{F}) \to E^{p,q+1}(\mathcal{F}) \text{ (respectively } d'' : E_c^{p,q}(\mathcal{F}) \to E_c^{p,q+1}(\mathcal{F}))$$

coincide with

$$(-1)^{p+q+1} d'' : K_c^{n-p,n-q-1}(\check{\mathcal{F}}) \to K_c^{n-p,n-q}(\check{\mathcal{F}})$$

(respectively

$$(-1)^{p+q+1} d'' : K^{n-p,n-q-1}(\check{\mathcal{F}}) \to K^{n-p,n-q}(\check{\mathcal{F}})).$$

Proof. Take $T \in K_c^{n-p,n-q-1}(\check{\mathcal{F}})$ and $\omega \in E^{p,q}(\mathcal{F})$. We shall verify that

$$\int d''\omega \wedge T = (-1)^{p+q+1} \int \omega \wedge d''T$$

and the lemma will be proved (the dual case is similar). Since $d'(\omega \wedge T) \in K_c^{n+1,n-1} = 0$, $d(\omega \wedge T) = d''(\omega \wedge T)$. Since $\int d(\omega \wedge T) = 0$, the conclusion will result provided that the equality

$$d''(\omega \wedge T) = d''\omega \wedge T + (-1)^{p+q}\omega \wedge d''T$$

holds.

This fact is local in nature; hence we may assume that X is an open subset of $\mathbb{C}^n$ where there exists an epimorphism $\mathcal{O}^r \to \mathcal{F} \to 0$. The map $E^{p,q}(\mathcal{O}^r) \to E^{p,q}(\mathcal{F})$ is surjective, hence there is a form $\omega' \in E^{p,q}(\mathcal{O}^r)$ whose image in $E^{p,q}(\mathcal{F})$ is ω. Denote by T' the image of T under the map $K_c^{n-p,n-q-1}(\check{\mathcal{F}}) \to \to K_c^{n-p,n-q-1}(\check{\mathcal{O}^r})$. Obviously, $\omega \wedge T = \omega' \wedge T'$, $d''\omega \wedge T = d''\omega' \wedge T'$ and $\omega \wedge d''T = \omega' \wedge dT'$. The conclusion follows from the equality

$$d''(\omega' \wedge T') = d''\omega' \wedge T' + (-1)^{p+q}\omega' \wedge d''T'$$

(in this case ω' and T' are systems of r forms in the usual sense...).

We consider topologies of strong dual on $K^{n-p,n-q}(\check{\mathcal{F}})$ and $K_c^{n-p,n-q}(\check{\mathcal{F}})$ obtained by means of lemma 4.9. Hence $K_c^{n-p,n-q}(\check{\mathcal{F}})$ is a DFS space and its strong dual is isomorphic to $E^{p,q}(\mathcal{F})$. The space $E_c^{p,q}(\mathcal{F})$ is a strict inductive limit of FS spaces; hence it is reflexive [68]. Accordingly, the strong dual of $K^{n-p,n-q}(\check{\mathcal{F}})$ is isomorphic to $E_c^{p,q}(\mathcal{F})$. From lemmas 4.9 and 4.10 one obtains the following:

PROPOSITION 4.11. *Let X be a complex manifold with countable basis, of dimension n. Then for any coherent analytic sheaf $\mathcal{F}$ on X the exterior product and the integration of the currents of maximal degree define a topological duality between the complexes*

$$\ldots \to E^{p,q-1}(\mathcal{F}) \to E^{p,q}(\mathcal{F}) \to E^{p,q+1}(\mathcal{F}) \to \ldots \textit{(respectively}$$

$$\ldots \to E_c^{p,q-1}(\mathcal{F}) \to E_c^{p,q}(\mathcal{F}) \to E_c^{p,q+1}(\mathcal{F}) \to \ldots)$$

and

$$\ldots \leftarrow K_c^{n-p,n-q+1}(\check{\mathcal{F}}) \leftarrow K_c^{n-p,n-q}(\check{\mathcal{F}}) \leftarrow K_c^{n-p,n-q-1}(\check{\mathcal{F}}) \leftarrow \ldots$$

$$\textit{(respectively} \ldots \leftarrow K^{n-p,n-q+1}(\check{\mathcal{F}}) \leftarrow K^{n-p,n-q}(\check{\mathcal{F}}) \leftarrow K^{n-p,n-q-1}(\check{\mathcal{F}}) \leftarrow \ldots).$$

(III) Actually, we pass to the study of the functors Ext. Recall the following result [54]:

For any point $x \in X$ the germs of distributions in x constitute an injective $\mathcal{O}_x$-module; in particular, the stalks $\mathcal{K}_x^{p,q}$ are injective $\mathcal{O}_x$-modules.

Consider the Dolbeault-Grothendieck resolution

$$0 \to \Omega^p \to \mathcal{K}^{p,0} \xrightarrow{d''} \ldots \xrightarrow{d''} \mathcal{K}^{p,n} \to 0.$$

Let $\mathcal{F}$ be a coherent $\mathcal{O}$-module. For each $x \in X$, $\mathcal{E}xt_{\mathcal{O}}^{\bullet}(\mathcal{F}, \mathcal{K}^{p,q})_x \simeq \operatorname{Ext}_{\mathcal{O}_x}^{\bullet}(\mathcal{F}_x, \mathcal{K}_x^{p,q})$. Hence $\mathcal{E}xt_{\mathcal{O}}^r(\mathcal{F}, \mathcal{K}^{p,q}) = 0$ for any $r \geqslant 1$. By using the spectral sequences which connect the local Ext's with the global Ext's, one can easily realize that

$$\operatorname{Ext}_{\mathcal{O}}^r(\mathcal{F}, \mathcal{K}^{p,q}) = 0, \ \operatorname{Ext}_{\mathcal{O},c}^r(\mathcal{F}, \mathcal{K}^{p,q}) = 0 \text{ for } r \geqslant 1.$$

Thereby the invariants $\mathrm{Ext}^{\bullet}(X; \mathcal{F}, \Omega^p)$ and $\mathrm{Ext}_c^{\bullet}(X; \mathcal{F}, \Omega^p)$ are isomorphic to the cohomology groups of the complexes

$$0 \to \check{K}^{p,0}(\mathcal{F}) \xrightarrow{d''} \check{K}^{p,1}(\mathcal{F}) \to \dots \xrightarrow{d''} \check{K}^{p,n}(\mathcal{F}) \to 0,$$

$$0 \to \check{K}_c^{p,0}(\mathcal{F}) \xrightarrow{d''} \check{K}_c^{p,1}(\mathcal{F}) \xrightarrow{d''} \dots \xrightarrow{d''} \check{K}_c^{p,n}(\mathcal{F}) \to 0.$$

In particular, $\mathrm{Ext}^{\bullet}(X; \mathcal{F}, \Omega)$ and $\mathrm{Ext}_c^{\bullet}(X; \mathcal{F}, \Omega)$ stand for the cohomology of the complexes

$$0 \to \check{K}^{n,0}(\mathcal{F}) \xrightarrow{d''} \check{K}^{n,1}(\mathcal{F}) \to \dots \to \check{K}^{n,n}(\mathcal{F}) \to 0,$$

$$0 \to \check{K}_c^{n,0}(\mathcal{F}) \xrightarrow{d''} \check{K}_c^{n,1}(\mathcal{F}) \to \dots \to \check{K}_c^{n,n}(\mathcal{F}) \to 0.$$

These complexes are topological and we will consider the induced locally-convex TVS topologies on $\mathrm{Ext}^{\bullet}(X; \mathcal{F}, \Omega)$ and $\mathrm{Ext}_c^{\bullet}(X; \mathcal{F}, \Omega)$.

PROPOSITION 4.12. *Let X be a complex manifold with countable basis and let $\mathcal{F}$ be a coherent analytic sheaf on X. Then the natural topology on* $\mathrm{Ext}^{\bullet}(X; \mathcal{F}, \Omega)$ *and* $\mathrm{Ext}_c^{\bullet}(X; \mathcal{F}, \Omega)$ *obtained by means of the Čech homology coincides with the topologies defined by means of the Dolbeault-Grothendieck resolutions.*

Proof. Let $\mathcal{U}$ be a locally finite covering of X by Stein open sets which are sufficiently small. The invariants $\mathrm{Ext}_c^{\bullet}(X; \mathcal{F}, \Omega)$ are isomorphic to the homology groups of the DFS complex of finite chains $C_{\bullet}(\mathcal{U}, \mathcal{E}xt_c^n(\mathcal{F}, \Omega))$.

Recall that $\mathcal{E}xt_c^n(\mathcal{F}, \Omega)$ is the precosheaf $U \mapsto \mathrm{Ext}^n_{\mathcal{O},c}(U; \mathcal{F}, \Omega)$. Consider the double complex $C_p(\mathcal{U}, \mathcal{H}om_c(\mathcal{F}, \mathcal{K}^{n,n-q}))$ of DFS spaces, where $\mathcal{H}om_c(\mathcal{F}, \mathcal{K}^{p,q})$ are the precosheaves $U \mapsto \check{K}_c^{p,q}(\mathcal{F}|\, U) = \mathrm{Hom}_c(U; \mathcal{F}, \mathcal{K}^{p,q})$. Let $C_{\bullet}$ be the associated simple complex. The link morphisms

$$C_{\bullet} \to C_{\bullet}(\mathcal{U}, \mathcal{E}xt_c^n(\mathcal{F}, \Omega)) \text{ and } C_{\bullet} \to \check{K}_c^{n,\bullet}(\mathcal{F})$$

are continuous quasi-isomorphisms (they are in fact transposed of the corresponding morphisms from the proof of proposition 4.5). Then the assertion from the proposition with respect to the invariants $\mathrm{Ext}_c^{\bullet}(X; \mathcal{F}, \Omega)$ follows from lemma 1.32.

Recall how the QFS topology on the invariants $\mathrm{Ext}^{\bullet}(X; \mathcal{F}, \Omega)$ is defined by means of the Čech homology. Consider a locally finite family $\mathcal{R} = (K_i)_i$ of Stein compacts of X whose interiors cover X. The invariants $\mathrm{Ext}^{\bullet}(X; \mathcal{F}, \Omega)$ are identified with the homology of the FS complex of infinite chains $C_{\bullet}^{\infty}(\mathcal{R}, \mathcal{E}xt^n_{\mathcal{R}}(\mathcal{F}, \Omega))$ where $\mathcal{E}xt^n_{\mathcal{R}}(\mathcal{F}, \Omega)$ is the system of coefficients $K \mapsto \mathrm{Ext}^n_K(X; \mathcal{F}, \Omega)$. The topology induced by this identification is just the required one.

In the following we obtain this topology by means of the open coverings. Let $\mathcal{U}=(U_i)_{i\in I}$ be a locally finite covering of X by relatively compact Stein open sets such that each $\overline{U}_i$ is a Stein compact. The homology of the complex of chains $C_{\bullet}^{\infty}(\mathcal{U}, \mathcal{E}xt_c^n(\mathcal{F}, \Omega))$ coincides with $\mathrm{Ext}^{\bullet}(X; \mathcal{F}, \Omega)$ (in order to prove this, consider an injective resolution of Ω and analyse the spectral sequence of a suitable bicomplex...). Consider the topology of the direct product on each $C_p^{\infty}(\mathcal{U}, \mathcal{E}xt_c^n(\mathcal{F}, \Omega)) = \prod \mathrm{Ext}_c^n(U_{i_0\dots i_p}; \mathcal{F}, \Omega)$. $C_{\bullet}^{\infty}$ becomes a TVS complex and we endow $\mathrm{Ext}^{\bullet}(X: \mathcal{F}, \Omega)$ with the induced topology, and show that it is the same as the already defined QFS topology. There exist two coverings $\mathfrak{K}' = (K_i')_{i\in I}$ and $\mathfrak{K} = (K_i)_{i\in I}$ as above such that the inclusions $K_i \subset U_i \subset K_i'$ are satisfied. The natural maps

$$C^{\infty}_{\bullet}(\mathfrak{K}, Ext^n_{\mathfrak{K}}(\mathcal{F}, \Omega)) \to C^{\infty}_{\bullet}(\mathfrak{U}, Ext^n_c(\mathcal{F}, \Omega)) \to C^{\infty}_{\bullet}(\mathfrak{K}', Ext^n_{\mathfrak{K}'}(\mathcal{F}, \Omega))$$

are continuous (they are in fact transposed from the maps $C^{\bullet}_c(\mathfrak{K}', \mathcal{F}) \to C^{\bullet}_c(\mathfrak{U}, \mathcal{F}) \to$ $\to C^{\bullet}_c(\mathfrak{K}, \mathcal{F})$ of the proof of 4.5). The morphisms induced to homology are therefore continuous, and since

$$H_{\bullet}(C^{\infty}_{\bullet}(\mathfrak{K}, Ext^n_{\mathfrak{K}}(\mathcal{F}, \Omega))) \to H_{\bullet}(C^{\infty}_{\bullet}(\mathfrak{K}', Ext^n_{\mathfrak{K}'}(\mathcal{F}, \Omega)))$$

is a topological isomorphism, the conclusion follows.

Compare now the topology from $\mathrm{Ext}^{\bullet}(X; \mathcal{F}, \Omega)$ thus obtained with the topology given by the complex $K^{n,\bullet}(\check{\mathcal{F}})$. We will consider the double complex of components $C^{\infty}_p(\mathfrak{U}, Hom_{\mathcal{O}}(\mathcal{F}, \mathcal{K}^{n,n-q})_c)$, endowed with the topology of the direct product. The proof proceeds exactly as in proposition 4.5, by means of two lemmas similar to 4.6 and 4.7 (we also notice that this bicomplex is the topological dual of the bicomplex $C^p_c(\mathfrak{U}, \mathcal{E}^{0,q} \otimes \mathcal{F})$ from the proof of 4.5 and this simplifies the proof...).

Remark. From the proposition it follows in particular that the topological structure on $\mathrm{Ext}^{\bullet}(X; \mathcal{F}, \Omega)$ obtained from the complex $K^{n,\bullet}(\check{\mathcal{F}})$ is of QFS type!

(IV) We now establish the connection with the duality theorems. Theorem 4.1 can be immediately reobtained from 4.11 and the topological lemmas of § 1.

As for theorem 4.2 one may proceed as follows (a little different way may be found in [28]). By proposition 4.11 one derives algebraic isomorphisms

$${}^{\sigma}H^p_c(X, \mathcal{F}) \rightleftarrows ({}^{\sigma}\mathrm{Ext}^{n-p}(X; \mathcal{F}, \Omega))', \quad {}^{\sigma}\mathrm{Ext}^{n-p}(X; \mathcal{F}, \Omega) \rightleftarrows ({}^{\sigma}H^p_c(X, \mathcal{F}))',$$

where σ means the associated separated space and the accent is, as usual, the topological dual; one may proceed exactly as in lemma 1.30 by using the reflexivity (in fact the semireflexivity!) of the spaces $\mathrm{E}^{0,p}_c(\mathcal{F})$, $\mathrm{K}^{n,n-p}(\check{\mathcal{F}})$.

The bilinear maps

$$E^{0,p}_c(\mathcal{F}) \times K^{n,n-p}(\check{\mathcal{F}}) \to \mathbb{C}, \quad (\omega, T) \mapsto \int \omega \wedge T$$

are continuous. In order to see this it is enough to prove that the restrictions to the spaces $\Gamma_K(X, \mathcal{E}^{0,p} \otimes \mathcal{F}) \times K^{n,n-p}(\check{\mathcal{F}})$ are continuous, where K is an arbitrary compact of X. If ρ is a C^{∞} function with compact support and which is equal to 1 on a neighbourhood of K, then the restriction to $\Gamma_K(X, \mathcal{E}^{0,p} \otimes \mathcal{F}) \times \mathrm{K}^{n,n-p}(\check{\mathcal{F}})$ coincides with the composition of the continuous map

$$\Gamma_K(X, \mathcal{E}^{0,p} \otimes \mathcal{F}) \times K^{n,n-p}(\check{\mathcal{F}}) \to E^{0,p}(\mathcal{F}) \times K^{n,n-p}_c(\check{\mathcal{F}}), \quad (\omega, T) \mapsto (\omega, \rho T)$$

with the map

$$E^{0,p}(\mathcal{F}) \times K_c^{n,n-p}(\check{\mathcal{F}}) \to \mathbb{C},\ (\omega, T) \mapsto \int \omega \wedge T.$$

However the latter is continuous (the reasoning may be made, for instance, in terms of sequences). Thus the induced bilinear maps

$$H_c^p(X, \mathcal{F}) \times \operatorname{Ext}^{n-p}(X; \mathcal{F}, \Omega) \to \mathbb{C}$$

are continuous. Thereby the above algebraic isomorphisms are continuous.

This is the part of theorem 4.2 which we can find here again. By means of propositions 4.5 and 4.12 (hence practically of theorem 4.2), we deduce that ${}^{\sigma}H_c^p(X,\mathcal{F})$ are DFS spaces and ${}^{\sigma}\operatorname{Ext}^{n-p}(X; \mathcal{F}, \Omega)$ are FS spaces. From the closed graph theorem it follows then that these isomorphisms are topological. Moreover, from 4.2 one derives (replacing $\mathcal{F}$ by $\mathcal{F} \otimes \Omega^p$) the following

COROLLARY 4.13. *The morphism* $E_c^{p,q-1}(\mathcal{F}) \xrightarrow{d''} E_c^{p,q}(\mathcal{F})$ *has closed image if and only if the morphism* $K^{n-p,n-q}(\check{\mathcal{F}}) \xrightarrow{d''} K^{n-p,n-q+1}(\check{\mathcal{F}})$ *has closed image.*

Remark. If we confine to locally free sheaves $\mathcal{F}$, then we have algebraic isomorphisms

$$\operatorname{Ext}_{\mathcal{O}}^{n-q}(X; \mathcal{F}, \Omega) \simeq H^{n-q}(X, \check{\mathcal{F}} \otimes_{\mathcal{O}} \Omega),\ \operatorname{Ext}_{\mathcal{O},c}^{n-q}(X; \mathcal{F}, \Omega) \simeq H_c^{n-q}(X, \check{\mathcal{F}} \otimes_{\mathcal{O}} \Omega).$$

In this case, the results from [54] are not necessary in order to obtain the duality.

We also remark that these isomorphisms are just homeomorphisms. Indeed, ${}^{\sigma}\operatorname{Ext}_{\mathcal{O}}^{n-q}(X; \mathcal{F}, \Omega)$ is in topological duality with ${}^{\sigma}H_c^q(X, \mathcal{F})$ and $H^{n-q}(X; \check{\mathcal{F}} \otimes_{\mathcal{O}} \Omega)$ in topological duality with $\operatorname{Ext}_c^q(X; \check{\mathcal{F}} \otimes_{\mathcal{O}} \Omega, \Omega)$. Since

$$\operatorname{Ext}_c^q(X; \check{\mathcal{F}} \otimes_{\mathcal{O}} \Omega, \Omega) \simeq H_c^q(X, (\check{\mathcal{F}} \otimes \Omega)^{\vee} \otimes \Omega) \simeq H_c^q(X, \mathcal{F}),$$

the conclusion for the first isomorphism follows by lemma 1.31 and similarly for the second one. As a matter of fact, the reasoning may also proceed directly as in the proofs of propositions 4.5 and 4.12.

(c) In this section we will prove a duality theorem with respect to the cohomology with supports in a closed set. The following facts are required:

LEMMA 4.14. *Consider the commutative diagram of complexes of abelian groups*

$$\begin{array}{ccccc} A^{\bullet} & \xrightarrow{\alpha} & B^{\bullet} & \xrightarrow{\beta} & C^{\bullet} \\ {\scriptstyle u}\downarrow & & {\scriptstyle v}\downarrow & & {\scriptstyle w}\downarrow \\ A'^{\bullet} & \xrightarrow{\alpha'} & B'^{\bullet} & \xrightarrow{\beta'} & C'^{\bullet} \end{array}$$

where $\beta\alpha = 0$, $\beta'\alpha' = 0$. *The morphism* v *induces a canonical morphism of complexes*

$$\operatorname{Ker}\beta/\operatorname{Im}\alpha \to \operatorname{Ker}\beta'/\operatorname{Im}\alpha'$$

and its cone is canonically isomorphic to the cohomology of

$$C^{\bullet}(u) \to C^{\bullet}(v) \to C^{\bullet}(w).$$

The proof results straightforwardly from definitions.

THEOREM 4.15. *Let* X *be a complex manifold with countable basis of dimension* n, $Y \subset X$ *a closed set,* Ω *the sheaf of germs of holomorphic forms in maximal degree,* $\mathcal{F}$ *a coherent analytic sheaf on* X, *and* p *an arbitrary integer.*

Then $H^p_Y(X, \mathcal{F})$ *has a QFS structure,* $\operatorname{Ext}^{n-p}_c(Y; \mathcal{F}, \Omega)$ *has a QDFS structure and the associated separated spaces are in topological duality.*

Moreover, $H^p_Y(X, \mathcal{F})$ *is separated if and only if* $\operatorname{Ext}^{n-p+1}_c(Y; \mathcal{F}, \Omega)$ *is separated* ($\operatorname{Ext}^{\bullet}_c(Y; \mathcal{F}, \Omega)$ *is by definition* $\operatorname{Ext}^{\bullet}_{\mathcal{O}|Y,c}(Y; \mathcal{F} \mid Y, \Omega \mid Y)$).

Proof. Let $\mathcal{U}$ and $\mathcal{V}$ be locally finite coverings of X and $X \setminus Y$ by Stein open sets such that $\mathcal{V} < \mathcal{U} \cap (X \setminus Y)$. Fix a function τ which defines the refinement $\mathcal{V} < \mathcal{U} \cap (X \setminus Y)$. For any $\mathcal{O}_X$-module $\mathcal{M}$ one obtains by restriction a morphism of complexes

$$C^{\bullet}(\mathcal{U}, \mathcal{M}) \xrightarrow{\tau^*(\mathcal{M})} C^{\bullet}(\mathcal{V}, \mathcal{M}).$$

Denote $C^{\bullet}(\mathcal{U}, \mathcal{V}; \mathcal{M}) = C^{\bullet}(\tau^*(\mathcal{M}))[-1]$, where $C^{\bullet}(\tau^*(\mathcal{M}))$ is the cone of $\tau^*(\mathcal{M})$. We will show that $H^{\bullet}(C^{\bullet}(\mathcal{U}, \mathcal{V}; \mathcal{F})) \simeq H^{\bullet}_Y(X, \mathcal{F})$. To this aim, let $\mathcal{I}^{\bullet}$ be an injective resolution of $\mathcal{F}$. Consider the double complex

$$E^{p,q} = C^p(\mathcal{U}, \mathcal{V}; \mathcal{I}^q).$$

For each q we have the exact sequence

$$\ldots \to H^p(C^{\bullet}(\mathcal{U}, \mathcal{V}; \mathcal{I}^q)) \to H^p(\mathcal{U}, \mathcal{I}^q) \to H^p(\mathcal{V}, \mathcal{I}^q) \to \ldots,$$

hence

$$H^p(C^{\bullet}(\mathcal{U}, \mathcal{V}; \mathcal{I}^q)) = \begin{cases} \Gamma_Y(X, \mathcal{I}^q) & \text{if } p = 0 \\ 0 & \text{if } p \neq 0. \end{cases}$$

By applying lemma 4.14 it follows that the spectral sequences associated to $E^{\bullet\bullet}$ are

$$'E_1^{p,q} = \begin{cases} \Gamma_Y(X, \mathcal{I}^q) & \text{if } p = 0 \\ 0 & \text{if } p \neq 0 \end{cases}, \quad \text{hence } 'E_2^{p,q} = \begin{cases} H^q_Y(X, \mathcal{F}) & \text{if } p = 0 \\ 0 & \text{if } p \neq 0 \end{cases}$$

and

$$''E_1^{p,q} = \begin{cases} C^p(\mathfrak{U}, \mathfrak{V}; \mathfrak{F}) & \text{if } q = 0 \\ 0 & \text{if } q \neq 0 \end{cases}, \quad \text{hence } ''E_2^{p,q} = \begin{cases} H^p(C^{\bullet}(\mathfrak{U}, \mathfrak{V}; \mathfrak{F})) & \text{if } q = 0 \\ 0 & \text{if } q \neq 0. \end{cases}$$

Our assertion follows.

Let now $\mathfrak{D}$ be a precosheaf on X. By corestriction one obtains a morphism of complexes (of chains)

$$C_{\bullet}(\mathfrak{V}, \mathfrak{D}) \xrightarrow{\tau_*(\mathfrak{D})} C_{\bullet}(\mathfrak{U}, \mathfrak{D}).$$

Denote $C_{\bullet}(\mathfrak{V}, \mathfrak{U}; \mathfrak{D}) = C_{\bullet}(\tau_*(\mathfrak{D}))\,[-1]$, where $C_{\bullet}(\tau_*(\mathfrak{D}))$ is the cone of the morphism $\tau_*(\mathfrak{D})$. Consider the precosheaf $Ext_c^n(\mathfrak{F}, \Omega)$ given by $U \mapsto \operatorname{Ext}_c^n(U; \mathfrak{F}, \Omega)$. We will show that

$$H_p(C_{\bullet}(\mathfrak{V}, \mathfrak{U}; Ext_c^n(\mathfrak{F}, \Omega))) \simeq \operatorname{Ext}_c^{n-p}(Y; \mathfrak{F}, \Omega)$$

for any p. In order to prove this, let $\mathfrak{J}^{\bullet}$ be an injective resolution of Ω. Consider the double complex of components

$$E^{-p,q} = C_p(\mathfrak{V}, \mathfrak{U};\ Hom_{\mathcal{O}}(\mathfrak{F}, \mathfrak{J}^q)_c).$$

For each q we have the exact sequence

$$\ldots \to H_p(\mathfrak{U}, Hom_{\mathcal{O}}(\mathfrak{F}, \mathfrak{J}^q)_c) \to H_p(C_{\bullet}(\mathfrak{V}, \mathfrak{U}; Hom_{\mathcal{O}}(\mathfrak{F}, \mathfrak{J}^q)_c)) \to$$
$$\to H_{p-1}(\mathfrak{V}, Hom_{\mathcal{O}}(\mathfrak{F}, \mathfrak{J}^q)_c) \to \ldots.$$

We also have the exact sequence

$$0 \to \operatorname{Hom}_c(X \setminus Y; \mathfrak{F}, \mathfrak{J}^q) \to \operatorname{Hom}_c(X; \mathfrak{F}, \mathfrak{J}^q) \to \operatorname{Hom}_c(Y; \mathfrak{F}, \mathfrak{J}^q) \to 0.$$

By applying lemma 3.9 one thus derives that

$$H_p(C_{\bullet}(\mathfrak{V}, \mathfrak{U}; Hom_{\mathcal{O}}(\mathfrak{F}, \mathfrak{J}^q)_c)) = \begin{cases} Hom_c(Y; \mathfrak{F}, \mathfrak{J}^q) & \text{if } p = 0 \\ 0 & \text{if } p \neq 0. \end{cases}$$

On the other hand, for any Stein open set U, $\operatorname{Ext}_c^p(U; \mathfrak{F}, \Omega) = 0$ if $p \neq n$. By these facts and by lemma 4.14, the spectral sequences associated to the bicomplex $E^{\cdot\cdot}$ have the terms

$$'E_2^{-p,q} = \begin{cases} \operatorname{Ext}_c^q(Y; \mathfrak{F}, \Omega) & \text{if } p = 0 \\ 0 & \text{if } p \neq 0, \end{cases}$$

$$''E_2^{-p,q} = \begin{cases} H_p(C_{\bullet}(\mathfrak{V}, \mathfrak{U}; Ext_c^n(\mathfrak{F}, \Omega))) & \text{if } q = n \\ 0 & \text{if } q \neq n. \end{cases}$$

The conclusion follows from the properties of spectral sequences.

The complexes $C^{\bullet}(\mathfrak{U}, \mathcal{F})$ and $C^{\bullet}(\mathfrak{V}, \mathcal{F})$ are FS and their topological duals are isomorphic to the complexes $C_{\bullet}(\mathfrak{U}, Ext_c^n(\mathcal{F}, \Omega))$ and $C_{\bullet}(\mathfrak{V}, Ext_c^n(\mathcal{F}, \Omega))$, respectively. In these dualities the maps $\tau^*(\mathcal{F})$ and $\tau_*(Ext_c^n(\mathcal{F}, \Omega))$ correspond to each other. Consequently, $C^{\bullet}(\mathfrak{U}, \mathfrak{V}; \mathcal{F})$ admits an FS structure, $C_{\bullet}(\mathfrak{V}, \mathfrak{U}; Ext_c^n(\mathcal{F}, \Omega))$ a DFS structure and these structures are in topological duality.

The proof of the theorem is concluded by applying 1.28 and 1.30.

§ 5. The dualizing sheaves

Let X be a complex space and $\mathcal{F} \in$ **Coh** (X). Denote by $\mathfrak{U}$ the family of the relatively compact Stein open sets. Let $U, V \in \mathfrak{U}, V \subset U$. Consider a closed immersion $U \xrightarrow{\pi} D$, where D is a Stein open subset of some numerical space. There exist topological isomorphisms

$$H_c^{\bullet}(U, \mathcal{F}) \simeq H_c^{\bullet}(D, \pi_*(\mathcal{F}))$$

(it is enough to use a locally finite covering of D by Stein compacts and to consider its inverse image under π ...). In particular, considering the above said, and by 4.4, the spaces $H_c^{\bullet}(U, \mathcal{F})$ are separated, hence DFS.

Let D' be an open subset of D such that $\pi(V) = \pi(U) \cap D'$. The spaces $H_c^{\bullet}(V, \mathrm{F})$ are topologically isomorphic to $H_c^{\bullet}(D', \pi_*(\mathrm{F}))$ and we have the commutative diagram

$$\begin{array}{ccc} H_c^{\bullet}(V, \mathcal{F}) & \simeq & H_c^{\bullet}(D', \pi_*(\mathcal{F})) \\ \downarrow & & \downarrow \\ H_c^{\bullet}(U, \mathcal{F}) & \simeq & H_c^{\bullet}(D, \pi_*(\mathcal{F})), \end{array}$$

where the vertical arrows are the natural morphisms obtained by the trivial extension of the sections. Taking into account the calculus of the cohomology by means of Delbeault resolutions, there results that the morphisms $H_c^{\bullet}(D', \pi_*(\mathcal{F})) \to \to H_c^{\bullet}(D, \pi_*(\mathcal{F}))$ are continuous. Then the morphisms $H_c^{\bullet}(V, \mathcal{F}) \to H_c^{\bullet}(U, \mathcal{F})$ are continuous.

Remark that this assertion is in fact true without any supplementary hypothesis on U and V: the reasoning may proceed either directly by means of Stein coverings or by applying the duality and lemma 1.31.

For any integer q we shall denote by $\mathfrak{D}^q\mathcal{F}$ the sheaf associated to the presheaf

$$U \in \mathfrak{U} \mapsto \text{the topological dual of } H_c^q(U, \mathcal{F}),$$

$V \subset U \mapsto$ the transposed of the morphism $H_c^q(V, \mathcal{F}) \to H_c^q(U, \mathcal{F})$.

$(\mathfrak{D}^q\mathcal{F})_{q \geq 0}$ are called *the dualizing sheaves of* $\mathcal{F}$ [4].

PROPOSITION 5.1. *The sheaves $\mathfrak{D}^q\mathcal{F}$ are coherent and for any* $\mathrm{U} \in \mathfrak{U}$, $\Gamma(U, \mathfrak{D}^q\mathcal{F})$ *is isomorphic to the strong dual of $H_c^p(U, \mathcal{F})$ (the topology on Γ is that deduced from the coherence).*

Proof. Let $U \in \mathfrak{U}$ and $\pi\colon U \to \mathbb{C}^n$ be a closed immersion for some integer n. We prove that

$$\mathfrak{D}^q\mathcal{F} \mid U \simeq \pi^*\,(\mathcal{E}xt^{n-q}_{\mathcal{O}_{\mathbb{C}^n}}(\pi_*\,(\mathcal{F}), \Omega)),$$

and the first assertion of the proposition will be completed.

Let $D \subset \mathbb{C}^n$ a Stein open set. By 4.4, there exist topological isomorphisms

$$H^q_c(\pi^{-1}(D), \mathcal{F})' \simeq H^q_c(D, \pi_*(\mathcal{F}))' \simeq \mathrm{Ext}^{n-q}_{\mathcal{O}_{\mathbb{C}^n}}(D; \pi_*\,(\mathcal{F}), \Omega)$$

$$\simeq \Gamma(D, \mathcal{E}xt^{n-q}_{\mathcal{O}_{\mathbb{C}^n}}(\pi_*(\mathcal{F}), \Omega)) \simeq \Gamma(\pi^{-1}(D), \pi^*\mathcal{E}xt^{n-q}_{\mathcal{O}_{\mathbb{C}^n}}(\pi_*(\mathcal{F}), \Omega)),$$

where we consider the strong topology on duals. These isomorphisms agree with the restrictions, and the required conclusion follows easily. Furthermore,

$$\Gamma(U, \mathfrak{D}^q\mathcal{F}) \simeq \Gamma(U, \pi^*\mathcal{E}xt^{n-q}_{\mathcal{O}_{\mathbb{C}^n}}(\pi_*(\mathcal{F}), \Omega)) \simeq (H^q_c(U, \mathcal{F}))',$$

as above.

THEOREM 5.2. *Let X be a complex space and $\mathcal{F} \neq 0$ a coherent analytic sheaf on X. Then*

(*i*) $\mathfrak{D}^q\mathcal{F} = 0$ *for* $q <$ proof $\mathcal{F}$ *and* $q > \dim \mathcal{F}$. *Moreover,* $\mathfrak{D}^q\mathcal{F} \neq 0$ *whenever* $q = \operatorname{prof} \mathcal{F}$ *or* $q = \dim \mathcal{F}$.

(*ii*) *For any* q, $\dim(\mathfrak{D}^q\mathcal{F}) \leqslant q$ *and if* $q = \dim \mathcal{F}$, *then* $\dim(\mathfrak{D}^q\mathcal{F}) = q$.

Proof. (*i*) follows by the previous proposition and by theorems I. 3.6 and I.3.7.

(*ii*) The problem is local in nature; hence we may assume that X is a Stein space and that there is a closed immersion $\pi\colon X \to \mathbb{C}^n$. If $\mathcal{F}^* = \pi_*(\mathcal{F})$, then

$$\dim \mathcal{F} = \dim \mathcal{F}^*,\ \pi_*(\mathfrak{D}^q\mathcal{F}) = \mathfrak{D}^q(\mathcal{F}^*),$$

hence $\dim(\mathfrak{D}^q\mathcal{F}) = \dim(\mathfrak{D}^q(\mathcal{F}^*))$.

We can thus assume X as a Stein manifold of dimension n and in this case,

$$\mathfrak{D}^q\mathcal{F} \simeq \mathcal{E}xt^{n-q}_{\mathcal{O}}(\mathcal{F}, \Omega).$$

Let $x \in X$. We will show that $\dim(\mathfrak{D}^q\mathcal{F})_x \leqslant q$ and that $\dim(\mathfrak{D}^{\dim \mathcal{F}_x}\mathcal{F})_x = \dim \mathcal{F}_x$ and the proof is completed.

There exist isomorphisms $(\mathfrak{D}^q\mathcal{F})_x \simeq \mathcal{E}xt^{n-q}_{\mathcal{O}_x}(\mathcal{F}_x, \mathcal{O}_x)$. Let $\mathfrak{p} \in \mathrm{Supp}(\mathrm{Ext}^{n-q}_{\mathcal{O}_x}(\mathcal{F}_x, \mathcal{O}_x))$, hence $\mathrm{Ext}^{n-q}_{(\mathcal{O}_x)_\mathfrak{p}}((\mathcal{F}_x)_\mathfrak{p}, (\mathcal{O}_x)_\mathfrak{p}) \simeq (\mathrm{Ext}^{n-q}_{\mathcal{O}_x}(\mathcal{F}_x, \mathcal{O}_x))_\mathfrak{p} \neq 0$. Since $(\mathcal{O}_x)_\mathfrak{p}$ is a regular ring (hence its injective dimension is finite and equals the Krull dimension), it follows that $\dim(\mathcal{O}_x)_\mathfrak{p} \leqslant n-q$. So $\dim(\mathcal{O}_x/\mathfrak{p}) = n - \dim(\mathcal{O}_x)_\mathfrak{p} \leqslant q$, therefore $\dim(\mathrm{Ext}^{n-q}_{\mathcal{O}_x}(\mathcal{F}_x, \mathcal{O}_x)) \leqslant q$.

It only remains to prove that $\dim(\mathrm{Ext}^{n-\dim \mathcal{F}_x}_{\mathcal{O}_x}(\mathcal{F}_x, \mathcal{O}_x)) = \dim \mathcal{F}_x$. Let us consider an ideal $\mathfrak{p} \in \mathrm{Supp}\,(\mathcal{F}_x)$ such that $\dim(\mathcal{O}_x/\mathfrak{p}) = \dim \mathcal{F}_x$; the support of the localization $(\mathcal{F}_x)_{\mathfrak{p}}$ is thereby reduced to the maximal ideal $\mathfrak{p}(\mathcal{O}_x)_{\mathfrak{p}}$. The ring $(\mathcal{O}_x)_{\mathfrak{p}}$ is regular and prof $(\mathcal{O}_x)_{\mathfrak{p}} = \dim (\mathcal{O}_x)_{\mathfrak{p}} = n - \dim (\mathcal{O}_x/\mathfrak{p}) = n - \dim \mathcal{F}_x$. By applying I. 1.16 we get

$$\mathrm{Ext}^{n-\dim \mathcal{F}_x}_{(\mathcal{O}_x)_{\mathfrak{p}}}((\mathcal{F}_x)_{\mathfrak{p}}, (\mathcal{O}_x)_{\mathfrak{p}}) \simeq (\mathrm{Ext}^{n-\dim \mathcal{F}_x}_{\mathcal{O}_x}(\mathcal{F}_x, \mathcal{O}_x))_{\mathfrak{p}} \neq 0.$$

Accordingly, $\mathfrak{p} \in \mathrm{Supp}\,(\mathrm{Ext}^{n-\dim \mathcal{F}_x}_{\mathcal{O}_x}(\mathcal{F}_x, \mathcal{O}_x))$, and the proof is concluded.

COROLLARY 5.3. *Let X be a complex space and $\mathcal{F} \neq 0$ a coherent analytic sheaf on X. Then $\mathcal{F}$ is Cohen-Macauley if and only if there exists an integer q_0 such that $\mathfrak{D}^q\mathcal{F} = 0$ for $q \neq q_0$.*

COROLLARY 5.4. *A complex space $(X, \mathcal{O})$ is perfect if and only if there is an integer q_0 so that the dualizing sheaves $\mathfrak{D}^q\mathcal{O}$ vanish for $q \neq q_0$.*

We give the connection with the dualizing complex:

PROPOSITION 5.5. *Let X be a complex space of finite dimension and $\mathcal{F} \in$* **Coh**(X). *Then*

$$\mathfrak{D}^q\mathcal{F} \simeq \mathcal{E}xt^{-q}_{\mathcal{O}}(\mathcal{F}, K^\bullet_X).$$

In particular, $\mathfrak{D}^q\mathcal{O}_X$ is isomorphic to the cohomology sheaf of the complex $K^\bullet_X$ in the dimension $-q$.

Proof. Let $\mathcal{I}^\bullet$ be an injective resolution of $K^\bullet_X$. Then $\mathcal{E}xt^{-q}_{\mathcal{O}}(\mathcal{F}, K^\bullet)$ is the cohomology sheaf of the complex $\mathcal{H}om_{\mathcal{O}}(\mathcal{F}, \mathcal{I}^\bullet)$ in the dimension $-q$. Accordingly, $\mathcal{E}xt^{-q}_{\mathcal{O}}(\mathcal{F}, K^\bullet_X)$ is the sheaf associated to the presheaf $U \mapsto \mathrm{Ext}^{-q}_{\mathcal{O}}(U; \mathcal{F}, K^\bullet_X)$.

If $U \in \mathfrak{U}$, then the spaces $H^\bullet_c(U, \mathcal{F})$ are separated. By the second theorem of absolute duality, the spaces $\mathrm{Ext}^\bullet_{\mathcal{O}}(U; \mathcal{F}, K^\bullet_X)$ are also separated and $\mathrm{Ext}^{-q}_{\mathcal{O}}(U; \mathcal{F}, K^\bullet_X)$ is canonically isomoprhic to the topological dual of $H^q_c(U, \mathcal{F})$. Moreover, if V is another relatively compact Stein open set, $V \subset U$, then the restriction map $\mathrm{Ext}^{-q}_{\mathcal{O}}(U; \mathcal{F}, K^\bullet_X) \to \mathrm{Ext}^{-q}_{\mathcal{O}}(V; \mathcal{F}, K^\bullet_X)$ corresponds to the transposed map of the corestriction $H^q_c(V, \mathcal{F}) \to H^q_c(U, \mathcal{F})$. This completes the proof.

Bibliographical indications

The construction of the dualizing complex and the theorems of absolute duality (§ 2 and § 3) are presented following Ramis and Ruget [61]. Theorems of separation of the invariants $H^\bullet_c(X, \mathcal{F})$, $\mathrm{Ext}^\bullet_c(X; \mathcal{F}, K^\bullet_X)$ can be found in [63]. The duality with respect to a morphism of complex spaces is studied in [62] and [63].

A detailed presentation of the first duality theorem on manifolds can be found in [87]. The assertion (*i*) of theorem 4.4 is due to Suominen [87] and the assertion (*ii*) to the authors [8]. The proof of the duality on manifolds by means of Dolbeault resolutions follows Serre [75] and Malgrange [53], [54]. Theorem 4.15 is due to Golovin [27].

The dualizing sheaves from § 5 are introduced in Andreotti and Kas' paper [4]. The same paper introduces the dualizing cosheaves and the homology groups of coherent analytic sheaves and provides another presentation of the duality in the singular case (namely, some spectral sequences which relate the dualizing sheaves to dualizing cosheaves are established).

Bibliography

1. ABHYANKAR, S., *Local analytic geometry*, Academic Press, New York, 1964.
2. ANDREOTTI, A., GRAUERT, H., *Théorèmes de finitude pour la cohomologie des espaces complexes*, Bull. Soc. Math. France **90**, 193–259, 1962.
3. ANDREOTTI, A., VESENTINI, E., *On the pseudo-rigidity of Stein manifolds*, Annali Sc. Norm. Sup. Pisa **16**, 213–223, 1962.
4. ANDREOTTI, A., KAS, A., *Duality on complex spaces*, Annali Sc. Norm. Sup. Pisa *27*, 187–263, 1973.
5. BĂNICĂ, C., *Un théorème concernant la séparation de certains espaces de cohomologie*, C.R. Acad. Sc. Paris **272**, 782–785, 1971.
6. BĂNICĂ, C., *Le complété formel d'un espace analytique le long d'un sous-espace: un théorème de comparaison*, Manuscripta math., **6**, 207–244, 1972.
7. BĂNICĂ, C., *Le comportement des faisceaux $\mathcal{F}(m)$ pour $m \to -\infty$*, Colloque sur les fonctions analytiques de plusieurs variables et l'analyse complexe, Paris, 1972.
8. BĂNICĂ, C., STĂNĂȘILĂ, O., *Des caractérisations topologiques de la profondeur d'un faisceau analytique cohérent*, Séminaire d'Espaces analytiques, Bucarest sept. 1969 (or C. R. Acad. Sc. Paris, **269**, 636–639, 1969; **270**, 1174–1177, 1970).
9. BĂNICĂ, C., STĂNĂȘILĂ, O., *Some results on the extension of analytic entities defined out of a compact*, Annali Sc. Norm. Sup. Pisa, **25**, 347–376, 1971.
10. BOURBAKI, N., *Algèbre commutative*, Hermann, Paris, 1961.
11. BREDON, G., *Sheaf theory*, Mc. Grow-Hill, New York, 1967.
12. BUCUR, I., DELEANU, A., *Introduction to the theory of categories and functors*, J. Wiley-Interscience, 1968.
13. CARTAN, H., *Sur le premier problème de Cousin*, C. R. Acad. Sc. Paris, **207**, 558–560, 1938.
14. CARTAN, H., *Quotients of complex analytic spaces*, Tata Institute, Bombay, 1960.
15. CARTAN, H., *Familles d'espaces complexes et fondements de la géométrie analytique*, Séminaire E.N.S., 1960–61.
16. CARTAN, H., EILENBERG, S., *Homological algebra*, University Press, Princeton, 1956.
17. CARTAN, H., SERRE, J. P., *Un théorème de finitude concernant les variétés analytiques compactes*, C. R. Acad. Sc. Paris, **237**, 128–130, 1953.
18. DOLBEAULT, P., *Sur la cohomologie des variétés analytiques complexes*, C. R. Acad. Sc. Paris, **236**, 175–177, 1953.
19. DOUADY, A., *Le problème des modules pour les sous-espaces analytiques compactes d'un espace analytique donné*, Ann. Inst. Fourier, **16**, 1–95, 1966.
20. DOUADY, A., *Flatness and privilege*, L'Enseignement math. **XIV**, 1, 1968.
21. FLONDOR, P., JURCHESCU, M., *Grauert's coherence theorem for holomorphic spaces*, Rev. Roum. math. **17**, 1199–1211, 1972.

22. Forster, O., *Zur Theorie des Steinschen Alegbren und Moduln*, Math. Z., **97**, 376–405, 1967.

23. Forster, O., Knorr, K., *Ein Beweis des Grauertschen Bildgarbensatses nach Ideen von B. Malgrange*, Manuscripta math., **5**, 19–44, 1971.

24. Frenkel, J., *Cohomologie nonabélienne et espaces fibrés*, Bull. Soc. Math. France, **83**, 135 – 218, 1957.

25. Frisch, J., *Points de platitude d'un morphisme d'espaces analytiques*, Inv. Math., **4**, 118–138, 1967.

26. Godement, R., *Topologie algébrique et théorie des faisceaux*, Hermann, Paris, 1958.

27. Golovin V. D., *On the spaces of local cohomology of analytic manifolds* (in Russian), Funkt. Analiz, **5**, 66, 1971.

28. Golovin, V. D., *Duality for the cohomology with compact supports* (in Russian), Dokl. Acad. USSR, **199**, 4, 751–753, 1971.

29. Grauert, H., *Ein Theorem der analytischen Garbentheorie und die Modulräume komplexer Strukturen*, Publ. IHES, N° 5, 1960.

30. Grauert, H., *Über Modifikationen und exzeptionelle analytische Mengen*. Math. Ann., **146**, 331–368, 1962.

31. Grauert, H., Kerner, H., *Deformationen von Singularitäten komplexer Räume*, Math. Ann., 153, 236–260, 1964.

32. Grauert, H., Remmert, R., *Bilder und Urbilder analytischer Garben*, Ann. Math., **68**, 393 – 443, 1958.

33. Grauert, H., Remmert, R., *Analytische Stellenalgebren*, Springer Verlag, Berlin, 1971.

34. Grothendieck, A., *Espaces vectoriels topologiques*, Societ. mat. Sao Paolo, 1954.

35. Grothendieck, A., *Sur quelques points d'algèbre homologique*, Tohoku Math. J., **9**, 119–221, 1957.

36. Grothendieck, A., *Séminaire de géométrie algébrique* (SGA), IHES, Paris.

37. Grothendieck, A., Dieudonné, J., *Eléments de géométrie algébrique* (EGA), IHES, Paris.

38. Gunning, R., Rossi, H., *Analytic functions of several complex variables*, Prentice Hall, New Jersey, 1965.

39. Hartshorne, R., *Residues and duality*, Lecture Notes 20, Springer Verlag, 1966.

40. Harvey, R., *The theory of hyperfunctions on totally real subset of a complex manifolds with applications to extension problems*, Am. J. of math., **91**, 853–873, 1969.

41. Hörmander, L., *An introduction to complex analysis in several variables*, Van Nostrand, Princeton, 1966.

42. Houzel, C., *Espaces analytiques relatifs et théorème de finitude*, Math. Ann., **205**, 13–54, 1973.

43. Jurchescu M., *On the canonical topology of an analytic algebra and of an analytic module*, Bull. Soc. Math. France, **93**, 129–153, 1965.

44. Kerner H., *Zur Theorie der Deformationen komplexer Räume*, Math. Z., **103**, 389–398 1968.

45. Kiehl, R., *Äquivalenzrelationen in analyt schen Räumen*, Math. Z., **105**, 1–20, 1968.

46. Kiehl, R., *Note zu der Arbeit von J. Frisch: "Points de platitude d'un morphisme d'espaces analytiques complexes"*, Inv. math. **4**, 139–141, 1967.

47. Kiehl, R., Verdier, J. L., *Ein einfacher Beweis des Kohärenzsatzes von Grauert*, Math. Ann, **195**, 24–50, 1971.

48. Knorr, K., *Der Grauertsche Projektionssatz*, Inv. math., **12**, 118–172, 1971.

49. Kodaira, K., Spencer, D. C., *On deformations of complex analytic structures* I, II Ann. Math., **67**, 328–466, 1958; III Ann. Math., 71, 43–76, 1960.

50. Kuhlmann, N., *Über die normalen Punkte eines komplexen Raumes*, Math. Ann., **146**, 397–412, 1962.
51. Laufer, H. B., *On Serre duality and envelopes of holomorphy*, Trans. Amer. Math. Soc., **128**, 414–436, 1967.
52. Łojasiewicz, S., *Triangulation of semi-analytic sets*, Annali Sc. Norm. Sup. Pisa, **18**, 449–474, 1964.
53. Malgrange, B., *Systèmes différentielles à coefficients constants*, Séminaire Bourbaki 15, N⁰ 246, 1962–63.
54. Malgrange, B., *Ideals of differentiable functions*, Univ. Press, Oxford, 1966.
55. Martineau, A., *Les hyperfonctions de M. Sato*, Séminaire Bourbaki, Exposé 214, 1960–61.
56. Narasimhan, R., *Introduction to the theory of analytic spaces*, Lecture Notes 25, Springer Verlag, 1966.
57. Narasimhan, R., *Compact analytic varieties*, L'Enseignement math., **XIV**, 1, 1968.
58. Narasimhan, R., *Grauert's theorem on direct images of coherent sheaves*, Univ. Montréal, 1971.
59. Popescu, N., *Abelian categories with applications to rings and modules*, Academic Press, New York, 1973.
60. Radu, N., *Local rings* I, II (in Romanian), Ed. Academiei, 1968.
61. Ramis, J. P., Ruget, G., *Complexes dualisants et théorème de dualité en géométrie analytique complexe*, Publ. IHES, N⁰ **38**, 77–91, 1970.
62. Ramis, J. P., Ruget, C., *Résidus et dualité*, Inv. Math., *26*, 89–131, 1974.
63. Ramis, J. P., Ruget, G., Verdier, J. L., *Dualité relative en géométrie analytique complexe*, Inv. Math. **13**, 261–283, 1971.
64. Reiffen, H. J., *Riemannsche Hebbarkeitssätze für Cohomologieklassen mit kompatken Träger*, Math. Ann., **164**, 272–279, 1966.
65. Remmert, R., *Holomorphe und meromorphe Abbildungen komplexer Räume*, Math. Ann., **133**, 328–370, 1957.
66. Riemanschneider, O., *Über die Anwendung algebraischer Methoden in der Deformationstheorie komplexer Räume*, Math. Ann., **187**, 4055, 1970.
67. Rothstein, W., *Die Fortsetzung vier und höherdimensionaler analytischen Flächen des* R_{2n} $(n \geq 3)$ (*Cousinsche Verteilungen 2. Art*), Math. Ann., **121**, 340–355, 1950.
68. Schaeffer, H., *Topological vector spaces*, Mc. Millan, New York, 1966.
69. Scheja, G., *Riemannsche Hebbarkeitssätze für Cohomologieklassen*, Math. Ann., **144**, 345–360, 1961.
70. Scheja, G., *Eine Anwendung Riemannscher Hebbarkeitssätze für analytische Cohomologieklassen*, Archiv der Math., 12, 341–348, 1961.
71. Scheja, G., *Fortsetzungssätze der komplex-analytischen Cohomologie und ihre algebraische Charakterisierung*, Math. Ann., **157**, 75–94, 1964.
72. Schneider, M., *Halbstetigkeitssätze für relativ analytische Räume*, Inv. Math., **16**, 161–176, 1972.
73. Schwartz, L., *Varietades analiticas complejas. Ecuaciones diferenciales parciales elipticas*, Universitad National di Colombia, Bogota, 1956.
74. Serre, J. P., *Quelques problèmes globaux relatifs aux variétés de Stein*, Coll. Bruxelles, 55–68, 1953.
75. Serre, J. P., *Un théorème de dualité*, Comm. Math. Helv., **29**, 9–26, 1955.
76. Serre, J. P., *Faisceaux algébriques cohérents*, Ann. Math., **61**, 197–278, 1955.

77. SERRE, J. P., *Géométrie algébrique et géométrie analytique*, Ann. Inst. Fourier, **6**, 1–42, 1956.
78. SERRE, J. P., *Groupes algébriques et corps de classes*, Hermann, Paris, 1959.
79. SERRE, J. P., *Algèbre locale. Multiplicités*, Lecture Notes 11, Springer Verlag, 1965.
80. SERRE, J. P., *Prolongement de faisceaux analytiques cohérents*, Ann. Inst. Fourier, **16**, 363–374, 1966.
81. SIU, Y. T., *Analytic sheaf cohomology with compact supports*, Comp. Math., **21**, 52–58, 1969.
82. SIU, Y. T., *Noetherianness of rings of holomorphic functions on Stein Compact subsets*, Proc. Amer. Math. Soc., **21**, 483–489, 1969.
83. SIU, Y. T., *Analytic sheaves of local cohomology*, Trans. Amer. Math. Soc., **148**, 347–366, 1970.
84. SIU, Y. T., TRAUTMANN, G., *Gap-sheaves and extensions of coherent analytic subsheaves*, Lecture Notes 172, Springer Verlag, 1971.
85. STEIN, K., *Analytische Zerlegungen komplexer Räume*, Math. Ann., **132**, 68–93, 1956.
86. STOIA, M., *Une remarque sur la profondeur*, C. R. Acad. Sc. Paris, **276**, 929–930, 1973.
87. SUOMINEN, K., *Duality for coherent sheaves on analytic manifolds*, Ann. Acad. Sc Fennicae I Math. 424, Helsinki 1968.
88. THIMM, W., *Untersuchungen über das Spurproblem von holomorphen Funktionen auf analytischen Mengen*, Math. Ann., **139**, 95–114, 1959.
89. THIMM, W., *Lückengarben von kohärenten analytischen Modulgarben*, Math. Ann., **148**, 372–394, 1962.
90. TRAUTMANN, G., *Ein Kontinuitätssatz für die Fortsetzung kohärenter analytischer Garben*, Archiv der Math., **18**, 188–196, 1967.
91. TRAUTMANN, G., *Cohérence de faisceaux analytiques de la cohomologie locale*, C. R. Acad. Sc. Paris, **267**, 694–695, 1967.
92. TRAUTMANN, G., *Ein Endlichkeitssatz in der analytischen Geometrie*, Inv. Math., **8**, 143–174, 1969.
93. VERDIER, J. L., *Topologie sur les espaces de cohomologie d'un complexe de faisceaux analytiques à cohomologie cohérente*, Bull. Soc. Math. France, **99**, 337–343, 1971.
94. VILLANI, V., *Cohomological properties of complex spaces which carry over to normalizations*, Am. J. of Math., **88**, 636–645, 1966.
95. WIEGMANN, K. W., *Strukturen auf Quotienten komplexer Räume*, Comm. Math. Helv., **44**, 93–116, 1969.